Environmental Protection in the European Union

Volume 3

Editors

Michael Schmidt and Lothar Knopp, Cottbus, Germany

Environmental Protection in the European Union

Michael Schmidt · John Glasson
Lars Emmelin · Hendrike Helbron
Editors

Standards and Thresholds for Impact Assessment

 Springer

Professor Dr. Michael Schmidt
Department of Environmental Planning
Brandenburg University
of Technology (BTU), Cottbus
P.O. Box 10 13 44
03013 Cottbus
Germany
michael.schmidt@tu-cottbus.de

Professor John Glasson
Oxford Institute for Sustainable
Development
School of the Built Environment
Oxford Brookes University
Gipsy Lane
Headington
Oxford, OX3 0BP
United Kingdom
jglasson@brookes.ac.uk

Professor Lars Emmelin
Department of Spatial Planning
Blekinge Institute of Technology (BTH)
SE-371 79 Karlskrona
Sweden
lars.emmelin@bth.se

Dipl.-Ing. Hendrike Helbron
Department of Environmental Planning
Brandenburg University
of Technology (BTU), Cottbus
P.O. Box 10 13 44
03013 Cottbus
Germany
helbron@tu-cottbus.de

ISBN 978-3-540-31140-9   e-ISBN 978-3-540-31141-6

DOI 10.1007/978-3-540-31141-6

Environmental Protection in the European Union ISSN 1613-8694

Library of Congress Control Number: 2007942755

Production: Jelonek, Schmidt & Vöckler GbR, Leipzig
Cover design: WMX Design GmbH, Heidelberg

Printed on acid-free paper

9 8 7 6 5 4 3 2 1

springer.com

Foreword

From the beginning of environmental regulation and environmental impact assessment, debates have raged about standards and thresholds. This book adds a great deal of light to such discussions. It will provide important guidance to those who must confront these issues.

The earliest pollution laws either banned activities outright (such as King Edward I's ban on the burning of sea coal in London in the fourteenth century) or provided only vague guidance to courts or administrators (such as opacity standards for smokestack pollution). The earliest environmental impact assessment law (the National Environmental Policy Act of the United States, 1969) simply applied to actions "significantly affecting the quality of the human environment," without more definition and used such terms as "the environmental impact of the proposed action," "long-term productivity," and "irreversible and irretrievable commitments of resources."

How far we have come! Now nearly every country has legislation providing specific standards for contamination or harm to the environment be used for environmental decisions. Most countries use numerical or similar thresholds for triggering an EIA and for evaluation of activities. Of course, the standards themselves may or may not be scientifically defensible or adequate to take into account the values of the society where they are used. The chapters in this book help to uncover the assumptions used in various contexts and offer a unique opportunity to view standards and thresholds in a comparative context. It is sure to become a standard reference work for environmental professionals of all kinds.

John E. Bonine, Professor of Law in the LL.M. and J.D. Programs, University of Oregon; Founder, Environmental Law Alliance Worldwide
Oregon, November 2007

Preface

Practitioners – responsible for decision-making in impact assessment in different sectors and at different planning levels – need on the one hand a lot of expertise, and on the other hand case studies as well as legal standards and scientific thresholds as benchmarks for decision-making. With new requirements for impact assessment and dynamically changing environmental conditions, there exists a sustained need for guidance to practitioners. Therefore the main motivation for editing this Handbook on standards and thresholds for impact assessment was to give guidance to practitioners for good practice on environmental impact studies (EISs), which are often very complex and comprehensive. The editors of this Handbook have attempted to partly fill the existing gap of scientific advice to practitioners in the field of assessment values and to meet the need for additional guidance. Standards and thresholds are applied in several stages of the environmental impact assessment (EIA) process – such as screening, scoping, impact prediction and assessment, as well as monitoring.

Many standards and thresholds are politically set to classify ranges of high risk or likely harm to human health and the environment. Assessment standards and thresholds are not defined in the EC EIA or SEA Directives, but national environmental policy, EIA Acts plus Spatial and Sectoral Planning Acts supply many such standards. However, legal obligations alone cannot guarantee high quality environmental assessments; expert knowledge and common efforts by all stakeholders involved in decision-making are needed. Non-binding assessment thresholds have to be derived from environmental objectives and operationalised as guidance values for the assessment of impacts affecting a specific area. Such case-by-case decisions at policy, plan, programme and project level require sophisticated knowledge on the significant effects of many development actions on the one hand and on a wide range of environmental media on the other hand. Practitioners need considerable expertise and high quality data to achieve an efficient and environmentally sound assessment process. Competent and licensing authorities additionally have to understand and approve quantitative and qualitative values, which are applied to and influence the results of assessments. They generally have to accept statistics from analyses and evaluations of environmental consultants or planners in charge of the environmental impact study (EIS). Delivered data, applied assessment methods and values are often not sufficiently transparent.

In EIS, estimates have to be made on the significance of impacts and the carrying capacity of the state of environment in the affected area. Prevention requires a long-term time schedule for future decades, which may not be achieved with the mitigation of the significance of impacts of one single project activity, but which requires an overall review of different types of environmental assessment in all sectors and at all planning levels – leading often to a cumulative and/or strategic approach. The book gives examples of the methodological derivation and practical application of environmentally relevant standards and thresholds in EIA. It seeks in particular to serve as guidance for competent authorities and licensing authorities to better: identify significant impacts in the scoping process, evaluate the quality of assessments of area-related environmental conflicts in the EIS, and to understand the effect quantitative and qualitative values have on final decisions.

The book is divided into five Parts, which present a wide variety of approaches from different technologies and sectors affecting different environmental media, future environmental issues and implementation processes. Part I introduces legal, procedural and political fundamentals, which deliver standards and threshold values of varying strengths and status. Part II discusses standard and threshold values for different types of projects, with examples for both site-specific and spatially dispersed projects. Part III evaluates thresholds and standards from the perspective of the environmental media and their carrying capacity. Part IV discusses emerging fields of application and Part V concludes with implementation steps. A variety of different case studies presented in the book link to possible practical application fields at different levels of planning and in different sectors. The book also includes some future oriented issues, where the implementation of new standard and threshold values will be necessary quite soon if good practice and high quality EIA is to be further promoted.

We wish to thank all authors from the various countries for their valuable articles, which made possible this comprehensive publication. We do also accept that despite this wide scope there are even more examples of project types and environmental media which might have been included. We also express our acknowledgment to the PhD students from the international network for Education and Research in Environmental and Resource Management (ERM) at BTU Cottbus, who were invited to contribute as future staff members in science and practice, and who will be soon in charge of promoting a sustainable use of our planet's resources. We also thank Mr. Dmytro Palekhov for his unwavering support in the final stages of formatting all manuscripts.

This Handbook is the result of cooperation between the Brandenburg University of Technology (BTU) Cottbus in Germany, the Oxford Brookes University in Oxford in the United Kingdom and the Blekinge Institute of Technology in Karlskrona in Sweden. We hope practitioners, researchers, academics, students and central and local government officials will find the content enlightening in both its practical application and its theoretical explanation of the function and importance of the use of standards and thresholds in impact assessment.

Michael Schmidt, John Glasson, Lars Emmelin and Hendrike Helbron
Cottbus, Oxford and Karlskrona, November 2007

Table of Contents

List of Contributors

Eike Albrecht is senior lecturer at the Centre for Law and Administration, Brandenburg University of Technology Cottbus (BTU) since 1999. In 2002 he has received his Doctorate Degree at the University of Leipzig, Germany, in soil protection law which is still his main research area beside international law and product related civil and administrative law.

Centre for Law and Administration, Brandenburg University of Technology (BTU) Cottbus, P.O. Box 101344, 03013 Cottbus, Germany
Tel: +49 (0)355 69 27 49; Fax: +49 (0)355 69 35 02; Email: albrecht@tu-cottbus.de;
Web: http://www-1.tu-cottbus.de/ZfRV/JuniorprofessorAlbrecht2006.pdf

Juliane Albrecht is research associate at the Leibniz Institute of Ecological and Regional Development in Dresden since 2005. In 2006 she has received her Doctorate Degree at the Technical University of Dresden, Faculty of Law. Her research interests are in environmental and planning law, especially in European water law. She developed her work on environmental requirements on spatial planning in the German Exclusive Economic Zone (EEZ) as a contribution to a research project of the Federal Environmental Agency.

Leibniz Institute of Ecological and Regional Development Dresden (IOER), Weberplatz 1, 01217 Dresden, Germany
Tel.: +49 (0)351 4679 223; Fax: +49 (0)351 4679 212; Email: j.albrecht@ioer.de

Amer Al-Ghorbany is the deputy general director of Environmental Policies and Programmes in the Ministry of Water and Environment in the Republic of Yemen. He holds a MSc degree in Environment and Resources Management from Brandenburg University of Technology Cottbus, Germany. His Master's thesis was on "Post EIA Indicator-based Monitoring System for Water Quality Assessment in Small Scale Dam Projects in Yemen".

General Directorate of Environmental Policies and Programme, Ministry of Water and Environment Sana'a, Republic of Yemen, P.O. Box 19204,
Fax: +967 1 418 296; Email: alghorbany_amer@hotmail.com

Fadhl Ali Al-Nozaily is an Associate Professor in Environmental Engineering at the Faculty of Engineering, Sana'a University. He is also working as a training director and a coordinator for IWRM MSc programme at the Water and Environment Centre (WEC) of the University. He is as an environmental consultant conducting EIA for the World Bank project executed by the Public Works Projects (PWP) in Yemen since 2006. His main career is on water and wastewater engineering. He awarded his PhD from TU University Delft, The Netherlands in 2001.

Faculty of Engineering (FoE) and Water and Environment Centre (WEC), Sana'a University, Sana'a, The Republic of Yemen. P.O. Box 13790, Sana'a, Yemen
Tel.: +967 777381627; Fax: +967 777381628; Email: d-fadl@maktoob.com

Amoah Benedicta Jacqueline is a PhD student. She was the first female African student from Ghana awarded a Master of Science Degree in Environmental and

Resource Management at the Brandenburg University of Technology – BTU Cottbus in July 2003. Mrs. Benedicta Amoah Fei-Baffoe also holds a BSc (Hons) degree in Food Science and Nutrition from the University of Ghana, Legon-Accra. She was a research assistant for the SUBICON project (BTU-Cottbus) "Succession of collembolan in the post mining landscape of Lower Lusatia". She is currently an Administrative manager at the Research and Business Development Centre-Ghana and a co-author of the Quarterly Cocoa Monitor series. Her expertise and research interest are in sustainability of natural resources, Biodiversity conservation and Ecosystem modeling along side Occupational health and women related issues.

Erich-Weinert-Str. 5, Zi 506, 03046 Cottbus, Germany or P.O.Box A/N 11231 Accra-North, Ghana
Tel.: +49 (0) 1792841252; Email: pastor1228@yahoo.com

Antwi Effah Kwabena is a PhD. Student Presently Working on the Topic, "Integrating GIS and Remote Sensing for Assessing the Impact of Disturbance on Biodiversity and Land Cover Change in a Post-Mining Landscape". Antwi was awarded a Master of Science Degree in Environmental and Resource Management at the Brandenburg University of Technology – BTU Cottbus in November 2003. He Supervised a Group of MSc. Students in a GIS and Remote Sensing Based Study Project at the Brandenburg University of Technology in summer 2005.

Department General Ecology, Brandenburg University of Technology (BTU) Cottbus, P.O. Box 101344, D-03013 Cottbus, Germany
Tel.: +49 (0) 3 55 86 91176; Fax: +49 (0) 3 55 69-2225; Email: antwieff@tu-cottbus.de

Cem Avci is Professor of Water Resources in the Department of Civil Engineering, Bogazici University Istanbul Turkey. He received his B.Sc. degree from Bogazici University Turkey. He received his M.Sc. degree from Princeton University U.S.A. and PhD degree from Purdue University U.S.A. His research interests include soil and ground water contamination investigation and remediation, landfill designs with emphasis on containment barriers. He serves as a consultant to many soil and groundwater remediation projects and participated in the design of major landfill projects in Turkey.

Civil Engineering Department, Faculty of Engineering, Bogazici University Bebek 34342 Istanbul Turkey
Tel: (90) 212 359 6410; Fax: (90) 212 287 2463; Email: avci@boun.edu.tr

Reinhart Bartsch is a senior agricultural economist working in development aid projects. His last assignment was for the GFA group in Yemen. He has a Doctorate Degree in Economics from the University of Hohenheim and works as a guest lecturer at the Brandenburg University of Technology Cottbus (BTU) since 2002, teaching mainly project planning methods.

Römerstrasse 23 in 72213 Altensteig, Germany
Tel: +49 (0)7453 8986; Fax: +49 (0) 7453 8986; Email: rr.bartsch@t-online.de

Chris Bennett is Chairman of the Noise Group of Stop Stansted Expansion (SSE). His special interest is the use of alternative methods of describing and illustrating noise impacts caused by aviation. He is the author of a paper on this topic given at the 2004 Airport Regions Conference in Brussels.

Email: hadstockmail-ssenoise@yahoo.co.uk

Ralf Buckley is Chair and Director of the International Centre for Ecotourism Research, and Research Director of the Climate Response Program at Griffith University, Australia. He has published over a dozen books and 200 articles (see: www.griffith.edu.au/centre/icer, >Publications); has contributed to several previous books on EIA; and has given evidence to a number of government inquiries into EIA and related topics.

Griffith University, Parklands Drive, Gold Coast 9726, Australia
Tel: +61 (0)7555 28675; Email: r.buckley@griffith.edu.au

Victor Ngu Cheo is a PhD candidate in the chair of Environmental Issues in the Social Sciences, BTU Cottbus, Germany. He is a holder of a Master of Arts Degree in Communication Studies from the University of Ibadan, Nigeria. Since 1999, he had been lecturing in the University of Buea, Cameroon where he rose to the position of Senior Lecturer in 2005. He has contributed book chapters and several scholarly articles in internal Journals. His current research interest is Communication and Environmental Sustainability.

Department of Environmental Issues in the Social Sciences, Brandenburg University of Technology (BTU) Cottbus, Postfach101344, 03013 Cottbus, Germany
Tel: +49 17620627143; Fax: +49 355693037; Email: vncheo@yahoo.com

Aleh Cherp is associate professor in environmental sciences and policy at Central European University, Budapest, Hungary and a guest professor at Lund University and Blekinge Institute of Technology, Sweden. He is also the Coordinator of the Erasmus Mundus Masters Course in Environmental Sciences, Policy and Management (MESPOM, www.mespom.eu). His research interests include environmental assessment, strategies for sustainable development and sustainability issues in countries in transition.

International Institute for Industrial Environmental Economics at Lund University, Box 196, 22100 Lund, Sweden
Tel: + 46 46 222 0280; Fax: +46 46 222 0240; Email: aleh.cherp@iiiee.lu.se

Tadeusz J. Chmielewski is Professor of Landscape Ecology and Nature Conservation in Lublin Agriculture University in Poland. His research interests include: analysis of landscape structure and function (especially peatland and lakeland landscapes); methods and techniques of nature and economy harmonization (in local and regional scale); National & Landscape Parks planning; ecosystem (especially wetland) restoration. His most important publication is two volumes book: *Spatial Planning Harmonizing Nature end Economy*, Lublin 2001. He is designer of the Polesie National Park, the West Polesie Biosphere Reserve, the Roztocze Biosphere Reserve, Ponidzie Region Landscape Parks Cluster and many others protected areas in Poland.

Lublin Agriculture University, Department of Landscape Ecology and Nature Conservation, Dobrzanskiego 37 St., 20-262 Lublin, Poland
Tel: +48 (0) 81 461 00 61 (321); Email: tadeusz.chmielewski@ar.lublin.pl

Ingrid Chorus is head of the section Drinking-Water Resource Protection and Treatment and deputy head of the Department Drinking-Water Hygiene at the Federal Environment Agency of Germany. Her focus is on protecting drinking-water resources (particularly from eutrophication) and on toxic cyanobacteria.

Federal Environment Agency, Corrensplatz 1, 14195 Berlin, Germany
Tel: +49 (0) 30 8903 1346 (1305); Email: Ingrid.chorus@uba.de

Stefan Dornack is senior advisor at the Department of Surface Waters and Flood Protection at the State Ministry of Environment and Agriculture in the Free State of Saxony. His focus is on dams and their multifunctional use in Saxony.

State ministry of environment and agriculture, Department of surface waters and flood protection, P.O. Box 10 05 10, D-01076 Dresden, Germany
Tel: +49 (0) 351 564 2317; Fax +49 (0) 351 564 2070; Email: stefan.dornack@smul.sachsen.de;
Web: http://www.umwelt.sachsen.de

Lars Emmelin is professor of environmental assessment in the Spatial planning programme, Blekinge Institute of Technology at Karlskrona. He has researched and taught on environmental issues in Sweden and Norway, worked with conservation and tourism in the Arctic and Scandinavian mountains and worked for major international organisations such as the UNESCO, UNEP and OECD on environmental education. At present he is programme director of a major research programme on tools for SEA funded by the Swedish Environment Protection Agency and directs a MSc on European Spatial Planning.

Spatial planning, BTH, SE 371 79 Karlskrona, Sweden
Email: lars.emmelin@bth.se

Jürgen Ertel is the founder and head of the Department of Industrial Sustainability at the Brandenburg University of Technology Cottbus, Germany since 1994. Previous to his university career, he worked for Siemens AG and established relations to various industry associations. His focus is on environmentally benign design of industrial goods, the rating methods and in particular the recycling properties

Head of Department of Industrial Sustainability, Brandenburg University of Technology (BTU) Cottbus, P.O. Box 101344, D-03013, Cottbus, Germany
Tel: +49 (355) 69 4385; Fax: +49 (355) 69 4700; Email: ertel@tu-cottbus.de;
Web: www.tu-cottbus.de/neuwertwirtschaft

John Glasson is Professor of Environmental Planning and Co- Director of the Oxford Institute for Sustainable Development (OISD) at Oxford Brookes University, where he also leads the Impact Assessment Unit (IAU). He is also a Visiting Professor in Planning at Curtin University of Technology in Perth, Western Australia. His research interests are in environmental impact assessment, strategic environmental assessment and sustainability appraisal, the socio-economic impacts of major projects (especially energy projects), impact monitoring, and tourism impact studies. He has managed over seventy major research and consultancy projects, largely in EIA / SIA, for a wide range of bodies including the European Commission, UK Government Departments and Agencies, energy companies, ESRC and various local authorities.

Oxford Institute for Sustainable Development, School of the Built Environment, Oxford Brookes University, Gipsy Lane, Headington, Oxford OX3 0BP
Tel: +44 (1865) 483401; Email: jglasson@brookes.ac.uk

Stanislaw Gruszczynski is Professor at the Department of Management and Protection of Environment, Faculty of Mining Surveying and Environmental Engineering, AGH University of Science and Technology in Kraków, Poland. His research interests include soil protection, land reclamation, geoinformatic, EIA (especially on mining terrains) and application of the artificial intelligence methods in spatial soil modelling.

Department of Management and Protection of Environment, Faculty of Mining Surveying and Environmental Engineering, AGH University of Science and Technology, Al. Mickiewicza 30, 30-059 Kraków, Poland
Tel: +48 (12) 617 22 89; Fax: +48 (12) 633 17 91; Email: sgrusz@agh.edu.pl

Erol Güler is Professor of Geotechnical Engineering in the Department of Civil Engineering, Bogazici University Istanbul Turkey. He received his undergraduate and graduate degrees from the Technical University of Istanbul Turkey. He has served as the Director of the Environmental Institute of Bogazici University and is currently the Chairmen of the Civil Engineering department. He is also the founding chairman of the Turkish Chapter of the International Geosynthetics Society (IGS). His research interests include geosynthetics, landfill design and liner designs. In addition to his academic research, he has participated in the design of major landfills in Turkey.

Civil Engineering Department, Faculty of Engineering, Bogazici University, Bebek 34342 Istanbul Turkey
Tel: +90 (212) 359 6452; Fax: +90 (212) 287 2463; Email: eguler@boun.edu.tr

Constantino Gutiérrez is a full time lecturer and researcher at the Department of Sanitary and Environmental Engineering in the Master and Doctoral Engineering Programme at National Autonomous University of Mexico. He was Environment Engineering Manager at National Found for Tourist Development.

Department of Sanitary and Environmental Engineering; Master and Doctoral Engineering Programme, National Autonomous University of Mexico (UNAM) Circuito Exterior Ciudad Universitaria, 04510, Mexico D.F. Mexico
Tel: +01 (555) 622 30 02; Fax: +01 (555) 622 30 00; Email: gupc@servidor.unam.mx, cgping@yahoo.com; Web: http://dgep.posgrado.unam.mx/ambiental/

Joachim Hartlik is Founder and Director of the Company for Environmental Assessments and Quality Management. He is Foundation member of the German EIA-Association, since 2004 member of the execution board and chairman of the EIA quality management working group.

Dr. Hartlik - Company for Environmental Assessments & Quality Management, Kreuzkamp 5s, D-32175 Lehrte, Germany
Tel: +49 (0)5175 9291003; Fax: +49 (0)5175 9291003; Email: j.hartlik@hartlik.de

Hendrike Helbron is a scientific research assistant and lecturer at the Brandenburg University of Technology of Cottbus, Department of Environmental Planning. She was responsible for the development of an assessment methodology and indicator system in a two year pilot project on strategic environmental assessment in regional planning. The topic of her PhD thesis involves an analysis, evaluation and application of environmental indicators and assessment thresholds in strategic environmental assessment of regional land use planning.

Department of Environmental Planning, Brandenburg University of Technology (BTU), Cottbus, P.O. Box 101344, D-03013 Cottbus, Germany
Tel:+49 (0) 3 55 69 23 52; Fax: +49 (0) 3 55 69 27 65; Email: helbron@tu-cottbus.de

Matti Johansson is Environmental Affairs Officer in the secretariat to the Convention on Long-range Transboundary Air Pollution in the Environment, Housing and Land Management Division, United Nations Economic Commission for Europe. His research has ranged over several air pollution modelling themes, in particular ecosystem critical loads and integrated assessment modelling, with later focus on particulate matter. He holds a Doctorate Degree in Technical Physics from Helsinki University of Technology.

United Nations Economic Commission for Europe (UNECE), Palais des Nations, CH-1211 Geneva 10, Switzerland
Tel: +41-22-917 2358; Fax: +41-22-917 0621; Email: matti.johansson@unece.org;
Web: http://www.unece.org/env/lrtap

Werner Kratz was a professor/lecturer for general ecology, soil ecology and ecotoxicology at the Free University Berlin and the Martin-Luther-University Halle. His research focuses on the impact of chemicals (heavy metals, PCP, BaP, Pesticides etc) on biota and eco-systematic processes primarily in terrestrial ecosystems. In 2000 he entered the Brandenburg State Office for Environment (LUA), Potsdam. From 2000-2004 he was the leader of the department Environmental Chemistry (a lab with 130 employees), Environmental Survey (12 employees) and Environmental Risk assessment (9 employees). Since 2004 he is the leader of the group Environmental Survey and Environmental Risk Assessment. He is also the research coordinator of the LUA and a member of several federal adviser groups for environmental chemistry, human-toxicology and ecotoxicology. He still occasionally lectures at the Free University Berlin in ecology, global climate change and sustainability.

Brandenburg State Office for Environment (LUA), Department of Environmental Survey and Ecotoxicology, Seeburger Chaussee 2, 14476 Potsdam
Tel: +49 (0) 33201 442 282/283; Fax: +49 (0) 33201 43677; Email: Werner.kratz@lua.brandenburg.de

Peggy Lerman is a legal counsel working in the field of environmental and planning law. She has been assistant judge at The Court of Appeal for southern Sweden. She was formerly head of legal affairs at the National Housing, Building and

Planning Agency and national expert in the EU commission expert group for EIA and SEA involved in the preparation of the SEA Directive. She has been a legal expert and consultant to a number of national government or parliamentary commissions on diverse topics such as the development of the Environmental Code, on sustainable development, environmental assessment, disposal of nuclear wastes, infrastructure, biodiversity, and served as coordinator for Nordic cooperation on EIA and planning.

Lagtolken AB, Jordosundsvagen 29, SE 370 24 NATTRABY, Sweden
Tel: +46 (457) 352 80; Email: peggy@lagtolken.se; Web: www.lagtolken.se

Robert Mayer is retired professor for Landscape Ecology and Soil Science, University of Kassel. His research focuses in environmental cycling and impacts upon soil, water and vegetation; soil science in planning and soil protection.

University of Kassel, Department of Architecture, Urban and Landscape Planning, Gottschalkstrasse 28, 34109 Kassel
Tel: +49 (0) 5 61 3160071 (private); Fax: +49 (0) 5 61 80 43 558; Email: rmayer@uni-kassel.de; Web: www.uni-kassel.de/fb6/fachgebiete/bodenkunde.htm

Ute Mischke is senior scientist at Leibniz Institute of Freshwater Ecology and Fisheries in Berlin. She received her doctor degree at Free University Berlin about microbial food web interaction in 1995. Further research areas focused on plankton communities in acidic mining lakes, drinking water reservoirs and hypertrophic lakes. Since 2000 she essentially contributed to the German assessment methods for lakes and rivers by means of phytoplankton for WDF and its intercalibration on European level.

Leibniz-Institute of Freshwater Ecology and Inland Fisheries, Dept. of Shallow Lakes and Lowland Rivers, Müggelseedamm 310, D-12587 Berlin
Tel: +49 30 64181690; Web: http://www.igb-berlin.de/abt2/mitarbeiter/mischke

Brigitte Nixdorf is the head of the Department of Freshwater Conservation at the Brandenburg University of Technology, Cottbus since 1993. Previous to her university career she worked for scientific institutions on basic research of freshwater ecology, esp. shallow lakes and eutrophication. Her focus is on plankton succession in natural and acidic mining lakes, water quality assessment acc. to the Water Framework Directive and lake restoration.

Head of the Department of Freshwater Conservation, Brandenburg University of Technology (BTU) Research Station Bad Saarow, D - 15526 Bad Saarow, Seestraße 45
Tel: +49 33631 8943; Fax: +49 33631 5200; web: http://www.tu-cottbus.de/BTU/Fak4/Gewschu

Edward K. Nunoo is an environmental scientist, economist and an instructor with a plethora of research interests in EIA, EMS, SFM, tourism and river basin management. Prior to his enrolment at BTU where he obtained his M.Sc. in Environmental and Resource Management, he was the project officer of Ghana United Nations Association, Cape Coast. During his PhD at the Department of Environmental Planning (BTU) he is currently researching measures of SFM in Ghana.

Papitzer Str. 4, WNR 315-2, 03046 Cottbus
Tel: +49 179 7775865; Email: Nunooedward@yahoo.com

Vincent Onyango is a PhD student in the Department of Environmental Planning, Brandenburg University of Technology, Cottbus, Germany. His PhD research is about "Evaluating the extent SEA process lends itself to quantitative analysis and behaves systematically". He completed his MSc in Environment and Resources Management with a thesis on "developing an SEA approach for assessing World Trade Organisation (WTO) rules in developing countries". He has also written several articles on SEA in Kenya.

Wilhelm-Kulz Str. 50, 03046 Cottbus, Germany
Email: vin_onyngo@yahoo.com

Dmytro Palekhov is currently doing PhD research at the Department of Environmental Planning, Brandenburg University of Technology, Cottbus, Germany. He holds a MA in Law from the National Mining University, UA (2003) and BSc in Environmental and Resource Management from the BTU Cottbus (2003). His research interests are on environmental assessment (EIA and SEA) and environmental law. His PhD thesis is about implementing SEA in Ukraine as a regional planning instrument.

Neue Str. 58, 03044 Cottbus, Germany
Tel: +49 179 120 8237; Email: palekdmy@tu-cottbus.de, dmitry_pal@mail.ru

Anastássios Perdicoúlis is an assistant professor at the University of Trás-os-Montes e Alto Douro (UTAD), Portugal, and a visiting research fellow at Oxford Brookes University, UK. He holds a PhD from the University of Salford, UK (1997) and a BS from the University of Washington, USA (1990). His research interests are on Sustainable Development, with particular emphasis on SEA, EIA, and Environmental Management Systems. He is currently the coordinator of the Landscape Architecture degree at UTAD, Portugal, and has been the founder and first leader of the Environmental SIG of the System Dynamics Society (SDS).

Department of Biological and Environmental Engineering, University of Trás-os-Montes e Alto Douro, Apartado 1013, 5001-801 Vila Real, Portugal
Tel: +351 259 350 728; Fax: +351 259 350 480; Email: tasso@utad.pt; Web: home.utad.pt/~tasso

Gennady Pivnyak is Rector of the National Mining University, Dnepropetrovsk, Ukraine, Academician of the National Academy of Science of Ukraine. He is a member of various national and international professional and academic organisations, including the Ukrainian Committee on State Prises in the Field of Science and Technology, State Commission for Academic Degrees and Titles, International Society for Engineering Education, European Society for Engineering Education, expert on sustainable energy at the United Nations Economic Commission for Europe. He holds Dr. h.c. titles from the AGH University of Science and Technology, Krakow and Moscow State Mining University.

Karl Marks Av., 19, 49005 Dnepropetrovsk, Ukraine
Tel: +380 56 7446219, +380 562 470766; Fax: +380 562 473330 ; Web: http://www.nmu.org.ua

Atis Rektins has completed his BSc from Brandenburg University of Technology (ERM, BTU) in Cottbus. Title of his Thesis: "Challenges to develop common freshwater ecological status assessment according to the requirements of the Water Framework Directive"

Ozolu 13, Jurmala LV – 2008, Latvia
Email: atis.rektins@gmail.com

Jürgen Ritschel has been an assistant in the division for Contaminated Sites and Soil Protection of the Brandenburg State Office for Environment in Potsdam. He mainly deals with extensive harmful soil changes, in particular with the development of protective and restrictive measures to avoid hazards as well as the definition and implementation of soil quality objectives within the framework of preventive soil protection.

Brandenburg State Office for Environment, Department for Technical Environmental Protection, Division for Contaminated Sites and Soil Protection, 2 Seeburger Chaussee, 14476 Potsdam, Germany
Tel: +49 (0) 33201 442 356; Fax: +49 (0) 33201 442 399; Email: juergen.ritschel@lua.brandenburg.de;
Web: www.brandenburg.de/lua

Agnieszka Rozej is a senior lecturer at the Department of Land Surface Protection Engineering, Lublin University of Technology, Faculty of Environmental Engineering. In 2004 she has received her Doctorate Degree in environmental engineering. Her scientific research and lecturing fields include sanitary and environmental microbiology and soil remediation.

Department of Land Surface Protection Engineering, Faculty of Environmental Engineering, Lublin University of Technology (LUT), ul. Nadbystrzycka 40B, 20-618 Lublin, Poland
Tel: +48 81 5384405, Fax: +48 81 5381997; Email: a.rozej@fenix.pol.lublin.pl
Web: http://wis.pol.lublin.pl/index.php

Michael Schmidt is the head of the Department of Environmental Planning at the Brandenburg University of Technology (BTU) Cottbus. His scientific research and lecturing fields include environmental planning, environmental assessment, strategies for sustainable development, techniques for combating desertification as well as monitoring and evaluation. He promotes long-term cooperation in research and education with environmental experts from Syria, Yemen and Jordan.

Head of Department of Environmental Planning, Brandenburg University of Technology (BTU) Cottbus, P.O. Box 101344, D-03013 Cottbus, Germany
Tel: +49 (0) 3 55 69 24 54; Fax: + 49 (0) 3 55 69 27 65; Email: schmidtm@tu-cottbus.de;
Web: www.tu-cottbus.de/environment

Hans-Ulrich Sieber is deputy director of the Dam Authority of the Free State of Saxony and head of the department of technology. He is member of the Steering Committee of the German Committee on Large Dams.

Dam Authority of the Free State of Saxony, Department of technology, P.O. Box 10 02 34, D-01782 Pirna, Germany
Tel: +49 (0) 3501 796 351; Fax: +49 (0) 351 796 101; Email: HansUlrich.Sieber@ltv.smul.sachsen.de;
Web: http://www.umwelt.sachsen.de

Martin Socher is head of the Department of surface waters and flood protection at the state ministry of environment and agriculture in the Free State of Saxony.

He is also honorary professor at the department of civil engineering and architecture at the University of Applied Sciences Dresden.

State ministry of environment and agriculture, Department of surface waters and flood protection, P.O. Box 10 05 10, D-01076 Dresden, Germany
Tel: +49 (0) 351 564 2273; Fax: +49 (0) 351 564 2070; Email: martin.socher@smul.sachsen.de;
Web: http://www.umwelt.sachsen.de/de/wu/umwelt/index.html

Marcos von Sperling is an associate professor at the Department of Sanitary and Environmental Engineering of the Federal University of Minas Gerais, Brazil. His lecturing, research and consultancy activities are in the field of wastewater treatment and water pollution control.

Department of Sanitary and Environmental Engineering, Federal University of Minas Gerais (UFMG). Av. Contorno 842 – 7 andar. 30110-060 – Belo Horizonte, Brazil
Tel: +55 31 3238-1935; Fax: +55 31 3238-1879; Email: marcos@desa.ufmg.br

Witold Stepniewski is the head of the Department of Land Surface Protection Engineering at the Lublin University of Technology (LUT) and Vice Rector for Science of LUT. His scientific research and lecturing fields include redox conditions of soil, landfill construction, exploitation and reclamation, greenhouse gases production and absorption in soil.

Department of Land Surface Protection Engineering, Faculty of Environmental Engineering, Lublin University of Technology, ul. Nadbystrzycka 40B, 20-618 Lublin, Poland
Tel: +48 81 5384413, Fax: +48 81 5381997; Email: w.stepniewski@pollub.pl
Web: http://wis.pol.lublin.pl/index.php

Harry Storch is a scientific research assistant at the Brandenburg University of Technology of Cottbus, Department of Environmental Planning. He is currently involved in the research project 'Sustainable Housing Policies for Megacities of Tomorrow, the Balance of Urban Growth and Redevelopment in Ho Chi Minh City, Vietnam' which is part of the research programme 'Sustainable Megacities of Tomorrow' funded by the German Federal Ministry of Education and Research (BMBF). In this project he is responsible for the development of the GIS-based Sustainability Indicator Framework.

Department of Environmental Planning, Brandenburg University of Technology (BTU), Cottbus, P.O. Box 101344, D-03013 Cottbus, Germany
Tel: +49 (0) 3 55 69 21 22; Fax: +49 (0) 3 55 69 27 65; Email: storch@tu-cottbus.de
Web: http://www.megacity-hcmc.org

Ernestine A. Tangang Yuntenwi was awarded a Master of Science Degree in 2003 by the Brandenburg University of Technology Cottbus, Germany in Environmental Resource Management. She is currently researching on Cooking Stove Technologies and Indoor Air Pollution for her PhD dissertation. Alongside her studies she coordinates the PhD Scientific Working Seminar and moderates the ERM-PhD e-group.

Department of Industrial Sustainability, Brandenburg University of Technology (BTU) Cottbus, P.O. Box 101344, 03013 Cottbus, Germany
Tel: +49 176 6231 0450; Fax: +49 355 69 47 00; Email: ernestinetangang@yahoo.co.uk

Aud Tennøy is researcher at Institute of Transport Economics (TØI) in Oslo, Norway. She holds an MSc in civil engineering with specialisation in urban and regional planning from The Norwegian University of Science and Technology (NTNU). Her research focuses on urban land use and transport planning in the environmental or sustainable development context, on cause-effect relations as well as the planning system and processes. This also includes how EIA and SEA are used, and can be used, as tools to achieve a more sustainable urban development.

Institute of Transport Economics (TØI), Department of Organisation and Implementation, Gaustadalléen 21, 0349 Oslo, Norway
Tel: +47 22 57 39 14; Email: ate@toi.no; Web: www.toi.no

Riki Thérivel is a partner of Levett-Thérivel sustainability consultants and a visiting professor at Oxford Brookes University's School of Planning. She specialises in SEA and environmental impact assessment. She has advised a wide range of organisations on SEA, has written some key guidance documents on SEA, and has (co-)authored three books on SEA. She is the 2002-3 recipient of the International Association for Impact Assessment's Individual Award for Contribution to Impact Assessment.

28A North Hinksey Lane, Oxford OX2 0LX, England
Tel/Fax: +44 (0) 18 65 24 34 88; Email: riki@levett-therivel.fsworld.co.uk; Web: www.levett-therivel.co.uk

Dirk Hein Westerveld is a senior rural engineer working for the GFA Group in Yemen as a team leader of a EC funded Food Security Project.

P.O.Box 5581 Sana'a, Yemen
Tel: 0967 1 250101; Fax: 0967 1 256083

Gerhard Wiegleb is professor of General Ecology at BTU Cottbus. His current research focuses on ecological and socio-economic driving forces of biodiversity change in disturbed landscapes. He is a head of the working group of Restoration Ecology of the German Ecological Society.

Department of General Ecology, Brandenburg University of Technology (BTU) Cottbus, P.O. Box 101344, D-03013 Cottbus, Germany
Tel: +49 (0)355 69 22 91; Fax: +49 (0)355 69 22 25; Email: wiegleb@tu-cottbus.de;
Web: www.tu-cottbus.de

Angkarn Wongdeethai is a lecturer and research associate at the Department of Industrial Sustainability. He holds a PhD from the Brandenburg University of Technology, Cottbus, Germany, since 2006. His research interests are on Sustainable Development, with particular emphasis on Industrial Ecology, Recycling and Design Analysis, and Environmental Protection.

Department of Industrial Sustainability, Brandenburg University of Technology (BTU) Cottbus, P.O. Box 101344, D-03013 Cottbus, Germany
Tel: +49 (355) 69 4794; Fax: +49 (355) 69 4700; Email: wongdeethai@hotmail.com; Web: www.tu-cottbus.de/neuwertwirtschaft

Part I – Environmental Policies and Key Drivers for Setting Standards and Thresholds

Standards and thresholds permeate all aspects of contemporary life – our lives at home, at work, at play and on the move as we travel between activities. Standards provide guidance that regulates the effect of an activity, normally a human activity, on a receptor. Thresholds refer to specific points above or below which there are effects with varying levels of significance and acceptability. Standards and thresholds are particularly numerous and increasing in the environment field; in Europe, the European Union is a major instigator through its Directives and other procedures. The focus of this handbook is on the nature and role of standards and thresholds for human health and the environment in Environmental Impact Assessment (EIA). The history of EIA dates back almost 40 years internationally, and it is almost 25 years since the advent of the EC EIA Directive in 1985. More recently the scope has widened to include a strategic level of activity, in the forms of Strategic Environmental Assessment (SEA) and Sustainability Appraisal (SA) and there will be reference also to these in some of the chapters in this text.

The aim of Part I is to set the exploration of standards and thresholds in Environmental Impact Assessment in the wider context of the key environmental policies and drivers which have led to their development. Chapter 1 outlines the nature of standards and thresholds and then explores their role in various stages of the EIA process – including scoping, description of the environmental baseline, impact prediction , decision making and monitoring. This is followed by a more in-depth examination of standards and thresholds in the important initial screening stage, which reveals considerable variations in practice in EU Member States, all operating under the same EIA Directive. This tailoring of standards and thresholds to fit particular national contexts is a recurrent theme in this text.

The next three chapters illustrate some of the key policies and drivers behind the evolution of standards and thresholds in particular countries. In Chapter 2, the case of Germany shows the importance of international environmental conventions, such as the Kyoto Protocol, on the national environmental legislation. In particular, it clearly illustrates the complications of transposing European standards into Member State environmental laws, and the risks involved in using a variety of approaches, including private norms.

Chapter 3 provides an example of how a major environmental problem, in this case air pollution in Ukraine, can be a driver of change in environmental standards. The Ukraine economy has been dominated by some of the most environmentally polluting industries—mining and quarrying, energy and chemicals constituting approximately 60 % of national GDP. Interrelated economic, social and environmental issues, plus a desire to harmonise environmental regulations with EU policies and standards, have provided key imperatives. Chapter 4 takes a different perspective in a different continent, exploring the impact of the UN Millennium Development Goals for target setting for the key social issues of HIV/AIDS remediation and poverty remediation in Kenya. Again the importance of setting targets in relation to national as well as international visions comes through as a key consideration.

Chapter 5 reverts back to the role of standards and thresholds in the environmental assessment process, but this time with reference to the emerging sustainability approach to environmental assessment. The case of Portugal illustrates again the importance of the EU as a driver, but also the importance of developing thematic groups of indicators to provide the information base to use, in a cascaded fashion, in the assessment process.

Chapter 6 concludes Part 1 with a discussion, using case studies from Europe, Africa and the Middle East, of some of the problems of setting thresholds which can result from not clearly defining project and policy objectives and envisaged outcomes'.

1 Principles and Purposes of Standards and Thresholds in the EIA Process

John Glasson

Oxford Institute for Sustainable Development, Oxford Brookes University, UK

1.1 Introduction

Standards and thresholds play an important role in the whole EIA process – they are the criteria against which the degree of significance of potential impacts of a development action (generally a project, but more recently also plans and programmes) are assessed. But what are standards and thresholds? They have many, and often overlapping dimensions. What roles do they play in the EIA process? They do play vital roles in many stages of the EIA process – ranging from initial screening as to whether a project/plan etc requires an assessment at all, through many other steps in the process, including decision making and the continuing but often underplayed monitoring of project/plan implementation. This chapter seeks to provide an introduction to standards and thresholds in EIA, by addressing such questions.

The chapter builds both on general concepts and practice, and on a more particular evidence base drawing on two pieces of research for the UK Government and the European Commission. Both pieces of research were undertaken by the Impacts Assessment Unit (IAU) in the Oxford Institute of Sustainable Development (OISD) at Oxford Brookes University. The first was a study for the UK Department of the Environment (DoE) on "Defining Thresholds in EIA" (DETR 1997); this was in the context of revising UK EIA procedures in response to the proposed changes in the amended EC EIA Directive. The second was for the EC Environment DG and was one element in the review of the implementation of the Amended EIA Directive across the EU Member States (IAU 2003). These two reviews are complemented by the findings of another IAU study for the UK Economic and Social Research Council (ESRC) on "Assessing Significance in the EIA Process" (see Wood and Becker 2005). These studies allow a more in-depth review of the role of standards and thresholds in one particular stage of the EIA process – the early, and vital, screening stage. They reveal considerable diversity in practice.

Standards and Thresholds for Impact Assessment. Edited by Michael Schmidt, John Glasson, Lars Emmelin and Hendrike Helbron. © 2008 Springer-Verlag

1.2 Thresholds and Standards – Definitions and Dimensions

1.2.1 Definitions

There are many ways to define both thresholds and standards. *Thresholds* refer to discrete points that must be exceeded to begin producing a given effect or result to elicit a response. They normally involve limits—upper or lower beyond which there will be an effect. In his pioneering work on Threshold Analysis, Kozlowski (1968) used thresholds, inter alia, in relation to steps in infrastructure provision (e.g. roads, sewerage capacity) needed to support various levels of urban growth. In his work on Thresholds of Concern, Sassaman (1981), identified thresholds for different environmental components (e.g. water quality) at which those assessing a proposal should become concerned with an impact.

Standards provide guidance that regulates the effect of an activity (normally human activity) on a receptor. More specifically, an environmental standard regulates the effects of human activity upon the environment. Standards may specify a desired state (e.g. noise levels from road traffic in residential areas at night) or acceptable changes (e.g. no more than 40 % of urban development shall be on 'green field' land). The UK Royal Commission on Environmental Pollution, in its 21st report "Setting Environmental Standards" (RCEP 1998), saw an environmental standard:

> *"to be any judgement about the acceptability of environmental modifications resulting from human activities which fulfils both of the following conditions:*
>
> *(a) it is formally stated after some consideration and intended to apply generally to a defined set of cases; and*
>
> *(b) because of its relationship to certain sanctions, rewards or values , it can be expected to exert an influence , direct or indirect, on activities that affect the environment."*

Standards can help development/business by clarifying in advance what criteria will be applied for environmental matters. They also help to clarify when sanctions may be applied for exceeding standards/thresholds and damaging the environment. They provide benchmarks of performance and, where the standards refer to a future date, they can provide important guidance for investment plans. Such roles tie directly into the EIA process as will be discussed below.

At this stage, it is worth re-iterating that although much of the work on thresholds and standards relates to the bio-physical dimensions of the environment, as highlighted in the UK Royal Commission report, there is also the important socio-economic dimension. The wider approach incorporating both bio-physical and socio-economic is reflected in the contents of this book.

1.2.2 Dimensions

There are many dimensions to thresholds and standards, and some of these are now discussed. They vary from the *legally enforceable* to *much vaguer guidance* and *general societal values and preferences.* For example a country may have specific and legally required building regulations in relation to structures, lighting and heating; on the other hand what are the relevant standards for levels of crime, fear of crime, health/ill-health and other socio-economic dimensions such as unemployment, and access to affordable housing? Such dimensions overlap with others relating to the *possible degree of quantification of the thresholds and standards.* Some are clearly measurable (e.g. levels of noise; air quality); but as noted here measurement is often more difficult and more *fuzzy* for socio-economic dimensions such as health, crime and some other quality of life indicators. Even where a numerical standard may be used, due regard must be given to the complexities of individual situations. The distinction between a hazardous physical impact and a safe one may also have a fuzzy element.

Standards and thresholds can also vary in their *degree of aggregation.* One weakness of EIA is sometimes the failure to disaggregate impacts on different groups affected by a project. The threshold may be an aggregate one highlighting that the project is likely to cause, for example, increased traffic in an area, or pressure on the housing market, but fail to identify the distributional impacts on affected groups. Other dimensions can be identified for example in relation to either *the nature of the project* (e.g. issues related to project over 'n' hectares in size); or in relation to the *nature of the environmental receptor* (e.g. air quality, water quality). Further, some thresholds and standards have a *spatial dimension;* others are *more a-spatial.* A spatial threshold may relate to a particular type of environmental designation; for example the often tighter controls on some types of development, or often any development at all, in designated natural areas, such as National Parks.

In the UK Royal Commission Report noted earlier (RCEP 1998), a typology of bio-physical environmental standards is presented. The classes of case covered "may be modifications to the environment, or the repercussions of such modifications, or activities that have the capacity to bring about such modifications." The typology categorises standards by reference to the pathways which substances follow until they meet a receptor which is susceptible to damage. Such standards are normally expressed as a particular level or concentration of a substance at a particular point on the pathway; for example quality standards for water, air or soil, and emission standards relating to releases of pollutants to the environment. Other forms of standards identified in the Royal Commission Report include process standards (e.g. providing criteria for deciding what level of emissions to the environment from industrial processes should be permitted); life-cycle standards (e.g. EC eco-labelling scheme); and management standards (e.g. International Organisation for Standardization standard ISO 14001 for environmental management systems).

1.3 Environmental Impact Assessment

1.3.1 EIA, SEA and SA

Any discussion of Environmental Impact assessment (EIA) must be set in the context of trends in sustainable development (SD). Sustainable development is a global movement, an over-arching approach for many national, regional and local levels of government worldwide. With its origins in the Brundtland Report (UNWCED 1987), the definition of SD is much wider than ecology and the natural environment. It entails social organisation of inter-and intra- generational equity. Importance is also assigned to economic and cultural aspects and to participation by all concerned stakeholders. Despite its global acceptance, SD is a contested and confused territory (see Faber 2005). There are numerous definitions, but a much used one is that of the triple bottom line, reflecting the importance of environmental, social and economic factors in decision making. There is also a growing acceptance of the holistic nature of SD, but the balance between the elements is also contested. One of the key roles of EIA is as an instrument for SD.

Definitions of EIA abound, but in essence EIA is a process, a systematic process that examines the environmental consequences of development actions (normally major projects), in advance (Glasson et al. 2005). The emphasis, compared with some other mechanisms for environmental protection, is on prevention. From the US National Environmental Policy Act of 1969, via the European Directive on EIA in 1985, EIA has spread and developed worldwide. By 1995, when more than 100 countries had EIA systems, the World Bank could note that "Over the past decade, EIA has moved from the fringes of development planning to become a widely recognised tool for sound project decision making" (World Bank 1995). The EIA tool has also changed over time, adapting to shifting environmental politics and managerial capabilities. Partly reflected in the various amendments to the European EIA Directive, there has been an attempt to have a more rigorous approach to all stages in the process, from initial screening through to monitoring of project implementation. More projects are now subject to EIA, and there is some (limited) shift in the range of impacts considered, with the inclusion of some socio-economic, as well as the bio-physical, impacts.

A particularly important development has been the widening of the scope of assessment with its application to the earlier, more strategic stages of development – at the level of policies, plans and programmes. Strategic Environmental Assessment (SEA) can be defined as "the formalised, systematic and comprehensive process of evaluating the environmental impacts of a policy, plan or programme and its alternatives, including the preparation of a written report on the findings of that evaluation, and using the findings in publicly accountable decision making" (Therivel et al. 1992). Again SEA origins date back to the US, and then were also given a boost by a European Directive – in this case in 2004 (EC 2004). The SEA Directive applies to a wide range of plans and programmes (including town and country planning, transport, tourism, energy, waste management, water management and others). However it does not yet include the very politically sensitive policy level.

An interesting innovation in UK, and in some other countries, is the streamlining, and complicating, of the task, in the UK for land use planning (regional and local), by requiring SEA to be undertaken as part of a wider Sustainability Appraisal (SA), which includes a more holistic approach to sustainability as noted earlier. Extensive guidance to the production of both SEAs and SAs has `been produced by relevant government departments in the UK (see for example ODPM 2005a, 2005b). A recent review of the first year of such activity in the UK by Therivel et al. (2005) shows that SEA/SA is likely to be a major growth area – with potentially as many assessments as EIAs per annum. Thus, whilst the focus of this book is on EIA, there will also be reference to the emerging SEA/SA activity.

1.3.2 The EIA Process

The EIA process can be seen as a series of steps, as outlined in Figure 1.1 – ranging from initial project screening (is an EIA required at all?) through to post decision monitoring and auditing (recording outcomes, and comparing them with predicted outcomes). However it must be stressed that EIA is rarely so logical and rational in approach –with just one key decision point. EIA should be more cyclical in approach with considerable interaction between the various steps. For example, public participation can be useful at most stages of the process; monitoring systems should relate to parameters established in the initial project and baseline description.

There is increasing recognition of EIA as a series of steps in decision- making with a mixed approach including both scientifically based rationality and community informed participation (see Lawrence 1997; Weston 2000; Glasson et al. 2005). Whilst there may be a formal decision point for a project, and EIA can have a key role to play at this stage, decision making occurs at many other stages and is undertaken by various parties. There are decisions for example, on: whether an EIA is needed at all (screening), the scope of the EIA, the alternatives under consideration, project design and re-design, the range of mitigation measures, and implementation and monitoring.

1.4 Standards and Thresholds in the EIA Process

Just as there are many decision points in the EIA process, so there are many stages where standards and thresholds can play an important role. A few of these – scoping, description of the environmental baseline, prediction and evaluation, decision making and monitoring – are now discussed briefly before a fuller discussion of the screening stage in the next section.

Scoping seeks to identify the types of impacts that should be investigated for a

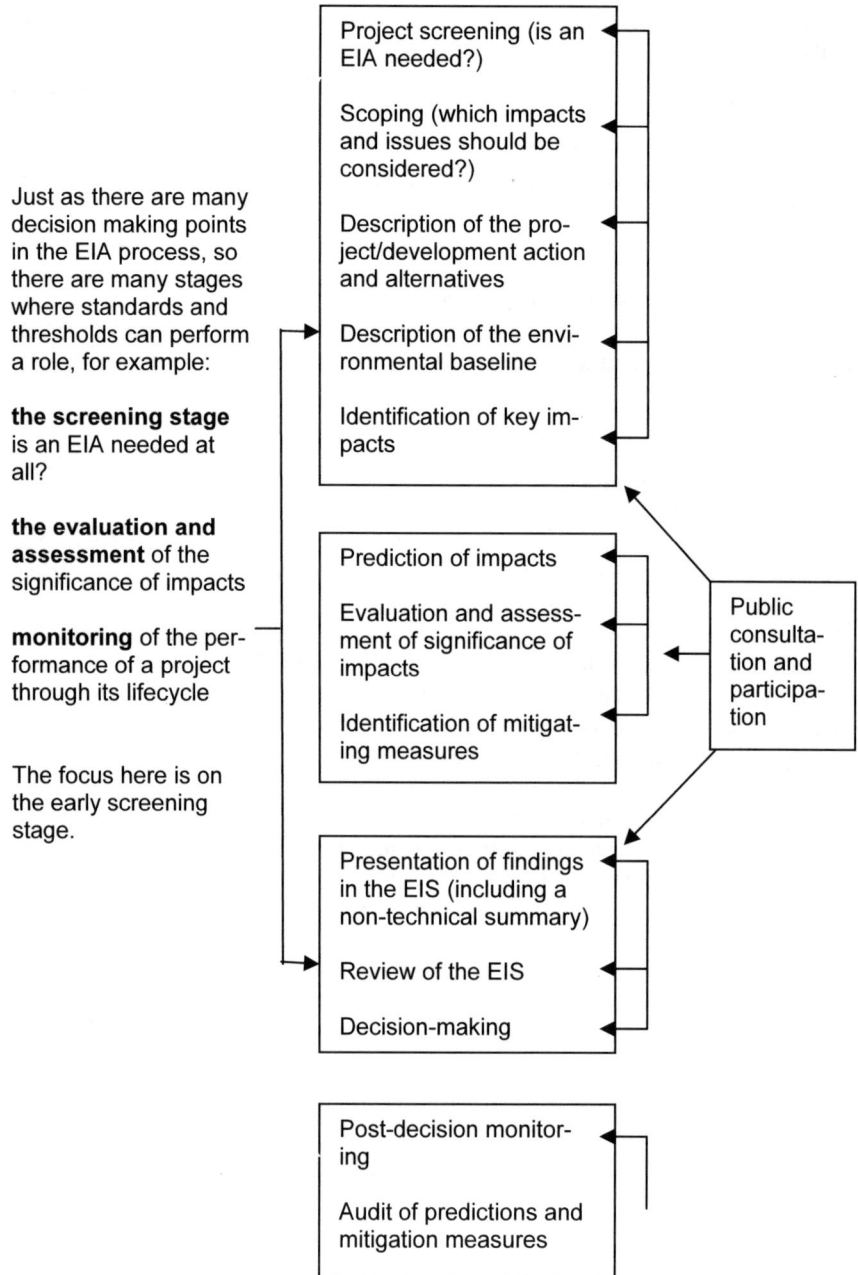

Just as there are many decision making points in the EIA process, so there are many stages where standards and thresholds can perform a role, for example:

the screening stage is an EIA needed at all?

the evaluation and assessment of the significance of impacts

monitoring of the performance of a project through its lifecycle

The focus here is on the early screening stage.

Project screening (is an EIA needed?)

Scoping (which impacts and issues should be considered?)

Description of the project/development action and alternatives

Description of the environmental baseline

Identification of key impacts

Prediction of impacts

Evaluation and assessment of significance of impacts

Identification of mitigating measures

Public consultation and participation

Presentation of findings in the EIS (including a non-technical summary)

Review of the EIS

Decision-making

Post-decision monitoring

Audit of predictions and mitigation measures

Fig. 1.1. Thresholds and standards and the EIA process (Glasson et al. 2005)

particular project, and to establish which of them are likely to be the key to the acceptability of the project. Many of the standards and thresholds used in the screening stage will also be relevant in scoping. For example, a spatial threshold, such as location in a sensitive area (e.g. a National Park, an area of archaeological importance) should indicate the direction of different types of impacts requiring study, such as landscape and ecological impacts. Scoping may also be driven by concerns from various stakeholders about the possibilities of a development crossing key thresholds (e.g. the capacity of a road), and breaching environmental standards (e.g. the air quality in a location). It is also at this stage that many of the more fuzzy considerations come to the fore, such as the impact on the quality of life in an area – including issues such as community health, cohesion and levels of crime. There is now guidance in many countries on scoping, plus EC guidance (see EC 2001).

The *description of the environmental baseline* includes the establishment of both the present and the future state of the environment, in the absence of the project, taking into account changes resulting from natural events and from other human activities. Baseline studies may include a range of data: 'hard data' from reliable sources which can be verified and which are not subject to short run change (e.g. geology); 'intermediate data' which are reliable but not capable of absolute proof (e.g. water quality and quantity, traffic flows); and 'soft data' which are more a matter of opinion or social values (e.g. landscape value, fear of crime). Standards and thresholds may be more apparent for the hard and intermediate categories. For example, baseline studies may reveal only one remaining case of a unique geological type, or already high levels of air pollution, in the study area.

Impact prediction seeks to identify the magnitude and other dimensions of identified change in the environment with a project or action, in comparison with the situation without that project or action. Prediction provides the basis for the *evaluation of significance of impacts*. Much of the evaluation activity in EIA is simple and often pragmatic, drawing on experience and expert opinion rather than on complex and sophisticated analysis. Box 1.1 provides an example of key factors used in Western Australia, where there is a particularly well developed EIA system. The softer factors, such as the degree of public interest in environmental issues likely to be associated with a proposal, are important. But the most formal evaluation method is the comparison of likely impacts against legal requirements and standards (e.g. air quality standards and building regulations). Such comparisons can highlight significant impacts requiring considerable mitigation, if the project is to proceed.

Influencing decisions on projects is at the very heart of EIA. As noted there are many decision–making stages in the EIA process, including the key stage of decision, usually by elected representatives, as to whether the project should go ahead or not, and with what conditions attached. A good EIA process and Environmental Impact Statement (EIS) should aid this decision stage by clearly highlighting any residual significant impacts associated with the proposal. The decision makers will need to assess the acceptability or otherwise of the proposal, and mitigation measures, in the context of both hard environmental standards and the softer, but equally important values and norms of the relevant stakeholders, including the lo-

---

Box 1.1. Determinants of environmental significance – Western Australia (Western Australia Environmental Protection Authority 1993)

Environmental significance is a judgement made by the Authority (West Australian Environmental Protection Authority) and is based upon the following factors:

    (a) character of the receiving environment and the use and value which society has as signed to it;
    (b) magnitude, spatial extent and duration of anticipated change;
    (c) resilience of the environment to cope with change;
    (d) confidence of prediction of change;
    (e) existence of policies, programmes, plans and procedures against which the need for applying the EIA process to a proposal can be determined;
    (f) existence of environmental standards against which a proposal can be assessed; and
    (g) degree of public interest in environmental issues likely to be associated with a proposal.

---

cal public who will have to live with the impacts.

Monitoring the outcomes of the decision is another vital, but often poorly covered, stage in the EIA process where standards and thresholds are very important. Monitoring involves the systematic collection of information on a regular basis over a long period of time. Ideally such information should include not only traditional indicators (e.g. ambient air quality, noise levels) but also causal underlying factors (e.g. the decisions and policies of the local authority and developer) which determine the impacts and may have to be changed if there is a wish/need to modify impacts. Monitoring can provide the data to audit performance of a project – including against relevant standards and legislation, and also against the predictions in the EIS. Unfortunately such monitoring still constitutes a weak link in the EIA process in many countries. It is still not mandatory under the EC EIA Directive, although there are examples of good practice in some Member States – for example in the Netherlands.

1.5 Thresholds and Standards in the Screening Stage – a More in Depth Review

1.5.1 Generic Approaches

In the European Union, Annex 1 of the EIA Directive uses a simple set of thresholds to screen those projects deemed to require mandatory EIA. For example, EIA is required for: thermal power stations of 300 MW or over; petroleum plants extracting at least 500 tonnes per day; and intensive poultry rearing projects with 60 000 places for hens or 85 000 places for broilers. For Annex 2 projects there is more discretion in the use of thresholds and standards to determine whether possible impacts on the environment might be significant. Indeed, should thresholds be used at all or should a 'case by case' approach be adopted? Some of the relative merits of the two approaches are set out in Table 1.1.

Where thresholds are used they may be of several types. The most common types are: inclusive, indicative and exclusive. *Inclusive thresholds* and criteria are those above which all projects will normally require an EIA. *Indicative thresholds,* or guidance thresholds and criteria, use a variety of criteria (often related to project size) to determine whether an EIA is required. Table 1.2 provides examples of some of the parameters commonly used for Annex 2 projects. *Exclusive thresholds,* or exemption thresholds and criteria, are those below which relevant projects will not normally require an EIA. This three-fold approach can be typified as a 'traffic lights' approach to screening (Fig. 1.2).

Table 1.1. Thresholds versus case-by-case approach to screening: advantages and disadvantages (Glasson et al. 2005)

Advantages	Disadvantages
Thresholds	
Simple to use	Arbitrary and inflexible
Quick, with more certainty	Less room for common sense
Consistent between locations	
Consistent between decisions	Difficult to set, and change
Consistent between project types	Proliferation of projects below threshold
Case by Case	
Allows more use of common sense	Can be complex and ambiguous
Flexible	Can be slow and costly
Can evolve (and improve) easily	Open to abuse by decision-makers
	Open to poor judgement of decision-makers

Table 1.2. Examples of screening parameters

Size		Time	Not between 18.00 and
Throughput	75 000 tonnes per years (waste disposal)		07.00 (industrial development
	500 000 tonnes p.a. (iron and steel production)	Designation	National Parks, AONBs
			Listed Buildings
Energy Capacity	50 MW	Legislation	Requirements for other authorisations
Height	15 metres (telecoms equipment)	Use	Previous uses (eg. contamination
Area	50 ha (mineral workings)	Materials and Emissions	Leachate Noise Odour

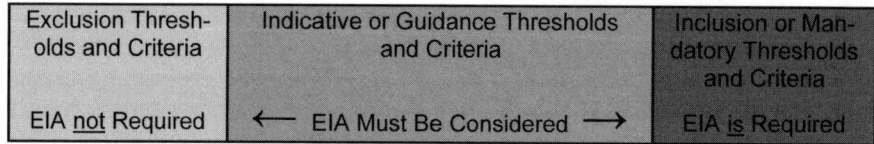

Fig. 1.2. The 'traffic lights approach' to screening (IAU 2002)

1.5.2 Recent Screening Practice in the UK

Under the original EC EIA Directive, from 1988-1998, the UK Government used inclusive thresholds for Annex 1 projects, and simple indicative thresholds for Annex 2 projects. From 1999, there was a change to the 'traffic lights' approach, introducing exclusive thresholds and more detailed indicative thresholds. Table 1.3 provides examples for some significant UK project types. This approach, and the detailed thresholds, draw on the research carried out for the UK Department of the Environment, Transport and the Regions by the IAU at Oxford Brookes University.

Further research by the IAU, for the UK ESRC (Wood and Becker 2006), reviewed the implementation of the use of thresholds in screening decisions in practice in the UK. Findings from a sample of 100 UK Local Planning Authorities (LPAs) showed that thresholds were identified as the main consideration in screening decisions by 44 % of the LPAs. 'Professional judgement' was seen as the second most important factor. Table 1.4 shows the criteria/issues considered to be important/very important in the LPAs' most recent screening decisions.

The use of the new approach to thresholds in project screening in the UK has brought more consistency of approach for particular project types, locations and decisions. But it has also brought a number of important issues. LPAs appear to be using the thresholds too easily as a short cut to making decisions , without considering fully enough the implications of particular cases. In addition, some developers are avoiding EIA by submitting projects which come just below the thresholds. There is also a need to be responsive to project types with fast moving technologies/innovations. Wind farms provide a good example of the latter in the UK –

Table 1.3. Examples of use of exclusive and indicative thresholds in the UK

Project	Exclusive threshold	Indicative threshold
wind farms	more than 2 turbines; hub height greater than 15m	visual and noise impacts are likely with 5 or more turbines, and more than 5 MW of generating capacity
industrial estate project	area of development exceeds 0.5 ha	increase in traffic, noise, and emissions more likely with development of more than 20 ha.
waste water treatment plant	area of development exceeds 1000 sq metres	site area more than 10 ha; discharges from more than 100 000 people; requires compliance with EC Urban Waste Water Directive

Table 1.4. Current issues of importance in screening decisions in practice in the UK (Wood and Becker 2006)

size/scale of project	87%	landscape impacts	61%
proximity to receptor	87%	cumulative impacts	46%
nature of project	74%	economic impacts	30%
traffic/access	65%	social impacts	30%
ecological impacts	63%	controversy/concern	25%
emissions	62%	risk of accidents	15%

with a major escalation in size from 'small' turbines (of perhaps 50 m) located inland in clusters of 10–20 to 'large' turbines (of perhaps 150 m) in very large wind farms about 10 km off the UK coast.

1.5.3 Diversity of Screening Practice in the EU

A key aim behind the Amended EIA Directive (EC 1997), was to reduce inconsistencies in the implementation of EIA in practice between the Member States. However the Five-Year Review of the implementation of the Amended Directive by the IAU (IAU 2003; Glasson et al. 2005) showed that there is still much diversity in the system. For example there is considerable variation in the annual throughput of what are regarded as EIAs. France undertakes over 7 000 each year; Germany over 1 000; the UK approximately 700; the Netherlands about 70; and Austria about 20. There is also major variation in the extent of public participation inn the EIA process. Mandatory scoping, commended by the EC in the Amended Directive, is still limited to a minority of the Member States, and the absence of systematic project monitoring is even more widespread. In addition, there is also considerable variation in the use of thresholds in screening. Table 1.5 shows some examples of screening mechanisms used in three Member States. There are also considerable variations in thresholds use by Member States for particular project types. Whilst some variation is to be expected by virtue of the differences in baseline conditions and relevant national policies between Member States, it is difficult to explain the extent of some of the variations illustrated in figures 1.3 to 1.7.

Table 1.5. Examples of screening mechanisms used in some Member States (IAU 2002)

Austria	Screening is based on a combination of thresholds and case-by-case examination. Some thresholds trigger mandatory EIA. Indicative thresholds are used with case-by-case examination; in sensitive areas the threshold values are usually halved. Exclusion thresholds are also used; new projects or modifications of existing projects that are less than 25% of the relevant threshold do not require EIA.
Greece	For Annex II projects a mandatory list is used which defines the thresholds and criteria above which EIA is always required. For projects that fall below these limits a "simplified EIA" procedure applies
Portugal	Mandatory thresholds are used for screening Annex II projects. Different thresholds apply in sensitive areas. There is no case-by-case screening.

14   John Glasson

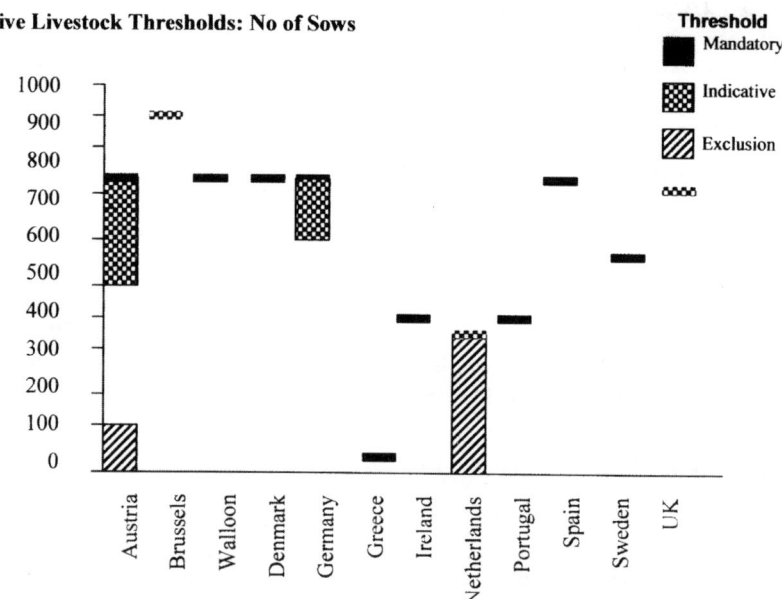

Fig. 1.3. Variations across Member States in the application of thresholds for intensive livestock projects

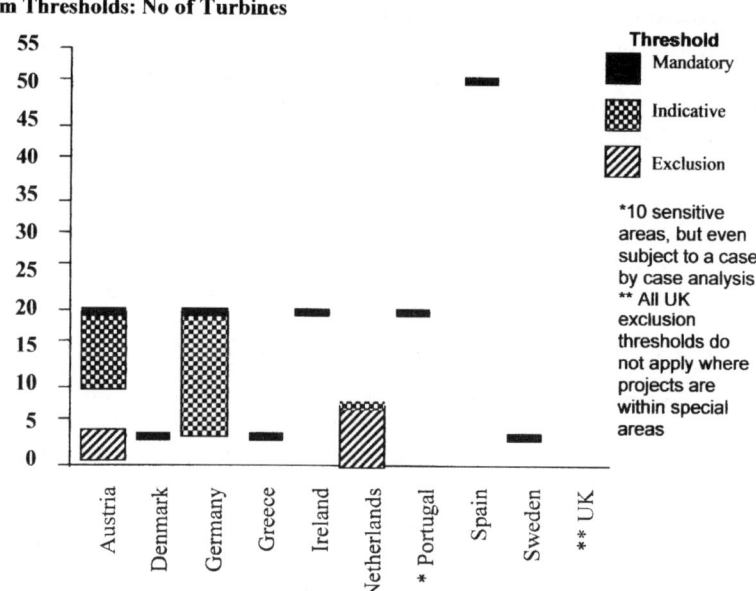

Fig. 1.4. Variations across Member States in the application of thresholds for wind farm projects

Hydroelectric Thresholds: Generating Capacity (MW)

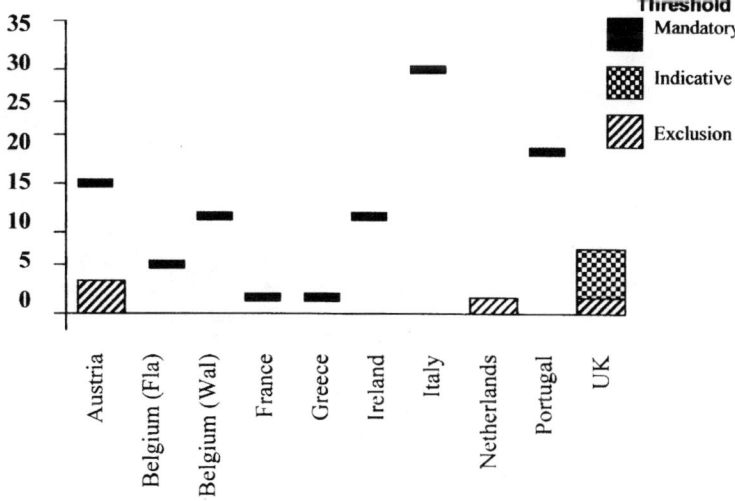

Fig. 1.5. Variations across Member States in the application of thresholds for hydroelectric projects

Industrial Estate Development Thresholds (ha)

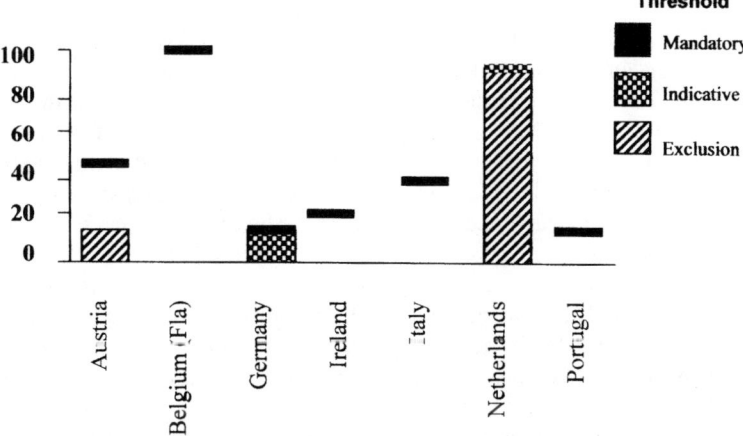

Fig. 1.6. Variations across Member States in the application of thresholds for industrial estate development projects

Afforestation Thresholds: Area (Ha)

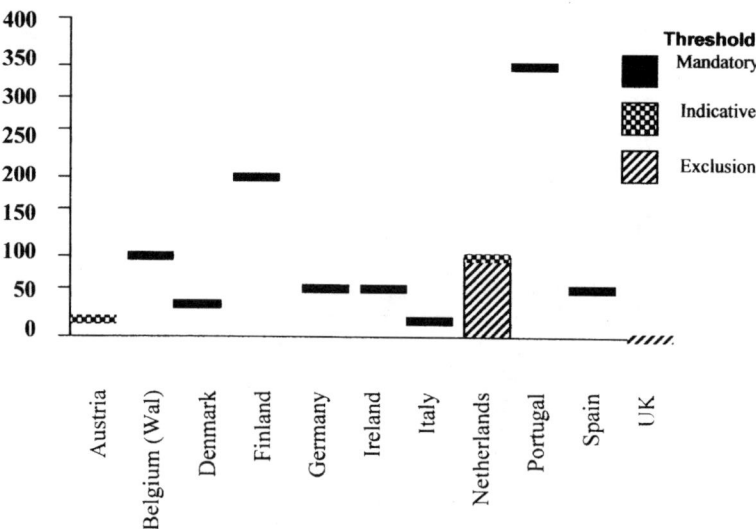

Fig. 1.7. Variations across Member States in the application of thresholds for afforestation projects

1.6 Conclusions and Recommendations

Thresholds and standards provide the very important criteria which underpin decision making in the many stages of the EIA process. They can range from those clearly embedded in legislation to those that generally represent the values and norms of a society. They can range from specific quantitative thresholds to those that are much more vague and fuzzy. They can relate to the characteristics of a project or to those of the receiving environment; they can be specific to a particular type of location, such as a National Park, or be of more general relevance.

Thresholds and standards can be used in almost all stages of the EIA process. They are important in the evaluation of the significance of impacts , decision makers must be mindful of them when finally deciding on whether to go ahead with a project, and they can provide the key criteria for monitoring and auditing project implementation. They are particularly visible in the initial screening stage, and this chapter has provided a more detailed review of evolving practice in the European Union, under the evolving EIA Directive. EU Member States have sought to rationalise their procedures in the use of thresholds (e.g. exclusive, indicative etc), but even in the 'converging EU system' there is evidence of considerable divergence in practice, especially with regard to thresholds for particular project types. This raises an important theme about thresholds and standards. They do and will vary between countries, and whilst there may be some general agreement on some thresholds and standards internationally, the experience of the recent history of the implementation of the EU EIA Directive illustrates that we should not be surprised when standards and thresholds are tailored to particular national contexts.

References

DETR – Department of the Environment, Transport and the Regions (1997) Consultation Paper on the Implementation of EC Directive 97/11/EC.Determining the Need for EIA. HMSO, London

EC – European Commission (1997) Council Directive 97/11/ec of 3 March 1997 amending Directive 85/337/EEC on the assessment of the effects of certain public and private projects on the environment. Official Journal L73/5, 3 March

EC (2001) Scoping checklist. Brussels

Faber N, Jorna R, van Engelen J (2005) A study into the foundations of the Notion of Sustainability. Journal of Environmental Assessment Policy and Management 7(1)

Glasson J, Therivel R, Chadwick A (2005) An Introduction to Environmental Impact Assessment: 3rd Edition. Routledge, London

IAU – Impacts Assessment Unit, Oxford Brookes University (2003) Five Years Report to the European Parliament and the Council on the Application and Effectiveness of the EIA Directive. Website of DG Environment, EC, Internet address:

http://europa.eu.int/comm/environment/eia/home.htm.m, last accessed on 05.03.2007

Kozlowski J (1968) Threshold theory and the sub-regional plan. Town Planning Review 39(2):99–116

Lawrence D (1997) The need for EIA theory building. Environmental Impact Assessment Review 17(2):79–107

ODPM – UK Office of Deputy Prime Minister (2005a) A Practical Guide to the Strategic Environmental Assessment Directive. ODPM, London

ODPM (2005b) Sustainability Appraisal of Regional Spatial Strategies and Local Development Documents. ODPM, London

RCEP – UK Royal Commission on Environmental Pollution (1998) 21st report: Setting Environmental Standards. HMSO, London

Sassaman RW (1981) Threshold of concern: a technique for evaluating environmental impacts and amenity values. Journal of Forestry 79:84–86

Therivel R, Wilson E, Thompson D, Heaney D, Pritchard D (1992) Strategic Environmental Assessment. Earthscan, London

Therivel R, Walsh F (2006) The SEA Directive in the UK. Environmental Impact Assessment Review (forthcoming)

UNWCED – UN World Commission on Environment and Development (1987) Our Common Future. OUP, Oxford

Weston J (2000) EIA, Decision-making Theory and Screening and Scoping in UK Practice. Journal of Environmental Planning and Management 43(2):185–203

Wood G, and Becker J (2005) Discretionary judgement in local authority decision making: screening development proposals for environmental impact assessment. Journal of Environmental Planning and Management 48(3):349-371

World Bank (1995) Environmental assessment: challenges and good practice. World Bank, Washington DC

2 Standards and Thresholds in German Environmental Law

Eike Albrecht

Junior partner Chair of Civil Law and Public Law with special references to the Environment and Law of Europe, Centre for Law and Administration, Brandenburg University of Technology (BTU), Cottbus, Germany

2.1 Introduction

This chapter aims to discuss the German practice of environmental regulation, in particular setting environmental standards and thresholds. In almost all fields in German Environmental Law, the use of standards and thresholds is commonly practiced. Therefore a selection is necessary, and the topic will be discussed with particular reference to the more European- influenced Emission Control Law and the typical national sector of Soil Protection Law.

Section 2.2 gives an introduction to the function of standards and thresholds in German Environmental Law including a discussion on the necessity of standards and thresholds. Section 2.3 provides an overview of the main limitations of regulating Environmental Law by standards and thresholds. In this section, the role of standards and thresholds in technical interferences and in the interplay between undefined legal terms and detailed standards is described. Section 2.4 describes the main sources of standards and thresholds in German Environmental Law and discusses the specific problems of each source. Finally, the problematic of transposing European standards into German Environmental Law is highlighted from the viewpoint of meeting the requirements of the European Court of Justice with respect to legal security and reliability. Section 2.5 illustrates the particular features of environmental regulations on the example of Emission Control Law and Section 2.6 discusses standards and thresholds in the field of Soil Protection Law. Section 2.7 provides some conclusions.

Standards and Thresholds for Impact Assessment. Edited by Michael Schmidt, John Glasson, Lars Emmelin and Hendrike Helbron. © 2008 Springer-Verlag

2.2 Functions of Thresholds and Standards in German Environmental Law

Standards and thresholds are necessary for the systematic and reasonable execution of German Environmental Law. Standards and thresholds make Environmental Law applicable and functional. Without standards and thresholds, the addressee of an environmental provision or of an administrative act (decision, permission or prohibition) simply does not know what is expected of him/her by the authority, society or the state. Provisions of Criminal Law or ethical codes of conduct, like the ten commandments, might be clear by relatively simple regulations, like "You shall not steal", "You shall not bear false witness against your neighbour" or "You shall not covet your neighbour's house", "you shall not covet your neighbour's wife, or male or female slave, or ox, or donkey, or anything that belongs to your neighbour" (Exodus 20:15-17). In Environmental Law, typically the case is not this simple. The general prohibition of using the environment is not possible, especially not in an industrialised world, and indeed may not be necessary. Different environmental sectors may be able to remedy or repair themselves. Of course this depends on the respective environmental sector and in particular on the respective use. Over exploited fish stocks can recreate themselves; poisoned or contaminated water or soil can be remedied; contaminants of the air can be washed out by the rain. This means soil contamination is not prohibited in general, but only in respect of a certain grade. Of course the introduction of substances into the water is allowed but – in the case of dangerous, toxic or noxious substances – only in very small doses. The emission of hazardous substances or even simple dust is not forbidden in general, and can be allowed by law or license for specific amounts, regulated by standards and thresholds and in specific cases.

Of course, an overuse of the environment bears the risk that it cannot be recreated or remedied. With respect to climate change, mankind might have already overused the environment irreversibly (Latif 2006). This means that although a total ban on specific harming substances or total prohibition of specific activities is in most cases not necessary, a general allowance of unlimited use of the environment is also not possible. Therefore, a system of standards and thresholds must be found to define the acceptable amount of use of the environment. The more detailed such a system, the more likely it is to meet the acceptable level of environmental usage. But a very detailed system can bring complications and difficulties for individuals, companies and the state itself in their use of the environment. When creating standards, a balance must be found between the optimum (e.g. zero emission or zero contamination) and practicable and economical measures. To find the right balance may be the most difficult task in the creation of standards and can be a sign of a successful environmental policy.

How can standards and thresholds be developed? At first, profiles of goods worthy of protection (e.g. soil, health) must be developed. Secondly, profiles of risks for these goods must be elaborated. This contains in particular the connection between legal values and goods and concrete dangers. One useful and well known tool is technology assessment. For specific cases of projects and plans, EIA, SEA or licensing procedures can also be used. The results of such assessments could be used for directing specific requirements, set by standards, to a specific plan or project. Authorisation procedures can be used also for licensing the production of dangerous goods, for examples of chemicals based on the coming European chemical policy REACH which will be set in force probably by the first half of 2007 (Büchler 2006; Knopp 2006; fur further details see Albrecht and Krause 2006).

2.3 Main Limitations of Standards and Thresholds in German Environmental Law

The difference between standards and thresholds in this context should be – as defined in the introduction – the degree of legal binding. A threshold in this sense is a criterion for assessing environmental impacts, for example in an EIA-procedure. A threshold is usually the result of use of different standards for specific environmental uses. A standard therefore shall be regarded as a legally binding and legally fixed value.

2.3.1 Environmental Law as Technical Law

In the modern, technical-industrial age environmental protection recurs often to technical cases. The use of advanced technical possibilities is on the one hand a source of environmental problems (use of cars, pollution by industries, contamination of waters and soil by the use of artificial fertilizers etc.). But, on the other hand it can help to provide the solution to environmental problems. The central objective of Environmental Law is therefore to provide the addressee of environmental provisions or administrative decisions (mainly industries, sometimes individuals) and the executive power (administration) with relevant and adequate standards and thresholds (Feldhaus 2000). Usually the law contains – as described before – only general principles or undefined legal terms. For example, Art. 5 of the Federal Emission Control Act obliges the operator of an installation

- to ensure that precaution is taken according to the best available technology,
- to protect (…) from harmful effects, and
- to ensure a high level of protection for the environment.

But these very open obligations are not sufficient with respect to the execution of the Federal Emission Control Act, because technicians cannot use such terms for developing machines, technical apparatus and procedures and such open terms do not meet the requirements of the principle of legal certainty (Art. 20 para 3 of the German Basic Law). On the other hand, technical standards cannot be included in the respective acts, because this would overload them, and overburden legislative and executive powers.

2.3.2 Standardisation in Environmental Law

In German Environmental Law provisions of a high number of undefined legal terms are used. Some of them contain no specific standard or a fixed and determined standard. But especially in the technically oriented Environmental Law, terms are used which have a certain dynamic. This has significant advantages in the process of creating a modern Environmental Law. For example, the term "best available technology", which originated from the IPPC Directive, is an undefined legal term and is used in Art. 3 para 6 of the Federal Emission Control Act or in Art. 12 para 1 of the Waste Management Act and in an annex to the respective acts with details (Kloepfer 2004 p.147). It is clear, that the best available technology twenty years ago is significantly different to the best available technology of today. Other examples for such dynamic terms are:

- Actual status of science and technology which is an improved standard in comparison to the best available technology in Art. 1 para 2 No. 5 of Product Liability Act or Art. 7 para 2 No. 3 of the Act on the Peaceful Use of Nuclear Power
- Commonly accepted rules of technology in Art. 3 para 1 S. 2 of the former Act on Apparatus Security

This opens a mainly static formation like a law to dynamic progresses (see German Federal Constitutional Court, BVerfGE 49 p.89). The main disadvantage of using undefined legal terms is the difficulty in the execution process. An undefined legal term has certain insecurity in respect of what is really meant, for example, by "current state of technology". The operator of an installation has a different opinion on this than, for example, an authority or an environmental NGO. Therefore it is necessary for a legally secure and effective execution of Environmental Law, to assist the deciding authorities, local, regional or federal, decide the usual cases of practice. The optimum would be to have measurable values, in particular standards and thresholds (Feldhaus, UPR 1982 p.137).

To use undefined legal terms in practice, sub law provisions or norms are regularly needed to clarify their meaning. There are cases in German Environmental Law where the lack of standards and thresholds, regulated by sub law or norms, led to the non-usability of environmental provisions. The German Federal Administrative Court decided that Art. 41 of the Federal Emission Control Act which regulates noise protection in road and railroad construction processes is unusable because of emission sub law standards (German Federal Administrative Court, BVerwGE 61 p.295 (299), later corrected by BVerwGE 71 p.150 (154f). The ordinance for setting standards and thresholds in such cases, based on Art. 43 para 1 sent. 1 no. 1 of the same Act, was simply never enacted. A similar case is the obligation of an operator of an installation, on the base of the Federal Emission Control Act, to use process heat for energy saving purposes (Art. 5 para 1 No. 4 of the Federal Emission Control Act, in the meantime revised to a general obligation for energy saving). This provision could not be used by the authorities in a single case, because the necessary sub law provisions to define this relatively open obligation of the operator of an installation does not exist (Feldhaus 2000 p.171).

These cases outline the importance of standards and thresholds set by sub law provisions or norms. A system of undefined legal terms in the law, filled out by sub law provisions and norms, leads to a dynamic law. The law itself is then an up to date system which guarantees a high level of environmental protection.

In German Environmental Law, the need for standards and thresholds is regarded as saturated (Feldhaus 2000 p.171). The last existing gap in soil protection law was filled by the Ordinance on the Federal Soil Protection Act from the 12th July 1999, though there was a major criticism with regard to its understandability and complexity (Higher Administrative Court Lüneburg, NVwZ 2000 p.1194 (1195); Ebermann-Finken 2000 pp.186 f)

Standards can be given a range of names. Typical names are: trigger values, action values, precautionary values, alert thresholds, limit values or target values. Different laws have different objectives and therefore different values and names. But, in a lot of cases, the different names should not hide the fact that the differences are in some cases marginal. This provides evidence that there is no convincing and consistent system with respect to standards and thresholds in German Environmental Law.

2.4 Sources of Standards and Thresholds

A standard usually contains a legal limitation for the use of the environment. Typically, a standard is set by the State. This can be the legislative power itself, but this is not very often the case. The setting of detailed provisions, in particular in technical and environmental law, like standards and thresholds are done mainly in provisions which rank below formal parliamentary acts.

2.4.1 Standards and Thresholds in Ordinances

Standards and Thresholds are often set by the executive power in the form of ordinances, but only if an allowance can be found in a formal law. More requirements for ordinances are stated in Art. 80 of the German Basic Law, such as the obligation to cite the legal allowance for enacting the ordinance. An ordinance, which is a sub-law regulation, is binding not only for the administration but also for third parties, in particular private individuals and companies. The specification of the addressee of an ordinance must be regulated in the ordinance itself. There are ordinances which have to be followed by everyone as far, like the Ordinance on Traffic Rules (*Straßenverkehrsordnung*) addressed to almost everybody. On the other hand the 13th Ordinance on the Federal Emission Control Act (*13. BImSchV über Großfeuerungs- und Gasturbinenanlagen*) is addressed only to operators of large power and heating installations.

2.4.2 Standards and Thresholds in Administrative Regulations

Another possibility for setting standards and thresholds is in the form of a general administrative regulation or administrative circular *(allgemeine Verwaltungsvor-schriften)*. General administrative regulations (like Technical Norms for Noise or Waste; *TA Lärm/TA Abfall*) are directly binding only to the administration. But, they are indirectly binding also to third parties and the courts, as far as these norms are accepted as general expert standards. The creation procedure of such regulations is partly regulated in the specific acts (like Art. 51 of the Fed. Emission Control Act), but it is mostly not regulated.

2.4.3 Private Standards and Thresholds

In a lot of cases, a detailed system of administrative standards does not exist. Quite often this gap is filled out by norms of private standardisation organisations, like the German Industrial Standardisation Organisation or the International Standardization Organization (ISO). Such private norms have no direct legal binding. But in a lot of cases these norms are used in addition to legal terms or for defining legal terms. For example, the question of whether a product is without fault can be answered by using private norms. A product which does not fulfil the requirements of a German Industrial Norm (DIN) is regarded as having a fault which opens the buyer to some legal possibilities (compensation, return, etc.).

In German Environmental Law, there are thousands of private norms used for specifying environmental obligations. For example with respect to noise reduction more than 100 guidelines exist on noise reduction methods and noise values. With respect to air pollution more than 100 guidelines exist for emission reduction methods, plus some hundred guidelines for analysing and measuring methods. With respect to working security closely connected with Environmental Law, more than 1000 German Industrial Norms exist as approved rules of security technology. Around 200 norms are in use with respect to water, sewage water and sludge analyses and with respect to soil protection, and there are about 70 international (ISO) norms and 40 national norms in use or in preparation (Feldhaus 2000 p.177).

In Environmental Law, private norms are often introduced into ordinances by setting a link. For example, measuring methods for soil protection investigations are regulated by the Ordinance to the Federal Soil Protection Act. In this ordinance there is a link to private German Industrial Norms for defining the requirements on specific investigation methods (location, duration, analysis methods etc.).

But could private norms be used as legal standards? An objection could be that they are not set by constitutional bodies. On one side, if a law or a jurisdiction recurs on norms and standards which are determined and fixed (and therefore are not dynamic, in the sense that a certain standard is to be seen in relation to the technical and scientific standard) this is allowed, as far as the state takes over the responsibility in terms of content and not only in a formal mode. This is in particular the case, if a binding law provision directs specific matters or a *modus operandi* to a private norm. In this case, the private norm is regarded as drawn into the legal provision and must be followed. Typical examples are German Industrial Norms (DIN) for remediation measures in case of contaminated sites or private norms for noise levels for employees, which describe in a binding way the detail of what is allowed.

But with respect to the fast moving area of technology, with more attractive dynamic (and therefore moving forward in relation to the current state of technical and scientific progress) private norms, the situation is different. To recur on private norms may conflict with the requirement for legal certainty of Art. 20 para 3 of the German Basic Law.

2.4.4 Transposition of European Standards into German Law

The main source of sub law standards and thresholds is in many cases European Law, in particular by Directives and/or Annexes. European Directives are directed to the Member States and have to be transposed into national law. The Member States are principally free in their decision how to transpose such provisions into their legal system, as long as certain standards with regard to legal security and reliability are followed. The European Court of Justice decided that the transposition of European standards into German Environmental Law by administrative circulars or recommendations or even by a stable judicial decision practice is not sufficient (Knopp and Albrecht 2005). The transposition of European standards by an ordinance is sufficient to meet these requirements. This means the transposition of European environmental standards into German Environmental Law must be done by a minimum of an ordinance. In fact, the fields of Environmental Law which are mainly influenced by European law, like emission control, waste or water law have a high density of ordinances to define standards and thresholds. For example, there are more than 30 ordinances to the Federal Imission Control Act, and most of them transposing European standards into German law.

On the other side, there are specific law fields which are mainly nationally influenced, in particular in the field of soil protection. Other than a soil protection strategy no direct European activities exist for the regulation of soil protection. Therefore the national legislation is a little bit freer to regulate standards and thresholds. There is one Ordinance on the Federal Soil Protection Act which contains a high number of different values. Furthermore, there is an administrative circular, published in the Federal Bulletin (*Bundesanzeiger*) 161a of 18th June 1999, which contains a system for defining further standards. Because this system is not clear, the Higher Administrative Courts of Lüneburg accepted the use of a different list with standards and thresholds for filling out soil protection obligations (Knopp 2001).

2.4.4 Private Standards and Thresholds for Assessing Environmental Effects

Standards set by the state, of course, are used for assessing (negative or positive) effects on the environment. But also, private standards could be used for this purpose. Some of the thousands of private standards known in German Environmental Law are being used for specifying environmental obligations. For example, with respect to noise reduction more than 100 standards and thresholds – regardless if enacted as private or administrative standards – are used for defining limits (Feldhaus 2000 p.177). But they can be used also for the assessment of environmental effects on different media. An example of private norms is the well known Handbook of the German Engineering Association (VDI) "Noise reduction – criteria for the evaluation of effects of noise and vibrations" which contains the standard 2058 (*VDI-Richtlinie*), page 1 "Evaluation of working noise in the neighbourhood". Such thresholds for assessing environmental effects are usually not measurable and therefore they are more difficult to set. If they have a legal value, for example as a basis for a license or a consideration in an EIA procedure, they have to be calculated somehow in a standardized way to make the results judiciable. Otherwise the results could be questioned very easily and the advantage of standards of threshold of clear and equal decisions would be eliminated.

2.5 Standards and Thresholds in Federal Emission Control Law

The German Federal Emission Control Act[1] is highly influenced by European Law and is therefore a good example of the functioning and transposition of European standards into national law. With respect to air quality, for example, the Council Directive 96/62/EC of 27 September 1996 on ambient air quality assessment and management, is a so called 'mother directive'. This mother directive is amended by 'daughter directives', for example:

- Directive 97/68/EC,
- Directive 98/69/EC,
- Directive 1999/30/EC, or
- Directive 2002/3/EC.

These Directives are transposed into German Environmental Law for example by the 7th amendment of the Federal Emission Control Act of 11.9.2002 (Fed. Law Gazette I p.3622) and the 22nd Ordinance for the Implementation of the Federal Emission Control Act, of 11.9.2002 (Fed. Law Gazette I p.3626). This regulation contains standards and thresholds which have to be found in a specific procedure, specified in Art. 4 of the Air Quality Directive (96/62/EC):

---

[1] A translation of the German Federal Emission Control Act is available in Mulloy M, Albrecht E, Häntsch T (2001) German Environmental Law, p.285ff.

- alert thresholds, and
- limit values.

An additional target value for ozone, introduced by the Directive 2002/3/EC of the European Parliament and of the Council of 12 February 2002 relating to ozone in ambient air, was transposed by the 33rd Ordinance for the Implementation of the Federal Emission Control Act, of 13.7.2004 (Fed. Law Gazette I p.1612). This Directive contains a target value for health protection in case of longer lasting pollution of 110 µg/m3 as average value during 8 hours. For the protection of vegetation a target value of 200 µg/m3 as average value during 1 hour and 65 µg/m3 as average value during 24 hours was set. Furthermore, a level for information obligations was introduced by Art. 6 of the Directive 2002/3/EC. So this could be defined as an information threshold for information of the public. This must be done if 180 µg/m3 ozone, as average value during 24 hours, is reached.

2.6 Standards and Thresholds in Federal Soil Protection Law

The German Soil Protection Act[2] is a good example for describing functions and the role of thresholds and standards in German Environmental Law because soil protection law is genuinely a national field of law (for details see Albrecht 2003). There are very few international or European attempts to regulate the use or the protection of soil in an international context.

2.6.1 Values in the Federal Soil Protection Act

The relevant law for this topic is the Federal Soil Protection Act. Art. 8 para 1 Sentence 1 of this Act defines:

- trigger values (no. 1) and
- action values (no. 2).

A *trigger value* is a value which, if exceeded, means that investigation with respect to the individual case in question is required, taking the relevant soil use into account, to determine whether a harmful soil change or site contamination exists. This means, if the trigger value is met or exceeded, the authority has to decide if further investigations, remediation or other measures are necessary.

An *action value* is a value for impacts or pollution which, if exceeded, shall normally signal the presence of a harmful soil change or site contamination, taking the relevant soil use into account, and means that measures are required. Thus, if the action value is met or exceeded, the authority has to start activities, because

---

[2] A translation of the German Federal Soil Protection Act is available in Mulloy M, Albrecht E, Häntsch T (2001) German Environmental Law, pp.254ff.

the level of having to face a danger for one or more of the protected legal components (like soil, water, groundwater or health). Such activities could include:

- warding off harmful soil changes; these shall include requirements relative to the handling of excavated, removed, and treated soil material;
- remediation of the soil and of contaminated sites, especially with regard to determination of the rehabilitation objective;
- the extent of decontamination measures and safeguarding measures that prevent spreading of pollutants in the long term; and
- protection and restriction measures.

Art. 8 para 2 of the Federal Soil Protection Act furthermore defines the so called precautionary value which is, from amongst the different values, the most important for EIA. A precautionary value is a soil value which, if exceeded, normally means there is reason that concern for a harmful soil change exists, taking geogenic or wide spread, settlement related pollutant concentrations into account. This value can be used for defining permissible additional pollution loads, and requirements for prevention or reduction of substance inputs. The competent authority shall order measures if the precautionary value is exceeded.

2.6.2 Values in the Federal Soil Protection and Contaminated Sites Ordinance

Art. 8 of the Federal Soil Protection Act just defines trigger, action and precautionary values. But, a measurable magnitude which makes such a value a standard or threshold is not regulated in the law itself. This is done in the Annex 2 of the Federal Soil Protection and Contaminated Sites Ordinance (Table 2.1).

Table 2.1. Federal Soil Protection and Contaminated Sites Ordinance: Action Values pursuant to Article 8 para 1, Sent. 2 no. 2 of the Federal Soil Protection Act for the direct intake of dioxins/furans at playgrounds, in residential areas, parks and recreational   facilities, and plots of land used for industrial and commercial purposes (in ng/kg dry matter, fine soil, analysis according to Annex 1)

	Action values [in ng/kg dry matter, fine soil, analysis according to Annex]*)			
Substance	Playgrounds	Residential areas	Parks and recreational facilities	Land used for industrial and commercial purposes
Dioxins/ furans (PCDD/F)	100	1,000	1,000	10,000

*) Sum of the 2,3,7,8-TCDD-toxicity equivalents (according to NATO/CCMS).

These values are divided into different paths of intake and differentiate between different uses of the ground. For example, in the soil-human health pathway – direct contact – (No. 1 of Annex 2) there are different standards set for playgrounds, residential areas, parks and recreational areas and land used for industrial and commercial purposes. Of course, the standards are high for playgrounds and relatively low for industrial sites. For the soil-plant pathway (No. 2 of Annex 2) the ordinance divides between agriculture, vegetable land and grassland.

There are standards for trigger values (direct intake), for action values (direct intake) and for precautionary values for metals (in mg/kg dry matter, fine soil, *aqua regia* decomposition). The main disadvantage of the Federal Soil Protection and Contaminated Sites Ordinance is that there are only a small number of different toxic substances defined as standards and thresholds. In particular, action values are only defined for dioxins/furans (Table 2.2).

Table 2.2. Federal Soil Protection and Contaminated Sites Ordinance: trigger values pursuant to Article 8 para 1, Sent. 2 no. 1 of the Federal Soil Protection Act for the direct intake of pollutants at playgrounds, in residential areas, parks and recreational facilities, and plots of land used for industrial and commercial purposes (in mg/kg dry matter, fine soil, analysis according to Annex 1)

Trigger values [mg/kg dry matter, fine soil]				
Substance	Playgrounds	Residential areas	Parks and recreational facilities	Land used for industrial and commercial purposes
Arsenic	25	50	125	140
Lead	200	400	1,000	2,000
Cadmium	10[1]	20[1]	50	60
Cyanides	50	50	50	100
Chromium	200	400	1,000	1,000
Nickel	70	140	350	900
Mercury	10	20	50	80
Aldrin	2	4	10	--
Benzo(a)pyrene	2	4	10	12
DDT	40	80	200	--
Hexachlorobenzene	4	8	20	200
Hexachlorocyclo-hexane (HCH-mix or β-HCH)	5	10	25	400
Pentachlorophenol	50	100	250	250
Polychlorinated biphenyls (PCP$_6$)[2]	0.4	0.8	2	40

[1] In back gardens and small gardens where children stay and food plants are grown, the trigger value 2.0 mg/kg TM must be applied in the case of cadmium.

[2] Where PCB total contents are determined, the measured values must be divided by a factor of 5.

The situation with regard to trigger values seems to be better. In fact, the number of dangerous substances is higher in comparison to the action values. But there are a lot of dangerous substances missing. Standards for those missing substances have to be calculated by the mentioned method, published in the Federal State Bulletin 161a of 1999 which is hard to handle, even for experts (Ebermann-Finken 2000 pp.186f). It must be mentioned that the Federal Soil Protection and Contaminated Sites Ordinance has not been amended since coming into effect. This is not a proof for a very well elaborated regulation; it gives evidence for a missing political will to improve the situation with regard to soil protection standards.

2.7 Conclusions and Recommendations

Standards and thresholds are regulated differently in German law. There are standards and thresholds regulated in ordinances, in general administrative rules and in private standards. Where transpositions of European standards are required at least an ordinance is necessary; general administrative regulations and private norms are not sufficient. The use of standards and thresholds in German Environmental Law for the different kinds of standards is not consistent. Each environmental sector uses different names for sometimes the same values.

Relevance for EIA mainly relates to for values concerning precautionary measures and planning measures. Trigger and action values are relevant mainly for administrative orders (e.g. order for a decontamination of a contaminated site) or licensing procedures. For the future, it can be expected that private norms will gain more importance in particular with respect to complex evaluation scenarios. Private norms are used for assessing negative effects on the environment, in particular for products, like life cycle assessments (LCA) or eco balances. With respect to industrial activities private norms like ISO 14000 for the organisation of enterprises, will gain importance when more and more control obligations of the states are transferred to the enterprise itself.

Though there are significant advantages in using private capacity for the creation of standards – industry has the respective capacity and knowledge to create such (partially very complicated and detailed) standardisation systems – there are specific risks for the state. The state as the original responsible and democratically legitimated body for defining standards and thresholds has to take care, that this constitutional obligation and responsibility is carried out or as a minimum carefully observed. The more this origin obligation to define standards is shifted to private bodies, the more the state loses the right to set standards. It has to be kept in mind that standardisation organisations are mainly influenced by the industry and that they are not democratically legitimated. This requires as a minimum, that the state, in situations where such private standards are used for defining standards and thresholds, takes care and assesses if the respective norms fulfil the legal requirements with respect to environmental protection or other state purposes.

References

Albrecht E (2003) Die Kostenkonzeption des Bundes-Bodenschutzgesetzes (The cost's conception of the (German) Federal Soil Protection Act). Peter Lang, Frankfurt/M

Albrecht E, Krause L (2006) Produktbezogener Umweltschutz und die neue europäsche Chemikalienpolitik (REACH) (Product-related environmental protection and the new European chemical's policy (REACH)). StoffR 2006, pp. 243–247

Büchler F (2006) Neues aus Brüssel (News from Brussels), EurUP 2006, pp. 273–275

Ebermann-Finken R (2000) Investitionssicherheit durch die neue Bundes-Bodenschutz- und Altlastenverordnung (Securing investments by new ordinance on soil protection). In: Knopp, L./Löhr, D., Bundes-Bodenschutzgesetz in der betrieblichen und steuerlichen Praxis (Federal Soil Protection Act in industrial practice). Recht und Wirtschaft, Heidelberg, pp. 168–195

Feldhaus G (1982) Entwicklung und Rechtsnatur von Umweltstandards (Development and legal nature of environmental standards), UPR 1982, pp. 137ff

Feldhaus G (2000) Umweltschutz und technische Normung (Protection of the environment and technical norms), In: Hendler R, Marburger P, Reinhardt M, Schröder M (eds) Jahrbuch des Umwelt- und Technikrechts 2000 (Yearbook of Environmental and Technical Law 2000). Erich Schmidt, Berlin, pp. 169–189

Kloepfer M (2004) Umweltrecht (Environmental law), 3rd Ed. Beck, Munich

Knopp L (2001) Bundes-Bodenschutzgesetz und erste Rechtsprechung (Federal soil protection law and first court decisions), DÖV 2001, p.441ff

Knopp L (2006) REACH – Eine Einführung in die Thematik (Introducton to the subject), in: Knopp, L./Boć, J./Nowacki, K., Aktuelle Entwicklungen europäischer Chemikalienpolitik (REACH) und ihre Auswirkungen auf deutsches und polnisches Umweltrecht (Actual development of the new European Chemical's Policy – REACH – and their impacts on German and Polish Environmental Law). Recht und Wirtschaft, Heidelberg, p.45–49

Knopp L, Albrecht E (2005) Transposition of the SEA Directive into national law. In: Schmidt M, João E, Albrecht E (eds) Implementing Strategic Environmental Assessment. Springer, Heidelberg, pp. 57–67

Latif M (2006) Globale und langfristige Strategie gegen den Klimawandel erforderlich (Global and long-term strategies necessary for climate change), EurUP 2006, pp. 267–270

3 Standards and Thresholds for EA in Highly Polluted Areas – The Approach of Ukraine

Dmytro Palekhov[1], Michael Schmidt[1] and Gennady Pivnyak[2]

1 Department of Environmental Planning, Brandenburg University of Technology
  (BTU) Cottbus, Germany
2 National Mining University (NMU), Dnepropetrovsk, Ukraine

3.1 Introduction

This chapter aims to discuss Ukrainian practice of environmental regulation (i.e. setting environmental norms) in the context of applying environmental assessment procedures – EIA and SEA. For highly polluted areas, which are plentiful in Ukraine, the development of standards and thresholds is of key importance for the successful application of modern methods of environmental assessment, since the adaptation of these methods can help provide a higher level of protection for the environment. The main focus of this chapter is to discuss the necessary changes in Ukrainian practice of environmental regulation, which can contribute to a more environmentally thoughtful management of the economy. It provides an example of how a major environmental problem, in this case air pollution, can be a driver of change in environmental standards. It also reflects a concern to improve environmental performance to meet international standards, in particular EU standards.

As a background for understanding the objectives for environmental regulation with regards to environmental assessment for highly polluted areas, Sect. 3.2 describes the present environmental situation in Ukraine and evaluates the current development of environmental policy since Ukraine's independence. Sect. 3.3 provides a legal definition of standards and thresholds, a detailed overview of their development principles and functions and presents the main deficiencies that limit the potential of environmental assessment for regulating impacts. Sect. 3.4 illustrates the particular features of environmental regulation for the example of atmospheric air protection. A description of the procedure for setting thresholds and some examples of threshold values will allow for a better understanding of the Ukrainian approach in this field. The chapter ends with conclusions and recommendations in Sect. 3.5 highlighting which reforms could contribute to an improvement of environmental assessment methods complying with environmental policy targets.

Standards and Thresholds for Impact Assessment. Edited by Michael Schmidt, John Glasson, Lars Emmelin and Hendrike Helbron. © 2008 Springer-Verlag

3.2 Basic Directions in Development of Ukrainian Environmental Policies

In Ukraine, the issue of the quality of the environment is high on the agenda. It should be noted that the Ukrainian economical model is based on the extensive utilisation of natural resources, with the economy dominated by some of the most environmentally polluting industries – mining and quarrying, energy, chemical – which together constitute approximately 60 % of the national GDP. A considerable majority of these industrial facilities were built in the 1950s-1970s and now do not comply with the current environmental requirements. A number of hazardous industries account for Ukraine's high level industrial waste per capita, and pollutant emissions into the atmosphere. As a result, Ukraine ranks poorly not only in Europe but also in the world. According to the Environmental Sustainability Index, released in Davos, Switzerland, at the annual meeting of the World Economic Forum 2005, among 146 counties Ukraine ranks the 108th (Esty et al. 2005).

Fig. 3.1 shows the results of the environmental pollution analysis for 2004, which was carried out by the Council for Productive Forces Studying of the National Academy of Sciences of Ukraine (Baranovsky et al. 2004). The atmospheric pollution was analysed based on the content of dust, sulphur dioxide, nitrogen dioxide, carbon monoxide and specific pollutants, typical for individual settlements (based on data from the environmental monitoring stations in 226 settlements). The surface water pollution was analysed based on an organoleptic and toxicological study of the sanitary condition of water bodies (based on data from 200 hydrometric stations). Soil pollution was characterised by the content of pesticides, mineral fertilizers and also by the environmental integrity of landscapes.

As it can be seen, the crisis situations are not localised; they affect the whole industrial agglomerations, mining and quarrying areas and adjacent territories – for instance, the Krivoi Rog iron-ore basin, the West-Donbass coal basin, and the Dnepropetrovsk-Dneprodzerzhinsk-Novomoskovsk agglomeration. High levels of different pollution types (dust, chemical, radiation, noise and others), especially in regions with high technogenic pressure, sharpen the necessity to develop environmentally sound development strategies (Schmidt et al. 2005). The modern environmental policy began to evolve in Ukraine after the country gained independence in 1991. For the first time, the Ukrainian Constitution of 1996 provided in Art. 16 that the state is obliged to "ensure ecological safety and to maintain the ecological balance on the territory of Ukraine", and proclaims in Art. 50 that "everyone has the right to an environment that is safe for life and health, and to compensation for damages inflicted through the violation of this right".

The "Complex programme for implementing at the national level decisions of the World Summit on Sustainable Development for 2003-2015" adopted by the Cabinet of Ministers Decree No. 634 specifies environmental policy targets according to principles of sustainable development. Absolute and relative objectives to be achieved by 2015 were set for different nature conservation objects and territories. For example, national parks in 2000 had an area of 600 000 ha;

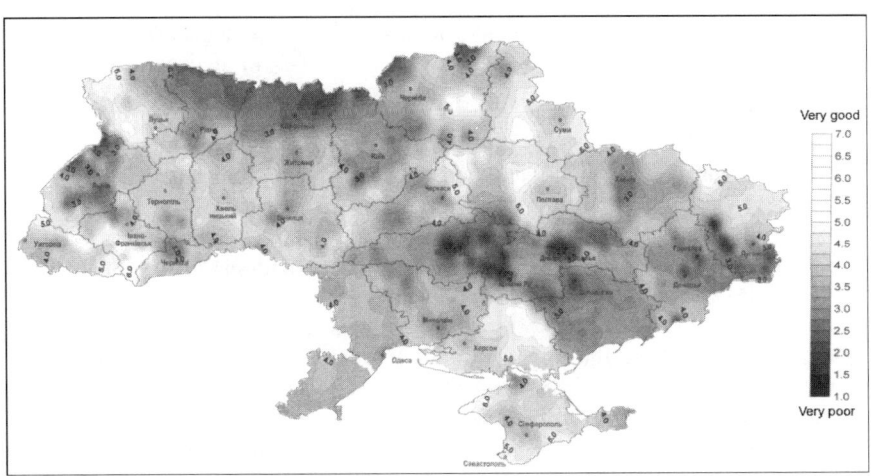

Fig. 3.1. Map of environmental pollution in Ukraine

it is planned to increase this gradually to 2 329 000 ha or 3.9 % of the total area of Ukraine by 2015 (MEU 2005).

Setting environmental targets and objectives presumes changes in the concepts of environmental management. The EU–Ukraine Action Plan developed by the European Commission in the framework of European Neighbourhood Policy, sets out a number of measures for conducting economic and social reforms. It defined the further development of Ukrainian environmental policy by two imperatives (EC 2005):

- complex correction of interrelated economic, social and environmental imbalances; prevention of environmental risks to health and well-being for the population, including future generations;
- harmonisation of the environmental regulations with the principles, tendencies and trends of environmental policy development in the EU.

Recognising an importance and necessity to reshape the environmental policies, Ukraine has taken concrete steps towards the reform. For example, in May 2003, Ukraine signed the Protocol on Strategic Environmental Assessment. As discussed by Palekhov et al. (2005), the implementation of governmental control procedures and methods, which are new for Ukraine are directly interlinked with the existing environmental protection institutions, including environmental standardisation and regulation.

3.3 Standards and Thresholds in Ukrainian Practice

The contents, types and practice in the application of environmental standards and thresholds may differ significantly in different countries, although their objectives have a lot in common. Ukrainian environmental protection legislation defines the following objectives for using standards and thresholds in environmental assessment procedures:

- ensuring environmental safety of the population;
- providing for the rational use and reproduction of natural resources.

3.3.1 Definition of Environmental Standards and Norms in Ukrainian Legislation

Ukrainian approach to the application of standards and thresholds in environmental assessment practice implies setting limit values (norms or *normatyvy* in Ukrainian) both for impacts and for state of the environment. Therefore, environmental standards and norms are the criteria for environmental quality, and are the tool for environmental management and legal regulation.

However, it should be noted that definitions used in Ukrainian legislation do not always correspond with the terms used in the EU legislation. According to the Law of Ukraine "On Standardisation", standards are the regulatory documents which establish rules, define basic principles or requirements for the different types of activities or their results (i.e. energy, oil and gas, environmental protection, construction, transportation, telecommunications, mining, food processing, and other industries).

According to the Cabinet of Ministers Decree "On standardisation and certification", Ukrainian standards are divided into the following categories:

- State standards of Ukraine (DSTU – *"Derzhavni Standarty Ukrainy"*) – state standards are compulsory and approved without validity period limitation. Standards of the former USSR (GOST) are also used as the state standards of Ukraine;
- Sectoral standards of Ukraine (GSTU – *"Galuzevi Standarty Ukrainy"*) – are developed for products or production processes failing the respective state standards, or in cases, where there is a necessity to supplement or extend requirements of the state standards;
- Standards of scientific-and-technical, engineering societies and unions of Ukraine (STTU) – are developed in cases, when the results of fundamental and applied research obtained in certain fields of knowledge or professional interests need to be disseminated;
- Technical specifications of Ukraine (TUU – *"Tehnichni Umovy Ukrainy"*) – contain requirements regulating relations between supplier (developer, producer) and consumer (customer);
- Industrial standards (STP – *"Standarty Pidpryemstv"*) – are developed for products and production processes that are used only at the specific enterprise. These standards must not be contrary to the obligatory requirements of the state and sectoral standards.

In the sense of the sustainable usage of natural resources, standards define the terminology and the methods used for the monitoring and protection of the environmental components (e.g. air, water, soil, ecological systems). GOST 17.2.3.02-78 "Environmental protection, Atmosphere. Rules for setting allowable pollutant emissions by industrial enterprises" (Gosstandard USSR 1979) is an example of environmental standards. Standards serve as a basis for developing environmental norms (e.g. norms of environmental safety of the atmospheric air, norms for maximum allowable concentrations). According to Art. 33 of the Law of Ukraine "On environmental protection", environmental norms set the maximum allowable emissions and discharges of pollutants into the environment, and permissible levels for the prevention of adverse effects from physical and biological factors (i.e. acoustic, electromagnetic, radiation, ionizing etc.). Ukrainian legislation divides environmental norms into the following groups (see Fig. 3.2):

1. norms of environmental safety;
2. norms for impacts of pollution sources.

Norms of the first group are often referred as the norms of environmental quality. They are the indicators for the state of the environmental components. Conceptually, these norms correspond to the German *immission standards* provided in the Federal Immission Control Act (Bundes-Immissionsschutzgesetz).

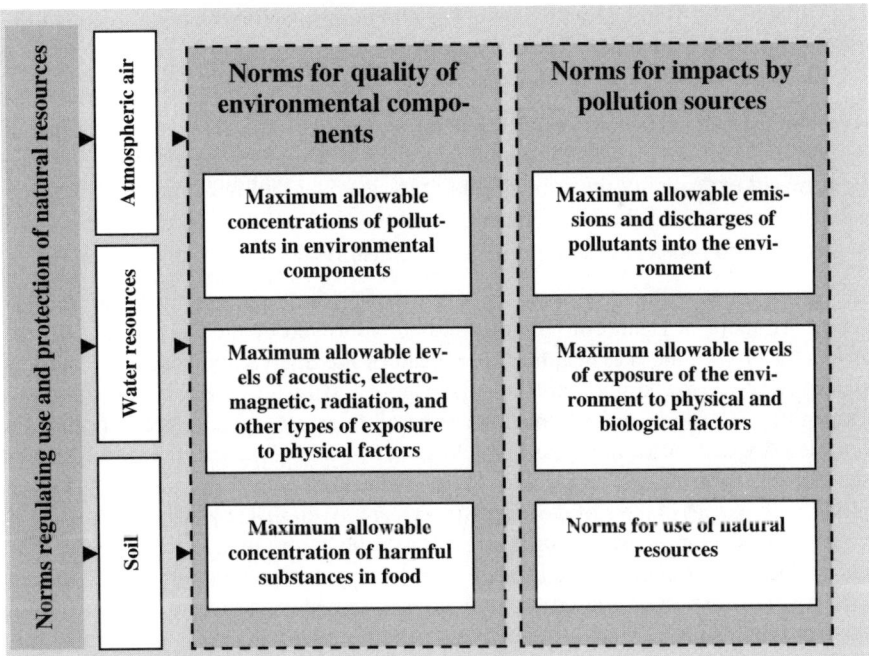

Fig. 3.2. System of environmental norms under Ukrainian legislation (compiled by authors)

Requirements for pollution sources themselves are reflected in the norms of the second group. Among these are the norms for emissions and discharges of pollutants. Norms of the second group are analogous to the German *emission standards* provided in the Technical Instructions on Air Quality Control (TA-Luft).

When carrying out an environmental impact assessment for different environmental components, the norms set forth in the specific environmental laws and regulations are used. For instance, norms provided by the Law of Ukraine "On protection of the atmospheric air" (Art. 5) are used in the assessment of impacts on the atmospheric environment, while norms used in water protection are set out in the Water Code of Ukraine (Art. 36, 38-40), and the Land Code of Ukraine defines norms used in assessment of soil pollution levels (Art. 165). Provided that the environmental norms are maintained by enterprises, i.e. main sources of environmental pollution in Ukraine, the content of any pollutants in water, air and soil should comply with environmental quality objectives.

3.3.2 Development of Environmental Norms

In Ukraine, the development of environmental norms is based on a concept of impact thresholds. These thresholds for adverse impact are the minimal effective dose which results in changes to the ecosystem exceeding the limits of the recovery capacity, i.e. limits at which adaptive and compensatory changes occur in the most vulnerable elements of the ecosystem.

The development of environmental norms is based on:

- technical and economic justification for maximum allowable concentrations of pollutants in environmental components (MAC), and for maximum allowable levels of exposure of the population to physical and biological factors (MAL);
- assessment of possible transboundary pollution;
- necessity to protect cultural heritage sites, historical sites and landscapes.

MAC are set based on long-term research studies done by institutions accredited by the Ministry of Health of Ukraine, and are approved by the Chief state sanitary inspector of Ukraine (i.e. health officer, head of the State Sanitary and Epidemiological Service of Ukraine and the First Deputy Minister of Health of Ukraine). Depending on intensity and characteristics of their biological effects, pollutants are also divided into four main hazard classes – 1,2,3,4, where hazard class 1 is the most dangerous. Examples of MAC values for some chemical substances along with hazard classes of these substances are given in Table 3.1 (Sect. 3.4.3).

In the case that the MAC norms are not set, the approximate safe exposure levels (ASEL) to a chemical substance, including chemical mixtures of constant composition, are used for environmental assessment. ASEL are the maximum dose or concentration which, on the basis of current knowledge, is likely to be tolerated by an organism without producing any adverse effect. ASEL are set based on short-term research studies according to the appropriate procedure, they are adopted as temporary sanitary norms, and are approved by the Chief state sanitary inspector of Ukraine. ASEL must be revised two years following their adoption or replaced by MAC taking into account the accumulated data on the impact of the chemical substances (or their mixtures) on health.

Norms for emissions and discharges are set for every pollution source (i.e. for every individual project on project development stage); if necessary, they are revised (with a minimal revision once every ten years). The reasons for revising emissions and discharges norms are:

- setting new targets and objectives aimed at decreasing environmental pressure;
- emerging new possibilities (technical, technological, organisational, economical) for decreasing environmental pressure;
- changes in requirements imposed by national legislation, and in international agreements on pollution control.

The procedure for developing, adopting and implementing environmental norms is regulated by the governmental acts (i.e. Cabinet's decrees, resolutions of the Ministry of Health of Ukraine).

3.4 Norms for Quality of the Atmospheric Air

In Ukraine, the problem of atmospheric pollution is one of the most difficult. In 2005, more than 11 000 Ukrainian industrial enterprises emitted over 4.4 million tonnes of pollutants. As reported by the State Statistics Service, the density of emissions from stationary pollution sources had totalled 7.4 tonnes of pollutants per km^2 of Ukraine's territory, or 94 kg per capita (Goskomstat 2005). In some regions air pollution indices exceed several times the guideline values.

Scientific justification for the air quality norms, their application in the practice of environmental assessment and state environmental monitoring, has an important place within the state system of environmental regulation in Ukraine.

3.4.1 Legal Regulation for Atmospheric Air Protection

The legal regulation for air protection is administered in accordance with the Law of Ukraine "On protection of the atmospheric air" (1992), and is based on the framework Law of Ukraine "On environmental protection" (1991) with a reference to the requirements of the Law of Ukraine "On sanitary and epidemiological welfare of the population" (see Fig. 3.3).

The majority of European countries have legal acts similar to the Law of Ukraine "On protection of the atmospheric air", which regulate, in one way or another, social relations with the aim of improving air quality. In Germany, for instance, such tasks are carried out by the Federal Immission Control Act (Bundes-Immissionsschutzgesetz), which regulates protection of the environment from adverse effects of air pollutants, noise, vibration and other processes (Jarass 2005).

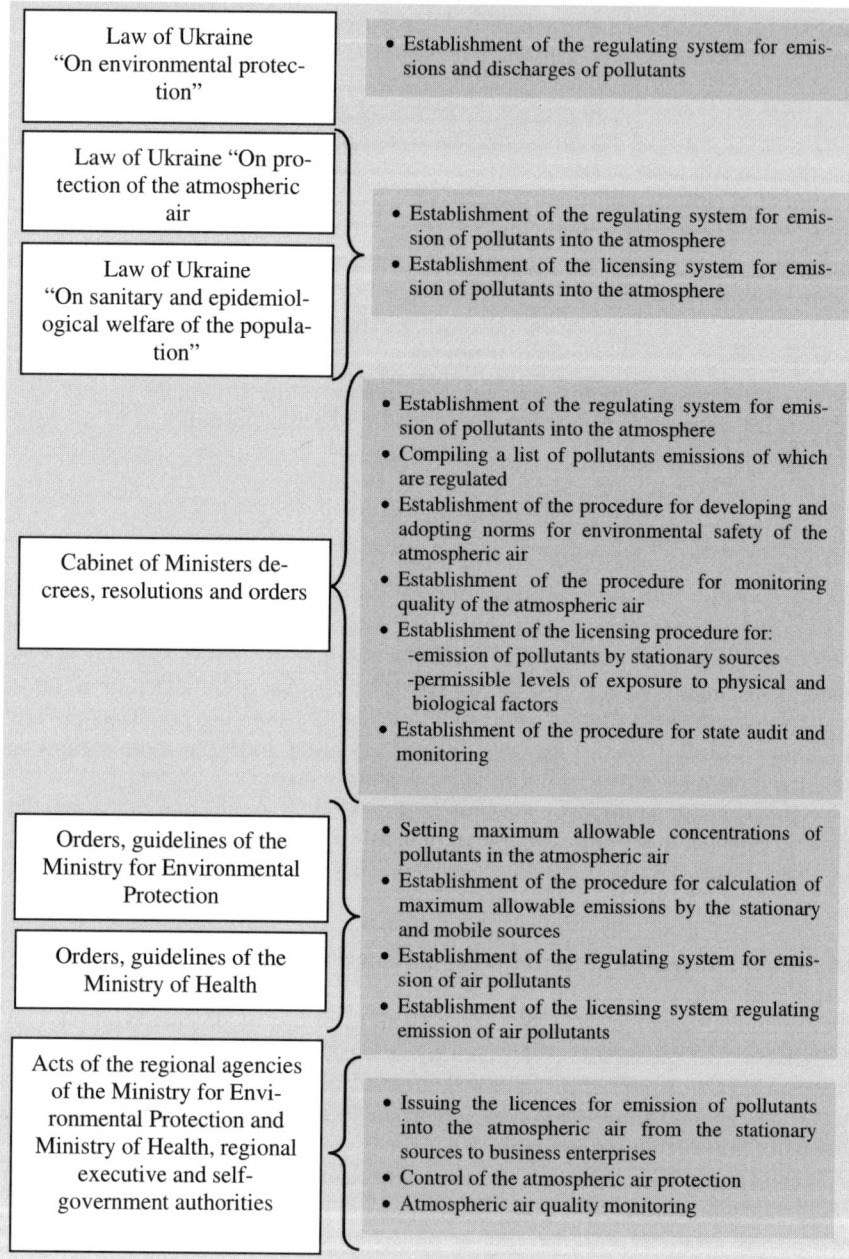

Fig 3.3. Legal regulation of the atmospheric air protection (compiled by authors)

3.4.2 Classification of the Air Protection Norms

Figure 3.4 shows the current system of environmental norms in the field of atmospheric air protection. These norms may be provisionally subdivided into:

- norms for the safety of the atmospheric air;
- norms set for emission sources that have or are likely to have an adverse impact on quality of the atmospheric air.

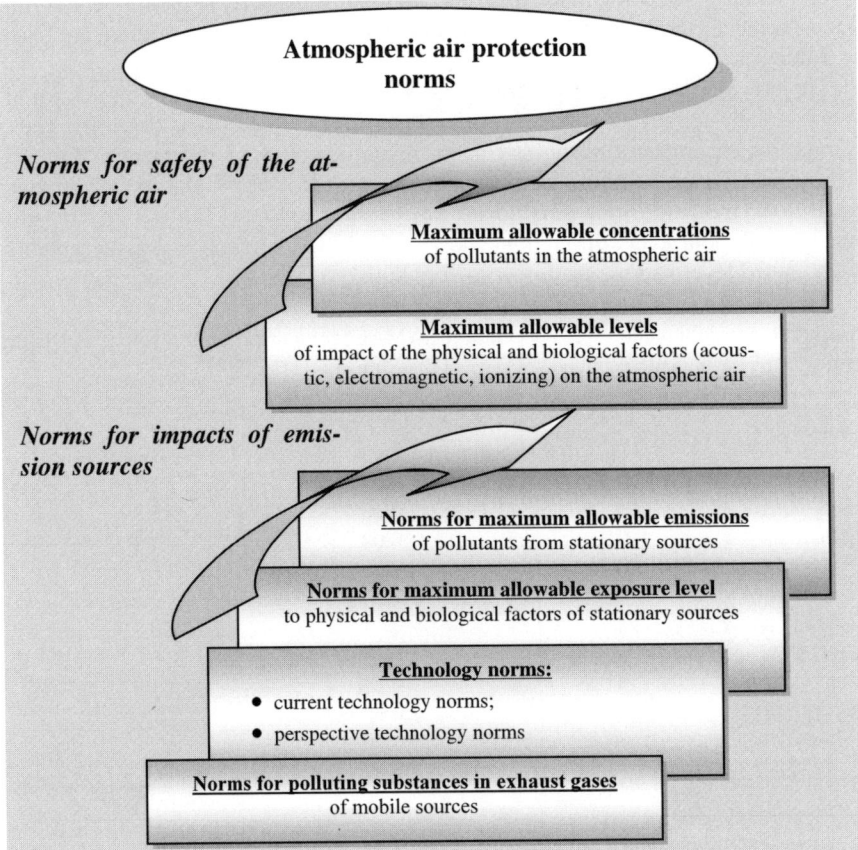

Fig. 3.4. Classification of norms in the field of atmospheric air protection (compiled by authors)

4.4.3 Norms for Safety of the Atmospheric Air

According to the Cabinet of Ministers Decree No. 299, norms for safety of the atmospheric air (MAC and MAL) are developed by the Ministry of Health for the assessment of impact of the planned or implemented development activities on public health. Additionally, in built-up areas, allowable limits on concentrations of pollutants are set separately for industrial areas and residential areas (i.e. intended for housing, public buildings and community facilities). These norms include also one-time maximum and daily mean limit values for the concentration of pollutants. The revision of these norms is carried out once every five years.

It is important to note that in Ukrainian practice, the norms of this group are used mainly as control (reference) values. When assessing the state of the environmental components, the ratio of the actual concentration to MAC/MAL is analysed. To date MAC for more than 500 pollutants are used in assessment of impacts on the atmospheric air of settlement areas. Some examples of them are listed in Table 3.1.

There is also an integral criterion that is used in project EIAs – maximum allowable pollution levels of the atmospheric air in settlement areas (MAPL). These norms, set the impact thresholds for the range of polluting substances. With the help of MAPL the pollution level (allowable, non-allowable), and its hazard level (safe, low, medium, high, very high) can be assessed (see Table 3.2). Unfortunately, pollution levels throughout most of Ukraine's territory are unacceptable, with hazard level varying from low to very high (see Fig. 3.1).

Table 3.1. Maximum allowable concentrations (MAC) for some chemical substances in the atmospheric air of settlement areas (Order of the Ministry of Health No. 201)

Substances	Maximum allowable concentrations (mg/m3)		
	One-time maximum	Daily mean	Hazard class*
Nitrogen dioxide	0.085	0.04	2
Nitrogen oxide	0.4	0.06	3
Ammonia	0.2	0.04	4
Acetone	0.35	0.35	4
Benzo(a)pyrene	-	0.1 µg/ 100 m3	1
Carbon oxides (CO and CO2)	5	3	4
Ethylbenzene	0.02	0.02	3
Cadmium	-	0.0003	1
Manganese	0.01	0.001	2
Formaldehyde	0.035	0.003	2

* The lowest number corresponds to the highest hazard class

Table 3.2. Assessment of the pollution level for the atmospheric air (Order of the Ministry of Health No. 201)

Pollution level	Hazard level	Ratio of actual pollution level to MAPL
Acceptable	Safe	< 1
Non-acceptable	Low	> 1 – 2
Non-acceptable	Medium	> 2 – 4.4
Non-acceptable	High	> 4.4 – 8
Non-acceptable	Very high	> 8

3.4.4 Norms for Impact of Stationary Sources on the Atmospheric Air

Thresholds for emissions from stationary sources are set during environmental assessment of construction and reconstruction projects, projects of technical documentation on new equipment, technologies and materials, which are likely to have significant impacts on the environment. The main task of these norms is to maintain norms for safety of the atmospheric air. During their development the following considerations are taken into account:

- economic feasibility and the level of technological processes;
- equipment and technical condition of the cleaning facilities;
- complying with the requirements of national and EU legislation.

The peculiar feature of Ukrainian practice of environmental regulation is that the thresholds developed for this group of norms are calculated individually for every stationary source when commencing operation (e.g. by its commissioning) or in the course of norms revision (on one's own or external initiative). According to the Cabinet of Ministers Decree No. 1780, these thresholds are calculated in unit emission mass per unit time and per unit output or used raw materials.

Norms for maximum allowable exposure level to physical factors are of key importance for the practice of environmental assessment. It is important to note that the Law of Ukraine "On protection of the atmospheric air" places noise among physical factors, which determine the quality of the atmospheric air (Art. 21 "Noise prevention and abatement"). According to the Cabinet of Ministers Decree No. 300 and similarly to the previous group of norms, norms for maximum allowable exposure level to physical factors (including noise) are set for every stationary source (and for every type of mobile sources) taking into account the current technologies. Norms for allowable exposure to physical factors are laid down in the State building regulations of Ukraine for the various objects and territories (see Table 3.3). If environmental monitoring reveals, that the statutory norms for impacts on the atmospheric air are violated, the business or other activities may be restrained, temporarily prohibited (halted) or discontinued.

Table 3.3. State standards for noise pollution (Order of the Ministry of Health No. 173)

Type of territory	Equivalent noise level, dB(A)		Maximum noise level, dB(A)	
	7a.m. - 11p.m.	11p.m. - 7a.m.	7a.m. - 11p.m.	11p.m. - 7a.m.
Residential areas	55	45	70	60
Residential areas being reconstructed	60	50	70	6
Residential areas adjacent to airports and airfields	65	55	75	65
Areas suitable for recreation and tourism	50	35-40	85	75
Resort areas	40-45	30-35	60	65
Natural and game reserves	< 25	<20	50	45

3.5 Conclusions and Recommendations

As illustrated in this chapter, Ukraine has a developed system of environmental regulation, which in general covers all environmental media and adverse impacts on human organisms. The methods for calculating and setting environmental norms and standards are, for the most part, adequate/comparable to the European approaches.

The environmental assessment of planned or current developments is conducted with a view to provide for the environmental safety of the population and for rational use of natural resources. In this context, Ukrainian legislation stipulates that the environmental norms are applied in two ways: firstly as quality criteria for the environmental components and secondly, as instruments for the legal and environmental regulation.

However, analysis of Ukrainian practice of environmental assessment revealed that it requires a serious reform, which, in its turn, involves appropriate changes to the current system of environmental regulation. The main reform directions may be defined as follows:

1. *Implementing SEA.* The current EA system has developed its procedures more or less consistently only for the project level and this is not sufficient for correcting the existing environmental imbalances. For this reason Ukraine expressed its intention to implement SEA as a new decision-making and planning instrument, which is able to supplement project EIA and overcome its deficiencies, by signing the Kiev SEA Protocol in 2003. The SEA procedure, as it is well known, is based on a complex use of the appropriate criteria and indicators (environmental, economic and social). The system of environmental norms and standards should be adjusted for its possible application within SEA procedure (Schmidt, Palekhov 2005).

2. *Strengthening screening procedure.* According to the Law of Ukraine "On Environmental Expert Review (Expertise)", environmental assessment is formally required for many strategic and almost for all project activities. As Cherp (2001) argues, this results in several hundred environmental assessments per million of population conducted annually, while only a fraction of the assessed project-level activities could be considered environmentally significant.

3. *Improvement* of the formal screening provisions will increase the quality of individual assessments by distinguishing between proposals requiring either full or simplified assessment, or even no assessment, depending on the environmental significance of their impacts. Screening procedure may be improved by introducing the significance thresholds and criteria. Such thresholds could be set similarly to the thresholds and criteria listed in Annexes I and II of the EU EIA Directives (85/337/EEC as amended by 97/11/EC). The indicated system includes exclusion thresholds, indicative or guidance thresholds, and inclusion or mandatory thresholds (see EC 2001).

4. *Introducing EA follow-up.* Ukrainian practice of environmental assessment does not include a procedure of post-EIA/SEA monitoring, although there is a good potential for its implementation. The developed system of the state environmental monitoring covering all major sources of technological pressure may provide with extensive environmental data necessary for post-EIA/SEA monitoring. However implementing post-SEA monitoring will additionally require the development of assessment indicators for the specific territories and/or certain types of strategic action.

As discussed in this chapter, the Ukrainian system of environmental regulation has a number of deficiencies impairing the efficiency of EA:

- Environmental quality norms used mainly as control (reference) values, reflecting national environmental targets (i.e. desired values vs. reality). These norms are used for all of Ukrainian regions, irrespective of their environmental conditions. The set norms are not mandatory and are used only as measurement units – the state of the environment is assessed as a ratio of the actual pollution level to the recommended (i.e. in environmental norms).
- Environmental pollution thresholds that are used in EA are calculated only in respect of various impacts on the human organism. However, it is well known that some other biota specimens may be more sensitive to pollution of environmental components. In Ukrainian practice, thresholds for sensitivity of individual species are used only in biological monitoring.
- Current environmental norms consider only impacts of individual factors and do not address possible cumulative or complex effects.

Thus, reform of the EA system is directly linked with transformation of environmental regulation. Such reform requires:

1. Setting realistic environmental quality norms for individual territories (i.e. administrative and territorial units) taking into account their geographic, landscape and other natural features.
2. Developing thresholds for impacts on the most sensitive species. Application of these norms during EA must become legally binding for certain territories and objects.
3. Introducing integrated norms that would take into account:

   - *Cumulative impacts* – concurrent or successive impacts of several substances on the same object;
   - *Complex effects* – reception of pollutants by different means and from different environmental media;
   - *Synergic impacts* by the whole range of physical, chemical and biological factors;
   - *Characteristics of diffusion and transformation* of pollutants in different media.

4. Developing integrated environmental norms that can be used in SEA procedure together with other assessment indicators – social, economic etc.

5. Developing legally binding thresholds necessary for the screening procedure, including mandatory thresholds, guidance thresholds, and exclusion thresholds.
6. Developing integrated environmental norms that can be used in post-EIA/SEA monitoring.

It is also worth noting that the proposed measures for improving the environmental regulation system in the context of reforming EA practice requires support from other mechanisms of state regulation – i.e. environmental control, environmental audit.

References

Baranovsky VA, Shishenko PG, Dmytruk OJ (2004) Ukraina. Technogenna nebezpeka, mashtab 1:3000000 (Ukraine. Techogenic Sefety, scale 1:3 000 000). National Academy of Sciences of Ukraine, Kiev

Bundes-Immissionsschutzgesetz in der Fassung der Bekanntmachung vom 26. September 2002 (BGBl. I S. 3830), zuletzt geändert durch Artikel 3 des Gesetzes vom 18. Dezember 2006 (BGBl. I S. 3180)

Cabinet of Ministers Decree No. 1780 of 28.12.2001 "On approval of the procedure for developing and adopting norms for the maximum allowable pollutant emissions from stationary sources". Official gazette of Ukraine (Ofitsiyny visnyk Ukrainy) No. 1 of 18.01.2002, p. 84, st. 9

Cabinet of Ministers Decree No. 299 of 13.03.2002 "Procedure for developing and adopting norms for environmental safety of the atmospheric air". Official gazette of Ukraine (Ofitsiyny visnyk Ukrainy) No. 12 of 05.04.2002, p. 55, st. 571

Cabinet of Ministers Decree No. 300 of 13.03.2002 "On procedure for developing and adopting norms for maximum allowable impact level to physical and biological factors of stationary pollution sources on the atmospheric air". Official gazette of Ukraine (Ofitsiyny visnyk Ukrainy) No. 12 of 05.04.2002, p. 57, st. 572

Cabinet of Ministers Decree No. 641 of 26.04.2003 "On adoption of the Complex programme for implementing at the national level decisions of the World Summit on Sustainable Development for 2003-2015". Official gazette of Ukraine (Ofitsiyny visnyk Ukrainy) No. 18 of 23.05.2003, p. 116, st. 847

Cherp A (2001) EA legislation and practice in Central and Eastern Europe and the former USSR. A comparative analysis. Environmental Impact Assessment Review 21(4): 335-361

EC – European Commission (2005) EU-Ukraine Action Plan, DGE VI, UE-UA 1051/05. Internet address: http://ec.europa.eu/comm/world/enp/pdf/action_plans/ukraine_enp_ap_final_en.pdf, last accessed on 10.04.2007

EC – European Communities (2001) Guidance on EIA. Screening. Office for Official Publications of the European Communities, Luxembourg

Esty DC, Levy M, Srebotnjak T, de Sherbinin A (2005) 2005 Environmental Sustainability Index: Benchmarking National Environmental Stewardship. Yale Center for Environmental Law & Policy, New Haven

Goskomstat – State Statistics Service of Ukraine (2005) Main indicators for atmospheric air protection and waste management (hazard class I-III), 1990-2005. Internet address: http://www.ukrstat.gov.ua, last accessed on 10.04.2007

Gosstandard USSR - State Committee of the USSR on Standardisation (1979) GOST 17.2.3.02-78 "Environmental protection, Atmosphere. Rules for setting allowable pollutant emissions by industrial enterprises". Official edition. Standards Publishing House, Moscow

Jarass HD (2005) Bundes-Immissionsschutzgesetz. Kommentar unter Berücksichtigung der Bundes-Immissionsschutzverordnungen und der TA Luft sowie der TA Lärm. Beck, München

Land Code of Ukraine, Law No. 2768-III of 25.10.2001. Bulletin of Verkhovna Rada (Vidomosti Verkhovnoi Rady), 2002, No. 3-4, st. 27, last amendment on 19.12.2006, Voice of Ukraine (Golos Ukrainy) No. 6 of 16.01.2007

Law of Ukraine "On environmental protection" No. 1264-XII of 25.06.1991. Bulletin of Verkhovna Rada (Vidomosti Verkhovnoi Rady), 1991, No. 41, st. 546, last amendment on 19.12.2006, Official gazette of Ukraine (Ofitsiyny visnyk Ukrainy) No. 52 of 09.01.2007, p. 54, st. 3477

Law of Ukraine "On protection of the atmospheric air" No. 2707-XII of 16.10.1992. Bulletin of Verkhovna Rada (Vidomosti Verkhovnoi Rady), 1992, No. 50, st. 678, last amendment on 03.06.2004, Bulletin of Verkhovna Rada, 2004, No. 36, st. 434

Law of Ukraine "On sanitary and epidemiological welfare of the population" No. 4004-XII of 24.02.1994. Bulletin of Verkhovna Rada (Vidomosti Verkhovnoi Rady), 1994, No. 27, st. 218, last amended on 09.02.2006, Bulletin of Verkhovna Rada, 2006, No. 22, st. 199

Law of Ukraine "On Standardisation" No. 2408-III of 17.05.2001. Bulletin of Verkhovna Rada (Vidomosti Verkhovnoi Rady), 2001, No. 31, st. 145, last amendment on 01.12.2005, Bulletin of Verkhovna Rada, 2006, No. 12, st. 101

MEU – Ministry of Economy of Ukraine (2005) Ukraine – Millennium Development Goals 2000+5. Materials for the 60th session of the UN General Assembly. Dija Publishing House, Kiev

Order of the Ministry of Health of Ukraine of No. 173 19.06.1996 "On adoption of the State sanitary regulations for planning and building of settlement areas". Internet address: http://www.legal.com.ua/document/kodeks/0000CH5A0379-96.html, last accessed on 10.04.2007

Order of the Ministry of Health of Ukraine of No. 201 09.07.1997 "On adoption of the State sanitary regulations for protection of the atmospheric air of settlement areas (from pollution with chemical and biological substances)", DSP-201-97. Internet address: http://www.legal.com.ua/document/kodeks/0CH560201282-97.html, last accessed on 10.04.2007

Palekhov D, Shapar A, Schmidt M (2005) Analysis of the state of environment and recreation potential in Dnepropetrovsk region. In: Mamutov VK (ed) Regional problems of tourism development and recreation. Collection of scientific papers. National Academy of Sciences (NAS) of Ukraine, Institute for Economic and Legal Research. Yugo-Vostok LTD, Donetsk, pp 169-177

Schmidt M, Palekhov D (2005) Implementing strategic environmental assessment into regional planning in Ukraine. In: Ecology and nature management: Collection of scientific papers of the Institute of Nature Management and Ecology Problems by the NAS of Ukraine, Dnepropetrovsk, Issue 8, pp 28-31

Schmidt M, Palekhov D, Shapar A (2005b) Criteria and requirements for the implementation of sustainable development policy in Ukraine. In: Ecology and nature management: Collection of scientific papers of the Institute of Nature Management and Ecology Problems by the NAS of Ukraine, Dnepropetrovsk, Issue 8, pp 18-22

Schmidt M, Pivnyak G, Palekhov D. (2006) Reformation of regional management in Ukraine as compared to experience of Germany. Scientific Bulletin of the National Mining University 5:9-13

Schmidt M, Shapar A, Pivnyak G, Palekhov D (2005a) Methodical approaches to elaboration of the balanced environmental-economic regional development plans in Ukraine. Scientific Bulletin of the National Mining University 7:96-100

TA-Luft - Technische Anleitung zur Reinhaltung der Luft vom 24. Juli 2002. GMBl. 2002, Heft 25 – 29, S. 511 – 605

Water Code of Ukraine, Law No. 213/95-VR of 06.06.1995. Bulletin of Verkhovna Rada (Vidomosti Verkhovnoi Rady), 1995, No. 24, st. 189, last amendment on 30.11.2006, Bulletin of Verkhovna Rada, 2007, No. 3, st. 31

4 Poverty and Disease Remediation in the Millennium Development Goals: Time for Kenya to Set Standards and Thresholds?

Vincent Onyango[1] and Michael Schmidt[2]

1 Brandenburg University of Technology (BTU) Cottbus, Germany
2 Department of Environmental Planning, Brandenburg University of Technology (BTU) Cottbus, Germany

4.1 Introduction

The Millennium Development Goals (MDGs) have become a key action-forcing agenda and baseline for performance, goals, targets and indicators (Annan 2005; Johnson 2005; UN 2003, 2001). Several policies, programmes, plans and projects have been created to meet the MDGs and others re-formulated to integrate the MDGs. This paper argues for a need and a conceptual framework to buttress MDGs in national standards and thresholds in order to provide a more comprehensive long-term framework for strategizing, planning and implementation beyond 2015. For convenience MDG targets on poverty and HIV/AIDS are used for illustration although the same argument applies to most MDGs. The key argument is anchored on the concept that standards and thresholds are critical to guiding the strategizing, planning and implementation processes; facilitate a more rationalized decision making process and a cost-effective allocation of resources in the long term. In the context of Part I, the chapter illustrates the importance of international goals in driving progress on key quality of life standards and thresholds.

The paper begins with a brief background on MDGs and targets. The second section shows the implementation of poverty and HIV/AIDS MDGS in Kenya, concentrating on the methodological aspects of setting targets. The attendant underlying weaknesses solvable through standards and thresholds are demonstrated in the third section. In the fourth, the need for standards and thresholds for comprehensive strategies, planning and implementation is explained. A conceptual framework on how Kenya can set MDG-related standards and thresholds is suggested, and a discussion on advantages concludes the chapter.

Standards and Thresholds for Impact Assessment. Edited by Michael Schmidt, John Glasson, Lars Emmelin and Hendrike Helbron. © 2008 Springer-Verlag

4.1.1 Aim and Objectives

The aim is to convince that MDGs implementation in Kenya should not rely only on set targets, but also embed the MDG targets in national standards and thresholds. This aim is accomplished through achieving the following objectives:

- Exposing the shortcoming of MDG targets in providing a comprehensive framework for strategic planning and action beyond the global targets
- Suggesting a conceptual framework for setting standards and thresholds related to the MDGS in Kenya
- Illustrating the advantages of standards and threshold.

4.2 The Millennium Development Goals

In September 2000, leaders from 189 nations agreed on a vision for the future: a world with less poverty, hunger and disease, greater survival prospects for mothers and their infants, better educated children, equal opportunities for women, and a healthier environment; a world in which developed and developing countries worked in partnership for the betterment of all (UN 2001). The UN General Assembly adopted a Declaration having goals that formulate a vision for global development. This vision took the shape of eight Millennium Development Goals monitored through 18 targets and 48 indicators, which provide a framework for development planning for countries, and time-bound targets by which progress can be measured (UN 2005, 2003; World Bank 2004). To facilitate implementation the UN established the Inter-Agency and Expert Group (IAEG) to advise on methodologies and technical issues in relation to the indicators, guidelines, analysis and reports relating to implementation of the MDGs. Each year, the Secretary-General presents a report to the United Nations General Assembly on progress achieved towards meeting the goals, based on data on indicators, aggregated at global and regional levels (see reports e.g. UN 2005, 2006). The baseline for assessment of MDG targets are based on the global situation in 1990. At the national level each country sets up its own mechanisms to translate and implement the MDGs.

4.2.1 Do the Targets Matter?

Though most MDGs hold a moral imperative (Bradshaw 2004) global targets have been known for making a difference by mobilizing actors and advocates at all levels (UN 2005; Vandermoortele 2005; Bradshaw 2004). Historically, global targets resulted in the elimination of smallpox by 1977 and in reduction of ozone-depleting substances to 10 % of 1990 levels (Vandermoortele 2005). The targets make governments accountable by stating end-points and making it easy for civil society to follow and measure progress (World Resources 2005). It is significant to note that there are strong inter-linkages between some of the MDG targets, for example achieving environmental sustainability (MDG7) targets is vital to reaching the other MDG targets on poverty, health and gender equality (Toepfer 2005; Johnson 2005; Hirsch 2005; World Resources 2005). Such linkage in MDGs can complicate target setting because of the synergistic cost-benefit stream across them; but can also help prioritization by concentrating on MDGs that confer greatest compound effects across the others. Targets are necessary and central to implementation, monitoring and decisions making and the MDG targets have been described by Annan (2005) as people-centred, time bound, measurable, achievable and enjoying global partnership and political support.

4.3 Implementation of MDGs in Kenya: Target Setting

In accordance with the UN guidelines for country-level implementation, Kenya undertook baseline studies and needs assessments which led to strategies and plans to achieve the MDGs (UN Millennium Project 2005; GOK 2004). An Action Plan was developed to 'Mainstream the MDGs within the Planning, Budgeting and Monitoring and Evaluation' processes through for example the Economic Recovery Strategy (ERS) and the budgetary facility called the Medium-term Expenditure Facility (GOK 2004; UNDP 2006). Once MDGs and targets were accepted and committed to, what followed at the national level can be summarized as an exercise in planning and setting of national targets aligned towards "feasible and immediate steps" for achieving the MDG targets (see NACC 2005a; Waithaka et al. 2003; see Gillis and Southey 2005). The key approach revolved around directing the available resources towards achieving the MDG targets. In this process, the MDG targets have become *de facto* standards at a national level. Of significance and germane to this chapter is the absence of Kenyan standards and thresholds on MDG-related issues on which the country can strategize and plan independently of, and beyond the global agenda. This begs the question whether there are any national "allowable levels" that are derived scientifically and independently from the expediency of MDGs, and consequently, how MDG targets are influenced by them. For Kenya to be singularly guided by MDG targets exposes several critical underlying shortcomings:

- An inherent assumption that the MDG targets are substitutable with nationally desired and allowable levels. For example, meeting an MDG target on HIV/AIDS is not necessarily equivalent to arriving at an HIV-AIDS prevalence level 'allowable', 'desirable' or 'conducive' to Kenya.
- The Kenyan implementation of MDGs is not adequately calibrated by context-based notion of "acceptable levels" or standards, either at national, regional or household scale.
- There is no methodically formulated framework for nationally acceptable levels of MDG-related issues.
- There is little strategic, conceptual and methodical tie-up between long term national vision and MDG targets.

- Baseline for assessment is 1990 data; yet there is no reason other than expediency why the status in 1990 is preferred as a referent.

Therefore, to transcend this limited perspective of the MDG targets and embed the remedial strategies in a broader context beyond 2015, standards and thresholds formulated independently of the MDGs are required. The standards shall indicate the national limits of 'allowability' and be key referents for strategizing, planning, implementation and legal enforcement if need be. Indicators can be calibrated, into lower, middle and upper thresholds in order to trigger respective levels of responses commensurate to the severity reflected by the indicators. To continue the discussion and put the argument within a relevant context the '*Kenya Vision 2030, Transforming National Development*' shall be used to represent Kenya's long term strategic aspirations (Daily Nation 2006; Muriuki 2006). Released in October 2006, this national vision targets an annual GDP growth rate of 10 % aiming to make Kenya a middle-income country with high standards of living. The Vision 2030 is a potential yardstick against which "allowable" national or regional levels of HIV/AIDS and poverty can be generated. In this context "allowable" means formally recognized levels or standards that will not adversely impact on the attainment of Vision 2030. Currently, the derivation of targets and remedial strategies in Kenya's MDGs can be characterized as an approach more embedded in "response mode" and less in "strategic long-term pro-action mode", and does not reflect standards and thresholds.

4.3.1 MDGs in Kenya: Target setting in HIV/AIDS Remediation

The MDG "Combat HIV/AIDS, malaria and other diseases" has the target "Have halted by 2015 and begun to reverse the spread of HIV/AIDS". The Government released the Kenya National Aids Programme 2005/6-2009/10 (KNASP) which set out an HIV/AIDS Strategic Plan to respond to the epidemic (NACC 2005a; NACC 2005b). KNASP set a target of national HIV/AIDS prevalence of less than 5.5 % by 2010. While admitting that setting a prevalence target in an era of expanded treatment is a complex task, KNASP's approach and methodology in arriving at this target only reflects the global commitment and does not integrate or refer to a national standard or threshold to internalize the national limits of "allowability". Box 4.1 provides a rendition by KNASP on how the target for women aged 15–24 was calculated.

The priority areas for KNASP 2005/06–2009/10 are three-fold:

- Priority Area 1: Prevention of new infections and to reduce the number of new HIV infections in both vulnerable groups and the general population.
- Priority Area 2: Improve the quality of life of people infected and affected by HIV/AIDS
- Priority Area 3: Mitigation of socio-economic impact.

Box 4.1. Approach and methodology used by KNASP to set its 5-year HIV/AIDS target (NACC 2005a)

This target reflects the emphasis of the strategy on young women, who are a highly vulnerable group. As death rates in this age group are low, reducing prevalence will require a major reduction in new infection rates. The current prevalence rates (KDHS 2003) are 3 % (ages 15–19) and 9% (ages 20-24). Assuming very low prevalence below 15, and assuming low death rates prior to age 25, this implies incidence rates of 3 % 15–19 and 6 % 20–24. Achieving the target prevalence rate of 4.5 % would require these incidence rates to be halved to 1.5% and 3% respectively by 2010 (i.e. this would yield average incidence rates for the five years of the KNASP of 2.3 % and 4.5 %, which would yield prevalence rates of 2.3% for 15–19 and 6.8% for 20–24, giving an overall female prevalence of 4.5 % for the 15–25 group.)

Prevalence is adult prevalence given by the total number of adults aged 15-49 living with HIV, divided by the total population of adults in this age range, expressed as a percentage. The baseline adult prevalence for the KNASP 2005/06–2009/10 is 7 % (range 6.1–7.5 %) based on a reconciliation of KDHS (2003) and sentinel surveillance data.

Incidence of HIV/AIDS infection is the number of new cases in a year.

Within the priority area for example "Prevention of new infections", specific targets are set out in target statements (Table 4.1). What is significant in these targets is the fact that none integrates any national standards or thresholds (see Box 4.1 for methodology). The same obtains for all targets in the other priority areas.

Table 4.1. Targets in priority area "Prevention of new infections" and the target statements (source: NACC 2005a)

Priority Area 1: Prevention of new infections		
Objective: Reduce the number of new HIV infections in both vulnerable groups and the general population		
Target Area	Targets for 2010	Baseline
Prevalence	Prevalence for young men in age range 15-24 less than 1% (a reduction of 20% on KDHS 2003)	1.2% (KDHS 2003)
	Prevalence for young women in age range 15-24 less than 4.5% (a reduction of 25% on KDHS 2003)	5.8% (KDHS 2003)

4.3.2 MDGs in Kenya: Target Setting in Poverty Remediation

The MDG "Eradicate extreme poverty and hunger" has a target "Halve, between 1990 and 2015, the proportion of people whose income is less than one dollar a day" (World Bank 2004). Poverty is defined as the proportion of population living below $1 purchasing power parity per day to mean those living on less than $1.08 a day at 1993 international prices.

Box 4.2. Examples of methods to measure poverty in Kenya (UN 2003)

The poverty headcount ratio which measures the proportion of the national population whose income is below the official threshold or thresholds set by the national Government.

Poverty gap ratio which measures the mean distance separating the population from the poverty line (with the non-poor being given a distance of zero), expressed as a percentage of the poverty line.

The one dollar a day poverty line is compared to consumption or income per person and includes consumption from own production and income in kind. Poverty was discussed by various Kenyan communities and given Kenyan contextual interpretation and definitions. Although several programmes have been set up to address poverty, there is neither an official poverty standard nor threshold in Kenya. Though several methods of measuring poverty are used in Kenya (Box 4.2) these methods do not integrate poverty standards in their conception and only a defined poverty line is used.

4.4 Need and Justification for Standards and Thresholds

This chapter provides a rendition of impacts and projections in order to drive home the oft severe implication of impacts, and hence demonstrate the need to have the concerns integrated into the strategic and remedial planning through standards and thresholds. In order to justify the need for standards and thresholds, two compelling issues come to bear: one- the need to achieve Vision 2030 despite the impacts of poverty and HIV/AIDS; and two- understanding the role of standards and thresholds in strategizing, implementation, evaluation and effective resource allocation in poverty and HIV/AIDS remediation. The exacerbated costs projected in future scenarios justify formal limits particularly when the costs have adverse impacts on vision 2030 (see facts and status in Box 4.3).

4.4.1 Impact of HIV/AIDS on Kenya

AIDS severely affects macroeconomic indicators such as GDP per capita, economic growth and level of investment (Nyaga et al. 2004). The costs are beyond Kenya's capacity to pay and were projected to reduce Kenya's GDP by 14.5 % in 2006 or by 25 % according to Nyaga et al. (2004) assuming there was no HIV/AIDS. At the same time, per capita income is projected to drop by 10 %. Impact on household economies would similarly be adverse, according to, for example, Steinberg et al. (2000), Arndt and Lewis (2000), Bertozzi and Aggleton (1995) and Lonwenson and Whiteside (2001). The loss in agricultural production due to HIV/AIDS represented 0.3 % of the total value of production in 1999 and is expected to rise to between 1.7 % and 2.4 % by the year 2010 (Leighton 1993).

Box 4.3. Facts, Status and Impacts of HIV/AIDS in Kenya (sources: NACC 2005a; Nyaga et al. 2004; NACC 2005b; KDHS 2003)

1. HIV/AIDS declared a national disaster in 1999.
2. Over 1.5 million had died from HIV/AIDS; 300 000 new cases in 2000.
3. By 2001 2.3 million adults and 0.22 million children were living with HIV/AIDS.
4. Urban areas prevalence of 17–18 %, some areas 20–30 %; 12 – 13% in rural areas; many sites with 20% of pregnant women having HIV.
5. Life expectancy dropped from 60 years in 1993 to 47 years in 2004.
6. In 2004: 90 000 new infections; 105 000 deaths; 120 000 new children infected; 1.4 million pregnant women with HIV; 2.3 million orphans
7. Aids prevalence 2000–13 %; 2001–14 %, 2002–10.2 %; 2003–6.7 %; 2004–5.7 %
8. 9 % women infected, against 4.6% males
9. 150 000 deaths per year, twice the figure in 1998

The impact of HIV/AIDS on economic growth and development has a particularly significant adverse impact on poverty reduction efforts (NACC 2005a, 2005b; Kwena 2004; KDHS 2003; Greener et al. 2000). HIV/AIDS has heavy cost on economy, spreads fast, and has been described as a grave danger to economic development (Nyaga et al. 2004; Gillis and Southey 2005; Barnett 2004; Amuyunzu-Nyamongo 2001; Lonwenson and Whiteside 2001). AIDS orphans were expected to rise to more than 1.5 million by 2002 (NACC 2005b) costing households and the Kenyan economy as a whole through crime and deviant behaviors (Steinberg et al. 2000; Bonnel 2000). Research across low income countries (Nichols et al 2000; Bonnel 2000; Arndt and Lewis 2000, WB 1997; Rugalema et al. 1998) has demonstrated that HIV/AIDS is the most serious impediment to economic growth, development and there is no reason to expect Kenya to be an exception. These numbers do have debilitating effects through any combination of direct, indirect, short term, long-term, synergistic and cumulative costs across Kenyan sectors (see GOK 2001; Bollinger et al. 1999; Nalo and Aoko 1993; Forsythe et al. 1993; Forsythe and Roberts 1995). If reducing prevalence from 6.7 % to 5.5 costs KShs million 179.452 (15.9 % GDP) (NACC, 2005) and if the same cost ratio is maintained, to eliminate HIV/AIDs will cost KShs millions 1202.3284 (107 % current GDP and 9.88 % of 2030 GDP) and will take 29 years. A reduction of one percentage point costs Ksh million 149.5, equivalent to 13.25 % GDP in 2005 (based on purchasing power parity).

4.4.2 Impacts of Poverty on Kenya

Bradshaw (2004) comprehensively collates universal reasons to fight poverty as:

- Moral imperative or religious duty to act on behalf of the poor
- Poverty is an intolerable unfairness and injustice

- Poverty is economic and social waste and leads to significant inefficiencies in the national systems
- Poverty has associated social problems and costs.

Studies from Britain (Bradshaw 2004, 2005; Hobcraft 2004; Sigle-Rushton 2004) have documented impacts of poverty which can be said to be universal. In the Kenyan context, justification for poverty standards arises from

- The need to set an indicator at when potential adverse impacts on Kenyan Vision may occur
- As a referent for strategizing, planning, assessment and evaluation
- For prioritising and decision making
- Having a formal indicator for poverty because it is a significant determinant of success on several other MDGs.

Status and impacts of poverty in Kenya are given in Box 4.4. An indicative instrument, a standard, can facilitate a systematic integration of "allowable" limits based on these impacts in formulating strategies for remedial action. Kenya can study and simulate the effects and impacts of poverty on its well being, especially in relation to its Vision 2030.

Box 4.4. Some figures and related impacts (source Nyaga et al. 2005; Wathaka et al. 2003)

1. Poverty rate increased from 48.8 % (1990) to 56 % (2003) 2. As a result of poverty, illiteracy and HIV/AIDS increased, life expectancy dropped
2. Infant mortality increased from 63 (1993) to 78 per 1000 (2000)
3. Mortality under five, increased from 96 (1993) to 114 per 1000 (2003)
4. % stunted children increased from 29 % (1993) to 31 % (2003)
5. % children 12–23 months fully vaccinated dropped from 79 % (1993) to 52 % (2003)
6. Urban unemployment increased from 7 % (1978) to 25 % (1999), national average of 14.6 %
7. Kenya recorded poorer standards of living and Human Development Index fell drastically through 2002, 2003 and 2004 (Shimoli 2006)

4.5 Justification for Standards and Thresholds

Aware of the costs and implications of poverty and HIV/AIDS for households, sectors and the Kenyan economy as a whole (Vision 2030 is relevant here), Kenya needs an established formal measure of extent, quantity, quality or value or comparator to indicate what level of poverty or HIV/AIDS is 'allowable'. Such standard is useful for both strategic and instrumental reasons. Strategically, it can guide on the levels of poverty and HIV/AIDS that Kenya can sustain while

achieving its developmental and welfare aspirations. This implies that formulation
of several strategic policies, plans, programmes and projects will be guided by the
standard. Instrumentally, the standard is a tool for enforcement, assessment, and
evaluation. Besides, national standards and thresholds will also extend the understanding and framing from the global agenda to a national one, for more effective
and long-term ownership. In practice, as illustrated in Figure 4.1, the MDG-related
target for HIV prevalence (M_T) can be achieved and yet the national level (S) still
not achieved, meaning that adverse impacts still continue ($V_T - V$). This means
that considerable resources would have been spent in achieving a target to a point
that is still significantly harmful to the Kenyan Vision and care is not taken to relate the MDG target and a national standard. In this context, achieving the MDG
target may lead to complacency and HIV/AIDS may relapse; besides such an investment is not cost effective in the long-term.

Although setting standards is both a technical and practical matter, scientific
undertaking must remain the essential basis for poverty HIV/AIDS and poverty
standards (RCEP 2004). In defining a standard, the question of how much should
be tolerated is a challenge to be overcome by strategists and planners, although the
operational definition is situational and dependent on several factors. More on setting of standards is provided among others by Harvard (2001), RCEP (2004),
Glennerster (2000) and Guttorp (2000). Rural, urban, pastoral, agrarian communities may have different standards based on how poverty and HIV/AIDS impact
their advancement towards Vision 2030.

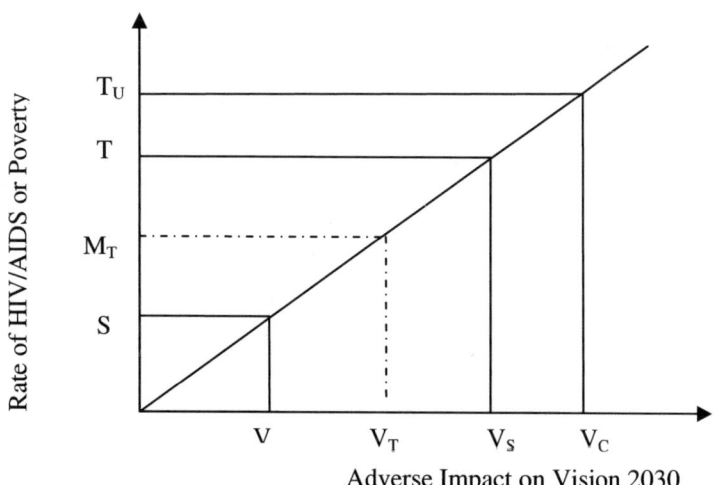

S - Standard and allowable level; no significant impact
T –Threshold; critical and worrisome impacts, urgent intervention needed
T_U – Upper threshold; Impact too great; vision completely derailed
M_T – MDG target level, BUT sill above the standard.
Rates above V_A are not desirable as they adversely impact the vision.

Fig. 4.1. Graph illustrating how an achieved MDG target may not be at a level that coin-
cides with that desired to achieve national vision e.g. Kenya's 2030 Vision

Besides, from the simulation of impacts on Kenya's 2030 vision, an upper or lower threshold at national, regional or household levels can be set and certain designated response strategies be triggered once a threshold is breached.

A threshold is the entrance point or beginning of anything; or the point at which a stimulus just produces a response (Webster's 1992). A threshold is a limit (high or low) placed on a specific monitored metric and is an analytical tool for judging significance. Adopting thresholds promotes consistency, efficiency, and predictability. Although thresholds may be either qualitative or quantitative, HIV/AIDS and poverty lend themselves to quantification and must therefore rely upon quantitative descriptions. Thresholds should be based on studies which can identify that point at which a given effect becomes significant, and various thresholds corresponding to various severities of MDGs on Kenya's Vision 2030 can be used to trigger various calibrated responses. This would create a systematic interlinkage and synergy between response, strategizing, planning and implementation to avoid remission of effects and impacts beyond what is bearable (e.g. does not adversely impact Vision 2030). A fully-fledged threshold should contain, in some form, the following elements:

- A brief definition of the potential effect
- Reasons for its significance
- Threshold criteria for significance
- Geographic scope of the criteria, if applicable
- References to the facts or data upon which the criteria are based.

Once poverty and HIV/AIDS thresholds are met with a menu of standardized mitigation measures the public will be assured that potential effects will be mitigated on a consistent basis particularly towards realization of Vision 2030. Tips for thresholds include:

- Enact only those thresholds with a basis in fact, upon which the thresholds can be enforced
- Do not adopt a threshold if a clear threshold does not exist
- Revise thresholds promptly upon the receipt of pertinent new information; and periodically to ensure their continued relevance and accuracy
- Adopt quantitative rather than qualitative thresholds whenever reasonably possible
- Base thresholds on existing standards and regulations.

4.5 Discussion

Although MDG targets in Kenya will not be met by 2015 (UN 2006; Nzioka 2006), considerable resources have and will be committed towards their achievement. While MDGs and their targets are significant in their own right, resources are limited and the need to be cost effective cannot be over-emphasized. Standards can be used to allocate resources in a cost effective way, e.g. by using distance to a standard as a criteria for prioritizing. Therefore not all MDGS will be pursued equally, particularly those that have lesser opportunity costs per unit or with lesser marginal returns *towards or in relation to* the standard. Since similar effort in many MDGs may end up producing dissimilar cost-effectiveness relative to Vision 2030, a methodical evaluation using standards can make a systematic and more objective differentiation possible, hence facilitating decision making. The possibility to incorporate moral aspects is still possible during the standard setting exercise. The standard becomes a tool to internalize and integrate "allowability" in the strategic planning process and tools such as Strategic Environment Assessment (SEA) and Environmental Impact Assessment (EIA) which are mandatory in Kenya will greatly benefit from such standards and thresholds (see Therivel 2004).

Benchmarking, which involves the analytical comparison of one's own performance with the performance of another (e.g. competitors) whose example is used as a benchmark, may be useful but is not always appropriate, particularly when the contexts are very different. Hence a standard is more appropriate. Such standards and thresholds can however only be realized when there is adequate data and information scientifically derived from thorough studies on the impacts of HIV/AIDS and poverty on Kenya Vision 2030. This is possible as KNASP has already prioritized such studies (NACC 2005a).

4.6 Conclusions and Recommendations

While achieving MDGs targets is significant, Kenya will benefit from basing its strategic planning and decision making processes in a wider longer-term context that integrates national visions. While poverty and HIV/AIDS costs are high, and while remedial action is achieving results, standards of allowability will significantly account for and integrate the strategic potential impact and influence of poverty and HIV/AIDS on the national vision. This will also facilitate methodical estimations of cost-effectiveness for resource allocation, particularly when several MDG targets are competing for the same resources. This is more so when standards are used *ex ante*, to strategize for future remedial plans. Kenya will gain significantly from the savings accruing from cost-effective implementation of remedial action on MDGS, which should be preceded by better strategies formulated using standards as guides. It is suggested that adoption of national standards and thresholds may be faster if the UN recommends it as part of the annual MDGs monitoring and indicator framework. Standards are an apt response to Nassiuma's (2005) suggestion for an "evidence-based policy-making to optimize resources" and also fits UNAIDS (2006) suggestion that HIV/AIDS fight should move from "episodic, crisis management to strategic response (...) for the long term (...)''. Although HIV/AIDS impacts are difficult to measure and poverty not easy to define and quantify, the advantages of setting standards and thresholds far outweigh the potential inaccuracies.

References

Amuyunzu-Nyamongo M (2001) HIV/ADS in Kenya: moving beyond policy and rhetoric, African Sociological Review 5(2) 2001

Annan K (2005) Foreword to The Millennium Development Goals Report 2005, UN, New York

Arndt C, Lewis JD (2000) The macro implications of HIV/AIDS In South Africa: a preliminary assessment. The South African Journal of Economics, 68(5):856-887

Barnett T 2004, The cost of an HIV/AIDS epidemic, volume 9 no 3 pp. 315–317 March 2004, Tropical Medicine and International Health, Blackwell publishing LTD, UK

Bertozzi SM, Aggleton P (1995) Report from a consultation on the socio-economic impact of HIV/AIDS on households, Chiang Mai,Thailand, 22-24 September 1995

Bollinger L, Stover J, Nalo D (1999) The economic impacts of AIDS in Kenya, Futures Group International

Bonnel R (2000) 'HIV/AIDS and economic growth: a global perspective' In The South African Journal of Economics, 68(5):820-855

Bradshaw J (ed) (2005) The Well-being of Children in the United Kingdom, Save the Children, London

Bradshaw J (2004) Poverty. presentation at the Centenary Conference, University of York, 13 December 2004, Social policy research unit, University of York, York

Daily Nation 2006, Editorial "The vision can become real" In The Daily Nation 13 October 2006. www.nationmedia.com/dailynation/printpage.asp?newsid=83283, last accessed 13 October 2006

Forsythe S, Roberts M (1995) The financial impact of HIV/AIDS on Africa's commercial sector: A case study from Kenya. AIDScaptions, Vol. 2, No.1

Forsythe S D, Sokal, Lux L, King TDN (1993) An assessment of the economic impact of AIDS in Kenya" Family Health International AIDSTECH/AIDSCAP

Gillis N and Southey S (eds) 2005, A community dialogue for meeting the Millennium development Goals, Fordham University press, New York

Glennerster H (2000) US poverty studies and poverty measurement: the best 25 years. Case paper 42 October 2000, Centre for Analysis of social exclusion, London school of economics

GOK – Government of Kenya (2001) AIDS in Kenya: background, projections, impact, interventions, policy, 6th edition, Ministry of Health, Nairobi

GOK – Government of Kenya (2004) Investments programme for the economic recovery strategy for wealth and employment creation 2003-2007, Revised March 2004, GOK, Nairobi

Greener R, Jeffris K, Siphambe H (2000) The impact of HIV/AIDS on poverty and inequality in Botswana. The South African Journal of Economics, 68(5):888-915

Guttorp P (2000) Setting environmental standards: a statistician's perspective, Technical report series,NRCSE-TRS No.048, 31 May 2000, EPA, USA

Havard (2001) The role of science and economics in setting environmental standards, Rapporteur's report, A Kennedy school policy workshop, Washington DC , 31 May 2001, JFK School of Government, Harvard University

Hirsch T (2005) Ecosystems protection, a key to development. In: Environment and poverty times 04. September 2005, UNEP/GRID-Arendal

Hobcraft J (2004) Parental, childhood and early adult legacies in the emergence of adult social exclusion: Evidence on what matters from a British cohort. In: Chase-Lansdale,
P.L. et al. (eds), Human Development across Lives and Generations· The potential for
change, Cambridge, University Press

Johnson HF (2005) Sustainable Development, a global challenge. In: Environment and poverty Times 04, September 2005, UNEP/GRID-Arendal

KDHS – Kenya demographic and health survey (2003) Central Bureau of Statistics website
http://www.cbs.go.ke/downloads/pdf/Kenya_Demographic_and_Health_Survey_2003
_Preliminary_Report.pdf last accessed 19 July 2006

Kwena ZA (2004) Politics etiquette and the fight against HIV/AIDS: negotiating for a common front in Africa Development Vol XXiX, No.4 2004 pp:113-131, Council for the
development of social science research in Africa, ISSN 0850-3907

Leighton C (1993) Economic and social impacts of the HIV/AIDS epidemic in African and
Asian settings: Case studies of Kenya and Thailand" Abt Associates

Lonwenson R, Whiteside A (2001) HIV/AIDS: implications for poverty reduction. Background paper prepared for the UNDP for the General Assembly, Special Session on
HIV/AIDS, 25-27 June 2001

Muriuki M (2006) Grand plan for wealth. In: The Daily Nation 13 October 2006
www.nationmedia.com/dailynation/printpage.asp?newsid=83325

NACC – National Aids Control Council (2005a) National HIV/AIDS Monitoring and
evaluation framework, Nairobi

NACC – National Aids Control Council (2005b) Kenya HIV/AIDS data booklet 2005, Nairobi

Nalo D, Aoko M (1993) Economic impact of AIDS in Kenya. Ministry of Planning and
Development, Long Range Planning Division, Nairobi

Nassiuma D Kenya Integrated Household Budget Survey (KIHBS), Joint Kenya poverty
assessment workshop, Mombasa, 19-20 May 2005

Nyaga TR, Kimani DN, Mwabu G, Kimenyi MS (2004) HIV/AIDS in Kenya: a review of
research and policy issues, KIPPRA Discussion Paper no 38, June 2004. KIPPRA,
Nairobi

Nzioka P (2006) Kenya unlikely to attain universal development goals. In: The Daily Nation www.nationmedia.com/dailynation/printpage.asp?newsid=55179

RCEP – Royal Commission on Environmental Pollution (2004) A summary of the 21st report of the RCEP: environmental standards and the pubic values, RCEP website
www.rcep.org.uk, last accessed 19 July 2006

Rugalema G, Weigang S, Mbwika J (1998) HIV/AIDS and the commercial agricultural sector of Kenya: impact, vulnerability, susceptibility and coping strategies. Consultancy
Report KE95/100. Rome: FAO

Shimoli E (2006) Kenya getting poorer, says UN report" In the Daily nation
www.nationmedia.com/dailynation/printpage.asp?newsid=56752

Sigle-Rushton W (2004) Intergenerational and Life Course Transmission of Social Exclusion in the 1970 British Cohort Study, CASE paper 78, London: STICERD

Steinberg M, Kinghorn A, Soderlund N, Schierhout G, Conway S (2000) HIV/AIDS facts
figures and the future: capacity of traditional coping mechanisms, Abt Associates,
SARH 2000

Therivel R (2004) Strategic environmental assessment in action, Earthscan, London

Toepfer K (2005) Summit must be red ribbon day for the environment' In Environment and
poverty Times 04, September 2005, UNEP/GRID-Arendal

UN (2001) Road map towards the implementation of the UN Millennium Declaration, UN, New York

UN (2003) Indicators for monitoring the MDGS: definitions, rationale, concepts and sources (ST/ESA/STA/SER.F/95), UN, New York

UN (2005) The Millennium Development Goals Report 2005, UN, New York

UN (2006) The Millennium Development Goals Report 2006, UN, New York

UN Millennium Project (2005) The Millennium Project Report: Investing in Development – A Practical Plan to Achieve the MDGs. www.unmillenniumproject.org

UNAIDS (2006) Report on global AIDS epidemic: executive summary A UNAIDS 10[Th] anniversary special edition, May 2006. www.unaids.org, last accessed on 12.11.2006

UNDP (2006) Mainstreaming MDGS: Action Plan developed 2006. UNDP website www.ke.undp.org/MDGS%20in%20Kenya.htm, last accessed on 7.07.2006

Vandermoortele J (2005) Do global targets matter?. In: Environment and poverty Times 04. September 2005, UNEP/GRID-Arendal

Waithaka JK, Anyona F, Koori A (2003) Ageing and poverty in Kenya: country report. for the regional workshop on ageing and poverty in sub-Sahara Africa, 29-31 October 2003, Dar es salaam, Tanzania

Webster's (1992) New illustrated Webster's dictionary, Pamco Publisging, New York

World Bank (1997) Confronting AIDS: Public priorities in a global epidemic. New York: Oxford University Press

World Bank (2004) Millennium Development Goals, World development indicators database, April 2004.World Bank website http://devdata.worldbank.org/idg/IDGProfile. asp?CCODE=KEN&CNAME=KENYA last accessed on 03.11.2006

World Resources (2005) The wealth of the poor – managing ecosystems to fight poverty, Washington DC, WR Institute in collaboration with UNDP, UNEP, WB 2005

5 Widening the Scope – Sustainability Indicators, Legal Thresholds and Standards in Portugal

Anastássios Perdicoúlis

Department of Biological and Environmental Engineering, University of Trás-os-Montes e Alto Douro (UTAD), Vila Real, Portugal

5.1 Introduction

This chapter widens the scope of legal thresholds and standards from EIA to sustainability, and illustrates this with the case of Portugal.

The Portuguese legal system is in tune with the respective EU Directives, which are being transposed in conformity with the existing national policies and specific legislation. The national legislation thus sets all the environmental legal thresholds and standards, sets the national environmental policy, and also regulates the EIA process as a "preventive instrument of the environmental policy" (Ministério do Ambiente, do Ordenamento do Território e do Desenvolvimento Regional 2005, Art. 2, It. e)). In parallel with the national environmental legislation, there exist a number of national indicator and information systems elaborated by various government offices, based on international references. These systems on one hand organise available information and guidance, and on the other they demonstrate a wider perspective towards sustainability. The co-existence of a complete environmental legislation (including EIA) and a number of relevant indicator and information systems marks an exceptional potential for synergy and mutual benefits.

The chapter outlines the Portuguese EIA and environmental legislation accompanied by examples of thresholds and standards (5.2), and provides an overview of the major indicator and information systems regarding environment and sustainability used in Portugal (5.3). Analysis (5.4) explores potential synergy options, discussed in Section 5.5. Section 5.6 summarises and makes recommendations.

Standards and Thresholds for Impact Assessment. Edited by Michael Schmidt, John Glasson, Lars Emmelin and Hendrike Helbron. © 2008 Springer-Verlag

5.2 Standards and Thresholds

Since Portugal's entry into the EU, in 1986, its legal system is being constructed with the transposition of the respective EU Directives into the national legal and physical context. Together with issues and regulations of national origin, the EU Directives function as the main drivers for the legal thresholds and standards in the Portuguese legislation. As for all aspects regulated by the central government, this also holds true for the particular case of the environmental legislation.

The fundamental reference of the Portuguese environmental legislation is the "Base Law of the Environment" (Assembly of the Portuguese Government 1987), which is considered as the national environmental policy. This has a strong anthropocentric character, recognising that "every citizen has the right to a human and ecologically equilibrated environment..." (Assembly of the Portuguese Government 1987, Art. 2, Par. 1), and frequently expresses concerns about the human well-being – e.g. "atmospheric emissions... that imply risk, damage, or grave inconvenience to the people and goods will be the object of special regulations" (Assembly of the Portuguese Government 1987, Art. 8, Par. 1). Besides the general national environmental policy, sector- or issue-specific policies are issued isolated – for instance: the national renewable energy policy (Box 5.1 (b)).

More specific than policy, the national legislation on the environment provides clear and unambiguous legal thresholds (Box 5.1 (c)) for a number of environmental indicators, as well as the standards for sampling, data processing, and reporting (Box 5.1 (g)(ii) and (f)(i)). The official publication of the central government, *Diário da República*, containing all national legislation, is openly accessible to all citizens, and thus makes widely available – *inter alia* – all environmental legal thresholds and standards set by the government.

As an instrument of the general national environmental policy, EIA is regulated by the legislation of the central government. The current, consolidated EIA legislation (Ministério do Ambiente, do Ordenamento do Território e do Desenvolvimento Regional 2005) is created on the standards of the EU Directives on EIA (European Council 1985, European Council 1997), including the Directive about public participation (European Council 2003) – Box 5.1 (e). The Portuguese EIA legislation defines legal thresholds for screening purposes, with clear and unambiguous values (Box 5.1 (a)). These thresholds provide an easy-to-use table for deciding whether a particular project proposal requires or not to undergo EIA and, like the EU Directives, the thresholds are absolute cut-off values, with no pre-defined margins of flexibility.

Box 5.1. Examples of thresholds and standards in Portugal

(a) **Thresholds** [*EIA – Project types*]: Hydroelectric power plants are subject to EIA when their nominal power rating is equal to or greater than 20MW in the general case, or at any power rating when they are proposed for installation in designated environmentally sensitive areas (Ministério do Ambiente, do Ordenamento do Território e do Desenvolvimento Regional 2005, An. II, Par. 3, It. (h))

Box 5.1. (cont.)

(b) **Thresholds** [*EIA – Project types*]: Hydroelectric power plants are subject to EIA when their nominal power rating is equal to or greater than 20MW in the general case, or at any power rating when they are proposed for installation in designated environmentally sensitive areas (Ministério do Ambiente, do Ordenamento do Território e do Desenvolvimento Regional 2005, An. II, Par. 3, It. (h))

(c) **Thresholds** [*Renewable energy policy*]: Based on EU Directive 2001/77/CE, the production of electricity in Portugal by means of renewable energy sources to be achieved by 2010 is defined at 39% (IA 2005, p.25)

(d) **Thresholds** [*Air Quality/ SO_2 concentration – Portuguese legislation*]: The SO_2 maximum limit is defined to 80 µg/m^3 as the median of the average daily values over a year, when TSP > 150 µg/m^3 measured gravimetrically; reference pressure and temperature for reporting volumes: 101.3 kPa and 293 K respectively; sampling and analysis methods are referenced (Ministérios da Indústria e Energia e do Ambiente e Recursos Naturais 1993, An. I, Tab. B)

(e) **Thresholds** [*Air Quality – SIDS indicator*]: "The WHO has stipulated norms for all the parameters of that indicator, in addition to the national and community (EU) norms" (DGA 2000a, p.48)

(f) **Standards** [*EIA legislation – References*]: The Portuguese EIA legislation is based on the EU Directives 85/337/CEE, 97/11/CE, and 2003/35/CE (Ministério do Ambiente, do Ordenamento do Território e do Desenvolvimento Regional 2005, Art. 1)

(g) **Standards** [*Sets of indicators – air pollution/ air quality*]: (i) Decreto-Lei Nº 276/99, 23 de Julho: SO_2, NO_2, fine particles, suspended particles, Pb, O_3; benzene, CO, polycyclic aromatic hydrocarbons, Cd, As, Ni, Hg; (ii) DGA 2000a (SIDS): greenhouse gas emissions, SO_2 emissions, NO_x emissions, NH_3 emissions, VOC emissions, consumption of substances that destroy the ozone layer, average air temperature, air quality, investment and expenses towards the reduction of air pollution; (iii) REA 2004 (air quality index): CO, NO_2, SO_2, O_3 and PM_{10} (particulate matter with diameter <10µm); (iv) EEA 2005 (air pollution and ozone depletion): emissions of acidifying substances, emissions of ozone precursors, emissions of primary particulates and secondary particulate precursors, exceedance of air quality limit values in urban areas, exposure of ecosystems to acidification, eutrophication and ozone, consumption of ozone-depleting substances

(h) **Standards** [*Sampling methodology – air quality*]: (i) DGA 2000a (SIDS): recommends normalised analytic methods to determine concentrations, and currently used statistical methods for the analysis of the data, according to the respective legislation (ii) Specific legislation: gravimetric method for TSP sampling: suspended particles are collected in a glass or membrane filter; the apparatus consists of a filter, a suction pump, a volumetric gas counter or flow meter; the duration of the sampling is 24 hours; the filter is protected from directly sedimentable particles and direct influence of weather conditions; the filters used must have an efficiency greater than 99% for an aerodynamic diameter of 0.3µm; the velocity of air through the filter must be between 33cm/s and 55cm/s; the reduction of velocity during sampling must not exceed 5% for glass fibre filters, and 25% for membrane filters; the number of samples must be at least 100 per annum, uniformly spread [over the national territory] (Ministérios da Indústria e Energia e do Ambiente e Recursos Naturais 1993)

(i) **Standards** [*National performance*]: the EU average values for certain indexes become the reference standards for Portugal for comparisons of performance – for instance, the energy intensity of the economy (energy consumption per GDP unit) indicates a rising efficiency for the EU average and a decreasing efficiency for Portugal (IA 2005, p.20)

5.3 Indicator and Information Systems

In parallel to the national environmental legislation, the Portuguese government –
mostly through the ministry of the environment[1] – has made several attempts in the recent years to organise the information about environment and sustainability with the
intent to increase the understanding, awareness, and information level of the public,
and also provide help to the professional community. This section presents the main
indicator and information systems for environment and sustainability used in Portugal.

One of the most notable efforts in this direction was the proposal of the General
Directorate of the Environment (DGA 2000a) for an indicator system about sustainable development, known as "SIDS". SIDS is a methodological guide for data-
collection, and thus contains no data *per se*. The philosophy of SIDS is based on
OECD's "Pressure-State-Response" (PSR) model, which is a simplified version of
the "Driver-Pressure-State-Impact-Response", or DPSIR model. SIDS contains
132 indicators in four groups: 72 environmental, 29 economic, 22 social, and 9 institutional indicators (Tables 5.1 to 5.4; Box 5.1 (f)(ii)). The references of SIDS to
methods, techniques, and legislation are in most cases general, and sometimes
vague (Box 5.1 (g)(i)). The reference to targets or thresholds is present in all indicators, although in a general manner (Box 5.1 (d)).

Table 5.1. The SIDS environmental indicators (Source: DGA 2002a)

Theme	Indicator Name	Type
Air	Greenhouse gas emissions	Pressure
Air	Sulphur oxide emissions (SOx)	Pressure
Air	Nitrogen oxide emissions (Nox)	Pressure
Air	Ammonia emissions (NH3)	Pressure
Air	Volatile organic compound emissions (VOC)	Pressure
Air	Consumption of substances that destroy the ozone layer	Pressure
Air	Average air temperature	State
Air	Air quality	State
Air	Investment and spending in the reduction of air pollution	Response
Marine and coastal environments	Population growth in coastal environments	Pressure
Marine and coastal environments	Evolution of the coast line	State
Marine and coastal environments	Constructed area	Pressure
Marine and coastal environments	Contamination of diffuse origin	State
Marine and coastal environments	Point discharges of effluents without treatment	Pressure

---

[1] The name (and jurisdiction) of the Portuguese ministry of the environment has changed
several times in the recent years; its current name is "Ministério do Ambiente, do Ordenamento do Território e do Desenvolvimento Regional".

Table 5.1. (cont.)

Marine and coastal environments	Environmental discharges of hydrocarbons	Pressure
Marine and coastal environments	Water quality in bathing zones	State
Marine and coastal environments	Bathing zones with blue flag	State
Marine and coastal environments	Quality of the aquatic system in coasts, estuaries, and lagoons	State
Marine and coastal environments	Fishing stocks	State
Marine and coastal environments	Fishing stocks below the biological limits of security	State
Marine and coastal environments	Fishing catches	Pressure
Marine and coastal environments	Investment and spending for environmental preservation and defence of coastal zones	Response
Inner waters	Water availability	State
Inner waters	Drawing of underground and superficial waters	Pressure
Inner waters	Water consumption	Pressure
Inner waters	Population with access to (regularly monitored) drinking water	State
Inner waters	Efficiency of the water supply systems	Pressure
Inner waters	Surface water quality	State
Inner waters	Underground water quality	State
Inner waters	Quality of the water for human consumption	State
Inner waters	Wastewater production	Pressure
Inner waters	Population served by drainage systems and wastewater treatment	Response
Inner waters	Efficiency of the drainage and wastewater treatment systems	Response
Inner waters	Reutilisation of treated wastewater	Response
Inner waters	Density of water networks	Response
Inner waters	Investment and spending for environmental preservation of inner water systems	Response
Soil	Use of soil	State
Soil	National Ecological Reserve (NER/ REN)	State
Soil	Area of irrigated agricultural soil	Pressure
Soil	Consumption/ utilisation of agricultural pesticides	Pressure
Soil	Consumption/ utilisation of commercial agricultural fertilisers (NPK)	Pressure
Soil	Contaminated soil	State
Soil	Soil area affected by desertification	State
Soil	Investment and spending ort he environmental protection of soil	Pressure
Nature conservation	Protected areas	State
Nature conservation	Marine protected areas	State
Nature conservation	Protected areas integrated in international networks	Response
Nature conservation	Degree of surveillance in protected areas	Response
Nature conservation	Protected areas included in land use plans	Response

Table 5.1. (cont.)

Nature conservation	Use of protected areas for environmental education	Response
Nature conservation	Endangered fauna and flora species	State
Nature conservation	Protected fauna and flora species	Response
Nature conservation	Maintenance of agricultural and forest systems with particular interest for nature conservation	Response
Nature conservation	Burned area in protected or sensitive areas	Pressure
Nature conservation	Public and private investment and spending ort he conservation of nature	Response
Forest	Type of forest cover	State
Forest	Total production of wood	Pressure
Forest	Forest production of material other than wood	State
Forest	Burned forest area	Pressure
Forest	Investment and spending ort he environmental preservation of the forest	Response
Biotechnology	Commercialisation of genetically modified organisms	Pressure
Waste	Production of waste	Pressure
Waste	Production of waste by sector of economic activity	Pressure
Waste	Production and final destination of wastewater treatment plant mud	Pressure
Waste	Treatment and final destination of waste	Response
Waste	Valorisation and reutilisation by class of waste	Response
Waste	Import and export of waste	State
Waste	Energy production from waste	Response
Waste	Investment and spending in waste management	Response
Noise	Population affected by external environmental noise	State
Noise	Measures for noise minimisation	Response
Noise	Investment and spending ort he control of noise pollution	Response

Table 5.2. The SIDS economic indicators (Source: DGA, 2002a)

Theme	Indicator Name	Type
Economy	Gross domestic product (GDP)	Pressure
Economy	Evolution of the gross added value by sector	State
Economy	National investment and spending on environmental protection and management	Response
Economy	Imports and exports	Pressure
Economy	Imports by goods type	State
Economy	Exports by goods type	State
Economy	Financial assistance to development, offered and received by the country	Response
Economy	Debt	State
Economy	Direct foreign investment	State
Energy	Energy consumption	Pressure
Energy	Production and consumption of renewable energy	State
Energy	Energy intensity	State
Energy	Energy intensity of the economy	State

Table 5.2. (cont.)

Energy	Evolution of the prices of fuels and electricity	Response
Transport	Average age of the vehicles	State
Transport	Vehicles in circulation	Pressure
Transport	Passenger transport, by type of transport	State
Transport	Traffic intensity	Pressure
Transport	Cargo transport, by type of transport	State
Transport	Structure of the road network	State
Transport	Real prices of passenger transport, by type of transport	Response
Transport	Road accidents	State
Agriculture	Agricultural production	State
Agriculture	National Agricultural Reserve (NAR/ RAN) areas	Pressure
Tourism	Tourism intensity	Pressure
Tourism	Tourist seasonality	Pressure
Tourism	Rural space tourism	State
Tourism	Lodging capacity	State
Industry	Industrial production	Pressure

Table 5.3. The SIDS social indicators (Source: DGA, 2002a)

Theme	Indicator Name	Type
Population	Population density	State
Population	Natality rate	State
Population	Infant mortality rate	State
Population	Maternal mortality rate	State
Population	Average life expectancy	State
Health	Children vaccinated against infectious diseases before they complete their first year of life	Response
Health	Hospitals and health centres	Response
Health	Medical doctors	Response
Health	Nurses	Response
Health	Health spending (total)	Response
Education	Illiteracy rate	Pressure
Education	Population with secondary school certificate	State
Education	Public spending on Education	Response
Social Security	Total public spending on social protection	Response
Social Security	Active beneficiaries and pensioners	State
Employment	Employment structure by sectors	State
Employment	Unemployment rate	Pressure
Culture	Public libraries and users	State
Justice	Criminality index	State
Justice	Convicted in criminal processes, below 20 years of age	State
Justice	Prisoners	State
Other	Complaints about Environmental issues	Response

Table 5.4. The SIDS institutional indicators (Source: DGA 2002a)

Theme	Indicator Name	Type
Institutions	Environmental accounting	Response
Institutions	Employment in the Environment sector	Response
Institutions	Local Agendas 21	Response
Institutions	University graduates	Response
Institutions	Research and development spending (R&D)	Response
Institutions	National implementation of the globally ratified accords	Response
Institutions	Access to global communication networks	State
Institutions	Newspaper consumption	State
Institutions	Environmental Management Systems/ Certification of Environmental Management Systems	Response

The Portuguese ministry of the environment collects environmental information on approximately the same indicators every year. These data are transmitted to the European Environment Agency, and are also published for the Portuguese citizens as State-of-the-Environment reports – known as "REA" – on paper and on the Internet. The philosophy of the REAs also follows OECD's PSR model and thus classifies information in three groups: environmental aspects of economic activities (Pressure), state of the environment (State), and environmental policy and management (Response). The more recent REAs make critical comparisons between the Portuguese data and external references, such as the performance indicators of other member states (Box 5.1 (h)). This constitutes references to relative values, which could function as relative thresholds.

In addition to the above, Portugal's National Institute of Statistics (INE) collects and publishes data on certain social, economic, and institutional indicators. Some of these data are duplicates of the REA, while for other data (e.g. demography, cultural aspects) INE is the main source of information. Aggregated information from the INE databases is available on the Internet free of charge, while detailed information is available for purchase from INE publications. The themes range from environment, people, finance, to sectoral economic activities – with a notable focus on econometric aspects. INE publications usually report facts, but may also contain processing of information in the form of indices, evaluations, or interpretations.

5.4 Analysis

This section explores synergy options between the national legal system (including EIA) and the indicator and information systems used in Portugal.

Since the transposition of the first EU EIA Directive, thematic groups of indicators (factors, aspects, or descriptors) such as water, soil, and biota are identified in the Portuguese EIA legislation for inclusion in the environmental impact statements (Ministério do Ambiente, do Ordenamento do Território e do Desenvolvimento Regional 2005, An. III, Par. 3.). This is an attempt to standardise the contents of "environment", but still leaves enough freedom to practitioners to consider specific elements of particular environments. After many years of EIA experience nationally and internationally, it could be possible to introduce more specification, or even standardisation of the indicators within the above groups. Relevant sets of indicators – even covering the wide spectrum of sustainability – are readily available either from the Portuguese government (e.g. DGA 2000a, 2000b; Instituto do Ambiente 2005; Instituto do Ambiente and University of Algarve 2005), or from external authorities such as the European Environment Agency (2005) and the Organization for Economic Co-Operation and Development (OECD 1993, 2003, 2004).

There are more circumstances in favour of the standardisation of indicators for EIA purposes. The new Portuguese EIA legislation requires the quantification of the previewable emissions, per project phase and per medium (Ministério do Ambiente, do Ordenamento do Território e do Desenvolvimento Regional 2005, An. III, Par. 4). This requirement is capable of facilitating the cross-referencing of indicators with legal thresholds in the national legislation for mitigatory (before implementation) or legal (after implementation) action, and also comparisons with other similar projects – either for raising issues of environmental performance or for forecasting cumulative impacts. In either case, i.e. cross-referencing with legal thresholds or comparing to other projects, the task would be significantly facilitated by a standard set of indicators.

EIA aims to forecast and mitigate environmental impacts of relatively small-scale development actions. Through indicators as the common thread, EIA can link seamlessly with larger-scope and -scale developments, thus reinforcing its role as an environmental policy (or sustainability) instrument. As the EEA (2005) suggests, standard sets of indicators (and, by extension, information systems) are stable but not static. Also, they may be specific to environment, or include social, institutional, and economic aspects and thus widen their scope towards sustainability. Standard sets of indicators can be also adjustable to suit particular requirements of detail or focus, such as projects, programmes, plans, or policies.

Besides the set of indicators *per se*, there is another type of standardisation that seems appropriate in EIA. As in many other countries, the description of impacts in EISs varies in the selection of impact parameters to consider (e.g. magnitude, probability, spatial pattern, time scale, recursivity, etc.) as well as in the choice of values for each of these parameters (e.g. numerical scale options, symbol options, text, etc.). The description of the likely impacts (Ministério do Ambiente, do Ordenamento do Território e do Desenvolvimento Regional 2005, An. III, Par. 5) raises the question of standard impact characterisation. The standardisation of the description of environmental impacts becomes more important when transcending EIA and approaching the field of Environmental Management Systems (EMS), thus making a potential bridge for the future (Eccleston and Smythe 2002, Sánchez and Hacking 2002, Ridgway 2005).

The SIDS and the REA, as well as the INE data collections, contain no legally binding government-defined thresholds or standards. Any reference they make to the national legislation is generic, and sometimes they make no reference at all. Conversely, the Portuguese national legislation transmits clear and unambiguous standards and thresholds for environmental indicators of concern, but makes no specific references to SIDS, REA, or the INE data collections. This situation raises the question of cross-referencing between the national indicator and information systems on one hand, and the national legislation on the other.

The national indicator and information systems on environment and sustainability do not overlap completely among themselves, nor with the international references (e.g. EEA, OECD). This provides a plurality of approaches, which is healthy for discussions. For practice, however, a multi-standard system can be confusing. Perhaps after a number of years of producing national and international discussions and sharing experiences, there could be a unique reference for a sustainability (including the environment) indicator and information system, capable of facilitating comparisons and evaluations at the national and international level.

5.5 Discussion

A list of environmental indicators, based on the national legislation and maintaining references to national or international indicator and information systems (e.g. SIDS, EEA, OECD), could serve as the recommended indicator set in the EIA legislation to include in each EIS. This could be divided in the already existing thematic sections (e.g. water, air, soil, biota).

The above environmental indicator and information system could be expanded with social, economic, and institutional indicators, marked with national government guidelines (not legal limits in this case), and would serve as a national reference for sustainability planning. Such a reference requires uniformisation with (a) the general and specific environmental legislation and sustainability guidance, and (b) with other national and international indicator and information systems such as the SIDS, REA, EEA, and OECD.

The marking of forecasted values for environmental indicators, as required by the EIA legislation, raises the question of national legal thresholds (including minimum, maximum, and tolerance levels). For comparison purposes, this in turn demands an accessible, searchable, easy-to-use, and always up-to-date legal reference of environmental thresholds. This need not be yet another compilation published on paper (which becomes outdated fast), but perhaps a database officially maintained by the ministry of the environment.

Since the national environmental legislation includes *inter alia* the methodology for data sampling, processing, and reporting, a second database of standards could be compiled from the environmental legislation. This could replace the SIDS methodological guidance, and become a unique reference handbook for EIA practitioners.

This investment of compiling databases with legal thresholds and standards, as well as standard sets of indicator and information systems, could be enhanced by a standardisation of the way to describe and report information such as environmental effects as forecasts and facts – i.e. before and after their occurrence.

5.6 Conclusions and Recommendations

The Portuguese legislation sets the national environmental policy, legal thresholds and methodological standards for a number of environmental indicators, and regulates EIA as an instrument of the environmental policy. Besides this legal framework, alternative systems of indicators and information are prepared in parallel by government offices with the intent of informing and assisting the national public and professionals. The co-existence of legal and support references presents a potential for synergy and mutual benefits, identified in this chapter. The following recommendations aim for the realisation of the identified benefits.

First, the compilation of two government-maintained databases is recommended: one for the legal thresholds and the other for the methodological standards regarding the environment, with the intent to facilitate consultation by the public and professionals. The format of this authoritative source could be inspired by the existing indicator and information systems, which they could eventually replace.

A standard way to describe impacts and a standard set of indicators that adequately describes the environment in EISs are recommended as the respective answers to the new national requirements of EIA for the description of forecasted impacts and the description of the environment. A unique reference set of environmental indicators and information could help EISs become more focused and efficient, furnishing important information in a common way that would also permit comparisons across projects for the purposes of measuring performance and/ or assessing cumulative effects.

Finally, appended with social, economic, and institutional information, the reference set is recommended to serve in sustainability assessment and planning. Such an indicator and information system needs to integrate with the national environmental (or sustainability) policy, the legal thresholds, as well as the methodological standards, and with international references such as those issued by OECD and EEA.

References

Assembly of the Portuguese Government (1987) Lei Nº 11/87, 7 April, Diário da República
Assembly of the Portuguese Government (1999) Decreto-Lei Nº 276/99, 23 July, Diário da República
European Council (1985) Council Directive 85/337/EEC, 27 June 1985, OJ L 175
European Council (1997) Council Directive 97/11/EC, 03 March 1997, OJ L 73
European Council (2003) Council Directive 2003/35/EC, 26 May 2003, OJ L 156
DGA (2000a) Proposta para um Sistema de Indicadores de Desenvolvimento Sustentável (Proposal for a System of Indicators of Sustainable Development). Direcção Geral do Ambiente, Lisbon
DGA (2000b) Relatório do Estado do Ambiente 1999 (State-of-the-Environment Report for 1999). Direcção Geral do Ambiente, Lisbon

Eccleston CH, Smythe RB (2002) Integrating Environmental Impact Assessment with Environmental Management Systems. Environmental Quality Management, Summer, 1-13

European Environment Agency (2005) EEA Core Set of Indicators – Guide. EEA Technical Report N° 1/2005, Office for Official Publications of the European Communities, Luxembourg

Instituto do Ambiente and University of Algarve (2005) Relatório do Estado do Ambiente 2004 (State-of-the-Environment Report for 2004). Instituto do Ambiente, Lisbon

IA (2005) Relatório do Estado do Ambiente 2003 (State-of-the-Environment Report for 2003). Instituto do Ambiente, Lisbon

Instituto Nacional de Estatística (INE) Internet address: http://www.ine.pt, last accessed 20.05.2006

Ministério do Ambiente, do Ordenamento do Território e do Desenvolvimento Regional (2005) Decreto-Lei N° 197/2005, 8 November 2005, Diário da República

Ministérios da Indústria e Energia e do Ambiente e Recursos Naturais (1993) Portaria N° 286/93, 12 March 1993, Diário da República

OECD (1993) OECD Core Set of Indicators for Environmental Performance Reviews, Environment – Monographs N° 83. Organisation for Economic Co-Operation and Development, Paris

OECD (2003) OECD Environmental Indicators: Development, Measurement and Use – Reference Paper. Organisation for Economic Co-Operation and Development, Paris

OECD (2004) OECD Key Environmental Indicators. Organisation for Economic Co-Operation and Development, Paris

Ridgway B (2005) Environmental management system provides tools for delivering on environmental impact assessment commitments. Impact Assessment and Project Appraisal, 23(4): 325-331

Sánchez LE, Hacking T (2002) An approach to linking environmental impact assessment and environmental management systems. Impact Assessment and Project Appraisal, 20(1): 25-38

6 Problems in Setting Thresholds

Reinhart Bartsch

Senior Agricultural Economist, Altensteig, Germany

6.1 Introduction

Economics "...studies human behavior as a relationship between ends and scarce means which have alternative uses" (Robbins 1932 p.15 cited in Baumgärtner et al. 2003 p.5). In analyzing the effects of environment conservation measures (e.g. to set a threshold), it is generally understood that the role of the natural sciences is to describe the ecological effects of measures. The economist then would use his tools to determine the optimal choice among the possible alternatives to achieve a planned output. If this sequence is accepted, the first task of the economist is to determine scarcity, or costs and benefits, as well as to define human ends or target functions.

A good is considered to be scarce if it has opportunity costs. Opportunity or shadow costs reflect the value that has to be given up to obtain one additional unit of the desired good. This value may be some amount of another good, or an opportunity to do or not to do something, or the payment of money. For example, for the price of one glass of wine one could buy two glasses of beer or pay the entrance ticket for a museum. Therefore, scarcity is defined relatively: a good is scarce in relation to other scarce goods. Absolute scarcity should be considered as a form of relative scarcity; in this case "a stronger form of scarcity". One could imagine that a person views his or her life as absolutely scarce. His daily submission to the risk of car traffic or aircraft travel proves this wrong. A good that does not exist (including that it has not even been imagined), is by definition not scarce (see Baumgärtner et al. 2003).

Human ends or desired outcomes (expressed in a target function, e.g. clean air, health, income or leisure time) are pursued by investing into a combination of scarce goods (to quell hunger, e.g., one does not eat two pieces of bread or two pieces of butter, but one piece of bread and one piece of butter). Economics tries to provide the tools to find rational, efficient or even optimum solutions for complex decision cases (e.g. by using the decision theory). It seeks to maximize the value of target functions with an amount of variables that cannot reliably be assessed by common sense judgment only.

Standards and Thresholds for Impact Assessment. Edited by Michael Schmidt, John Glasson, Lars Emmelin and Hendrike Helbron. © 2008 Springer-Verlag

A threshold describes by definition a value limit that requires action[1]. Setting a threshold requires examining what the benefits and costs for setting a certain threshold value would be for a given target function. E.g., if the reduction of air pollution is the target, the implementation of a threshold should not cause more pollution than the pollution value exceeding the threshold value represents, or, if the conservation of a certain resource is the objective, the implementation of the threshold should not cost more than the value of the resource saved.

Of course, scarcity and human ends are, as a rule, difficult to quantify, define exactly or to prove correct. And so "optimization" is sometimes considered a vague notion, if not one that can be manipulated easily by the profession.

Also, in the decision-making processes for any action, measure (e.g. threshold) or project, underlying assumptions on scarcity and human objectives are made, mostly in the form of political declarations on the benefits and costs of ecological measures. However, the argument of the vagueness of the scarcity or the human end notions or political influence should not be an argument to circumvent the analysis of the effects of ecological measures (and eventually the setting of thresholds). At least a tentative comparison of what had been declared as objective and planned as outcome of a measure and what was invested in the quest to achieve it, should be carried out.

To highlight the need of this, case studies on the effects of selected environmental conservation measures are presented in the following. I have chosen the examples from my professional life and contacts. In the cases studied, the struggle for correct economic decisions were and still are complicated by

- A general and specific lack of data,
- Disagreement on the nature and quantification of costs, and
- The character and tangibility of benefits.

This lead to deficient assessment of costs, benefits or objective functions resulted in wrong decisions (i.e. in the setting of wrong thresholds).

6.2 Assessing Target Functions, Benefits and Costs of Ecological Measures Incorrectly

6.2.1 The Case of the Reforestation of the Black Forest – Defining the Human Ends or Assessing the Benefits Wrongly

As has been tried to prove, the assessment of costs and benefits of a measure is directly related to the objective or the objectives of a measure, or, the scarcity of goods is related to the objective or target function established. Therefore, an incor-

---

[1] A threshold is a special indicator. An indicator describes by quantity, quality, temporal and spatial parameters that should be reached. The threshold is the indicator value that should not be reached (EC, 2004).

rect definition of the objective function can lead to a wrong assessment of the costs and benefits involved.

In Germany, a rather famous example for this is connected to the invention of the Cost-Benefit Analysis (the Rentabilitätsformel, Faustmann 1849, see Mantel 1990 p.414) by foresters trying to move from the traditional sustainable forestry management to maximize profit from forestry in the Black Forest (Weidenbach, 1994 pp.9f and pp.15ff).[2] The Forest Net Benefit Analysis had been extended to include a time axis and an interest rate was introduced to take into account the position of costs and benefit over time (Bodenreinertragslehre) (Mantel 1990 p.413). As the Net Present Value (or the Internal Rate of Return) was found to be negative or below the minimum of 2.5% (Rau et al. 2000 p.46) the newly invented method justified lower times of turnover and the creation of a "Forest Reserve Funds". These Funds were to be invested in industrial projects to generate higher returns than the forest. The Faustmann formula was shelved (Wurz 2003 p.11) and the sustainable forest management given preference over the profitability aspect. Ex post, it is commonly agreed that this was very useful and nobody would call the present Black Forest seriously a waste of public funds. The Cost-Benefit Analysis method could not be successfully suppressed and it is at present a method required to be used by law for a range of projects (Scholles 2004).

What had happened? The foresters had defined monetary profit from cutting and selling the trees as the objective of the reforestation measure and an interest rate higher than 2,5 % (or eventually the saving bank rate of 3 %) was set as a threshold for the decision to invest (Rau et al. 2000 p.46) The main benefit for the economy as a whole, the existence of the forest itself, was not taken into account in the analysis and in defining the threshold.

The value of leaving an inheritance ("I will leave my children something"), the values of biodiversity, the protection of drinking water and soil resources and the recreational value of the forest were not accounted for, due to an incorrect (or too limited) definition of the human end or the target function. These benefits may have been difficult to quantify at that time, and except for the recreational value[3] economists still struggle in quantifying these correctly, but their omission, justified by the objective defined, led to an incorrect analysis.

Correct target functions differ e.g. in relation of the sociological level of the person or group who establishes it, and geographically. Wealthy people would have given a higher value to the sustainable management of the Black Forest, than those who were food insecure, so the influence of the "class" on human ends and

---

[2] The Black Forest had been largely deforested for timber in Holland, mining and as a pasture for sheep and goats. When the demand for Black Forest timber declined, reforestation was systematically started in 1833 by the Forest Law of Baden (Stächele 2003 p.1) and on large scale in 1890 (Rau et al. 2000 p.45). By the beginning of the 20th century the states of Baden and Württemberg were sitting on a huge forest capital due to the sustainable policy of reforestation in the century before.

[3] Here the concept of the expenditures made to reach a recreational target, to spend there some time, is used to reflect the value of the recreational benefits.

scarcity assessments remains to be further researched as well as the geographical and cultural framework.

The German hyperinflation of 1923 annihilated The Forest Reserve Funds as far as these were generated. The value for an economy as a whole to reduce investment risks by diversifying them to secure and low interest options was therefore also omitted, and might ex post have been the strongest argument for the sustainable approach.

6.2.2 Erosion Control in Morocco – the Use of an Ideological Target Function

If the human end is not too narrow as in the case of the Black Forest reforestation but too broad, vast or "absolute" due to ideologically challenged objectives, the analyses may be equally wrong. For decades erosion control had been set as a goal in itself, i.e. the threshold for the level of erosion tolerated was defined to be zero. If arguments were needed, silting of dams, loss of farmland and forests needed to generate oxygen were at hand. A detailed economical analysis (Bartsch et al. 1994) of erosion prevention measures showed their function in delaying migration into cities by increasing the income of the rural population by subsidies, but they did not contribute to the objective declared.

For the Wadi Srou area in Morocco (near Khenifra), a number of erosion control measures were examined from the economic point of view. The costs of the measures were established and their effect on erosion calculated (Table 6.1). The benefits of erosion control were not only the reduction of the sediment load of water flowing downstream but also the loss of farmland and in the case of e.g. reforestation, also other benefits. Table 6.2 shows such a calculation for the biomechanical stabilization of ravines with Jujube branches and stones. The calculation was made on the basis of a typical 1 to 5 m deep ravine of 50 running meters length. The stabilization of this ravine length reduces the erosion by 7 MT/ha and year. The costs of stabilizing ravines were calculated to be 37.5 €/ha.

Table 6.1. Costs and effects of erosion control measures in the Wadi Srou in Morocco

Action	Costs €/ha	Quantity of sediments stabilised MT/year/ha
Reforestation	6,000	40
Stabilisation of bad-lands	350	12.6
Contour stone walling 1)	168	4.6
Stabilisation de ravines 2)	38	7
Improved fallow 3)	120	8
Fruit tree planting 4)	450	4

1) 0.5 €/m
2) 750 €/running km or 37.5 €/ha (50 running m/ha)
3) Investment costs
4) 1.5 €/tree at 300 trees/ha

Table 6.2. Costs and benefits of stabilising 50 m ravine or of 1 hectare farmland

Average loss on the edge of the ravine per year	3 cm
Loss in farmland (200% of the ravine length)	3 m2)/year
Costs of initial stabilisation	37.5 €
Costs of maintenance (1/3 of investment costs, 2 years after construction)	12.5 €
Value of farmland lost every year additionally 1)	0.15 €
Reduction of sediment load downstream	7 MT/year

1) The Gross Margin of wheat (500 €/ha) for one m2 multiplied by the m2 saved accumulated.

Taking the lifetime of the erosion control measures to be 15 years and the opportunity costs of capital 8 %, the value of one MT of sediment prevented to reach a dam should be at least as given in Table 6.3 (the loss in farmland is included in the calculation). In the case of the stabilization of the ravines, this value was about 0.7 €/MT.

Table 6.4 shows the costs of dredging for different dams of Morocco. At all dams, except one, dredging is cheaper than to prevent the sediment from reaching the dam. Moreover, research on site has shown that sediments travel only about 10 km. This would mean that sediments stemming from erosion in the Wadi Srou, rarely ever reached a dam. The conclusion is, that not only the stabilization of ravines, but also all the other erosion control measures, are far too expensive to justify erosion control measures upstream. This relation must have been detected also in other cases and might have contributed to the outmoding of erosion control (as no other reasons were found).

Of course, some of the erosion control measures have other benefits, like e.g. reforestation has that of wood production or others (see section 6.2.1). But if this is the case, the target function focusing on erosion control is wrong.

Table 6.3. Value of one MT sediment load prevented in the Wadi Srou area should have at least at the dam site

Action	Minimum value required in €/MT
Reforestation	18.0
Stabilisation de bad-lands	3.3
Contour stonewalling	4.3
Stabilisation of gullies	0.7
Improved fallow	1.8
Fruit tree plantation	13.5

Table 6.4. Cost of dredging in different dams in Morocco

Dam name	Costs of dredging (€/MT)
Mohammed V	1.17
Mansour Eddahbi	0.14
El Kansera	0.54
Moulay Youssef	0.16
Ibn Batouta	0.24

6.2.3 Classical Biological Control Concepts – Assessing Benefits and Costs Incorrectly

Any definition of costs and benefits implies the determination of a point of view, i.e. answering the question of to whom, which and to what extent the effects examined are costs and benefits. In classical biological control, costs and benefits occur partly on different levels of point of view and therefore the assessment of the profitability of such measures is frequently incorrect (as well as of thresholds set, if any are set).

Without the intervention of man, pest antagonists control innumerable amounts of pests. If a pest is imported to a country without its antagonist, this system loses balance. Classical biological control is the importation of natural enemies or antagonists of a pest and their establishment in the environment of the pest. If this is successfully accomplished, the parasite or predator population grows with the pest population, reducing the pest eventually to insignificance and obliterating all future control measures.

As a rule, biological control it is not a farmer-implemented action. This is because he, the individual farmer, would not like to carry the costs of a measure from which many others, in cases the whole society, would benefit. Moreover, environmental conservation is not a direct benefit to the farmer, and he cannot control it. Therefore, the economy as a whole represented by its administration, implements biological control measures and bears its costs.

Commonly, there are no technical problems in identifying the costs of biological control. Table 6.5 below shows the costs of biological control measures as they were carried out in Yemen (Bartsch 1994). The target function for biological control is mostly the reduction in the use of chemical pesticides. Frequently even an indicator is formulated, such as "reduction of the use of chemical pesticides by 5% annually, starting in year X". Thresholds, defining where, when and how biological control should be initiated are not formulated, except occasionally in the form of "always" or, as in the case below, only in the form of profitability thresholds.

Table 6.5. Costs of establishing parasites and predators in Yemen

Cost item	Copidosoma koehleri	Aphelinus mali	Neoseiulus idaeus
Problem definition and search for parasite sources (USD, 10,000 USD/ MM)	10,000	10,000	20,000
External consulting expenses (USD)	14,573	-	
Travel expenses (USD)	-	5,000	5,000
Parasite acquisition costs	7,947	4,000	1,000
Releasing and establishing (USD)	600	600	600
Travel and transport (USD)	1,000	1,000	1,000
Total	34,120	20,600	27,600

The identification of the benefits of biological control is more difficult than the assessment of the costs. They can be measured by (Tisdell 1985):

- Indicators of change, as the reduction in the population of the target pest or yield increases;
- Cost savings like the value of losses in production saved and savings over alternative pest controls, such as pesticides;
- Profit increases or variations, such as price increases for the products;
- Variations of land values and
- Production cost reductions.

In the cases examined, the benefits are calculated as savings over alternative pest control measures. The costs of biological control are calculated and these figures are compared with the costs of chemical pest control. The ecological effects of the biological measures, such as the reduction of carcinogenic substances in the food chain, would have to be added to the benefits (Pesticide induced cases by MT of pesticide used multiplied by the average treatment costs). In the cases examined, thresholds of profitability are calculated and then analyzed for the probability of reaching them.

From the point of view of the society, the benefits of this action are e.g. the reduction of environmental pollution (this implies that using less chemicals is a value in itself, a target function hardly shared by African subsistence farmers, see section 6.2.1) and the savings in foreign exchange, when imported chemicals are substituted, if subsidizing agriculture indirectly is not considered a benefit.

In the case of introducing the parasite Copidosoma koehleri to control the potato tuber moth in Yemen, the costs of establishing the parasite are 2.6 USD per hectare of potato grown (13 221 ha) or 22 % of the costs of one conventional treatment. If assumed that the parasite is efficient for 10 years, the costs of the parasite are 03. USD/ha/year and the IRR of this measure would be above 500 %.

The control of the woolly apple aphid with the parasite Aphelinus mali, has costs of 73.6 USD per hectare due to the small apple growing area (280 ha). Still, if it is assumed that the apple orchard acreage increases by 5 % a year and that one conventional treatment per year is saved; the IRR of the measure would be 55 %.

The control of spider mites with predatory mites in apples and peaches has costs of 27.4 USD per hectare (Peaches are grown on 790 ha). If the apple growing area would grow by 5 % yearly and one spray would be saved every year, the IRR would be 217 %. With one spay saved every second year, the IRR is 50 %.

These examples show that if the crop and the pest are economically important, classical biological control can be extremely profitable. If the crop is not important, successful biological control can still be profitable. It is the character of the non-occurrence of damage as the benefit of an investment, which leads mostly to overestimates regarding the costs of the measure.

From the farmer's point of view, the benefit of a successful biological control action is the reduction of his variable costs and thereby the increase of his gross margins and income from farming. The individual farmer, as a rule, does not bear any of the costs of classical biological control. It is the government administration, frequently in the form of the Ministry of Agriculture or the Faculty of Agriculture of a University, which has to meet the costs of implementing the control measure, e.g. of establishing a parasite or predator.

The costs of all the failed control attempts are born by these entities and funded through taxes, by the economy as a whole. The best-known example of a failed measure is the control of pests of sugarcane by Agas in Australia[4]. It has, characteristically, gained much more publicity than the very successful biological control of the Australian rabbits. However, the benefits of one successful biological control measure on the macro-economic level are high enough to more than compensate for many failed attempts in this field[5] (See Norgaard 1988). The problem is that the benefits occur on a different level than the costs.

This problem of the different level appears also in another context. If a pest is reduced to insignificance by the classical biological control measure, no further costs occur, while the benefits continue. If the pest is not reduced by biological control below the Economic Control Threshold (ECT), a conflict at farmer level arises:

• If he does not treat his crop chemically, the pest might grow over the ECT level and reduce production and his income from farming.
• If he controls the pest with a chemical, he not only affects the pest, but mostly also its antagonist, i.e. the biological control agent, reducing the effect of the biological control and increasing the chemical pesticide doses required.

In abstaining from chemical control in such a case, the farmer would have to trade the long-term benefits of a low pest population level combined with yield losses for the short term benefit of a higher Gross Product or Gross Margin and increasing variable costs of chemical pest control. He is reluctant to do this for several reasons:

• Chemical plant protection functions as an insurance for the farmer. A certain yield is required to maintain the farmer family. This yield cannot be substituted by future benefits.
• Benefits in the future have to be discounted to obtain their present value. Even if the farmer does not know the concept of discounting, he will know that money tomorrow is of less value than money today. To sacrifice profits now for higher profits in future is considered risky by the farmer.

---

[4] The popular example of such a failed measure is the introduction of dingoes to control rabbits in Australia. The wild dog is reported to have become a pest itself and caused very expensive control measures. However, the dingoes are an (rather) indigenous animal species and their population grew with the introduction of cattle and sheep on the subcontinent.

[5] In 1951 the research institute CSIRO started the release of the Myxoma virus to control rabbits in Australia. 99 % of the population of 400 million animals were eradicated, the costs of the damage caused by the rabbits was calculated to be 300 million € annually. Although resistance or immunity developed in the course of 20 years and the rabbit population increased again, the profitability of the measure is obvious (see ARTE 1999).

The society, aiming at the reduction of chemicals, mainly for environmental reasons, would have to compensate the farmer for his losses. This is not done because huge costs for the taxpayer are to be expected. This leads to the requirement to evaluate realistically the effect of classical biological control measures and abandon less successful measures, without abandoning research into biological measures.

6.2.4 Release of Beneficials in Greenhouses – Calculating the Costs Incorrectly

Differently from the classical biological control concept, the release of parasites or predators, called beneficiaries, does not establish a permanent natural balance between pest and his antagonist but only a temporary one, similar to conventional pest control. A number of such beneficials have been tested worldwide for their efficiency in greenhouse vegetable growing. Of these, two have shown promising results in Jordan in greenhouse crops, i.e. the red spider mite control on tomato and cucumber with phytoseiullus persimilis.

The costs of releasing these beneficials (Table 6.6) were calculated for a 500 m² greenhouse (monospan tunnel).

Table 6.6. Costs of red spider mite control in tomato and cucumber production

Item	Value	Unit	US $
Control organism	Phytoseiullus persimilis		
Character of organism	Predator		
Rate of application (min.)	2000	adult	
Rate of application (medium)	3000	adult	
Rate of application (max.)	4000	adult	
Price per unit	0.0076325	JD	0.01075
Costs of beneficial (min.)	15.265	JD	
Costs of beneficial (medium)	22.8975	JD	
Costs of beneficial (max.)	30.53	JD	
Manpower requirements			
Preparation	10	min.	
Treatment	40	min.	
Other	10	min.	
Total	1.0	h	
Manpower costs 0.5 JD/h	0.5	JD	
Machinery costs:	0		

The price for the beneficials is the market price Amman. The benefits of the measure were tested in the context of IPM trials in case studies and found to be positive (Table 6.7).

Table 6.7. Tomato production in multispan (1500 m²) greenhouses

Item	IPM trial average (A)	Control (B)	Difference (A-B)	Difference on 500 m²
Yield (kg)	17038,000	16758,000	280,000	93,333
Gross Benefit (JD)	5775,063	5289,780	485,283	161,761
Gross Benefit at +15 % for IPM products (JD)	6641,322	5289,780	1351,542	450,514
Costs of plant protection (JD)	167,278	162,138	5,140	1,713
Gross Benefit - costs of plant protection (JD)	5607,785	5127,642	480,143	160,048
Gross Benefit +15 % - costs of plant protection (JD)	6474,044	5127,642	1346,403	448,801

The trials show the profitability of the measure, not only in Jordan, but also in other countries where the supply of greenhouse beneficiaries is commercially developed. The target function seems to be the increase of the gross margin of green house production. But the target function is rather the elimination of a certain rage of (toxic) pesticides (not allowed in "integrated" production) from the green houses. The threshold could then be defined as a zero level of the banned chemicals. This target function is illustrated by a long history of government subsidies in introducing "Integrated Pest Management".

But why has the use of beneficials in greenhouse production declined in recent years? This is, because the costs of the measure were not calculated correctly. Firstly, research on the subject was not valuated but more importantly, the costs of the pest management and the costs of the increased risk were not taken into consideration. The use of beneficials in greenhouse production requires highly specialised personnel and the detection of the pests as well as the timing of the use of the beneficiaries creates costs, mostly shadow costs of labour. Then, the risk of a failed treatment is higher than in conventional pest control. From time to time the production of whole greenhouses is lost. These costs of the higher risk were not calculated. The maintenance of a widespread integrated plant production or a ban of certain toxic pesticides would have required much more government subsidies than anticipated. Or in other words: a zero threshold level for certain chemicals would have cost more than the society was ready to pay.

6.3 Conclusions and Recommendations

The case studies presented try to highlight the role of the clear definition of human ends for the correct economic assessment of ecological measures. Objectives and human ends can be misjudged and wrongly evaluated for ideological reasons (as in the case of erosion control in Morocco) or unintentionally (as in the case of the discovery of an ultimate formula to maximize profits in the Black forest). If the target function is correct, it may be difficult to quantify the benefits of these measures, leaving important benefits out of the calculation (as e.g. the occurrence and effect of hyperinflations, wars etc., is most difficult to foresee). Equally, in the cost calculation of these measures important factors are often omitted (as in the case of biological pest control). Costs may be overstated or underestimated, either because of a lack of sound calculations – as in the case of the use of Beneficiaries in greenhouses - or with intention. All these mistakes resulted in setting erroneous thresholds.

To establish reliable thresholds, more data are needed. Threshold level/benefit and threshold level/cost function should be established. If such data are not available, calculations should still be made to assess at least orders of magnitude of the costs and benefits of threshold levels.

References

ARTE (1999) 360° – Die GEO Reportage: Der Kaninchen-Krieg

Bartsch R (1994) Economic Evaluation of Integrated Pest Management Concepts, GTZ Eschborn

Bartsch R, Driouchi A (1994) Appui en agro-économie et conception d'un système de subvention adapté, Projet Oued Srou, Maroc

Baumgärtner S, Becker C, Faber M, Manstetten R (2003) Relative and absolute scarcity of biodiversity. In: Verhandlungen der Gesellschaft für Ökologie 33

EC – European Commission (2004) Project Cycle Management Guidelines

Mantel K (1990) Wald und Forst in der Geschichte, Hanover

Robbins L (1932) An essay on the nature & significance of economic science, London

Rau H, Brandl H (2000) Zwei Jahrhunderte Forstgeschichte Baden-Württemberg, Forstliche Versuchs- und Forschungsanstalt Baden-Württemberg, Freiburg

Scholles F (2004) Planungsmethoden, Die Kosten-Nutzen-Analyse, Institut für Landesplanung und Raumforschung, Universität, Hannover

Stächele W (2003) Die Erhaltung der Kulturlandschaft, Landinfo 5/2003

Weidenbach P (ed) (1994) Wald , Ökologie und Naturschutz, Heilbronn

Wurz, Antje (2003) Sustainable Forest Management. In: Seminarbeiträge des Schweizerischen Forstvereins

Part II – Thresholds and Standards for Different Types of Projects

Part IIa – Thresholds and Standards for Site-Specific Projects

Environmental impact assessment is required for different types of projects and spatially-relevant developments as defined in national legislations of different countries worldwide, including Germany, Poland from the European Union, Turkey as a neighbouring European state, Mexico and Brazil in Middle and South America, Yemen in the Middle East, Cameroon and Ghana in Africa as well as Thailand in Asia and Australia in Oceania. Part II of the Handbook focuses on traditional sectors from waste management to forestry, which are all linked to assessment of impacts on human health and the environment at project level. Each chapter combines methodology combined with best practice in the specific country of application.

Part II was divided into the two Sub-Parts IIa Thresholds and Standards for Different Types of Projects and IIb Thresholds and Standards for Spatially Dispersed Projects. The objective was to include typical examples from practice on one hand for man-made developments on clearly defined and limited sites and on the other hand for designated land uses, which generally spread on a larger area and whose impacts are generally mitigated in a less technical way and are closely linked to human health and societies.

Sub-Part IIa starts with Chap. 07 by Hartlik which deals with standards for the quality of the EIA procedure, contents and its implementation in practice. It functions as a 'lighthouse' for future EIA and SEA to come. Chap. 08 by Güler and Avci discusses EU and US Standards for municipal waste landfills including quantitative values and prohibited land uses.

Chap. 09 by Gutierrez looks at sewage treatment standards and thresholds for pollution of surface and groundwater in Mexico. The following Chap. 10 by von Sperling documents similar standards for the Brazilian situation with other values and references. Both authors state, that major problems remain a high percentage of untreated municipal sewage being flooded into natural waters and a general insufficient compliance with current standards.

Chapter 11 by Schmidt et al. investigates impacts of dam projects specifically on soil and water quality and local populations in Yemen. Current standards for dams and water quality of reservoirs are not well implemented and further regulations are needed in the future.

Wongdeethai and Ertel (Chap. 12) look at environmental standards of products and their life-cycle on a larger project scale of recycling industries in Thailand. The authors recommend the implementation of new thresholds of recycling rate of products comparable to EU standards on one hand and to foster communication and education between politicians, entrepreneurs and the public on the other hand.

The relatively new issue of strategic environmental assessment of offshore wind energy use in marine spatial planning in the German exclusive economic zone is discussed by Juliane Albrecht in Chap. 13. She comes to the conclusion that the legal and methodological framework of SEA, including adequate standards and thresholds at different planning stages, are essential for an efficient SEA. However as there is still a lack of objectives, standards and data concerning the marine environment, further research on cause-effect-relationships and the derivation of environmental indicators is needed. Tangang Andandoh et al. (Chap. 14) also study renewable energy technologies (RET), in this case biomass use for cookers in Cameroon. Their focus lies especially on standards for indoor air quality as a major component that influences human health. The authors see an urgent need for Cameroon to re-think it's RET policy standards and to be more domain- specific rather than a "shallow" generalization.

Chap. 15 by Gruszczynski presents adverse effects of mining activities and their wastes and strictly set standard for the operation itself. However, the large-scale land degradation remains an unsolved environmental conflict mainly affecting the sectors of agriculture and forestry.

7 Requirements on EIA Quality Management

Joachim Hartlik

Dr. Hartlik Consultants, Hanover, Germany

7.1 Introduction

Quality management as an integrated part of EIA procedures is an important current challenge to focus on. Even those countries, where EIA is already implemented in national law and has become a standard instrument to assess the environmental effects of projects, plans or programmes, need requirements to ensure good EIA quality. For example, the European Commission stresses the need of quality assurance in their third 5-years report, reviewing the operation of the EIA directive, adopted in 1985 (EU Commission 2003):

> "The quality of the EIA process, and especially the EIS, are the key for an effective EIA. The Commission urges those Member States that have yet to do so to introduce formal provisions for the review of the environmental information supplied by the developer to ensure strict compliance with the terms of the EIA Directive. Such measures could comprise the establishment of expert pools, guidelines on the coordination of experts, clear instructions about responsibilities, the use of independent external expert review etc. Another tool of quality control could be the introduction of an efficient post-decision monitoring system."

Furthermore, the EIA Directive Guidance Group emphasis the need of quality management in their "Topic – Quality assurance - best practices", published in April 2005. An essential conclusion is, that good and efficient EIAs need both:

- an efficient EIA procedure comprising a systematic participation of all concerned authorities, experts, stakeholders, NGOs and the public and
- an understandable Environmental Impact Study or Environmental Impact Statement considering all significant likely effects.

Standards and Thresholds for Impact Assessment. Edited by Michael Schmidt, John Glasson, Lars Emmelin and Hendrike Helbron. © 2008 Springer-Verlag

7.2 Quality Requirements on EIA-Procedure

Overview

The typical stages of an EIA procedure are shown in Figure 7.1. Quality management procedures are not limited on reviewing just the Environmental Studies but have to comprise all major steps in the EIA process. The competent and decision making authority is responsible for integration of such measures. Depending on the national EIA basic conditions, this authority should involve all other concerned authorities, bodies or - like in some countries - a board of national EIA experts in the quality review tasks.

The consultation process, where stakeholders, the public and NGOs have the opportunity to give comments and complaints about the environmental information, has an important role in quality assurance. Often, there is a big knowledge about the affected environment and the ecological interactions within these organisations, so neither the competent authority nor the developer can afford to miss this information to prevent from cost intensive prolongations in the EIA process. The next chapters deal with the quality requirements of the main stages of the EIA process:

- screening,
- scoping,
- preparing environmental impact study (EIS),
- review of the EIS,
- consultation,
- decision making.

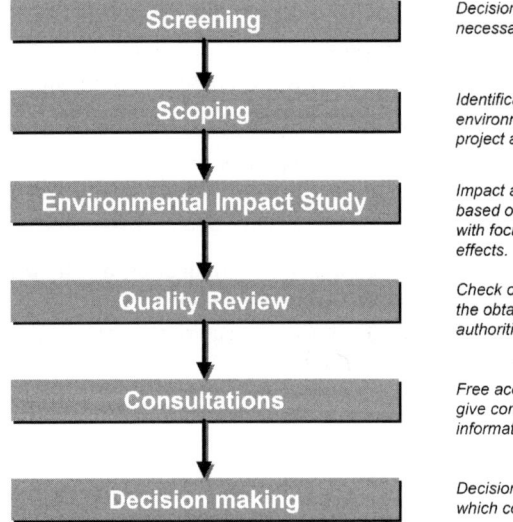

Screening — Decision by the competent authority (CA), if EIA is necessary.

Scoping — Identification of the matters to be investigated like environmental components, project impact factors, project alternatives, prediction and evaluation methods.

Environmental Impact Study — Impact analysis usually prepared by the developer, based on the terms of reference from scoping process with focus to the likely and significant environmental effects.

Quality Review — Check of completeness, plausibility and adequacy of the obtained environmental information, involving other authorities, independent panels or experts.

Consultations — Free access to all interested parties and the public to give comments about the gathered environmental information and the project.

Decision making — Decision by the CA, comprising which alternative and which conditions and mitigation measures are to be

Fig. 7.1. Key stages in the EIA process

Requirements on Screening

Within Screening, the authority has to decide, whether an EIA should be provided or not, considering the likely significant effects on the environment based on a developers project outline. Quality requirements are:

- inclusion list of projects
- exclusion list of projects
- support of case by case screening decision due to the use of a general set of criteria or set of criteria for specific projects; this set should comprise environmental criteria (ecological sensitive areas, areas protected by law, natural resource etc.), characteristics of project and potential impacts
- involvement of other concerned authorities or independent experts (primarily in doubtful cases)
- documentation und publication of the decision and the reasons.

Requirements on Scoping

The scoping process has to identify both the important project impacts and the characteristics of the affected environmental components. Based on that, the significant environmental conflicts have to be estimated in order to set the "terms of reference" for the detailed impact analysis. That includes the determination of the projects reasonable alternatives to be considered. Quality requirements are:

- carrying out a consultation process
  - meeting with the developer, who makes a suggestion about things to be investigated ("scoping paper")
  - involvement of all concerned and environmentally related authorities, bodies and independent experts, based on the "scoping paper"
  - participation of all concerned stakeholders, NGOs and the interested public, if the projects kind or size is likely to cause serious effects on several environmental components
- documentation of the findings in a "scoping document" or "terms of reference" with
  - determination of extent and detail of description of the environmental characteristics
  - determination of extent and detail of description of the project characteristics
  - determination of extent and detail of resulting environmental effects
  - determination of methods to be used for impact analysis if required
- submission of the scoping document to the developer and publication of the document for free access to the public
- opportunity to adapt or complete the terms of reference according to attain unexpected facts and information during the EIA process.

Requirements on Environmental Impact Study

The Environmental Studies, generally called "Environmental Impact Study" (EIS), also known as "Environmental Impact Statement" or "EIA report", are usually be prepared by the developer. Nevertheless, there are existing EIA procedures and the authority is responsible for it. Quality requirements concerning contents and methodology on EIS are described in section 7.3.

Requirements on Quality Review

Unfortunately, a comprehensive quality review of the obtained environmental information submitted by the developer is not yet a common standard step in the national EIA procedures. Nevertheless, there is a need to check the developer's information on adequacy in any kind before publishing it and starting the consultation. Quality requirements are:

- involvement of other concerned authorities or independent experts, review bodies etc.
- comprehensive and systematic check of compliance with terms of reference
- check of fulfilment of all legal provisions
- check of comprehensibility
- check of plausibility
  - use of methods which are independent of the assessor
  - consistency and adequacy of methods
- record of the results published in a "quality review report", listing the deficits and ways to improve quality.

Requirements on Consultation

Consultation is a very important part within the EIA process to achieve acceptance and understanding in the public. Therefore, the main findings of the gathered information should be understandable in a broad sense. Quality requirements are:

- non technical summary of the environmental impact study's important results as a base for the consultation
- involvement of all concerned authorities, stakeholders, NGOs etc.
- opportunity of participation for the public (no exclusions) considering
  - the spatial range of the likely impacts to identify the potential affected public
  - broad information of the public to ensure, everyone is informed about the project and its environmental effects and the consultations
- fair conduction of the consultation without preference towards any stakeholder or the developer
- information about all changes in project planning or in the obtained environmental information during the EIA process and carrying out additional consultations if necessary
- collation and analysis of all responses and feedback to all commentators

- documentation of all reasonable comments
- adequate consideration of all comments into the EIA process.

Requirements on Decision Making

In the decision making process, the competent authority has to consider all gathered information. The decision must be made available to the public including the reasons and the measures that will be required to mitigate negative environmental effects. Quality requirements are:

- clear outcome of the decision
    - proposal approved
    - proposal approved with conditions
    - proposal rejected
- free access to the decision for the public
- publishing of all important aspects of the decision
    - reasons of the decision
    - conditions concerning measures to prevent, reduce or compensate significant effects
    - monitoring measures in cases of uncertainty according to the expected environmental effects
- access of the concerned public to a review procedure before a court of law or another independent and impartial body established by law to challenge the substantive or procedural legality of decisions.

7.3 Quality Requirements on EIS

Overview

In some countries, there are provisions for licensing in order to prevent, that everyone can offers EIS services without proof of special knowledge. In the member states of the European Union, a certification of special EIA consultants is not common, because of the 'free market principle' in any member state. Legislation of European Union guarantees free access to national markets, so private consultants need neither license nor diploma. Regarding this, there are different ways to ensure 'Good Practice' on EIS:

- regular implementation of quality reviews by independent bodies or experts into the EIA process (see Chaps. 35 and 37)
- regarding official or other guidance from professional bodies.

Therefore, in the following chapters are given notes and recommendations on contents and methodology from a professional point of view, the German EIA Association and the experiences of the author as an EIA quality management consultant. The environmental impact study is the document, including the results of the impact analysis. It is a central source of the environmental information. Usually prepared by the developer's consultant, the EIS is build up on five major activities, shown in Fig. 7.2:

- project description
- environmental description
- prediction of environmental effects
- evaluation of environmental effects
- comparison of alternatives.

In general, all applied methods of the different EIS activities should be well described including the reasons of selection. All methods should be replicable, so that different assessors at different times using same methods, obtain identical or at least similar results.

Project Description

The function of project description is, to identify all factors like objectives and physical characteristics which are able to cause significant environmental effects. In order to meet this, the description should comprise

- all phases of the project itself as
  - construction phase
  - operating phase
  - decommissioning phase, closure or shut down of project
- all other existing or proposed projects, plans or programmes, which are likely to cause cumulative effects on the environment with the project to be authorized,
- the risk of accidents arising from breakdowns of processes or facilities, traffic accidents, natural disasters etc.

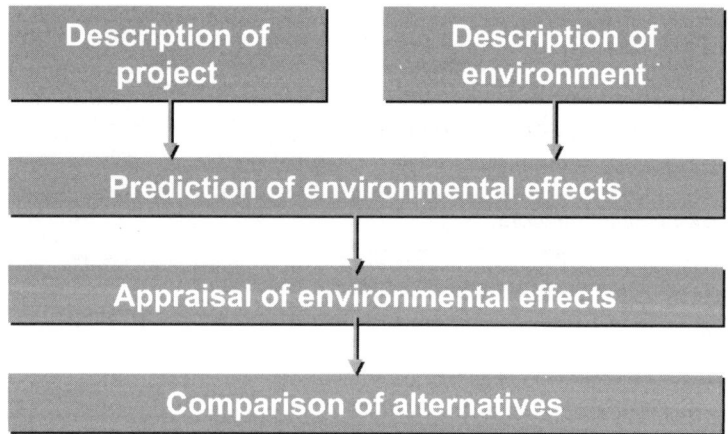

Fig. 7.2. Major activities of an environmental impact study

Box 7.1. Information concerning the description of project

Objectives and physical characteristics
- need and objectives
- programme for implementing the project
- estimated dates of construction, operating and decommissioning phases
- all activities of construction, operating and decommission phase
- induced activities, developments or other projects as a consequence of the project which are likely and are able to cause cumulative impacts on the environment

Size and kind of project
- description of all main components by site, design and size
- temporal and permanent land-use requirements during all project phases
- temporal and permanent need of environmental resources during all project phases (energy, materials etc.)
- main characteristics of the production processes
- outline of the main alternatives and reasons for this choice with respect to the environmental effects

Residues and Emissions
- expected residues and emissions (water, air and soil pollution, noise, vibration, light, heat, radiation etc.) during all project phases
- requirements of solid waste management, locations of final disposal
- requirements of waste water treatment, locations

Risks of accidents and hazards
- use of any hazardous or toxic materials
- risks from breakdowns or failure of processes or facilities
- risks from natural disasters (flood, earthquake, landslide), the project can be exposed

---

In Box 7.1, main project factors are listed, usually needed as baseline information for the EIS. The description of project is focussed on the aim, to identify all components which are likely to be a starting point of significant impacts on the environment.

Environmental Description

The description of the environment is of fundamental meaning of the EIS. Only if the environmental aspects are described well and adequate, the baseline conditions for the prediction of environmental effects are given. The question 'what is the environment?' is answered by the European Union EIA Directive as follows:

- human beings, fauna, flora,
- soil, water, air, climatic factors, landscape,
- material assets, including the architectural and archaeological heritage,
- inter-relationship between this factors.

Box 7.2 shows typical environmental aspects to be described within an EIS. Depending on the role of EIA in the national legislation, the special requirements in each state and the socio-economic and cultural conditions, EIA can be extended by health impact and social impact assessments. Especially in countries, where big investments or projects can cause migration of people, social impact assessment is

---

Box 7.2. Information concerning the description of environment

Physical Characteristics
- demographic, social and socio-economic situation of likely affected resident
- habitats of flora and fauna
- quality of soil
- geology features
- quality of running and surface water
- hydrology and groundwater situation
- quality of air
- local climate and regional meteorological conditions
- landscape, topography, viewpoints
- locations of archaeological, historical, cultural, architectural meaning

Use of natural resources
- location of ecological sensitive areas (wetland, estuary)
- location of areas protected by law (national parks, biosphere reserves, nature conservation areas etc.)
- location of recreation areas
- location of areas for mineral or water resources

Existing environmental loads
- actual loads to all environmental aspects due to existing projects
- actual loads to all environmental aspects due to natural reasons

---

a very important task to face. Migrations can result in much bigger environmental effects than the actual project processes and facilities itself, because people are pulled from their normal environment and confronted with other conditions, they are not used to. The description of the environment should always be focussed on the aspects, which are really likely to be affected. Depending on the kind and of the spatial range of the identified impacts, the environmental area to be regarded can differ considerably. For instance, an EIS of a wind energy project does not usually need a wide description of ground water or geology conditions. Major aspects to be considered in this case, are the factors fauna and landscape.

Prediction of Environmental Effects

Leaving the level of description, prediction of the likely environmental effects is much more demanding. Aim of this step is the identification of links between the project impact factors and affected environmental goods to estimate the likely effects on the environment. Methods to be used have to consider the quantity and quality of baseline data. Therefore, methods can be complex like computer aided calculation of spreading noise or emissions or less exact if describing the likely behaviour of species affected by fragmentation and disturbance of their habitats with the scenario technique.

Basing on a description of the environmental aspects subdivided in different values or ranking zones to fulfil ecological functions, methodological approaches are very common, that calculate the loss of areas due to the projects location (direct loss) or the projects impacts like noise, pollution (loss of ecological functions). Predicting the environmental effects, following aspects are to be considered:

- characteristics of the effects like direct, indirect, secondary, cumulative, short, medium and long-term, permanent and temporary, reversible/irreversible, positive and negative effects
- estimation of location and magnitude of the impacts and the likelihood
- description of the prediction methods used and the adequacy of application
- regarding the developers intended mitigation measures.

Box 7.3 shows typical environmental effects of project impact factors caused by the projects different phases. Box 7.4 contains prediction methods with a brief description and application cases. Although the prediction methods are listed separate, in the EIS practice there is always a mixture of these techniques.

Evaluation of Environmental Effects

The prediction of environmental effects and its magnitude is one step, the evaluation of the significance and importance is another and should always be handled separate. The appraisal of the identified environmental effects within the EIS may not be mixed up with the evaluation due to the competent authority in the decision making process. In this process, other concerns than environmental aspects are to be considered, too.

Box 7.3. General environmental effects

Impacts during construction occur temporal while projects constructions for a limited time. Because of this limit, they aren't that important like constantly lasting impacts. Nevertheless, impacts for example like construction noise can cause irreversible effects, if sensitive species are disturbed permanent.
For example:
- temporal claimed areas because of construction works,
- temporal visual effects because of construction facilities,
- temporal emissions of noise and pollutants because of construction work and construction vehicles
- temporal groundwater abstraction or redirection.

Impacts during construction occur temporal while projects constructions for a limited time. Because of this limit, they aren't that important like constantly lasting impacts. Nevertheless, impacts for example like construction noise con cause irreversible effects, if sensitive species are disturbed permanent.
For example:
- temporal claimed areas because of construction works,
- temporal visual effects because of construction facilities,
- temporal emissions of noise and pollutants because of construction work and construction vehicles
- temporal groundwater abstraction or redirection.

Impacts because of project existence are lasting constantly and therefore they are usually much more important.
For example:
- constantly claimed areas because of project components,
- constant visual effects because of project components,

Box 7.3. (cont.)

- changes in spatial and functional relations on the environment, for example because *of cutted ecological habitat networks.*

Impacts during operation are lasting also constantly and therefore they are of general importance.
For example:
- constant emissions of pollutants, noise, vibrations, light, heat, radiation etc.
- changes in spatial and functional relations on the environment, for example because of cutted ecological habitat networks because of induced traffic.

Impacts during project decommission occur temporal while projects shut down or closure for a limited time.
For example:
- temporal claimed areas because of construction works,
- temporal visual effects because of construction facilities,
- temporal emissions of noise and pollutants because of construction work and construction vehicles
- temporal groundwater abstraction or redirection.

Evaluating the predicted likely environmental effects, there should be considered as a scale

- the legal frame (international, national, local requirements) as legal limits, quality standards, environmental objectives and targets
- plans, programmes or policies with statements and recommendations according to sensitive areas or limits of use of natural resources
- scientific thresholds, environmental load capacities etc.

Box 7.4. Description and Application of Prediction methods

Scenario
Identification of effects on fauna, flora, landscape, cultural heritage
- A Scenario enables to describe possible states of a system with the help of factual knowledge or other techniques of prediction, even though the basic knowledge might be incomplete. Scenarios have the advantage of taking different aspects into account at the same time. They are especially suitable for situations, that don't contain quantifiable elements, as it is often the case in estimations of environmental impact concerning fauna, flora or landscape.
- In contrast to usual methods, the purpose of a scenario is not an accurate and precise forecast of future developments, but rather to find out if there is only one, virtually inevitable development, or if other developments are conceivable and possible. The variety of possible future states of the system is supposed to make citizens, politicians and planners think and avoid false feelings of security.
- Scenarios are usually backed up by statistic material, but at the core, they are nevertheless primarily verbal. They are a blend of intuition and creativity, systematic scientific work and experience. Correspondingly, this is also the character of their instruments and methods. In the making of a scenario, there are usually a few premises from which it is attempted to sketch a logical sequence of hypothesis of the change from an actual to a future state.
- A scenario consists of three main elements:
    (1)   General description of the examined object and its behaviour

---

Box 7.4. (cont.)

    (2)   Premises for the examined object

    (3)   Description of the expected changes, consequences and impacts

Overlay technique

General method for identification of effects on the environment

- Using more and more geographical information systems (GIS), the overlay of maps with environmental sensitive areas and impact factors and their spatial range becomes a standard technique for all environmental components.

As a result of overlay, calculations about direct loss of environmentally important areas or indirect effects on special ecological functions are made.

Simulation

Identification of effects on soil, water, air/climate

- Complex, computer-based technique of prediction working with mathematical simulation-models are contextually and technically rather demanding procedures. Usually these techniques are based on models, that try to describe the examined system with a smaller or larger amount of parameters.
- Based upon the manipulative changing of certain parameters, developing changes in the system are observed in order to draw conclusions about the consequences which can be expected to take place (see also the following illustration).
- The simulation of complex models computer aided. They can be applied as spreading-models for air pollutants, noise emission, changes in the ground water at the extraction of water and infiltration, transport of pollutants in the ground water, erosion processes for instance.

Expert judgement

Identification of effects on fauna, flora or interactions.

- In special fields, environmental effects may uncertain and not predictable with normal methods. IN such cases, experts or an expert panel can give their opinion and expectations by own experiences.

This method is known as "delphi-method" after the "oracle of Delphi" (Greek mythology).

Checklists, Matrices

General method for identification of effects on the environment

- Based on experiences in comparable cases, checklists and matrices are often used as a simple approach to get a first impression about the main effects likely to be expected.

In simple cases, these tools can be adequate, in more complex cases it's usually not sufficient to meet the aims of a detailed impact analysis.

Analogues

General method

- Experiences from case studies with similar environmental conditions and projects are taken to predict the effects in the current case.

EIA is an instrument of precaution to prevent from adverse environmental effects whose elimination would be cost-intensive. So the only application of legal limits is not satisfying, to meet the real aims of EIA. Table 7.1 shows a general approach on an ordinal scale for evaluating environmental effects. This scale has to be adopted case by case to any single environmental effect.

Comparison of Alternatives

The comparison of alternatives is normally a routine within EIS concerning projects like big infrastructure investments. In cases of three or more alternatives to be assessed by a set of criteria of about forty to fifty, the comparison can't be done just by arguing. Decision-making tools have to be used, to prevent from a non transparent and arbitrary way of selection the best alternative. Therefore, the process of comparison should

- use an understandable approach of a step by step aggregation with respect to any single assessed effect,
- determination of the best alternative covering all identified environmental effects in an overall view.

To meet these requirements, there should be at least two steps of aggregation:
- aggregation on the level of the different environmental aspects to enable a statement about the best alternative concerning human beings, fauna, flora, soil, water, air etc.
- aggregation on the more general, comprehensive level, to enable a statement about the best alternative considering all environmental effects in a global view.

There are a lot of different methods for decision making, so the selected technique should be adequate to the special case. Table 7.2 shows an example of a simple ranking-scoring technique using weights to differ between the meaning of the affected environmental components. The ranking is based on a ordinary scale, allowing just the statement, which alternative is better than the others. Weighting is always a case by case decision and cannot be determined in advance without

Table 7.1. Ordinal scale for evaluating the significance on environmental effects

Significant Impact	Scoring	Description
Very High	4	Legal limits and environmental standards exceeded. Serious irreversible and continuous effects on environmental goods. Affection of protected areas and health impacts on local residents. Loss of protected or ecological sensitive areas.
High	3	Legal limits and environmental standards almost achieved. No compliance with environmental objectives and targets of policies, programs and plans. Protected or ecological sensitive areas could be affected. Sustainable use of natural resources not ensured.
Moderate	2	Limited effects on environmental components, usually reversible and with restricted spatial range. Sustainable use of natural resources can be threatened.
Low	1	No serious effects but slightly impairment on environmental goods.

Table 7.2. Comparison of alternatives

Significant Impact	Weight	Alternative A		Alternative B		Alternative C	
		rank weight	rank x	rank weight	rank x	rank x weight	rank
Recreation areas	3	1		2			2
		3		6			6
Noise emissions	3	2		3			1
		6		9			3
Air pollution	2	1		2			3
		2		4			6
Fauna	2	3		1			2
		6		2			4
Landscape	1	2		1			3
		2		1			3
Total		9		9			11
		19		23			22
Overall ranking		1		3		2	

knowledge based on the detailed impact analysis. The use of weighting implies always a clear and consistent foundation. In complex cases, there should be applied more than one aggregation scheme to ensure the result is stable in sense of a sensitivity analysis of the aggregation model:

- using different weights or don't using weights at all,
- change rankings or scorings of alternatives concerning those environmental impacts, where are very close or controversial decisions,
- using a total different approach for aggregation.

7.4 The Role of Standards and Thresholds within Quality Management

There are two main relations between standards respectively thresholds and EIA quality management requirements:

- Standards or thresholds concerning projects capacity

Standards or thresholds can limit the range of EIA application if they are defined too high in the national regulations (compare Annex I and II of the EIA-Directive). In Germany, existing thresholds for projects concerning Annex II are criticized as not adequate by the EU Commission, because the risk of significant environmental effects can also result by very small projects if the affected ecological area is very sensitive. Therefore, it is recommended, to make a case by case examination if significant effects on the environment cannot be excluded safely.

- Standards or thresholds concerning environmental quality

Standards or thresholds for air, water, soil etc. influence the scope and deepness of environmental investigation. If there are no fixed standards (for example in terms climate, material assets or the cultural heritage), things will possibly not be considered properly. On the other hand, if there are only legal standards and no thresholds comprising ecological precaution, there is no reason to unlimit these values and to think about minimizing pollution or other environmental effects.

7.5 Conclusions and Recommendations

The implementation of an efficient EIA procedure as an instrument of sustainable environmental policy can help seriously to avoid irreversible damage on environmental goods and inefficient use of resources. It should be carried out at an early stage of project authorization when all options are still open.

To achieve these demands, the EIA process should comprise extensive provisions on quality management tools to ensure a broadly acceptable fair procedure and an adequate impact analysis, which is reviewed by independent bodies or experts to guarantee "good practice".

References

EIA Directive – Council Directive 85/337/EEC of 27 June 1985 on the assessment of the effects of certain public and private projects on the environment, OJ L 175 of 05.07.1985 p. 40, corr. OJ L 216 of 03.08.1991, p. 40, amended by Directive 97/11/EC of 03.03 1997, OJ L 73, p. 5, and Directive 2003/35/EC of 26.05.2003, OJ L 156, p. 17

8 Environmental Impact Assessment Standards and Thresholds for Sanitary Landfills

Cem B. Avci and Erol Güler

Civil Engineering Department, Boğaziçi University, Istanbul, Turkey

8.1 Introduction

Municipal waste landfills represent a viable and quite commonly used method for domestic waste disposal even though the method has been used for centuries. Municipal waste landfills, at the same time, pose potentially adverse environmental impacts that need to be considered during the site selection process, landfill design, operational and post closure phases. An Environmental Impact Assessment (EIA) study represents an effective tool built into the legislative process of many countries and institutions that can be used to identify potential adverse effects on human health, environment impacts and ecologic risks. The decision makers are able to review all the necessary data on potential impacts prior to project approval. In this manner appropriate remedial measures can be identified and implemented prior to the project start-up.

The present study identifies various standards and associated thresholds for the site setting selection, landfill design and operations that may be incorporated in the EIA process. These components were identified following the review of legislative framework in the European Union and the USA.

8.2 Regulatory Overview

The basis for the EIA standards were derived from regulations governing the solid waste management practices in the European Union countries and the USA. The European Council Directive 1999/31/EC (Landfill Directive) was enacted in 1999 on landfill waste encompassing inert, non-hazardous waste (including domestic waste) and hazardous waste. The directive provides operational and technical requirements on the waste categories and landfills.

Standards and Thresholds for Impact Assessment. Edited by Michael Schmidt, John Glasson, Lars Emmelin and Hendrike Helbron. © 2008 Springer-Verlag

It proposes measures, procedures and guidance to prevent or reduce as far as possible negative effects of landfilling operations on the environment. Specifically, the protection of surface water, ground water, soil and air quality and impacts on the global environment were considered. The specified criteria represent minimum guidelines that European Union members need to implement in their respective regulatory framework.

Municipal waste landfilling requirements in the U.S. are governed by the 40CFR 258 (Landfill Regulations) Criteria enacted in 1991. The purpose of the Landfill Regulation is to establish minimum national criteria under the Resources Conservation and Recovery Act (RCRA) and the Clean Water Act so as to ensure the protection of human health and the environment.

An EIA study (regulated by European Council Directive 85/337/EEC enacted in 1985) may be necessary prior to the permitting and construction of landfills as stated in the Landfill Directive. EIA requirements are not explicitly stated to be a requirement in Landfill Regulations since the documents sets specific thresholds set as minimum national guideline criteria. On the other hand, EIA type studies may be implemented for the regulatory agency framework if the landfill facility needs conditions to be placed below the national criteria.

8.3 EIA Standards and Thresholds

Standards and threshold conditions that may be used in the EIA studies were identified and linked to the existing regulatory conditions. These standards were developed for the setting conditions chosen for a landfill site, design development, operational and post-closure phases of the landfill.

8.3.1 Site Setting Standards

A number of conditions may be specified in the EIA process related to the setting of the landfill. The proposed landfill setting may have adverse impacts on residential and recreational areas as well as agricultural or urban sites, floodplains, wetlands, airports, water bodies and drinking water resources, unstable areas. These items should, therefore, be part of the standards being considered. The Landfill Directive does not specify threshold values for the site setting except to indicate that the site selection should review these components and, if necessary, ensure corrective measures should be implemented. Member countries are allowed set distance limitations or other thresholds for these setting conditions. The Landfill Regulations on the other hand sets the following limitations for the landfill setting:

- *floodplains*: prohibited unless it can be shown that a 100 year flood waters are not restricted, water storage capacity is not reduced or washout of solid wastes causing adverse impact on human health and environment does not occur

- *wetlands*: prohibited unless it can be shown that the landfill system will not adversely impact the wetland conditions and all applicable environmental regulations are met
- *seismic zones*: prohibited unless it is demonstrated that all infrastructure and upper structure are designed to resist earthquake loads
- *unstable areas*: prohibited (suspect to natural and events capable of impairing integrity of some of the containment systems) prohibited unless demonstrated that engineering measures can be incorporated into the landfill design

Both regulatory frameworks consider site setting conditions as standards that should be considered in the approval and permitting process. Limitations as stated above and minimum distance settings for airports, recreational areas, urban and agricultural setting, water bodies and drinking water supply should be identified in the EIA process. Both limitations and thresholds should be allowed to be reviewed and reassessed on risk based assessment if the need develops.

8.3.2 Design Standards

Design conditions related to the landfill infrastructure need to be incorporated into the EIA standards. These conditions can be grouped as water control and leachate management, soil and ground water quality protection and gas management measures.

Water Control and Leachate Management

The applicable standards encompass water from precipitation entering the landfill body, surface water runoff from the landfill waste, collection and treatment of contaminated water and leachate.

The rainfall run-on runoff systems should be designed and subsequently constructed to certain operational threshold. One type of threshold as given by the Landfill Regulations is to set a rainfall event such as a peak discharge, recurrence period and duration for rainfall event and require the control of the water volume run-on into the active landfill phase. Similarly, the runoff control system of the active portion of the landfill should be able handle the water volume from a specified rainfall duration and recurrence storm event. The surface water quality threshold must be in line with the applicable water pollution control regulation parameters.

The Landfill Directive requires that the leachate collection design standards should prevent pollution of soil and groundwater media. The minimum leachate drainage layer thickness is specified to be a minimum of 0.5 m. The Landfill Regulations, on the other hand, require that the leachate collection design will restrict the maximum leachate to 0.3 m above the liner in the landfill. In addition, it is also mandated that the leachate collection and the containment barrier will ensure pre-specified ground water quality limits at a point of compliance with within the upper aquifer below the landfill.

The EIA standards for surface water run-off and run-on should be based on pre-specified rainfall events established in view of the meteorological events and site setting conditions. Protection of soil and ground water can be minimized by an appropriate leachate collection system. A pre-specified point of compliance for ground water quality based on site setting conditions can be used for threshold of the leachate management system. The leachate treatment system thresholds should be based on the wastewater quality control regulations of the regulatory body.

Soil and Water Protection

The Landfill Directive indicates that a combination of geologic barrier, a constructed bottom liner should be used to protect the soil and ground water quality during the operational and post closure phases of the landfill. A similar requirement is put forward in the Landfill Regulations. The minimum design requirements have been set as presented in Table 8.1.

The Landfill Directive requires that the bottom liner design should be based on the geologic and hydrogeologic conditions of the site and should have adequate attenuation capacity to prevent potential risk to soil and groundwater. The Landfill Regulation requires that the containment barrier design should meet ground water quality limits at a point of compliance with specified thresholds for chemical parameters within the upper aquifer below the landfill location.

It is unlikely that an engineered containment barrier together with a geologic barrier can be leak-proof and, therefore a certain amount of leakage should be expected in the active and post operational phase of the landfill. The EIA threshold for the artificial liner thickness and the technical specifications of the mineral barrier should therefore take into account the potential for leakage or a level of acceptable leakage from the bottom of the landfill. Under these conditions an acceptable limit should be set on the allowable pollutant load in the soil and ground water media underneath the landfill. One means of setting threshold limits would be the use of a risk based analysis taking into account the site setting, landfill design and potential receptors.

Gas Management

The decomposition of municipal waste will create gases that are harmful to human health and environment. The Landfill Directive specifically indicates that the landfill gas should be collected and treated either by means of flaring or use in energy

Table 8.1. Minimum design requirements of Landfill Directive and Regulations

Standard\Thresholds	Landfill Directive	Landfill Regulations
Geologic barrier	Presence Accounted	None Described
Mineral Layer thickness	$>= 1$ m	$>= 0.60$m
Mineral Layer permeability	$<= 1x10^{-7}$ cm/sec	$<= 1x10^{-7}$ cm/sec
Artificial Sealing Liner	Required	$>= 2$ mm thick HDPE Geomembrane

production. The criteria for the gas management system are based on minimization of the damage or deterioration to human health and environment. The Landfill Regulations on the other hand require that landfill gas be monitored so as to not create explosive levels within and around the landfill site and if needed a collection system should be implemented.

The EIA standards for gas management should be taken as an active collection system design and implementation of a gas collection system and treatment if the potential for landfill gas generation is present in the landfill. The threshold values for the allowable gas discharge should be based on applicable air quality control regulations set in the regulations as well as development a risk based approach for landfill gas emissions to the atmosphere.

8.3.3 Operating Standards

An appropriately operated landfill must be able to take precautions against nuisance and hazards (odours and dust, wind blown material, fires, vectors, noise and traffic). The EIA standards set for operating standards and thresholds should include best management practices for these potential problems based on the landfill setting and governing environmental regulations (air quality, noise, odour (…)).

Wastes accepted at the landfill have a direct impact on environmental protection measures (liners, leachate collection system, treatment system, gas generation). The Landfill Directive has set guidelines for waste properties to be accepted in each type of waste (inert, non hazardous and hazardous waste), general procedures for testing and waste acceptance and sampling of waste. The EIA process should include standards for the waste identification, testing and acceptance criteria based on the landfill classification and operational requirements. The thresholds should be based on waste characterization based on existing regulations.

8.3.4 Monitoring Standards

Despite having taken into account all appropriate precautions during the site selection and design phases and operational conditions of the landfill, there is always the possibility that adverse impacts may develop to human health and environment. Monitoring standards must therefore be in place during the operational and post closure phases to ensure that adverse impacts may be detected and dealt with appropriately.

Monitoring programme to be undertaken at the landfill should include discharge values for surface water, ground water, gas generation ground water and meteorological data. These should be considered as monitoring standards to be requested within the EIA process. The monitoring standards with appropriate threshold conditions can be as follows:

- *Surface water*: sampling at appropriate discharge locations within the landfill for run-on and runoff discharge volume and surface water quality. The information is to be used for assessing whether discharge conditions are met, operational difficulties are present. The sampling locations, sampling frequency (peak rainfall periods) and quality testing parameters should be selected appropriately based on site setting, regulatory requirements and best management practices.

- *Leachate*: sampling of leachate volume and composition at appropriate intervals to provide information on the efficiency of the collection system, loads incurred to the wastewater treatment plant system and biodegradation of the wastes. At the same time, treated water samples should be taken at appropriate frequency to ensure compliance with regulatory requirements.
- *Landfill gases*: monitoring of the landfill gases to establish the gas generation rate and composition, potential emissions to the atmosphere, discharge from the flare or energy conversion unit.
- *Groundwater*: a ground water monitoring network establishment to detect up-gradient and down-gradient ground water quality in the landfill site setting. The number and depth of the monitoring wells should be based on the site geology and hydrogeology. The sampling test protocol should include all of the chemical parameters that might have been leached from the barrier material or surface water runoff impacting the uppermost water bearing zone. The sampling protocol should also be in line with the regulatory requirements for ground water protection zones.
- *Meteorological data*: data to be used in the water balance of landfill operations and post closure phases together with leachate and surface water generated at the site to review the leachate collection system performance. Discrepancies in the water balance would be indicative of leachate accumulation or leakage from the landfill.

8.3.5 Post Closure Operation Standards

The closure standard of the landfill or active portions of the landfill encompass the construction and maintenance of a cover system to prevent rainfall infiltration into the landfill. The criteria that must be developed for the post closure operations include the specifications of the various layers of the landfill cover, surface water drainage conditions and the gas collection component of the landfill cover. The Landfill Directive and Landfill Regulations criteria are given in Table 8.2.

Table 8.2. Closure Standards of Landfill Directive and Regulations

Standard	Landfill Directive	Landfill Regulations
Artificial Sealing Layer	Not required	Requirement equivalent to bottom sealing layer
Mineral Layer thickness	Required	Requirement equivalent to bottom liner
Mineral Layer permeability	Not specified	Requirement equivalent to bottom liner
Drainage Layer	Required	Required

Duration of the post closure operation should be specified in the EIA process. The threshold may be an upper limit quality of the leachate and gas generation occurring in the landfill following construction of the landfill cover which would indicate the cessation of the biodegradation activities and leaching occurrence within the landfill. Activities should also include proper investigation and upgrade if necessary of cover integrity and continued monitoring as specified in the monitoring programme for landfills.

8.3.6 Remediation Standards

It is conceivable that at one point in time during the operation or post operational phase of the landfill operations adverse impact may occur to the environment. It is therefore important to be able to define the standards and thresholds for imposing remediation activities to circumvent the adverse impact. The monitoring data collected (gas, surface water, leachate etc.) should be assessed and remediation requirement threshold levels should be established. These conditions thresholds could be specified as follows:

- *Water balance*: meteorological data, coupled with surface water runoff and leachate generation data, field capacity of the landfill should be used to establish water balance components of the landfill. Discrepancies in the mass balance should be reviewed and decision to investigate the leachate collection system efficiency should be undertaken.
- *Surface water*: surface water quality at appropriate locations should provide information on compliance with regulatory discharge requirements. If levels are found to be above the regulatory limits, surface water drainage system of the landfill should be improved.
- *Leachate*: large amounts of leachate discharge from the landfill would be indicative of poor surface water ingress control at the site. Efforts to minimize surface water ingress into the active phase of the landfill should be undertaken. Wastewater treatment system discharge analysis would provide indication of compliance with regulatory requirements.
- *Landfill gases*: monitoring of the landfill would provide information on the adequacy of the gas collection system. Inadequate efficiency or health and safety hazard development could be monitored and appropriate measures can be taken to improve collection efficiency.
- *Groundwater*: ground water quality data should be reviewed using an appropriate statistical analysis procedure to review adverse impact on the ground water quality. Trigger levels for the statistical procedures should be set to start implementing additional ground water investigation and remediation process.

8.4 Conclusions and Recommendations

It should be remembered that the present EIA standards discussed in this article dealt with site setting, design, operation, closure and remediation aspects of sanitary landfills only. Use of sanitary landfilling is just part of the solid waste management system that would include collection system, transport schemes, recycling systems and transfer stations. These components together with their potential adverse impacts would have to be reviewed and incorporated into the decision regarding the site setting and size of the landfilling component. The interaction between the various components as it would impact the site setting selection and size of landfill has not been incorporated in the present study.

The standards and thresholds that can be used in the EIA process for solid waste landfill are summarized in Tables 8.3 to 8.8:

Table 8.3. Site setting conditions

Standard	Threshold Conditions
Floodplains	Prohibited unless risk study done
Wetlands-archaeological setting	Prohibited unless risk study done
Unstable areas	Prohibited unless risk study done
Seismic zones	Prohibited unless risk study done
Impact areas: airports, recreational areas, urban and agricultural setting, water bodies, drinking water supply	Minimum distance set or risk study done

Table 8.4. Design standards for water control and leachate management

Standard	Threshold Conditions
Rainfall run on control	Specific Rainfall storm event control
Rainfall runoff control	Specific Rainfall storm event control
Surface water quality	Discharge point testing for regulatory framework
Leachate collection system	Minimize impact on ground water point of compliance standards
Leachate treatment system	Treatment plant discharge testing for regulatory framework

Table 8.5. Design of containment barrier

Standard	Threshold Conditions
Mineral layer	Thickness >= 1 m Permeability <1x10-7 cm/sec More stringent conditions based on risk assessment on groundwater protection criteria
Artificial sealing layer	Thickness >= 2 mm thick HDPE geomembrane More stringent conditions based on risk assessment on ground water protection criteria

Table 8.6. Design standards for gas management

Standard	Threshold Conditions
Gas collection System	Flare or Energy condition so as to meet air quality discharge regulatory overview Acceptable explosive gas levels within the landfill and environment

Table 8.7. Operating and post-closure standards

Standard	Threshold Conditions
Nuisance and Hazard Control	Best management practices implemented for mitigation of operational hazards
Waste Acceptance	Characterization testing and acceptance criteria based on regulatory definition of waste
Final Cover	Minimization of rainfall infiltration based on containment barrier conditions and maintenance

Table 8.8. Monitoring and remediation decision standards

Standard	Threshold Conditions
Surface water discharge quality	Discharge quality testing programme based on regulatory framework and remedial actions when exceeded
Gas emission quality	Gas emission testing programme based on regulatory framework and remedial action when exceeded
Ground water quality	Ground water monitoring programme testing based on regulatory framework and remedial action when statistically significant increase
Leachate quality	Composition testing prior and after wastewater treatment based on regulatory framework and remedial action when exceeded
Water balance	Review of mass balance in landfill and remedial action when mass balance indicates accumulation in landfill or leakage

References

40 CFR Ch I Part 258 Criteria for Municipal Solid Waste Landfills 7-1-96 edn, pp. 378–416

Council Directive 1999/31/EC of 26 April 1999 on the landfill of waste 399L0031, Official Journal L 182, 16/07/1999 pp. 0001–019

Council Directive 1985/337/EEC, EIA Directive, 27 June 1985 on the assessment of the effects of certain public and private projects on the environment Official Journal L 175 , 05/07/1985 pp. 0040–0048

9 Standards and Thresholds for Waste Water Discharges in Mexico

Constantino Gutiérrez

Department of Environmental Engineering, Faculty of Engineering, National Autonomous University of Mexico (UNAM), Mexico D.F., Mexico

9.1 Introduction

The aim of this chapter is to discuss the standards and thresholds that are effective to control wastewater discharges in Mexico. The Mexican Official Standards (NOMs) to the control of wastewater discharges are:

(a) NOM-001-SEMARNAT-1996. It establishes the maximum pollution limits for municipal and industrial wastewater discharges to water bodies.
(b) NOM-002-SEMARNAT-1996. It specifies the maximum pollution limits for wastewater discharges to urban or municipal sewer systems.
(c) NOM-003-SEMARNAT-1997. It establishes the maximum pollution limits for the reuse of treated wastewater.
(d) NOM-004-SEMARNAT-2002. This regulation specifies the maximum pollution limits for the use and final disposition of sludge resulting from the wastewater treatment.

This chapter is divided as follows: section 9.2 presents the historical evolution of standards on wastewater discharges in Mexico, section 9.3 introduces the NOM-001-SEMARNAT-1996, section 9.4 presents the NOM-002-SEMARNAT-1996, section 9.5 examines NOM-003-SEMARNAT-1997, section 9.6 studies NOM-004-SEMARNAT-2002, section 9.7 analyses the results of ten years of application of these technical regulations, and finally, section 9.8 includes some conclusions and recommendations.

Standards and Thresholds for Impact Assessment. Edited by Michael Schmidt, John Glasson, Lars Emmelin and Hendrike Helbron. © 2008 Springer-Verlag

9.2 Standards Antecedents

Mexican official standards for wastewater discharges had been developed since 1973. Their last version was published in 1996. Before 1973 the Mexican legislation only comprised certain kind of general prohibitions for discharging untreated wastewater into soil and rivers. *The Federal Water Law*, published in 1972, prohibited among other actions: to discharge industrial wastewater without permission into streams or impoundments of national competence. The first Mexican official standard appeared in 1973, together with the *Regulation for the Prevention and Control of Water Pollution*, which was part of the *Federal Law to Prevent and to Control the Environmental Contamination (FLPCEC)*, promulgated in March of 1971 as the first Mexican law oriented to environmental protection. The implementation of the FLPCEC resulted in the creation of basic standards and thresholds designed to control wastewater discharged outside municipal sewer systems. Table 9.1 shows the quality parameters and thresholds that were applied at that time.

At the same time, the Ministries of Water and Health proposed a classification of water bodies according to assimilation, dilution capacity, and water use: municipal, industrial, agriculture, recreation and conservation. Table 9.2 shows the classification of water bodies according to water uses and quality characteristics. Table 9.3 resumes the thresholds of some of the toxic substances permitted in surface water bodies. The Water Ministry also started a program to establish particular thresholds for wastewater discharges according to their characteristics and the water bodies' classification.

This control policy continued until the beginning of the 90's, when new standards and thresholds were established to control the wastewater discharges of the 31 more preponderant industries in Mexico, inter alia: sugar cane, petroleum, plastic and synthetic polymer, brewery, iron and steel, etc., and the municipal wastewater discharges for agriculture irrigation. In 1993 thirty three new Mexican Official Norms were applied (NOM-CCA-N-ECOL/1993??).

For each industry it was specified the maximum allowable limits for wastewater discharges. As an example, the limits for conventional thermoelectric power station are shown in table 9.4. All of the regulations presented before were abolished and they were replaced with the following Official Mexican Norms.

Table 9.1. Maximum Allowable Limits included in the Federal Water Law

Quality parameter	Threshold
Settle able solids	1,0ml/l
Grease	70 mg/l
Floating mater	None that can be retained by mesh of 3 mm of clear free square
Temperature	35°C
pH	4,5 – 10,0

Table 9.2. Classification of surface water bodies in function of the water uses and quality characteristics

Class	Uses	pH	Tempe- rature (°C)	D.O (mg/l) Max. limit	Coliform Bacteria MNP (Organisms/100 ml) Max. limit	Grease (mg/l) Max. limit	Dissolved Solids (mg/l) Max. limit	Turbidity (J.T.U.) Max. limit	Colour (Platinum Cobalt) Max. limit	Odours and Flavour	Nutrients: Nitrogen Phos- Max. limit	Floating Matter	Toxic substances
			2	3	4	5	6	7	8	9	10	11	12
DA	Supplying for systems of drinking water and meals industry with disinfection solely. Recreation (primary contact) and free for the uses I, DII and DIII.		C.N. plus 2,5 (a)	4,0	200 fecal (b)	0,76	No more than 1000	10	20	Absent	(c).	Absent	(d)
DI	Drinking water supply with conventio-nal treatment (coagulation, sediment-tation, filtra-tion and disinfection) and industry	6,0 to 9,0	C.N. plus 2,5 (a)	4,0	1000 fecal (b)	1,00	No more than 1000	N.C.	(f)	(g)	(c).	Absent	(d)
DII	Water suitable for recreational use, conservation of flora, fauna and industrial uses.	6,0 to 9,0	C.N. plus 2,5 (a)	4,0	10 000 total coli-forms as monthly average; none va-	Absence of visi-ble layer	Not higher than 2000	N.C.	C.N.	C.N.	(c).	Absent	(d)
DIII	Water for agricul-ture and industrial uses.	6,0 to 9,0	C.N. plus 2,5 (a)	3,2	1000 (j) and free for the	Absence of visible (i)	(i)	N.C.	C.N. plus ≤0		(c).	Absent	(d)
DIV	Water for industrial use (except meals proc-essing).	5,0 to 9,5		3,2									(d)

pH   = POTENTIAL HYDROGEN   J. T. U. = JACKSON TURBIDITY UNITS
D. O.   = DISSOLVED OXYGEN   MG/L   = MILLIGRAMS PER LITER
M.P.N   = MORE PROBABLE NUMBER   N.C.   = NATURAL CO

Table 9.3. Maximum allowable limits for toxic substances in natural water (in milligrams per litre)

Classification	DA	DI	DII	DIII
Arsenic	0,050	0,05	1	5
Barium	1,000	1	5	-
Boron	1,000	1	-	2
Cadmium	0,010	0,01	0,01	0,0005
Cupper	1,000	1	0,1	1
Hexavalent chromium	0,050	0,05	0,1	5
Mercury	0,005	0,005	0,01	-
Lead	0,050	0,05	0,1	5
Selenium	0,010	0,01	0,05	0,05
Cyanide	0,200	0,2	0,02	-
Phenols	0,001	0,001	1	-
Active substances to blue methylene (detergents)	0,500	0,5	3	
Extractable with Chloroform				
Chloroform	0,150	0,15	-	
Pesticides				
Aldrin	0,017	0,017		
Chlordane	0,003	0,003		
D.D.T.	0,042	0,042		
Dieldrin	0,017	0,017		
Endrin	0,001	0,001		
Heptachlorine	0,018	0,018		
Epoxic of Heptachlorine	0,018	0,018		
Lindane	0,056	0,056		
Metoxichlorine	0,035	0,035		
Organic Phosphates with carbamates	0,100	0,1		
Toxaphene	0,005	0,005		
Total herbicides	0,100	0,1		
Radioactivity		picuries per litre		
Beta	1,000	1	1	
Radio-226	3,000	3	3	
Strontium	10,000	10	10	

Table 9.4. Maximum allowable limits for conventional thermoelectric power station

Parameters	Maximum allowable limits	
	Daily Average	Instantaneous
pH (pH units)	6 – 9	6 – 9
Total Suspended Solids (mg/L)	150	180
Settle able Solids (ml/L)	1,0	1,2
Grease (mg/L)	30	36
Biochemical Oxygen Demand (mg/L)	150	180

9.3 Standards for Discharges into Surface and Groundwater

The Mexican Official Standard NOM-001-ECOL-1996 came into force in January 7th 1997. It determines the maximum limits of polluting agents in the municipal and industrial wastewater discharges into natural water and soil.

In this NOM two types of parameters are established: a) basic contaminants and; b) heavy metals and cyanides. The maximum limit of polluting agents for each type are presented in Tables 9.5 and 9.6 respectively. The permissible rank of potential hydrogen (pH) is of 5 to 10 units.

The Most-Probable-Number (MPN) of faecal coliforms was specify to determine contamination by pathogens. The permissible maximum limit for wastewater discharges into waters bodies and soil (used in agricultural irrigation) are of 1.000 and 2.000 MPN of faecal coliforms per 100 millilitres for the monthly and daily average, respectively. Helmint eggs are indicator for parasites. The permissible

Table 9.5. Maximum limits for basic pollutants

Parameters (milligrams per lit-re, except when it is specified)	Rivers — Agriculture Irrigation M.A	Rivers — Agriculture Irrigation D.A	Rivers — Urban Public (B) M.A	Rivers — Urban Public (B) D.A	Natural and Artificial Impoundments — Aquatic Life Protection (C) M.A	Natural and Artificial Impoundments — Aquatic Life Protection (C) D.A	Natural and Artificial Impoundments — Agriculture Irrigation (B) M.A	Natural and Artificial Impoundments — Agriculture Irrigation (B) D.A	Coastal Waters — Urban Public (C) M.A	Coastal Waters — Urban Public (C) D.A	Coastal Waters — Fishing Operation Navigation, others (A) M.A	Coastal Waters — Fishing Operation Navigation, others (A) D.A	Soil — Recreation (B) M.A	Soil — Recreation (B) D.A	Soil — Estuaries (B) M.A	Soil — Estuaries (B) D.A	Soil — Agric ulture irrigation (A) M.A	Soil — Agric ulture irrigation (A) D.A	Soil — Natural Wetlands (B) M.A	Soil — Natural Wetlands (B) D.A
Temperature °C (1)	N.A	N.A	40	40	40	40	40	40	40	40	40	40	40	40	40	40	N.A	N.A	40	40
Fats & oils (2)	15	25	15	25	15	25	15	25	15	25	15	25	15	25	15	25	15	25	15	25
Flouting matter (3)	Absent	Absent	Absent	Absent	Absent	Absent	Absent	Absent	Absent	Absent	Absent	Absent	Absent	Absent	Absent	Absent	Absent	Absent	Absent	Absent
Settle able solids (ml/l)	1	2	1	2	1	2	1	2	1	2	1	2	1	2	1	2	N.A	N.A	1	2
Suspended Solids Totals	150	200	75	125	40	60	75	125	40	60	150	200	75	125	75	125	N.A	N.A	75	125
Biochemical Oxygen Demand₅	150	200	75	150	30	60	75	150	30	60	150	200	75	150	75	150	N.A	N.A	75	150
Total Nitrogen	40	60	40	60	15	25	40	60	15	25	N.A.	N.A	N.A.	N.A.	15	25	N.A	N.A	N.A	N.A
Phosphorus Total	20	30	20	30	5	10	20	30	5	10	N.A	N.A	N.A.	N.A.	5	10	N.A	N.A	N.A	N.A

(1) Instantaneous; (2) Average single sample; (3) Absent according to the Method of Test defined in the NMX-AA-006; D.A. Daily average; M.A. Monthly average; N.A. Non applicable; (A),(B) and (C)Type of receptor body according to Federal Rights Law.

maximum limit for wastewater discharges into soil (used in agricultural irrigation), is of one helminth egg per litre for non restricted irrigation (irrigation of agricultural products in limitless form like forages, grains, fruits, and vegetables) and five eggs per litre for restricted irrigation (irrigation of agricultural products except vegetables that are consumed crude).

The dates of fulfilment were established in tables 9.7 and 9.8 for municipal and industrial discharges respectively. These dates of fulfilment could be advanced by the National Commission of the Water for a receiving body in specific, as long as the corresponding study exists that validates such modification. This norm has been more applied in Mexico as most of the wastewater discharges were lead into natural water bodies: springs, rivers, lakes, ponds and even into the sea.

Because this NOM-001-ECOL-1996 is more flexible than the former 33 NOMs discussed in Sect. 9.2, industries and municipalities adopted it as the reference for their discharges control. This new norm was designed according to the Mexican economy and in a more rational way that can ensure water pollution control. Most small towns could not afford the high costs of construction and operation of efficient treatment systems for a more complete pollutants removal.

Table 9.6. Maximum allowable limits for heavy metals and cyanides

Parameters (*) (milligrams per litre)	Rivers						Natural and Artificial Impoundments				Coastal Waters						Soil			
	Agriculture Irrigation (A)		Urban Public (B)		Aquatic Life Protection (C)		Agriculture Irrigation (B)		Urban Public (C)		Fishing Operation, Navigation and others (A)		Recreation (B)		Estuaries (B)		Agriculture irrigation (A)		Natural Wetland (B)	
	M.A	D.A	M.A	D.A	M.A	D.A	M.A	D.A	M.A	D.A	M.A	D.A	M.A	D.A	M.A	D.A	M.A	D.A	M.A	D.A
Arsenic	0.2	0.4	0.1	0.2	0.1	0.2	0.2	0.4	0.1	0.2,	0.1	0.2	0.2	0.4	0.1	0.2	0.2	0.4	0.1	0.2
Cadmium	0.2	0.4	0.1	0.2	0.1	0.2	0.2	0.4	0.1	0.2	0.1	0.2	0.2	0.4	0.1	0.2	0.0	0.1	0.1	0.2
Cyanide	1.0	3.0	1.0	2.0	1.0	2.0	2.0	3.0	1.0	2.0	1.0	2.0	2.0	3.0	1.0	2.0	2.0	3.0	1.0	2.0
Copper	4.0	6.0	4.0	6.0	4.0	6.0	4.0	6.0	4	6.0	4	6.0	4.0	6.0	4.0	6.0	4	6.0	4.0	6.0
Chromium	1	1.5	0.5	1.0	0.5	1.0	1	1.5	0.5	1.0	0.5	1.0	1	1.5	0.5	1.0	0.5	1.0	0.5	1.0
Mercury	0.01	0.02	0.005	0.01	0.005	0.01	0.01	0.02	0.005	0.01	0.01	0.02	0.01	0.02	0.01	0.02	0.005	0.01	0.005	0.01
Nickel	2	4	2	4	2	4	2	4	2	4	2	4	2	4	2	4	2	4	2	4
Lead	0.5	1	0.2	0.4	0.2	0.4	0.5	1	0.2	0.4	0.2	0.4	0.5	1	0.2	0.4	5	10	0.2	0.4
Zinc	10	20	10	20	10	20	10	20	10	20	10	20	10	20	10	20	10	20	10	20

Table 9.7. Dates of fulfilment of the NOM-001-ECOL-1996 for municipal discharges

Date of fulfilment:	Population range (According to 1990 Census)	Locations number (According to 1990 Census)
January 1st., 2000	Larger than 50 000 inhabitants	139
January 1st., 2005	From 20 001 to 50 000 inhabitants	181
January 1st., 2010	From 2 501 to 20 000 inhabitants	2 266

Table 9.8. Dates of fulfilment of the NOM-001-ECOL-1996 for non municipal discharges

No municipal discharges		
Date of fulfilment:	Biochemical Oxygen Demand (t/day)	Total Suspended Solids (t/day)
January 1st. 2000	Higher than 3,0	Higher than 3,0
January 1st. 2005	From 1,2 to 3,0	From 1,2 to 3,0
January 1st. 2010	Less than 1,2	Less than 1,2

9.4 Standards for Discharges into Municipal Sewerage

In December 9th 1997 it was promulgated the Mexican Official Standard NOM-002-ECOL-1996. This regulation was oriented to control the wastewater discharges into the municipal sewerage. Table 9.9 shows the maximum allowed limits. The permissible maximum limits for polluting substances of the wastewater discharges to urban or municipal sewerage, do not have to be superior to the indicated ones in Table 9.9, for fats and oils it is the average weighed based on the volume, resultant of the analyses practiced to each one of the simple samples. The established permissible maximum limits in the column "instantaneous", are solely values of reference, in case the value of any analysis exceeds the instantaneous one, the person in charge of the discharge is forced to present to the competent authority in the time and form that establish the local legal orderings, the averages daily and monthly, as well as the results of laboratory of the analyses that endorse them. The permissible interval of pH (potential hydrogen) in the wastewater discharges is of 10 (ten) and 5.5 (five point five) units, determined for each one of the simple samples. The units of pH will not have to be outside the permissible interval, in none of the simple samples. The permissible maximum limit of the temperature is of 40°C. (Forty degrees Celsius), measured in instantaneous form to each one of the simple samples. It will be allowed to discharge wastewater with higher temperatures, as long as it is demonstrated to the competent authority by means of a sustained study that it does not cause damage to the system. The floating matter must be absent in the wastewater discharges.

The permissible maximum limits for Biochemical Oxygen Demand (BOD) and Total suspended solids (TSS), that must fulfil the responsible of the discharge to the urban or municipal sewerage system, are established in Table 9.5 of the NOM-001-ECOL-1996 referred to in Sect. 9.3 or to the own particular conditions of discharge that the authority establishes. The person in charge that does not give fulfilment to the BOD and TSS thresholds by his own wastewater treatment can choose to remove these pollutants, through the municipal wastewater treatment, for which he has to:

a) Present to the competent authority a viability study that assures that damage to the urban or municipal sewerage system will not be generated.
b) Support the costs of investment, when therefore it is required, as well as the ones of operation and maintenance that correspond to him in accordance with their volume and loads polluting substances in accordance with the applicable local legal orderings.

It is forbidden to discharge hazardous waste into the sewerage system. The competent authority will be able to fix particular conditions of to the people in charge of the wastewater discharge to the sewerage system, of individual or collective way, that establish the following items:

a) New permissible maximum limits for polluting substances.
b) Permissible maximum limits for not contemplated additional parameters in this Standard. This action will have to be just by means of a technically sustained study, presented/displayed by the competent authority or the person responsible of the discharges.

The dates of fulfilment of the NOM-001-ECOL-1996 were established in table 9.10.

Table 9.9. Standards for discharges into sewerage system

Maximum allowed limits			
Parameters (milligrams per litre, except when other is specified)	Monthly Average	Daily Average	Instantaneous
Oil and grease	50	75	100
Settle able solids (mililiter per liter)	5	7,5	10
Total Arsenic	0,5	0,75	1
Total Cadmium	0,5	0,75	1
Total Cianyde	1	1,5	2
Total Cupper	10	15	20
Chromium hexavalent	0,5	0,75	1
Total Mercury	0,01	0,015	0,02
Total Nickel	4	6	8
Total Lead	1	1,5	2
Total Zinc	6	9	12

Table 9.10. Dates of fulfilment for maximum allowed limits for discharges into sewerage system

Date of fulfilment since:	Range of population
January 1st. 1999	Larger than 50.000 inhabitants
January 1st. 2004	From 20.001 to 50.000 inhabitants
January 1st, 2009	From 2.501 to 20.000 inhabitants

Beside the maximum allowed limits shown in table 9.9, the water authority can be able to fix particular conditions for a specific discharge based on:

a) New permissible maximum limits of polluting agents
b) Permissible maximum limits for not contemplated additional parameters in this Norm.

9.5 Standards for Reuse Treated Wastewater

The Mexican Official Standard that establishes the standards to control the quality of the treated wastewater for reuse was promulgated on 14th of August, 1998. These standards are presented in table 9.11 Additionally it is required the absence of flouting matter and it will not have concentrations of heavy metals and cyanides greater to the established permissible maximum limits presented in the column that corresponds to natural and artificial dams with use in agricultural irrigation of table 9.6 of the Official Mexican Norm NOM-001-ECOL-1996 presented in Sect. 9.3

Floating matter must be absent in the treated wastewater. The treated wastewater for public services will not contain heavy metals and cyanides greater to the established permissible maximum limits in the column that corresponds to natural and artificial dams with use in agricultural irrigation of Table 9.6 of the Mexican Official Norm NOM-001-SEMARNAT-1996.

Public organisations responsible for the treatment of the residual waters that use them in public services, have the obligation to sample and analyze the treated water in the terms of the present Mexican Official Standard and to conserve, at least for the next three years, the registries of the resulting information of the sampling and analysis.

This norm has been very useful to regulate the quality of treated wastewater for reuse purposes. In fact, many projects of wastewater treatment plants fix their water quality goals to be used for reuse. The more common types of treated wastewater reuse in Mexico are: a) irrigation; b) industry; c) recreation; d) municipal; e) fisheries; f) car washing.

Table 9.11. Maximum allowed limits for treated wastewater reuse

Type of reuse	Fecal Coliforms MPN/100 ml	Helmint eggs (he/l)	Oil and grease mg/l	BOD$_5$ mg/l	TSS mg/l
Public services with direct contact	240	[1	15	20	20
Public services with indirect or occasional contact	1.000	[5	15	30	30

9.6 Standards for Sludge Disposal

The objective of the NOM-04-SEMARNAT-2002 is to control the quality of residual sludge and bio solids that are produced in wastewater treatment plants and in the sewerage system cleaning. Sludge is defined as solids with variable humidity content, coming from: a) urban or municipal sewerage cleaning; b) potable and wastewater treatment plants that have not been stabilized. Bio solids are sludge that has been stabilized and that due their content of organic matter, nutrients and characteristics acquired after their stabilisation can be susceptible of use. It was considered that those sludge and bio solids, because of their own characteristics after a stabilisation process, can be used as long as they comply with maximum limits permissible in polluting agents according to the present Mexican Official Standard. Otherwise, they are treated as no dangerous residues in order to attenuate their polluting effects on the environment and to protect the population.

Table 9.12. Maximum allowed limits for heavy metals in bio solids

CONTAMINANT (determined in total form)	EXCELLENT mg/kg in dry base	GOOD mg/kg in dry base
Arsenic	41	75
Cadmium	39	85
Chromium	1 200	3 000
Copper	1 500	4 300
Lead	300	840
Mercury	17	57
Nickel	420	420
Zinc	2 800	7 500

Table 9.13. Maximum allowable limits for pathogens and parasites in sludge and bio solids

Class	Bacteriological indicator of contamination Faecal Coliforms MPN/100 ml in dry base	Pathogens Salmonella spp. MPN/g in dry base	Parasites Helmint eggs (he/l)
A	Less than 1 000	Less than 3	Less than 1(a)
B	Less than 1 000	Less than 3	Less than 10
C	Less than 2 000 000	Less than 300	Less than 35

(a) Viable Helmint eggs; MPN Most probable number.

Table 9.14. Bio-solids use

Type	Class	Use
EXCELENT	A	• Urban uses with direct public contact during its application • The established ones for class B and C
EXCELENT or GOOD	B	• Urban uses without direct public contact during its application • The established ones for class C
EXCELENT or GOOD	C	• Forest use • Soil improvement • Agricultural use

Table 9.12 classifies the bio solids as excellent and good according their maximum concentration of heavy metals. In table 9.13 the maximum allowable limits for pathogens and parasites in sludge and bio solids are presented. Table 9.14 presents the use of bio solids after type and class defined in tables 9.12 and 9.13.

9.7 Experiences in Ten Years of Application

Before 1970s, there were minimal standards and thresholds for wastewater control in Mexico. The first real standards published in the 90s which were composed by specific regulations to control wastewater discharges of the 31 preponderant groups of industries and for municipal discharges were strict because they were designed basically according to the United States Environmental Protection Agency, USEPA, criteria. The American standards corresponded to a high level wastewater treatment technology. Mexican economy was not enough to fulfil the gap of wastewater treatment demand in most of large, medium size cities and even in small towns. Poor villages were not able to carry out programs to design and construct wastewater treatment facilities. Nevertheless, there were an increasing environmental market for wastewater technology; many private investors offered equipments that they imported from industrialised countries. Despite of this increasing wastewater treatment technology, the increment of governmental investment were very low because of insufficient budget. The cities that could construct facilities which treatment processes required imported mechanisms had problems to operate them properly because of lack of money for maintenance and equipment replacement. Many plants were abandoned or operated with low efficiency.

This situation was obligated to the Mexican authorities responsible of water and wastewater management to analyse better strategies to reduce water pollution in an economical way. The strict regulations were modified in order that municipalities could fulfil with the standards. These new standards were designed after Mexican economy and in a more rational way ensuring water pollution control.

(a) Standards were divided into two groups a) basic contaminants and b) heavy metals (Tables 9.1 and 9.2)

(b) BOD and Solids thresholds became flexible in order that they can be reached with economic wastewater treatment processes.

(c) Standards limits for discharges were established according to the use of water bodies (municipal water supply, agriculture, environmental conservation, recreation, fishery, power generation)

(d) It was established a fulfilment programme along the time according to the number of inhabitants (Table 9.3)

(e) Responsible of industries discharges can elect to continue using the standards for specific industries or adopt the modified regulations.

This new policy helped to have better conditions to fulfil the standards to control water pollution. Recent data illustrate the current situation of wastewater facilities in Mexico. In 2000 existed 1,018 wastewater treatment plants with a capacity of

75,9 m^3/sec (The Water Management in Mexico. Advances and Challenges; National Water Commission 2006). Just 793 of them were reported in a regular operation, treating 45,9 m^3/sec of the total wastewater in the country that reached 200 m^3/sec, which means 23 %. Regarding the type of treatment processes, around 50 % were composed of stabilization lagoons. In 2006 wastewater treatment capacity grew up to 74,2 m^3/sec which represented 36 % of the total. Between 2000 and 2006 period wastewater treatment coverage increased 13 %. This increment was achieved in part due to the programmes of CNA (Water National Commission) that helped to the municipalities to make investments on wastewater facilities. CNA pardoned municipalities that had debts because the payments of the rights for wastewater discharge into federal rivers.

9.8 Conclusions and Recommendations

The Official Mexican Standards presented are the basic regulations to establish the standards for wastewater control. It is necessary to review continuously the maximum allowable limits to require more rigorous values in order to have greater security for human health and environmental protection. It is advisable to incorporate other parameters that are representative of industrial discharges additional to the basic pollutants and heavy metals into the standards. Although an important advance in the number of wastewater facilities and in the percentage of treatment coverage in Mexico was achieved, there is still 76 % of the wastewater not treated before being disposed.

References

Official Mexican Norm NOM-001-ECOL-1996; O.D. January 7th 1997
Official Mexican Norm NOM-002-ECOL-1996; O.D. December 9th 1997
Official Mexican Norm NOM-003-ECOL-1997; O.D. August 14th, 1998
Official Mexican Norm NOM-004-ECOL-1996; O.D. September 24th, 2002
National Water Commission (2006) The Management of Water in Mexico. Advances and Challenges
National Water Commission (2005) Resume of Wastewater Plants 2005

10 Standards for Wastewater Treatment in Brazil

Marcos von Sperling

Department of Sanitary and Environmental Engineering, Federal University of Minas Gerais, Brazil

10.1 Introduction

The setting up of an adequate legislation for the protection of the quality of water resources is a crucial point in the environmental development of all countries. The transfer of written codes from the paper into really practicable guidelines, which are used not merely for enforcement, but mainly as an integral part of the environmental protection policy, has been a challenge for most countries.

Many developed nations have already surpassed the basic stages of water pollution problems, and are currently fine-tuning the control of micro-pollutants or the impacts of pollutants in sensitive areas. However, developing nations are under constant pressure, from one side observing or attempting to follow the international trends of frequently lowering the limit concentrations of the standards, and from the other side being unable to reverse the continuous trend of environmental degradation. The increase in the sanitary infrastructure can barely cope with the net population growth in many countries. The implementation of sanitation and sewage treatment depends largely on political will and, even when this is present, financial constraints are the final barrier to undermine the necessary steps towards environmental restoration and public health maintenance. Time passes, and the distance between desirable and achievable, between laws and reality, continues to enlarge (von Sperling and Fattal 2001).

Figure 10.1 presents a comparison between the current status of developed and developing countries in terms of actual water quality concentrations of a particular pollutant and its associated standard. In developed countries, compliance occurs for most of the time, and the main concern relates to occasional episodes of non-compliance, at which most of the current effort is concentrated. However, in developing nations the concentrations of pollutants in the water bodies are still very high, and the efforts are directed towards reducing the distance to the standards and eventually achieving compliance. Brazil is certainly not an exception in this scenario.

Standards and Thresholds for Impact Assessment. Edited by Michael Schmidt, John Glasson, Lars Emmelin and Hendrike Helbron. © 2008 Springer-Verlag

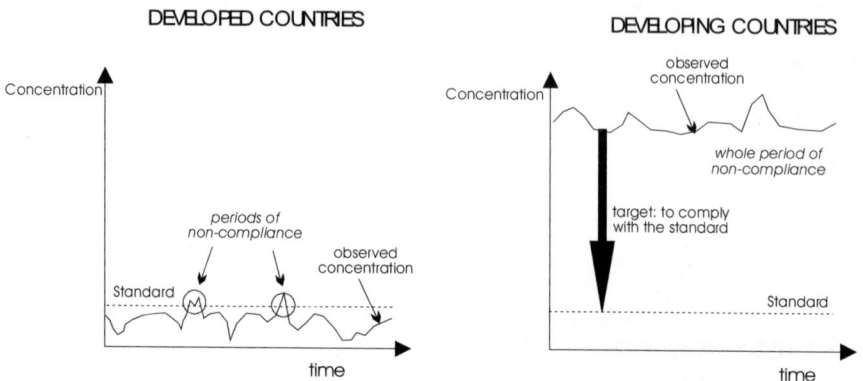

Fig. 10.1. Comparison between developed and developing countries in terms of compliance to standards (source: von Sperling and Chernicharo 2005)

Wastewater treatment deficit in Brazil is around 75 % of the collected sewage flow, and huge efforts to decrease it are in many cases insufficient to compensate for the net population increase in the areas without sanitation. The problem is more significant in urban areas, where the large concentration of population results in very small dilution ratios in the receiving water and due to the scarcity of the substantial funds required to treat the pollution loads. These problems have caused a general status of non-compliance with former Brazilian water quality standards.

Brazil is a federative union, divided into states. Setting up and controlling the environmental legislation is at the federal level a task of the National Environmental Council (CONAMA) and its executive branch (IBAMA). The federal law applies at national level. Each state has also a state environmental council and an executive agency. Besides dictating the state environmental policy, the council has the responsibility of licensing and controlling polluting activities, with the technical support of the environmental agency.

Whereas the environmental control policies are established as a function of political boundaries (states), water resources management follows geographical limits (river basins). If the river basin is completely confined within one state of the union, its water resources are managed by the state water resources council, which may assign a river committee and agency for the basin. However, river basins that cross two or more states or national boundaries are managed by a federal agency (ANA - National Water Agency). There are no fixed rules for defining the minimum size of the basin that is required for having a committee. Basin committees, together with the state water resources council, define the general policies of water resources management in the catchment area, including establishment of classes of use for the water body, licensing for water abstraction and wastewater discharge and the definition of the application of the revenue arising from the user-polluter-pay principle. The basin agencies are the executive branch of the committee.

These principles are set by the National Water Resources Policy (Law 9433 of 1997).

The chapter describes and comments on the major points of the Brazilian national standards for water quality and effluent discharge (CONAMA Directive No. 357/2005). These standards are used as a basis for licensing new polluting activities, since they are a reference for Environmental Impact Assessment (EIA) studies. A permit may only be issued if has been demonstrated by EIA that the legislation will be complied with. Also, for existing polluting activities, enforcement based on the legislation may be put into practice by the state environmental councils. It should be recognized, however, that putting these principles into real practice is a difficult task and a challenge for many state environmental systems. A closer control on private polluters (industries) seems to be easier to apply than on public polluters (municipalities). This is one of the reasons that, in many places, pollution by domestic sewage is a more serious problem than pollution by industrial wastewater.

10.2 Quality Standards for Water Bodies in Brazil

The Federal directive CONAMA (National Environmental Council) No. 357/2005 updated a directive which was established in 1986 (CONAMA Directive 20/1986). Both structures are similar, but the 2005 legislation revised water uses, classes, constituents and parameter values. Besides the federal legislation, there are also state legislations. Each state must comply with the federal law, and has also the option in the state legislation of including specific parameters or more stringent standard values.

CONAMA Directive 357/2005 divided the waters in the national territory into *fresh* (salinity $\leq$ 0.05 %), *brackish* (0.05 % < salinity < 3 %) and *saline* (salinity $\geq$ 3%) waters. For each of these categories, there are different classes. Each class is associated with a grouping of intended uses for the water. For fresh waters there are five classes, for brackish water there are four classes and for saline waters there are four classes. The chapter concentrates mainly on fresh water, and Table 10.1 presents a summary of the main potential water uses assigned to each class.

The Special Class is associated with the most important uses, and Class 4 is designated for less critical uses. The Special Class is intended for the preservation of the environment under natural equilibrium. However, abstraction of water for supply is accepted, but no wastewater (even treated) is allowed to be discharged into a Special Class water body. Wastewater discharge may only take place into water bodies with classes 1 to 4.

Each river basin, through its committee and agency, together with the environmental agency, must have their waters classified according to the system shown above. In the absence of any specific classification, the legislation specifies that the water body will automatically remain as Class 2.

Table 10.1. Classification of fresh waters as a function of their intended uses (CONAMA Directive Nr. 357/2005)

Use	Class				
	Special	1	2	3	4
Domestic drinking water supply	X (a)	X (b)	X (c)	X (d)	
Preservation of natural equilibrium of aquatic communities	X				
Preservation of aquatic environment in special protection units	X				
Protection of aquatic communities		X	X		
Recreation with direct contact (*)		X	X		
Irrigation		X (e)	X (f)	X (g)	
Breeding of species (aquaculture) and fishing activities			X		
Amateur fishing				X	
Animal water supply				X	
Recreation with indirect contact				X	
Navigation					X
Landscape harmony					X

(a) water supply after disinfection.
(b) water supply after a simple treatment.
(c) water supply after conventional treatment.
(d) water supply after conventional or advanced treatment.
(e) irrigation of vegetables eaten uncooked or low-growing fruits eaten unpeeled.
(f) irrigation of fruits and vegetables, and also parks, gardens and sports fields with which the public may have direct contact.
(g) irrigation of trees, cereals and fodder.
(*) a specific bathing directive applies (CONAMA Directive 274/2001).

The classification of water bodies has already been undertaken for some catchment areas in Brazil, but the majority still remains classified as Class 2.

Because of the grouping of water uses, each of the classes is associated with a certain water quality to be maintained in the water body, expressed in terms of *receiving water standards*. Besides these, there are general *discharge standards*, which are independent of the class of the receiving water body. Discharge standards are dealt with in Section 10.3.

The federal standards set by the CONAMA directive are summarised in Table 10.2 for some of the main water quality constituents. The complete list of parameters covered by the directive is of course much larger, and the full legislation should be consulted, if necessary. The setting up of the limiting values is based on international experience, and is driven mainly by the protection of human health and aquatic species. Besides the list of the several parameters and their limit values clearly established in the legislation, the CONAMA directive specifies that the quality of the aquatic environments may be evaluated, when appropriate, by biological indicators using organisms or aquatic communities.

Table 10.2. Brazilian water quality standards for fresh water bodies according to the class (CONAMA Directive 357/2005)

Parameter	Unit	Fresh water class			
		1	2	3	4
Colour	mgPt/L	natural	75	75	
Turbidity	NTU	40	100	100	
Total dissolved solids	mg/L	500	500	500	
pH	-	6.0 to 9.0	6.0-9.0	6.0-9.0	6.0-9.0
Thermotolerant coliforms	MPN/100mL	200 (a)	1000(a)	(b)	
Chlorophyll a	µg/l	10	30	60	
Cyanobacteria density	cells/mL (or mm3/L)	20,000 or 2	50,000 or 5	100,000 or 10(c)	
Biochemical oxygen demand	mg/L	3	5	10	
Dissolved oxygen	mg/L	≥ 6	≥ 5	≥ 4	≥ 2
Total ammonia (pH$\leq$7.5)	mgN/L	3.7	3.7	13.3	
Total ammonia (7.5<pH$\leq$8.0)	mgN/L	2.0	2.0	5.6	
Total ammonia (8.0<pH$\leq$8.5)	mgN/L	1.0	1.0	2.2	
Total ammonia (pH>8.5)	mgN/L	0.5	0.5	1.0	
Nitrate	mgN/L	10.0	10.0	10.0	
Nitrite	mgN/L	1.0	1.0	1.0	
Total P (lenthic environment)	mgP/L	0.020	0.030	0.050	
Total P (intermediate environment and direct influent to a lenthic environment)	mgP/L	0.025	0.050	0.075	
Total P (lotic environment and direct influent to an intermediate environment.)	mgP/L	0.10	0.10	0.15	

Only some parameters are listed – for a full list consult the legislation CONAMA 357/2005.
Intermediate environment: residence time between 2 and 40 days.
(a) See bathing water directive (CONAMA 274/2000).
(b) Class 3 – thermotolerant coliforms: water supply for breeding of animals under con
    finement: 1000 MPN/100mL; indirect contact: 2500 MPN/100mL; other uses: 4000
    MPN/100mL
(c) Class 3 – Density of cyanobacteria: for animal water supply 50.000 cel/mL ou 5 mm3/L.

Also, the possible interactions of substances and the presence of contaminants not specified in the legislation and that are potentially harmful to living beings must be investigated using toxicological, ecotoxicological or other scientifically recognized methods.

These standards are to be met under the so-called reference flow for the river at the point of the discharge. Reference flow is usually a flow characterizing dry-weather periods and low dilution capacities. Each state environmental agency must decide upon the reference flow to be adopted. Common criteria are: Q90 (flow value expected to be exceeded 90% of the time), Q95 (flow value expected to be exceeded 95 % of the time) and Q7,10 (flow value associated with a minimum of seven consecutive days and a return interval of 10 years).

10.3 Discharge Standards in Brazil

As mentioned in the previous section, in the Brazilian legislation there are two types of standards: water body standards and discharge (effluent, emission) standards. This concept is also adopted in many countries. The main reason is that for the environmental agency it is very difficult to control the river water quality when there are multiple discharges and, in case of infringement of the law, assign responsibilities. As a result, the agencies may concentrate their efforts on a more systematic basis on controlling mainly the *discharge standards*.

Table 10.3. Brazilian discharge standards according to CONAMA Directive 357/2005

Category	Parameter	Condition
General conditions	Floating matter	Absent
	pH	5 to 9
	Temperature	$\leq 40\ oC$
	Settleable solids	$\leq 1\ mL/L$
	Mineral oils	$\leq 20\ mg/L$
	Vegetable oils and animal grease	$\leq 50\ mg/L$
	Parameters	Maximum value
Inorganic parameters	Ammonia (total)	20.0 mg/L N
	Arsenium (total)	0.5 mg/L As
	Barium (total)	5.0 mg/L Ba
	Boron (total)	5.0 mg/L B
	Cadmium (total)	0.2 mg/L Cd
	Chromium (total)	0.5 mg/L Cu
	Copper (dissolved)	1.0 mg/L Cr
	Cyanide (total)	0.2 mg/L CN
	Fluoride (total)	10.0 mg/L F
	Iron (dissolved)	15.0 mg/L Fe
	Lead (total)	0.5 mg/L Pb
	Manganese (dissolved)	1.0 mg/L Mn
	Mercury (total)	0.01 mg/L Hg
	Nickel (total)	2.0 mg/L Ni
	Selenium (total)	0.30 mg/L Se
	Silver (total)	0.1 mg/L Ag
	Sulphide	1.0 mg/L S
	Tin (total)	4.0 mg/L Sn
	Zinc total	5.0 mg/L Zn
Organic parameters	Carbon tetrachloride	1.0 mg/L
	Clorophormium	1.0 mg/L
	Dichloroethene	1.0 mg/L
	Phenols (total)	0.5 mg/L C_6H_5OH
	Trichloroethene	1.0 mg/L

In Brazil, this is also the case, and there are general discharge standards specified, which are independent from the class of the water body the effluent is being directed to. The relationship between both standards is:

- An effluent, besides complying with the general discharge standards, must also allow compliance of the receiving water with the specific standards for its class.
- If the compliance with the receiving water body standards is demonstrated by environmental studies, the polluter may apply for the environmental agency for relaxation of its discharge standards.

Table 10.3 presents the discharge standards adopted in the Brazilian federal legislation (CONAMA 357/2005). Some parameters of broad interest are not included in the national standards, but are left for possible inclusion in the standards from the states, in order to better reflect local reality. For BOD_5, some states adopt the discharge standard of 60 mg/l. This value may be exceeded if the efficiency of BOD removal in the treatment is greater than or equal to 80 % (in some states) or 85 % (in other states). For SS, some states adopt the value of 60 mg/l. Few states apply discharge standards for nitrogen, phosphorus and coliforms.

10.4 Progressive Implementation of the Standards

Usually the stepwise implementation of a wastewater treatment plant is through the *physical expansion* of the number of treatment units. A plant can have, for instance, two reactors built in the first stage, and another reactor built in the second stage, after it has been verified that the influent load has increased, frequently due to the population growth. This stagewise implementation is essential, in order to allow reduction in present value construction costs. However, another concept of stagewise implementation, which should be put in practice, especially in developing countries, is the *gradual improvement of the treated effluent quality or of the receiving water quality*. It should be possible, in a large number of situations, to implement in the first stage a less efficient process, or a process that removes fewer pollutants, transferring to a second stage the improvement towards a system more efficient or more wide-reaching in terms of pollutants. If the planning is well structured, the environmental agency could make allowances in the sense of permitting a temporary small violation in the standards in the first stage. Naturally a great deal of care must be exercised in not allowing that a temporary situation becomes permanent, which is a very common occurrence in developing countries. This alternative of stagewise development of the effluent or of the water quality is undoubtedly much more desirable than a large violation of the standards, whose solution is often unpredictable over time (von Sperling and Fattal 2001).

Brazil has adopted this approach in the 2005 revision of the legislation, allowing a progressive improvement towards the target (standard) values. This is considered to be an advancement in terms of environmental control, instead of what could be regarded by a few as some sort of relaxation or postponement of the solution. The new legislation states clearly that the environmental licenses, discharge consents, polluter-pay principle and terms of adjustment must be based on the intermediate and final targets approved by the river basin committee. Therefore, commitment from the polluter is, in principle, guaranteed.

10.5 Conclusions and Recommendations

The setting of standards should be based on sound, logical, scientific grounds and should be aimed at achieving a measured or estimated benefit or minimising a given risk for a known cost (Johnstone and Horan 1994). The standard values to be implemented should be also interlinked with the availability of wastewater treatment technologies.

Whereas it is recognised that receiving water quality standards should be based on quality criteria for the intended uses of the water, it should not be forgotten that they should also take into account existing capable and affordable technologies. Otherwise, the standards will remain confined to official papers, without reaching reality and without helping the country in its path towards environmental protection. In a developing country such as Brazil, the technologies must be affordable in order to be really implemented and maintained. The relationship between effluent quality and treatment technologies as a basis for standards setting up is discussed by von Sperling and Chernicharo (2002).

It is felt that the current Brazilian legislation provides a suitable framework for environmental pollution control. However, the setting up of the legislation must be accompanied by funding availability, strengthening of environmental institutions, commitment from private (industries) and public (water and sanitation companies) and a police of incentives for improvements in effluent and river water quality.

References

CONAMA Directive 274/2000 Criteria for bathing in Brazilian waters, www.mma.gov.br/port/conama/res/res00/res27400.html, last accessed 13.10.2006
CONAMA Directive 357/2005 Classification of water bodies; environmental guidelines for their classification; establishment of conditions and standards for effluent discharge www.mma.gov.br/port/conama/res/res05/res35705.pdf, last accessed 13.10.2006
Johnstone DWM, Horan NJ (1994) Standards, costs and benefits: an international perspective. J IWEM 8:450–458
National water resources policy, law 9433/1997; National System for water resources management, www.mma.gov.br/estruturas/srh/_arquivos/lei9433.pdf, last accessed 13.10.2006
Von Sperling M, Chernicharo CAL (2002) Urban wastewater treatment technologies and the implementation of discharge standards in developing countries. Urban Water, 4: 105-114
Von Sperling M, Chernicharo CAL (2005) Biological wastewater treatment in warm climate regions. vol 1, IWA Publishing, p. 1496
Von Sperling M, Fattal B (2001) Implementation of guidelines: some practical aspects. In: Fewtrell L, Bartram J (eds) Water quality: guidelines, standards and health, assessment of risk and risk management for water-related infectious disease. IWA Publishing, World Health Organisation Water Series, pp. 361–376

11 Standards for and Evaluation of Small-Scale Dam Projects in Yemen

Michael Schmidt[1], Fadhl Al-Nozaily[2] and Amer Al-Ghorbany[3]

1 Department of Environmental Planning, Brandenburg University of Technology (BTU) Cottbus, Germany
2 Faculty of Engineering and Water and Environment Centre, Sana'a University, Yemen
3 General Directorate of Environmental Policies and Programmes, Ministry of Water and Environment, Yemen

11.1 Introduction

This chapter presents current standards for establishment and evaluation of small dams in Yemen. Starting with a brief introduction on the needs for a construction of small dams in Yemen, the Sect. 11.2 describes some natural and anthropogenic causes of water quality deterioration in Yemen and effects of water quality deterioration on human and animal health. Sect. 11.3 presents available guidelines that are related to dams' water quality in connection with the common uses of dams' water in Yemen. Before ending with conclusion and recommendation, the role of Environmental Impact Assessment (EIA) as a standard for establishment and evaluation of small dams is highlighted with a case study for EIA in practice at Public Works Projects (PWP) in Yemen.

11.2 Need for Dams in Yemen

Yemen is a non-riverine country that is located in the arid and semiarid region. Water scarcity is one of the main challenges that face Yemeni government due to dramatic increase of population that has one of the highest annual growth rates in the world with 3.2 % (NIC 2005). The annual gap of available water resources and uses has increased from 400 million m^3 in 1990 to 900 million m^3 in 2000 (PRSP 2002). In rural areas where 73 % of the population live, immense efforts are being made to meet the increasing demand for water. Among other efforts to use renew-

Standards and Thresholds for Impact Assessment. Edited by Michael Schmidt, John Glasson, Lars Emmelin and Hendrike Helbron. © 2008 Springer-Verlag

able water resources, the government has started in restoring existing water harvesting structures. Construction of small dams is widely seen as a viable and reliable solution to meet water demand for agricultural and domestic purposes. There are also many flood control dams, which are not intended to store water, but to divert spate floods immediately to the adjacent irrigation network (AL-Hemiary 2002). In addition to a recharge of the depleting groundwater, small dams are preferable due to their affordable price, easy construction and multiple-use in rural areas. Therefore, the construction of small dams is strongly supported by local communities, the government and international donors.

11.3 Dams' Water Uses and Guidelines

Recommended guidelines and criteria have been published by a number of countries and organizations and set maximum values for different variables of water used for different purposes. However, higher levels for some variables may be tolerated in these guidelines in some areas where water is scarce (Chapman 1996). Yemen has national guidelines for some water uses like drinking water quality guidelines and irrigational water quality guidelines. Guidelines for irrigational water include water used for animals watering. Drinking water quality guidelines are mainly based on other guidelines for some Arab countries and also the World Health Organization (WHO) guidelines, which are used to insure water quality of urban water supply against restricted range of constituents (WB 2000).

On the contrary guidelines for some uses of water such for bathing do not exist in Yemen yet. It is also mentionable that data about water quality is not known for the entire Yemen (WB 2000). Lack of information undermines the development of current guidelines and the introduction of new guidelines.

Water for Irrigation

Irrigation in Yemen needs more than 90 % of the overall water consumption (Ward et al. 2000). Agriculture employs more than half of the working population (54.2 %) in 2003 and constituting over 20 % of gross domestic product (GDP) since 1990, 22.4 % in 2003. Agriculture is the mainstay of Yemen's economy (LOC 2006). Accordingly, dams' water quality used for irrigation in Yemen has a great importance for local people and the government. Water that is contaminated with pathogens or toxic compounds becomes a potential health risk, particularly when the contaminated water is used to irrigate food and crops. However, the risk of contaminating crops increases if water is not rinsed on the soil around the plant and rather sprayed directly onto the crop from above (Chapman 1996). Yemen has a national guideline for water quality that is used for irrigation. The guideline which was prepared in 1999 is based on other international guidelines mainly the guidelines of WHO (Table 11.1).

Table 11.1. Yemeni Standards for Drinking Water Quality

Parameter	Unit	Maximum Value
Physical Characteristics		
Colour		15
Turbidity	NTU	5
Temperature	°C	25
PH		6.5-9
Conductivity	(μmhos/cm)	2500
Chemical Characteristics		
Total Dissolved Solids	mg/l	1500
Total Hardness	- -	500
Bicarbonate	- -	500
Chloride	- -	600
Sulphate	- -	400
Calcium	- -	200
Magnesium	- -	30-150
Sodium	- -	400
Potassium a	- -	12
Iron	- -	1
Manganiz	- -	0.2
Copper	- -	1
Zink	- -	15
Aluminum	- -	0.2
Nickel	- -	0.02
Toxic Organic Pollutants		
Parathion	mg/l	0.035
Endrin	- -	0.0002
Lindane	- -	0.004
Methoxychlor	- -	0.1
Toxaphene	- -	0.002
Malathion	- -	0.19
Dimethoate	- -	0.002
Diazinon	- -	0.02
HexachlorCyclohexan	- -	0.04
Acetic Acid	- -	0.1
Propionic Acid	- -	0.01
Toxicants		
Lead	mg/l	0.05
Selenium	- -	0.01
Arsenic	- -	0.01
Chromium	- -	0.05
Cyanide	- -	0.1
Cadmium	- -	0.005
Mercury	- -	0.001
Barium	- -	0.5-1.0
Silver	- -	0.01-0.1
Fluorine	- -	1.5-0.5
Antimony	- -	0.005
Barium	- -	0.3-0.1
Total trihalomethanes c	μg	150

Table 11.1. (cont.)

Parameter	Unit	Maximum Value
Chemical Pollutants		
Nitrate	mg/l	10-50
Ammonium	- -	0.3-0.05
Nitrite	- -	2
Phosphate	- -	0.5
Radioactive material		
Beta emitters	Becquerel (Bq)	0.1
Alfa emitters	- -	1
Microbiological contaminants		
Faecal Coliforms	Cell/100 ml	10-15

a 30 mg/l if sulphate is 250 mg/l and more and 150 mg/l if sulphate is less than 250 mg/l.
bTotal pesticides must not exceed 0.1 mg/l.
c Includes Chloroform, Bromoform, Bromide, Chloromethane, Dibromochloromethane.

Water for Drinking

In areas that suffer from sever scarcity of water, dams' water is used for drinking. However, some measures are taken to make sure that water is of adequate quality even though its quality is not of high standard. Such measures can include the construction of sand filters that help filtering water before leaving the dams. Such dams are also protected by fences to restrict access of children and animals and trespassers to the dam body. Some other measures are imposed by the local people themselves on rights of access and uses of the dam, which are fixed in a written official document that is agreed by all users of the dam. This document prohibits using dams for swimming or leaching livestock, which would render it unfit for drinking and other domestic uses. A national standard for drinking water quality exists since 1999.

Livestock Watering

"In principle, water used for livestock watering should be of high quality to prevent livestock disease, salt imbalance or poisoning from toxic compounds. Nevertheless, higher levels of suspended solids and salinity may be tolerated by certain livestock than by humans. Many of the variables included in monitoring the quality of livestock water are the same as for drinking water sources" (Chapman 1996 p.122). As was mentioned in Sect. 11.2, using polluted water in animals watering may form a health risk to humans. Accordingly providing quality criteria for water used for livestock watering aims at protecting both the livestock and the consumer (Helmer and Hespanhol 1997). In Yemen national guidelines for water quality used for animals watering are included in the guidelines of water quality used for irrigation. International guidelines were also prepared by different organizations e.g. Food and Agriculture Organization (FAO).

Recreational and Hygiene Use

Dams' water is also used in rural areas for recreational purposes i.e. swimming and bathing. Such usage for water can hold a risk for the water user in case the quality of water is inadequate for such usage. The potential risk comes from the possibility of ingesting polluted water when swimming or bathing. In Yemen, there is no national guideline for water that is used for recreation. Nevertheless, international guidelines for water quality used for recreational purposes were developed by organizations like WHO. Helmer and Hespanhol (1997) indicated that "the main concern behind developing such guidelines is to protect human health by preventing water pollution from faecal material or from contamination by microorganisms that could cause gastro-intestinal illness, or infections of ears, eyes or the skin. Indicators are therefore usually set for faecal pollution such as faecal coliform as an indicator of pathogens".

11.4 Dams and Water Quality Deterioration

Beside water scarcity, water pollution is one of the main growing environmental problems in Yemen. Water quality gained attention in recent years due to its role as a limiting factor for water usage for certain purposes. The health risk caused by polluted water attracted attention of national and international sponsors of small dams. Dam projects in Yemen, particularity those which are sponsored by international donors, undergo Environmental Impact Assessment (EIA). In EIA potential factors or processes that may affect the environment caused by the project are studied. Any impact that may affect the water quality of the dam shall be identified and mitigated or minimized. However, after the construction of the dam and during the operational phase, the monitoring of the water quality in the dam is left to take care of for the beneficiaries and most of the time, no regular tests and continual check of the water quality of the dam is carried out. This means that monitoring which is a crucial part of EIA is not applied in most of the cases, which contributed to the current deterioration of water and lack of data on water quality. Due to water scarcity in arid and semiarid areas, it is always the purpose of any dam construction project to increase water availability in the area. However, making water available does not necessarily mean to achieve a good quality of water. There may be changes in the water quality of the dam due to contamination from natural sources or pollution that mainly result from human activities.

11.4.1 Natural Contamination of Dams' Water

Natural contamination may be caused when runoff passes through a ground water phase in limestone areas before emerging with more calcium carbonate content. Other sources of water contamination can be suspended sediments of erosions (Carpenter 2001) which can occur naturally or as a result of the activities that accompany dam construction, e.g. constructing access roads or excavation works.

Table 11.2. Important processes affecting water quality (Bartram and Balance 1996 p.14)

Process type	Major process within water body	Water body
Hydrological	Dilution	All water bodies
	Evaporation	Surface waters
	Percolation and leaching	Groundwaters
	Suspension and settling	Surface waters
Physical	Gas exchange with atmosphere	Mostly rivers and lakes
	Volatilisation	Mostly rivers and lakes
	Adsorption / desorption	All water bodies
	Heating and cooling	Mostly rivers and lakes
	Diffusion	Lakes and groundwaters
Chemical	Photodegradation	All water bodies
	Acid base reaction	All water bodies
	Redox Reaction	All water bodies
	Dissolution of particles	All water bodies
	Precipitation of minerals	Ground waters
Biological	Primary production	Surface waters
	Microbial die-off and growth	All water bodies
	Decomposition of organic matter	Mostly rivers and lakes

The eroded soil containing clays and silts is washed away through the catchment's area and ends in the dam body, causing turbidity of the water and lowering the quality of water. Some other natural conditions can be irreversible. Salinization of surface waters is a common example of permanent natural conditions, where evaporation in arid and semiarid region can make water unfit for irrigation or animals watering (Bartram and Balance 1996). At the same time there are other natural processes that affect the balance of some elements in the water. Various types of natural processes, i.e. physical, chemical, hydrological and biological affect the concentration of chemical elements and compounds in the water body. Important processes affecting water bodies are listed in Table 11.2. This is also the case in coastal areas in Yemen, where most of the flow of the main wadis ends in the mountainous region. The quality of surface water that reaches coastal regions is deteriorating. As surface water flows, it is fully exploited in the upstream of the watershed. Consequently, the limited amount of water that reaches coastal regions has a very low quality (WB 2000).

Changes in Water Quality due to Storage

Other processes can occur in the dam during the storage period, which contributes to water quality deterioration such as the decomposition of organic material. This process occurs in the dam due to the formation of deep zone in the water body. The presence of anaerobic conditions creates the required conditions for anaerobic micro-organisms to decompose organic matter e.g. dead aquatic plants or their parts resulting in a reduction of water quality. Deposition in the water body which removes some sediments, e.g. silt and attached nutrients, from the stream flow make the outflow less suited for irrigation. The dam can also become a source for health hazards by providing a breeding ground for snails carrying bilharzia, mosquitoes and other water bone diseases (Carpenter 2001). Sediments accumulation in the dam plays also an important role in changing water quality in the dam. Sediments are carried when water flows through the watershed and settle in the dam. According to Ongley (1996) sediment affect water quality in two dimensions: first physically, when turbidity of water increases due to sediments; second chemically, when the silt and clay fraction (< 63 mm fraction) function such as a primary carrier of adsorbed chemicals (e.g. pesticides and metals) cause pollution and deteriorate the water quality.

11.4.2 Changes in Water Quality due to Human Activities

In Yemen, pollution types that result from human activities and affect water quality are microbial and organic pollution. Microbial pollution in dams occurs at rain events when a large quantity of faecal material is washed from the catchment area into the water body, polluting the water with pathogens. Human activities and pollution in the watershed, e.g. the presence of domestic waste, is important in the determination of the microbiological quality of water flowing to the dam. Organic pollution has become a great concern in Yemen due to the wide and uncontrolled application of chemicals like pesticides and fertilizers in agriculture. The main effect of organic pollution is that it reduces the dissolved oxygen concentration in the water body. These two parameters are key indicators of water quality (Carpenter 2001). Pollution may also occur when people come to the dam to fetch water. Water may be polluted by dirt from clothes, shoes, hands and buckets. Other possible sources for pollution are domestic trash which may be dumped close to the dam. The presence of domestic or wild animals in the area of the dam or when animals get in direct contact with the water body form a potential risk of contaminating the water body. Some pathogens, which are infectious to humans, can be transmitted to the water via animals' bodies or animals' excrements being washed into the water basin during heavy rain events or storms.

11.4.3 Water Quality and Impacts on Health

The creation of a water body is always associated with human health risks as it can become a habitat for water-related diseases. Some insects that transmit diseases to humans need water as a fundamental part for their life cycle, e.g. malaria. In arid and semiarid areas, the construction of dams means the provision of the "missing link" for these insects to start active proliferation and disease transmission to humans and animals. In arid and semi arid areas, the most common waterborne diseases related to irrigation projects are malaria and schistosomiasis (FAO 1997).

Another risk potentially affecting health of local communities appears when dam's water is used for bathing or swimming. If the water is contaminated with pathogens, these will easily find their way into the human body. Another potential pathway for pathogens transmission to humans is the contact of water with the eyes, nose, ears or open wounds (WHO 2002). Uses of small dams in Yemen – especially for animals drinking and recreational purposes – present main pathways for the transmission of microbiological organizations. In some areas of high poverty of the population, inadequate water quality has led to outbreaks of diseases such as cholera, bacterial dysentery, infectious hepatitis, salmonellosis and typhoid. Estimations show that about 70 % of infant mortality (or 107 deaths per 1 000 life-births) is due to waterborne diseases. For instance, typhoid has recorded an increase in the number of cases in Yemen from 7 811 cases in 1998 to 8 287 in 2000. Schistosomiasis, which is also a water-based disease, has spread with 17 000 cases in 2000 (NBSAPY 2004).

11.4.4 Water Quality and Animal Disease in Yemen

In Yemen, one of the main uses of dams' water is for livestock watering. Alabasi, (2001) mentioned that livestock are estimated to contribute by about 20 % to agricultural GDP. At the time where animal production in rural areas is controlled by women, it is an essential food source and also provides financial security. According to Ayers and Westcot (1994), livestock in arid and semiarid regions use poor or marginal quality drinking water. This is the case for several months of the year. The water, which is brought from small wells, canals, streams or 'water holes', occasionally has a high content of salt. Consequently, it may lead to physiological upset of animals and even death in livestock. Water imbalance is reported to cause depression of appetite for animals. Such impact is not attributed to the presence of certain ion in water. One common exception, which is known to cause scouring and diarrhoea, is a high level of magnesium contained in water. Using poor water quality for livestock watering causes sickness, impaired growth or death (Helmer and Hespanhol 1997). Such effect may also form a potential risk to human health as some substance or their products that exist in the water used for livestock watering may occasionally be transmitted to humans.

11.5 EIA Practice at PWP in Yemen

The major steps of EIA for small dams will be elaborated using the EIA practice for small dams projects which are mainly supported by World Bank through the public work project (PWP). These steps are:

An *EIA screening checklist* is filled as part of the baseline form for the village(s) under study that requests a source of water for drinking, irrigation, livestock and/or other purposes. From such checklist, the *source of impact, receptors of impact, environmental impacts* and *mitigation measures* are extracted and finally *recommendation for decisions* are made either with or without further environmental analysis. Projects with limited EIA being required are category B projects. The purposes of such small ponds are:

- Storage of storm water and runoff from springs
- Irrigation
- Recharge of groundwater

- Drinking
- Recreation site

The *mitigation measures* for the vector disease transmission in small dams are: introducing a catchment basin upstream of the small dam; removal of grasses from the boundaries of the dam, avoiding 'dead spots' by making straight boundaries around the dam; avoiding shallow depths of less than one meter; introducing a biological measure of removing the vectors, e.g. certain fish species such as *Nothobranchius virgatus* or *Sarotherodon Niloticus*.

Table 11.3 presents environmental impacts, mitigation measures and monitoring activities for small dams. The main expected impacts are the proper selection of the site, which should have a tributary area free from human settlement which could conflict with sewage and solid waste disposal. This area should also be protected against animal faeces. The solid waste and sewage would introduce nutrients into the water basin, leading to algae proliferation and making the water inadequate for drinking (Fig. 11.1).

Fig. 11.1. Algae as a result of garbage and nutrients disposed from the tributary area of the small dam of Bani Matar, Sana'a

A sedimentation pond prior to the small dam is constructed to collect the sediments and garbage from entering the small dam (Fig. 11.2.).

Fig. 11.2. Road and dam construction material covering the site of the dam

Table 11.3. Environmental management plan (EMP) for small dams project

Item	Potential negative impact	Mitigation measure	Implementation responsibility and cost	Monitoring responsibility
Design Phase				
	Vector breeding sites	Closed conduit outlets hygienic conditions around public collection points by paving at least 1 m^2 apron concrete slab under the water taps with proper drainage and fencing; discuss the extent of local malaria problem with local health-care officials to emphasize the importance of implementing their preventive and curative plans for vector control and malaria roll back.		
	Leakage of the water from the dam	Enough geotechnical study underneath the dam.		
Health	The water is not suitable for drinking	Cleaning the tributary area form the solid waste and sewage; directing the runoff away from agricultural area.	PWP-3 design engineers	
Land use	Disputes about designed project site on privately owned land, or disconcerting areas of public, touristic interest, disturbing wildlife	Discuss the planned site with communities and landowners to get approval or purchase land, or change the design to communal owned land or to land with less expected conflicts; consider drop of sub-project if problems are not resolved.	The communities; local authorities; PWP-3 design engineers	
Construction Phase				
Air	Increased dust during excavation and preparing of drainage bed	Inform nearby houses. Protect excavation works with proper shielding scaffolds; spraying water during excavation might reduce the dust; workers wear protective masks.	Implementing contractors; local authorities, the community	PWP-3; environmental specialist
Nois	Increased level of noise and vibration	Inform nearby houses; avoid work during night hours.	Implementing contractors	PWP-3; environmental specialist
Safety	More possibility of accidents	Protect construction site from unauthorised persons; provide proper support for excavations to protect against their collapse; improve the readiness of health facilities in the region to deal with emergency cases; provide workers with protective clothing.	Implementing contractors; local authorities with MoPHP	PWP-3; environmental specialist
	Getting clean water	Executing the sand filter either at the site or as house level.		
	Awareness for protecting the tributary area	Conducting awareness program to keep the tributary area clean from the solid waste, sewage, animal waste; constructing the canals away from the agriculture area.		

Table 11.3. (cont.)

Operation Phase				
Health	The formation of vector breeding sites	avoid malaria-infested areas	Local community	Environmental specialist, EPA Local NGOs
Equal water distribution	The water is not equally distributed	Male (or preferably Female) should be employed for equal water distribution among beneficiaries; the village should be kept clean form man and human excreta; runoff is directed towards the canals away from agriculture area; proper operation and maintenance of sand filter.		
Consultation and training Components				
Capacity building	The possibility of failure due to low capacity in O&M, administrative or financial management of the project.	Support training of local authority, local NGOs and members of the community on O&M of the system; support training on the administrative and financial management of the project	PWP to contract specialized local consulting firms	Environmental Specialist EPA Local NGOs

The social impact is very important in order to stop tribal tension and in one way compensate for the land acquisition and on the other hand equal water distribution among the beneficiaries. Finally the water in the dam should be monitored and chemically and biologically tested as well the tributary area should be monitored.

11.6 Conclusions and Recommendations

In Yemen recently water quality issues have been raised because water quality plays a crucial role as limiting factor for water use and for potential risks poor water quality imposes on human health. In general dams' water quality in Yemen is not area-wide known. However, waterborne diseases indicate that microbiological pollutants affect water quality in dams. Other type of pollution like organic pollution has not been studied yet. Data on the extent and severity of this type of pollution is unknown on local as well as on national level. National guidelines which are based mainly on international guidelines exist for drinking water quality and irrigational water quality. These guidelines are applied irregularly; accordingly insufficient data about water quality in Yemen hinders efforts to develop the existing national guidelines. Guidelines for recreational use do not exist in the country. To insure that guidelines of water quality are met, the gap of insufficient data about dams' water quality in Yemen should be filled. Regular testing and collection of data should be implemented at project level. Data which is collected and documented locally should be communicated on national level to insure compliance with national guidelines or international guidelines, if national guidelines do not exist. This would facilitate the establishment of proper management for water quality including national guidelines that are based on reliable data.

References

Al-Hemiary AbdulMaged (2002) Gateway to land and water information, Yemen national report. Internet: http://www.fao.org/AG/AGL/swlwpnr/reports/y_nr/z_ye/ye.htm, last accessed on 06.12.2006

Alabsi A (2001) Country Pasture/Forage Resource Profiles, Yemen Internet: http://www.fao.org/AG/AGP/agpc/doc/Counprof/Yemen/yemen.htm, last accessed on 02.06.2006

Ayers RS, Westcot RS (1994) Water quality for agriculture, experiences using water of various qualities, Food and Agricultural Organization, Rome

Bartram J, Balance R (1996) Water quality monitoring - A practical guide to the design and implementation of freshwater quality studies and monitoring programmes, World Health Organization, E&FN Spon, London. Internet: http://www.who.int/entity/water_sanitation_health/resourcesquality/wqmchap2.pdf, last accessed 06.01.2007

Carpenter TG (2001) Environment, Construction and sustainable Development, vol 2, Sustainable Civil Engineering, Wiley & Sons Ltd, Chichester

Chapman D (1996) Water quality assessments - a guide to use of biota, sediments and water in environmental monitoring. 2nd edn, E&FN Spon, Cambridge, Internet: http://www.who.int/water_sanitation_health/resourcesquality/watqualassess.pdf, last accessed 18.02.2007

FAO (1997) Food and agriculture organization of the United Nations, Irrigation in the near east region in figures. Internet: http://www.fao.org/documents/show_cdr.asp?url_file=/docrep/W4356E/w4356e0z.htm, last accessed 21.10.2006

Helmer R, Hespanhol I (1997) Water pollution control: A guide to the use of water quality management principles, 1st Edition, E&FN Spon, Cambridge. Internet: http://www.who.int/water_sanitation_health/resourcesquality/watpolcontrol/en/, last accessed on 15.12.2006

LOC (2006) Library of Congress, Country Studies-Yemen, Federal Research Division , United States of America, Internet: http://lcweb2.loc.gov/frd/cs/profiles/Yemen.pdf), last accessed on 17.06.2006

NBSAPY (2004) Yemen First National Report To The Convention On Biological Diversity. http://www.biodiv.org/doc/world/ye/ye-nr-01-en.pdf, last accessed on 06.12.2006

NIC (2005) National Information Center, Profile of Yemen, http://www.nic.gov.ye/English%20site/SITE%20CONTAINTS/about%20yemen/quick%20view/aboutyem.htm, last accessed 18.04.2007

PRSP (2002) Poverty Reduction Strategy Paper, Republic of Yemen. Internet: http://www.mpic-yemen.org/dsp/PRSP2003_2005_2.pdf, last accessed 04.05.2007

Ward C (2000) Water Resources Management in Yemen Internet: http://siteresources.worldbank.org/INTYEMEN/Overview/20150274/YE-Water.pdf, last accessed 26.03.2007

WB – World Bank (2000) Land and Water Conservation Project, PROJECT, Internet: http://lnweb18.worldbank.org/oed/oeddoclib.nsf/DocUNIDViewForJavaSearch/BAF2C70A3DD8F6BD8525711D00746F30/$file/ppar_35004.pdf, last accessed 13.03.2007

WHO (2002) Guidelines for drinking-water quality recommendations, 3rd edition, Geneva, World Health Organization, Internet: (http://www.who.int/water_sanitation_health/dwq/GDWQ2004web.pdf), last accessed 24.10.2006

12 The Need for Developing Thresholds for the Recycling Rate of Products in Thailand

Angkarn Wongdeethai and Jürgen Ertel

Department of Industrial Sustainability, Brandenburg University of Technology, (BTU), Cottbus, Germany

12.1 Introduction

Achieving sustainable development requires greater responsibility of each single person to take care the ecosystems on which all life depends, for the generations that will follow our own (Annan 2005). We can wait until a crisis forces us to do something. Or we can commit to working together, and start by asking the tough questions: How do we meet the needs of the developing world and those of industrialised nations? What role will recycling-oriented product/society play? What is the best way to protect the environment? Whatever actions we take, we must look not just to next year, but to the next 50 years (O'Reilly 2005).

Thailand has taken measuring approach for monitoring and evaluating the quality of human health and environment focusing on: reducing and recycling waste; improving treatment and safe disposal of solid and hazardous waste; enhancing the supporting institution, regulatory, and financing framework; and expanding public and civil society participation. Thailand currently produces nearly 22 million tonnes of solid waste annually from residences, industries, businesses and hospitals. If current consumption growths prevail and recycling rates remain low, it is likely that this could bring potential risks to human health and the environment. While industries have effectively harnessed the market for recyclables such as glass, paper, metal and plastic, each year over 4.5 million tons of recyclables are thrown away by households and businesses (World Bank 2003).

Each year, the country generates 58 000 tonnes of electronic scrap or 'e-waste', which is now accumulating in the junkyard nation wide. The raise of demand for appropriate measures, treatment facilities, competitive markets and globalisation, and the recycling profit of end-of-life (EOL) products catch the attention in the country. Strengthening national laws, standards on producers' responsibility and implementation could facilitate these challenges. After UNEP (2005), each year 20 to 50 million tonnes of e-waste are generated worldwide. The European Union,

Standards and Thresholds for Impact Assessment. Edited by Michael Schmidt, John Glasson, Lars Emmelin and Hendrike Helbron. © 2008 Springer-Verlag

Japan and the United States have taken action and preparing to encounter with EOL products by enforcing legislation on producers' responsibility and implementation e.g. EU Directives: ELV, WEEE, RoHS. But such laws are not directly enforced to the Thai producers or exporters. Although many Thai manufacturers are international players, the majority of domestic players rely on national standards.

Discussion in this chapter is restricted to the thresholds of recycling rate of products (target values in percentage of recyclable materials) and what can be said from the EU Directives. It particularly focuses on the need for developing thresholds of recycling rate for Thai products. Pollution controls such as water quality standards, air quality standards, soil quality standards, noise standards provide thresholds and standards usually solve problems at the end-of-pipe. Therefore a long-term strategy such as 3R (initiative by the G8 nations), recycling-oriented society, lifecycle thinking, as well as producers responsibility and implementation are the tools that could be set as milestones for planning and combating with future problem of EOL products. The information was taken from various sources including published reports of government agencies, international agencies, newspapers and meeting with the relevant government agencies including Electrical and Electronics Institute.

Currently, there is no legal requirement for recycling rate target of Thai products. Only the recycling target for solid waste is proposed in the Ninth National Economic and Social Development Plan (2002-2006). According to Thailand environment monitor 2003, the actual solid waste recycling rate in 2003 is 11 %, the country has to take dynamic effort in solid waste management to reach the recycling rate target of 30 % in year 2006. The effective introduction of economic and other incentives for waste recycling could potentially reduce the amount of waste disposal, approaches such as: taxes, fees, thresholds can be used. The main challenge of the program is ensuring that finances are adequate, and regulations are enforced. Considering a process of implementing a new standards and thresholds, the socio-economic facets must be taken into account. This chapter therefore wants to contribute to the further discussion about the need, requirements and methods for developing thresholds (target values) of recycling rate of products. It investigates different questions, in the following sections.

- What are the chances of new thresholds for recycling rate of Thai products, which factors may exacerbate its realization?
- What products are concerned for implementing thresholds for recycling rates?
- Who are involved for the implementation?
- Which methods exist to fulfil the requirement for developing the thresholds for recycling rate of products?

Although the stringent international products recycling rates are enforced in member countries, however it is not necessary for other countries having different socio-economic infrastructures and policies must be followed. To demonstrate the need for developing the national products' recycling target values, first the study of current situation should be carried out to see whether the society is aware of recycling-oriented products. International experiences can be well-resourced which concurrently displays efforts made on global environmental protection.

12.2 Challenges and Opportunities

The development of new thresholds for recycling rate of Thai products offers a variety of opportunities for improving overall products performance, which, however, can be jeopardized by different subjections and obstacles. The opportunities can be described as follows:

- reduce the use of virgin materials
- reduce the waste arising from EOL products
- holistically improve performance of all involved in the life cycle of products

Evidently, these opportunities are beneficial to human health and the environment. However, supports from relevant actors e.g. planning agencies, planners, authorities, producers could be insufficient. In consequence the chances of developing the new thresholds for recycling rate of Thai products may not be expatiated and the possible quality of the results may not be obtained. The following apprehensions and objections to the opportunities are frequently quoted.

- higher work load
- higher cost of compliance
- higher requirements concerning new standards for implementation
- higher need of adjustment and modification of the strategic action while implementation take place
- higher time required of the planning process and capacity building

It is necessary to bear in mind those apprehensions in order to mitigate the subsequent obstacles of developing new thresholds for recycling rate of Thai products. Wider involvement of policy makers, industries, stakeholders and infrastructures are crucial for successful implementation of new thresholds.

In relation to the current situations for waste management in Thailand the following cooperation and activities among the involving people is illustrated, see Fig. 12.1. The flow chart displays the materials flow through the system from the cradle-to-grave perspective, starting from products manufacturing until the EOL products linking with responsible persons. Currently there are no standards/ thresholds for recycling rate for Thai's products. The control of the government, on waste management, is focusing on the recycling rate. It has set the recycling rate of solid waste to 30 % in year 2006. With the Decentralisation Plan, the country has divided the solid waste collection and treatment responsibility of 76 provinces with over 1 000 municipalities and 60 000 industries into five main local governments: 1) Bangkok, 2) Nakorn Municipalities & Pattaya, 3) Muang Municipalities, 4) Tambon Municipalities, and 5) Tambon Administration. The solid waste officers are loaded with works. There is only one solid waste officer for every 142 municipalities and one inspector for every 180 industries (World Bank 2003).

Industrial waste has become more problematic because a number of local municipalities have operated hazardous waste disposal illegally and dumped the waste on their garbage landfills or public land for more than twenty years despite the lack of proper facilities and experience (Danish Trade Council 2004). In order to play their respective roles effectively, skills and capacity building need to be established. Department of Industrial Works (DIW) also tries to enforce the industries to dispose the waste through authorised operators.

'Local governments, taxpayers and communities are bearing the burden for dealing with discarded materials, and yet manufacturers who profit from the production and deployment of these products into the market are getting away scot-free without any kind of liability,' says Kittikhun Kittiaram, toxic campaigner for Greenpeace Southeast Asia (Taipei Time 2005). Share of solid waste collection for recycling can be classified as: tricycle garbage collectors 'sa-leng' 67 %, factories and stores 14 %, municipal collectors 13 %, and waste pickers 6 % (PCD recycling study 1998). According to World Bank (2003), it is estimated that

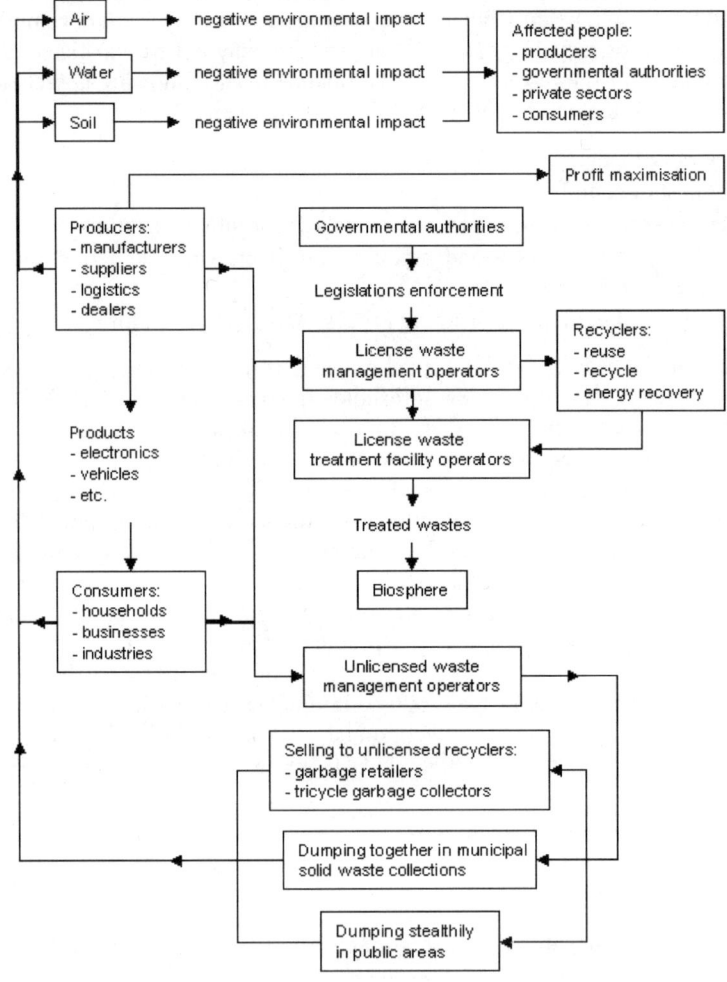

Fig. 12.1. Case without standards/ thresholds for recycling rate of products

there are 25 000 people involved in informal recycling. This includes more than 15 000 sa-lengs, 2 000 waste agents, and just fewer than 4 000 scavengers who collect waste from disposal sites and transfer stations. In addition, all tricycle garbage collectors, and over 3 000 municipal collectors in the country supplement their income by selling recyclables on an informal basis.

The development of the target values of recyclable materials of products would benefit the people in the long-run, as illustrate by the ideal model (see Fig. 12.2), when the system reaches the optimum stage, where everybody wins. It could reduce demand of virgin materials, makes products more recycling friendly, improves products performance, increases products competitiveness, reduces EOL waste, increases recyclers income, saves disposal cost. However, to reach this optimum sustainability stage, the overall impact assessment and the development plans have to be prepared carefully. The cooperation and collaboration of the government, relevant authorities, agencies, public and private sectors have to be strengthening even tougher. However to implement a tougher legislation, effective policies, financial supports and technical know-how are crucial which can bring success or failure.

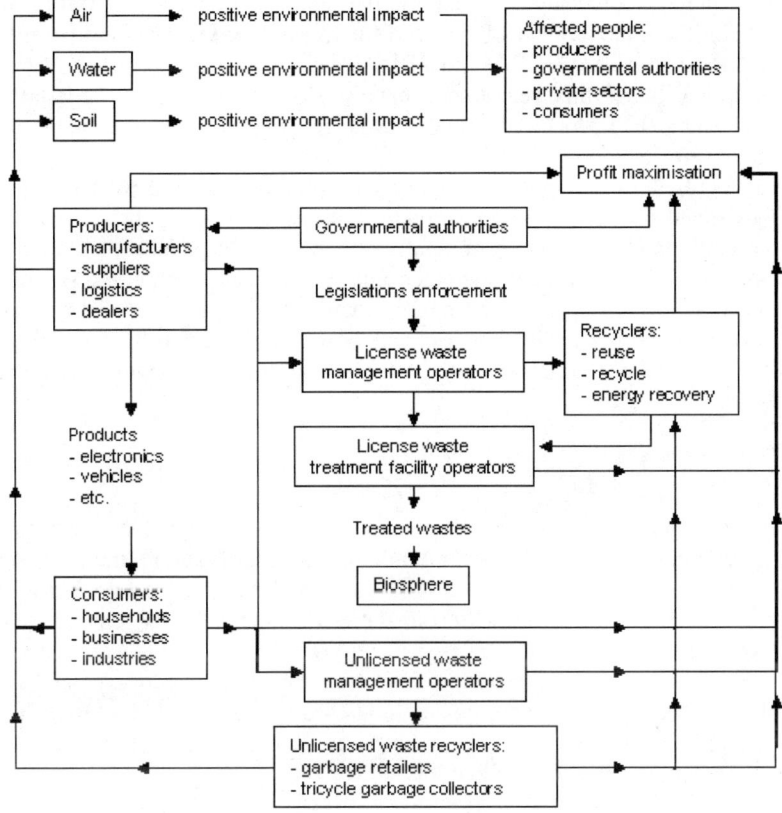

Fig. 12.2. Case with new standards/ thresholds for recycling rate of products

12.3 Which Products Should be Concerned?

It was suggested that to determine which product categories should specify the recycling rates, current national products consumption and consumers' behaviours have to be carefully measured. Products which have priority for implementing the recycling standards/ thresholds should basically:

- contain hazardous materials, toxic heavy metals (e.g. lead, mercury, cadmium, chromium 6+, PBB, and PBDE)
- comprise high compositions of recyclable and/or valuable parts/ materials (e.g. ferrous and non-ferrous metal, precious material, plastic, and glass)
- have high consumption rate (quantity and/or value) in the market

The electronics consumption in Thailand reached 2.5 Billion USD, in 2003 (Bank of Thailand 2003). After the financial crisis in 1998, the trend for electronics consumption has grown steadily. Electronic is Thailand's leading export, account for 20 percent of all manufacturing exports from the Kingdom (World Bank 2003). In 1999, Electronics export reached 14.4 Billion USD (The Customs Department, Ministry of Finance). Thailand's automotive industry produced 928 081 vehicles in 2004. In 2006, the number of cars and trucks produced in Thailand from January to August increased by 22 % to 710 889 units and seems on track to exceed one million vehicles this year (Runckel 2005). According to the Federation of Thai Industries (FTI), 62 % of vehicles produced were consumed in the country, and 38 % were for export.

Future growth of commercial product consumption and solid waste generation largely depend on the population growth, reuse and recycling behaviour. Commercial equipments ranging from small appliances to large machines have high recycling potentials, electrical and electronics equipments often comprise hazardous materials e.g. battery, printed circuit board, etc., when recycled by informal recyclers or improperly disposed off can pose health hazards. Concerning the above issues, product categories with high potential for implementing requirement for recycling rate are:

- electrical and electronics equipments
- vehicles

The two product categories contribute significantly to export values; many industries show interests to comply with international standards and regulations. World Bank (2003) reports: 'certification of Thai firms under the International Organization for Standardisation 14001 (ISO14001): environmental management standard has been risen seven fold to 700 firms since 1998. The largest 712 certified companies are from the largest exporting industries, including electronics and motor vehicles'. Electronics and automotive industries are among the most effective manufacturers which voluntarily prepare themselves for strict international environmental regulations, and sustainability.

12.4 Who will be Affected from the Implementation?

Who will be involved in the implementation? All levels of people take the responsibility. The responsible national-level ministries are 1) the Ministry of Natural Resources and Environment (MoNRE), 2) the Ministry of Public Health (MoPH), 3) Ministry of Industry (MoIND), and 4) Ministry of Interior (MoInt). They primarily set the national policy. Department and authorities under the ministries are responsible for implementing the provisions of the law through regulations and technical guidelines. These agencies include the PCD, DIW, Industrial Estate Authority of Thailand (IEAT), Office of Natural Resources and Environmental Policy and Planning (ONEP), and the Local Administration Office.

Not only in the ministry level, but also other corresponding sectors such as industries, consumers, recyclers, waste treatment facilities are the significant public and private sectors that shares the responsibilities. The following questions are frequently quoted when new strategic-decisions are made for better performances:

- Is there a need or demand for implementing new things?
- What, who will benefit or affected from the decision?
- What duties are necessary and how-to execute them to achieve goals?
- Who will be responsible for those duties, consequences, and costs?
- What special issues and alternatives should be concerned, and evaluated?
- Whether or not to start the implementation? if yes, where and when?

The new legislation on recycling rate of products will make planning process more important and will furthermore provide a procedural framework for determining the recycling performance provision of product categories e.g. electronics, vehicles.

Figure 12.3 illustrates the three main phases for implementing new regulations/thresholds for recycling rate of products. Phase 1 - Planning: the study of environmental, technological, economic situations has to be addressed in order to visualize trends and alternatives. Document draft is required to provide scope, objective, and assessment methodology. Affected organisations and environmental impacts need to be studied and specified. Phase 2 - Data collection, public hearing, and research study: positive and negative impacts on organisations e.g. governmental authorities, private sectors, consumers, and governmental authorities and private sectors as well as the environment have to be adequately assessed via environmental report. Several instruments, e.g. life-cycle assessment, cost-benefit analysis, can be used to evaluate impacts on organisations. Alternatives, and influenced factors should be taken into account. The study of sensitivity analysis should be carried out. Discussion, recommendation and conclusions on the strength and weaknesses of the study should be given in the final report. - Consideration for proposed regulations/ thresholds and requirements is made in the last phase. Phase 3 - Executive decision- making[1]: the decision is made by the respect-

---

[1] Hierarchy of laws: 1) 1997 Constitution 2) Acts passed by the Parliament 3) Regulations and Notifications enacted by the respective Ministries (Ministry of Justice 2003).

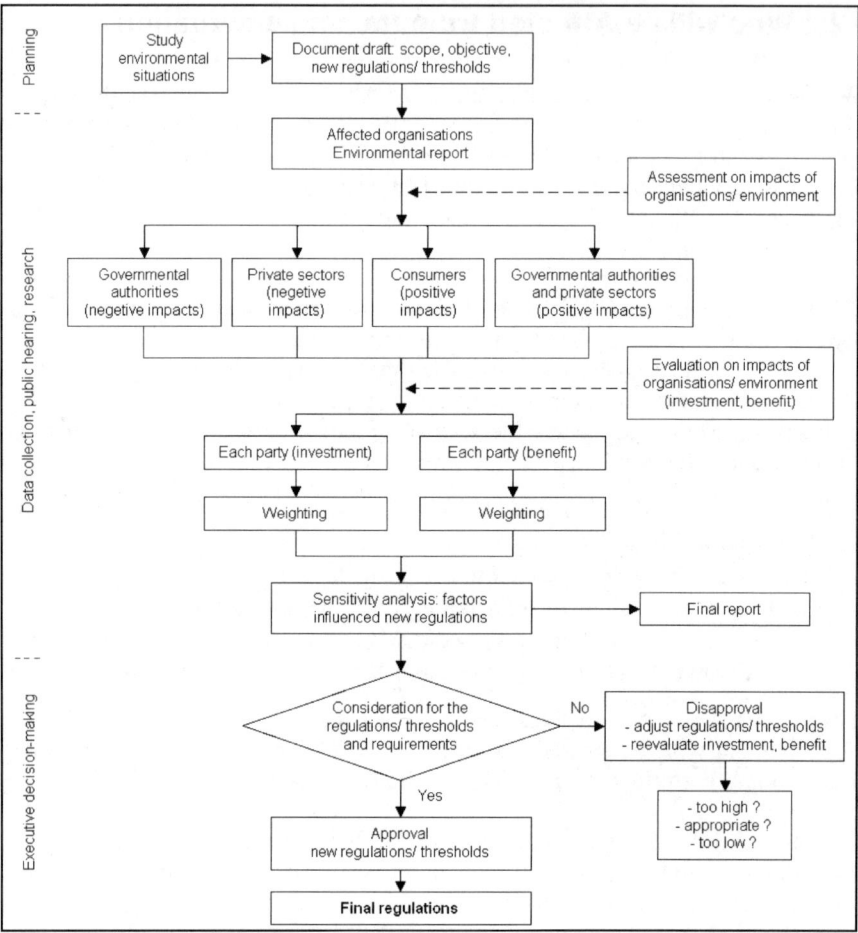

Fig. 12.3. Steps for implementing new regulations/ thresholds for recycling rate of products

tive Ministries. If the proposed regulation is rejected, i.e. too high, too low, or inappropriate; re-evaluation or adjustment of the regulations can be made. Finally, if the regulation is well elaborated, having good structure and fitted with environment, the final approval can be granted, published and enter into force.

12.5 Methods to Fulfil the Requirements for Developing Thresholds for Recycling Rate of Products

Several methods that can help to fulfil the requirement for developing thresholds for recycling rate exist including:

- Disassembly analysis

- Recycling analysis
- Life cycle assessment

Disassembly analysis helps to investigate the product profile. Recycling analysis is used to evaluate design for recycling, identify material compositions. Life cycle assessment could illustrate where are pollutions, how serious of human health and environmental impacts are, and trade-offs of different end-of-life scenarios. For calculating the recycling rate of a product, most importantly, is to know the ratio of recyclable materials. How to get the data is sometimes troublesome for both novice and experienced practitioners. Usually, the data of material composition are not provided in the client product manuals. The data of product profile from producers are frequently not adequate and incomplete, which is also hard to collect and obtain. Reusable parts or components are often found in variety of product models and categories. Recycling activities can only be sustained when total incomes, mainly from selling recyclables (parts, components, fraction materials) is greater than total investments, such as overhead cost, equipments, dismantling cost, maintenance cost, and disposal cost. It is not necessary to disassemble the product until reach the last part that constitutes single material. For example, if the benefit of selling part as a reusable part, is greater than disassemble and selling them as fraction materials, then the part shall not be dismantling, rather to sell as the reusable part. At formal recycling plants, prior to obtain different materials from EOL products and feed them back into a system of circular economy[2], EOL products are processing by the following steps:

- Sorting
- Disassembling
- Shredding
- Separating (plastic, metal, aluminium, glass, etc.)

The recycling rate of each product category can be obtained by carry out disassembly analysis (by robots, machines, and/or human) of the same EOL products category. Further study of other conditions and alternatives (see also Fig. 12.3) has to be considered. When overall processes have been carried out, the initial thresholds for recycling rate of the product category can be recommended.

The EOL products' recycling process, in large scale, (see Fig. 12.4) starts from sorting for reusable, refurbishable, repairable items. The remaining will transfer to disassembly process for separating hazardous and reusable parts. Non-reusable parts will be sent for shredding. Later on, major scraps fractions outputs: plastic, metal, aluminium, glass, and the residues (wastes) can be sorted out. The residues have several EOL destinations. If the residues still potentially have valuable materials, no impacts, economy and meaningful, then it might have to put in the recy-

---

[2] Circular economy refers to an economy that features closed-circuited material flow. It is an economic model where materials and energies are utilised in an aggressive, closed-circuited way, which features low emission or even zero emission. Protection on exploitation of biological resources is the key foundation for a circular economy (UNEP 2004).

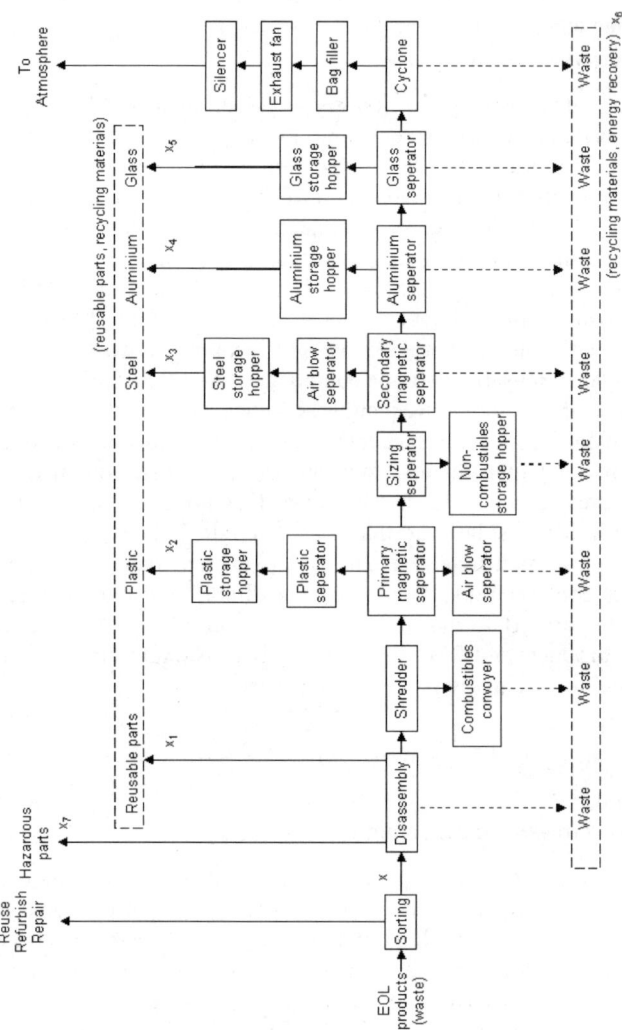

Recovery rate = $(x_1 + x_2 + x_3 + x_4 + x_5 + x_6)/x$.
Recycling rate = $(x_1 + x_2 + x_3 + x_4 + x_5)/x$.

Fig. 12.4. The EOL products' recycling process, large scale

cling again. Otherwise it may be utilized as inputs for energy recovery, or need to be treated, or landfills. Additional LCA study for each EOL scenario is useful, in order to understand the environmental trade-offs and help for the decision making (e.g. virgin raw material VS. recycle material; landfills VS. incineration; material recycle VS. energy recovery). Outputs of one system always are the inputs for other systems. Dumping the waste elsewhere, but not at my backyard never be the real solution, rather shift the problem elsewhere. There are three types of outputs:

- Products (secondary raw materials, energy)
- By-products
- Residues (wastes, which can be convertible to energy, or send to treatment fa-
cilities, or deposition in landfill sites)

A following example shows how to avoid environmental impact from output
(waste) of metal recycling activities, i.e. by-product of secondary aluminium industry. Disposal of highly saline industrial by-products in landfills is not permitted
in member states of the European Union, e.g. Germany. Large amounts as such
by-products thus have to be disposed of in alternative ways. Hermsmeyer et al.
(2002) has found that a fine-granular saline aluminium recycling by-product
(ALRP), when mixed with flue gas desulphurisation by-product (FGDP) of a coal
combustion power plant, as a soil substitute in a surface barrier, can improve the
soil quality.

Although EOL products in Thailand are collected and sold by formal and informal operators to second hand market, repairing shops, recycling plants, or landfills, but waste still increasing. Rapid change of new design, innovation, and smart
technologies push old products (i.e. with old technologies) out of service sooner.
The amount of EOL products is increasing rapidly, but the number of the collection system and treatment facilities, is inadequate and these are challenges for the
recycling business. Effective strategies and approaches, such as new standards for
recycling rate of products could solve this problem at the begin-of-pipe.

12.6 Conclusions and Recommendations

Understanding human and ecological impacts of products during the entire life-
cycle is crucial for making proper environmental prevention and protection plans.
Moreover, an integration of different principles, e.g. industrial ecology, industrial
symbiosis, eco-industrial estate[3], material flows analysis, life-cycle-assessment,
eco-design, design for recycling, and cost-benefit analysis is very useful, helping
enrich information and ideas, identifying strengths, weaknesses, opportunities, and
solutions. If sustainability needs communication, then those who are able to give a
meaning to the term 'sustainability' are obliged to do so (Tobias 2004). The impacts that we and our predecessors have caused, unless we deal with them otherwise, will be passed on to our children. Therefore, it is our responsibility to take
care of what we have produced. New thresholds of recycling rate of products shall
gradually be developed and give enough time for the adaptation of industries. Arguably, this could bring a new constraint for reorganization of governmental authorities, industries, waste collectors, and recyclers. This, on another hand, will
enable Thailand to meet the higher standards and contribute more responsibly to

---

[3] Eco-Industrial-Estate (EIE) concept is formulated first by the Industrial Estate Authority
of Thailand (IEAT). EIE is a group of productions and services that are oriented toward
raising the standards of environmental quality and business performance through collaboration on natural resource and environmental management.

the environment. Numbers of people have already expressed their opinion to support this mechanism, as reported by Bank of Thailand (2003): There has been interest in introducing laws similar to the EU, set out standards/ thresholds for recycling rate. This effort could increase competitiveness of Thai industries and prevent dumping of used electronics in Thailand from countries with more stringent electronic waste laws.

References

Annan K (2005) The UN works to secure our future for sustainable development. Internet address: http://www.un.org/works/sustainable, last accessed on 19.05.2006

Clements V (2004) Doing Business with the EU - Waste management regulations: implications to electrical and electronic industries in Thailand, EU Environmental Policy. Bangkok, Thailand

Hermsmeyer D, Diekmann R, Van der Ploeg RR, Horton R (2002) Physical properties of a soil substitute derived from an aluminum recycling by-product. Journal of hazardous materials, The Netherlands, Nov 11; 95(1-2):107-24

Lee K (2004) Discussion paper: environmental regulations on the electrical and electronic equipment by European Union (EU). Republic of Korea

Thai Industrial Standards Institute (2006) Modernization of Thai standardization system. Ministry of Industry. Bangkok, Thailand

Taipei Times (2005) Business Quick Take, Recycling. Taipei City, Taiwan. June 19, p. 11

Danish Trade Council (2004) Sector Overview, Environment in Thailand, Royal Danish Embassy, Bangkok, Thailand, Aug 16

World Bank (2003) Thailand environment monitor 2003, The World Bank, Thailand

Bank of Thailand (2003) Thailand environment monitor 2003

EPA (1997) Solid waste and emergency response: measuring recycling, a guide for state and local governments, United States Environmental Protection Agency. September

O'Reilly (2005) Time, Time Warner Publishing, Amsterdam, The Netherlands, Oct 31, vol 166, 18:9

Runckel CW (2005) Thailand automotive industry update 2005. Business-in-Asia, Internet address: http://www.business-in-asia.com/auto_article2.html, last accessed on 19.05.2006

Tobias M (2004) Sustainability needs communication: German ICT- industry's strategy to promote sustainability, Proceedings of the Electronics Goes Green 2004+, Berlin, pp.1007-1012

UNEP (2004) Panel discussion paper for the 20th Anniversary of UNEP SASAKAWA Environmental Prize Internet address: http://www.unep.org/sasakawa/XX_aammoversaru /shrestha.asp#_ftn1, last accessed on 08.06.2005

13 Guidelines for SEA in Marine Spatial Planning for the German Exclusive Economic Zone (EEZ) – with Special Consideration of Tiering Procedure for SEA and EIA

Juliane Albrecht

Leibniz Institute of Ecological and Regional Development Dresden (IOER), Germany

13.1 Introduction

The economic use of the oceans for marine mining, for extracting gravel and sand, laying cables and pipelines and, more recently, installing wind energy plants is increasing. The call to implement spatial structure plans for marine areas has therefore become louder (Molitor 2005 p.66 et seq.). In taking account of various economic, social and ecological interests, such plans aim to identify and resolve potential conflicts at an early stage. Spatial planning is needed not only within the 12 nautical mile zone in the North and Baltic Seas (i.e. in territorial waters), but also within the Exclusive Economic Zone (EEZ), where, for example, most German offshore windfarms are planned (BSH 2006). For this reason, a legal basis for spatial planning in the EEZ (Section 18a Federal Spatial Planning Act, Raumordnungsgesetz, ROG) was established in Germany in July 2004. The applicability of the ROG has been extended to the EEZ, and the Federal Ministry of Transport, Building and Urban Affairs (Bundesministerium für Verkehr, Bau- und Stadtentwicklung, BMVBS) has been empowered to set targets and principles for the German EEZ in the Baltic and North Seas. According to Sections 18a, (7) and 5 ROG, spatial structure planning in the EEZ includes strategic environmental assessment (SEA) as an important component. SEA is an integral part of official procedures implemented by public authorities, governments, and legislative bodies for establishing and changing plans and programmes. It is the purpose of the SEA to ensure that the implications of the plan or programme are comprehensively determined, described and evaluated by means of standardized principles at an early

Standards and Thresholds for Impact Assessment. Edited by Michael Schmidt, John Glasson, Lars Emmelin and Hendrike Helbron. © 2007 Springer-Verlag

stage and that the results of the evaluation are taken into consideration in establishing and changing plans. Standards and thresholds play an important role in this process because they are the criteria for assessing the significance of the plan's potential impacts. There are many stages in the SEA process where standards and thresholds are needed, especially in scoping, in describing the state of the environment, in evaluating impacts on protected items and in monitoring (see below). It should be noted, however, that legal environmental regulation is much weaker for marine areas than for land areas. For this reason, existing environmental legal standards for marine territory in general and the German EEZ in particular are very general and difficult to apply. This is due partly to a lack of knowledge about the marine environment. Furthermore, SEA as well as marine spatial structure planning are new legal instruments in Germany. Little experience has therefore been gathered in implementing these procedures.

This chapter describes the application of spatial planning SEA of in the German EEZ of the Baltic and North Seas to provide guidelines for developing standards and thresholds in this area. Emphasis is placed on the tiering of the assessment of plans and the environmental assessment of projects (EIA) because of its influence on the degree of aggregation and detail for the standards and thresholds used at the level of SEA.

13.2 Legal Peculiarities of Marine Spatial Planning in the German EEZ

The EEZ is established and codified by the United Nations Convention on the Law of the Sea of 10 December 1982 (UNCLOS). According to Art. 55, 57 UNCLOS coastal states are entitled to establish an EEZ of a maximum breadth of 200 nautical miles from the baseline. Within this zone, the coastal state enjoys sovereign rights for the purpose of exploring and exploiting, conserving and managing the natural resources, whether living or non living, of the waters superjacent to the sea-bed and of the sea-bed and its subsoil, and with regard to other activities for the economic exploitation and exploration of the zone (Art. 56 para. 1 UNCLOS). Since the EEZ does not form part of the sovereign territory of the Federal Republic of Germany, spatial planning is carried out within the framework of Art. 56 (1) UNCLOS. Accordingly, only selected uses (economic and scientific use, safety and efficiency of navigation, conservation of the marine environment) are addressed in the German EEZ (Section 18a (1) ROG). In comparison to spatial structure plans for land areas, spatial structure plans for the German EEZ in the Baltic and North Seas exhibit a number of practical and legal peculiarities. For example, they are not part of a multi-stage planning system, but directly influence the approval of specific projects (Schomerus et al. 2006 p.206). Furthermore, they can deal with only certain aspects of plans and are not accompanied by landscape planning. Their effect, however, has the same basis in Section 4 ROG as plans for land areas. The targets and principles laid down in the plans are required to be taken into account in subsequent projects relating to spatial structure. In the EEZ this mainly involves the approval of marine facilities such as offshore windfarms under Section 3 Marine Facilities Ordinance (Seeanlagenverordnung, SeeAnlV), approval for the exploration and exploitation of mineral resources under Sections 6 et seq., 51 et seq. of the Federal Mining Act (Bundesberggesetz, BBergG), the laying of submarine cables and pipelines (Section 133 BBergG) and marine scientific research (Section 132 BBergG).

13.3 The Importance of Offshore Wind Use for Marine Spatial Planning in the EEZ

Among the various economic uses of the EEZ, offshore wind energy generation is especially important today. To limit harmful emissions of carbon dioxide, the contribution of renewable energy sources to total energy generation is intended to increase to 12,5% by 2010 and to 20% by 2020 minimum (Section 1 (2) Renewable Sources Act, Erneuerbare Energien Gesetz[1]). The potential of water power is widely believed to be fully exploited. The development of land-based wind energy farms is limited to repowering owing to the lack of further sites. Consequently, major progress in generating renewable energy requires new offshore wind energy sites to be developed. In general, offshore areas are characterized by more stable and more powerful winds. Typically, wind velocities are twice those of land areas (Risch 2006 p.15). In this respect, the EEZ is especially important, because alternative areas closer to shore have a number of disadvantages. They are already used intensively for various conflicting activities such as shipping, fishing, tourism, pipeline routing etc. Furthermore, the visual effects of wind energy sites close to the coast meet with opposition from coastal dwellers (Risch 2006 p.15). For these reasons, the German Federal government favours wind energy sites in the EEZ (BMU 2002 pp. 4 et seq.). Despite the technological difficulties of developing such installations posed by considerable depths and long distances from shore, more than 30 project applications have been filed with the Federal Maritime and Hydrographic Agency (BSH) responsible for approving offshore windfarms in the German North and Baltic Seas. As the impact of offshore windfarms on navigational safety and the marine environment has not yet been finally assessed, the BSH has hitherto only approved pilot scale projects with a maximum of 80 wind turbines. The purpose of these smaller scale projects is the detailed investigation of the impact of offshore windfarms on the marine environment and navigation. As the instruments for marine spatial planning in the EEZ are now available, it will in future be possible to approve the construction of larger numbers of wind turbines in preferred areas for wind energy use, which implies that other, more sensitive areas can be kept free from installations (BSH 2006).

---

[1] Gesetz über den Vorrang erneuerbarer Energien vom 21. Juli 2004 (Act on Granting Priority to Renewable Energy Sources of 21 June 2004), Federal Law Gazette I, 2004, p.1918.

13.4 The Requirements of SEA of Spatial Structure Planning in the EEZ

The SEA is based on Directive 2001/42/EC of the European Parliament and of the Council of 27 June 2001 on the assessment of the effects of certain plans and programs on the environment[2] which has been implemented in Germany by the European Law Adaptation Act for the Construction Sector (Europarechtsanpassungsgesetz Bau)[3] and the Act adapting the Environmental Impact Assessment Act (Gesetz über die Umweltverträglichkeitsprüfung, UVPG).[4] The content and procedure of the SEA are regulated.

13.4.1 Content Requirements

SEA includes the identification, description and evaluation of direct or indirect impacts of plans and programmes on the population, especially human health, fauna, flora, biodiversity, soil, water, air, climatic factors, landscape, material assets, cultural heritage including architectural and archaeological heritage, and interrelationships between the above factors (Section 2 (4) and (1) UVPG). In the EEZ, the assessment of the significant effects first of all addresses issues such as marine flora and fauna (benthos, fish, birds, marine mammals), biodiversity, water, bottom soil and landscape (Köppel et al. 2004 p.12; Schomerus and Busse 2005 p.48). Also to be considered are secondary, cumulative, and synergistic effects, whether short, medium or long-term, permanent or temporary, positive or negative (cf. Annex I lit. f SEA Directive). The SEA permits the impact of a development to be assessed long before specific projects have been realised and prior to planning and decision-making processes. The broader scope of assessment makes it possible to summarize the impact of a number of individual projects. Fundamental alternatives, which may not be available at a later stage in the approval process for individual projects, have to be taken into consideration (Section 7 (5) sent. 2 ROG). With respect to the EEZ, there are multiple alternatives because the spatial structure plan applies for the total geographic extent of the EEZ and for all allowable uses. Furthermore, far-reaching compensatory measures for the entire area of the German EEZ in the Baltic and North Seas can be provided for.

---

[2] Official Journal 2001 L 197 p.30.

[3] European Law Adaptation Act for the Construction Sector (Gesetz zur Anpassung des Baugesetzbuchs an EU-Richtlinien vom 24. Juni 2004), Federal Law Gazette I, 2004, p.1359.

[4] Act on the Introduction of Strategic Environmental Assessment and Implementation of the Directive 2001/42/EC of 25 June 2005 (Gesetz zur Einführung einer Strategischen Umweltprüfung und zur Umsetzung der Richtlinie 2001/42/EG vom 25. Juni 2005), Federal Law Gazette I, 2005, p.1746.

13.4.2 Procedural Requirements

Section 7 (5 – 10) ROG and Sections 14f et seq. UVPG in accordance with the SEA Directive stipulate a set of procedural requirements for environmental assessment. Having imposed a SEA for a plan or programme ("screening"), the competent authority together with the other authorities involved defines the scope of the evaluation ("scoping"). An environmental report must then be prepared, describing and evaluating the impact of the plan on the environment. Other authorities, the general public and third states affected can express their opinion on the environmental report and the draft plan ("consultations"). The competent authority considers the environmental report, the opinions expressed, and the results of transboundary consultations during the preparation of the plan or programme and before its adoption or submission to the legislative procedure. Once the decision-making process has been completed, the authority prepares an explanatory memorandum on the plan. At a later stage, the significant environmental effects of implementing the plan or programme have to be monitored to identify unforeseen adverse effects and permit appropriate remedial action.

13.5 Methodology and Standards for Application of the SEA in the EEZ

Unlike procedure, the methods of the SEA are not defined precisely. As far as offshore windfarms are concerned, Section 3 SeeAnlV requires only that they not be detrimental to the marine environment, especially to bird migration. Similar restrictions apply for marine scientific research (Section 132 (2) no. 3 BBergG), the laying of submarine cables and pipelines (Section 133 (2) sent. 2 BBergG) and the exploration and exploitation of mineral resources (Section 55 (2) no. 11 BBergG). Beside such general rules, there are no legal norms for assessing the impacts of marine spatial planning on the environment. For this reason SEA needs to be operational in marine areas. To this end, the methods and techniques developed for SEA in regional planning can be adopted for marine spatial planning. This methodology is in principle transferable to other types of plan (Stratmann et al. 2007 p.239). However, the technical and legal peculiarities of the spatial structure plan in the EEZ have to be taken into account, for instance the differences in the data base, the biogeographic conditions of the sea and the specifically marine contents as well as scale of the plan.

13.5.1 Selection of Relevant Content for the Spatial Structure Plan

According to Section 3 (1) UVPG, the subject matter of the SEA is the plan as a whole. This formal definition of the scope of the SEA does not, however, exclude focusing on specific elements, depending on the importance of the decisions to be made. Aspects of spatial structure plans in the EEZ that require special attention could, for instance, be priority areas for mineral extraction wind energy. Both the targets and principles of spatial planning can fall within the scope of the study (MKRO 2004 p.6; Jacoby 2005 p.28). Furthermore, special attention has to be paid in the EEZ to matters requiring assessment pursuant to the Habitats Directive. Since environmental planning is required to provide a comprehensive analysis, any potential positive effects, as well as preparatory measures to compensate negative effects should be noted at least briefly (cf. Jacoby 2005 p.28.).

13.5.2 Definition of Impact Factors that Record the Effects of Spatial Planning Decision on the Environment

In the SEA, the environmental effects of planning decisions on a given area are described in terms of impact factors (Heiland et al. 2006 p.11). Applying this concept to marine areas, the following examples for impact factors can be found: assignment of areas to shipping routes, spatial demands for other purposes such as cable or pipeline routes or mining, separation, partition or barrier effects on fauna (e.g. barrier effects on birds during migration or blocking of paths between different resting and/or feeding areas), harmful emissions, noise, or visual impairment of the environment (Janssen et al. 2007 6.2.2.2). Individual impact factors may have a significant effect on the environment such as scaring birds, destroying an ecologically valuable sea bottom or disturbing a landscape. Such impact factors will usually be effective even beyond the geographic limits of the EEZ. For example offshore wind energy plants can endanger marine protection areas within the 12 nautical mile zone or affect the landscape as viewed from the shore. SEA has to take such potential influences beyond the geographical limits into consideration.

13.5.3 Operationalisation of Protected Properties to be Considered in the Course of SEA for Protected Concerns

Protected properties to be considered in the SEA are listed in Section 2 UVPG. However, the scope of the investigation pursuant to Section 2 (1) UVPG is so broadly couched as to almost all-encompassing. The characteristics and qualities of protected properties to be submitted to SEA therefore need to be defined. This depends on their functions within the environment, their rareness and stress situation. Such characteristics are called protected concerns (Heiland et al. 2006 p.13). They describe the relevant aspects of protected properties. With respect to animals and plants, such protected concerns could, for example, be protection areas according to the Habitats Directive, a function as habitat or habitat system, or the protection of especially threatened species. Soil as a protected property, to take another example, can be considered in terms of protected concerns such as rareness, documentation function or function as a habitat for animals and plants. As far as landscape or "seascape" is concerned, the recreation function and the visual quality of the landscape can be addressed as protected concerns (Janssen et al. 2007 6.2.2.3).

13.5.4 Identification of Environmental Objectives Related to Protected Properties as a Basis for Evaluating the State of Environment and Impacts on the Environment In the Framework of Environmental Assessment

Evaluation of the state of and changes in the environment or corresponding protected properties requires not only knowledge about that state, but also standards against which to assess the state. Such standards arise from environmental objectives, which are derived from legal or planning stipulations, scientific standards or policy (Heiland et al. 2006 p.16). The objectives have a hierarchical structure ranging from abstract, high-level objectives with broad scope and little detail to mid-level objectives focusing on certain technical, spatial or sometimes temporal aspects, and, finally, at the lowest level, quantified environmental quality standards (in detail Heiland et al. 2006 pp.16 et seq. and Albrecht 2007 Sections 2 and 3). In practice, the hierarchical structure sometimes becomes diffuse owing to the difficulty of clearly assigning objectives to individual levels. In the framework of the EEZ, environmental objectives are derived from international treaties like UNCLOS, and from European or national regulations. The relevant objectives are established in many fields of regulation, to some extent with different, even opposing aims. This conflict is particularly evident with regard to wind energy, because this kind of energy not only limits CO_2 emissions to mitigate climate change on a global level but also adversely affects nature and the landscape. This is also why corresponding environmental objectives are established at various levels: internationally (UNCLOS, MARPOL[5], OSPAR[6], HELCOM[7], Kyoto-Protocol), at the Community level (Directive on the promotion of electricity produced from renewable energy sources[8], Habitats Directive[9], Birds Directive[10]), and at the Member State level (Strategy for Sustainable Development of the Federal Government, Strategy for Wind Energy Use in Coastal Waters of the German Advisory Council on the Environment) (Schomerus et al. 2006 pp.92 et seq.).

---

[5] International Convention for the Prevention of Pollution from Ships.

[6] Convention for the Protection of the Marine Environment of the North-East Atlantic.

[7] Helsinki Convention on the Protection of the North Sea and North-East Atlantic.

[8] Directive 2001/77/EC of the European Parliament and of the Council of 27 September 2001 on the promotion of electricity produced from renewable energy sources in the internal electricity market, Official Journal 2001 L 283 p.33.

[9] Directive 92/43/EC of 21 May 1992 on the conservation of natural habitats and of wild fauna and flora, Official Journal 1992 L 206 p.7.

[10] Directive 79/409/EEC of 2 April 1979 on the conservation of wild birds, Official Journal 1979 L 103 p.1.

13.5.5 Derivation of State Indicators for Describing the State of the Environment (in Respect of Specific Protected Items) and Evaluating it (in Respect of Environmental Objectives)

Besides the effects of planning decisions on certain protected items, the relevance of the planning area for these items has to be evaluated by means of state indicators. Such indicators have to be defined for each protected item as describing selected properties of the area. By preference, quantitative indicators should be used. In some cases, however, indicators are only qualitative in nature and provide verbal criteria for description. For example, the state indicator for water as a protected item, could include pollutant concentrations or temperature. Likewise, indicators for the state of the sea bottom could be the suitability of soil as a habitat for certain animals or plants. This qualitative indicator could, for example, be classified as high, medium or low (Janssen et al. 2007 6.2.2.6).

13.5.6 Determining the Availability of Data for Assessing the Practicability of State Indicators

The state of the environment in the area under study, as well as the changes caused by the spatial structure plan, can be described and evaluated only to the extent allowed by the available data. This is of particular importance as regards the EEZ, because there is still a considerable lack of information on the marine environment. The necessary data is therefore often not available (Köppel et al. 2004 p.31; Dahlke 2002 p.479). In the course of SEA, the availability of data must accordingly be ascertained in judging the practicability the proposed state indicators. During assessment, the detailed evaluation of the data sources can reveal an insufficiency of data for certain indicators. In such cases, the indicators in question need to be adjusted or replaced. Likewise, the available data may permit reasonable conclusions beyond the scope of the original indicators. The up-to-datedness of the data needs to be taken into account in accordance with the nature of the indicator and the dynamics of the environmental aspect. Whereas initial data on the morphology and sediment composition of the sea bottom can be assumed to remain accurate over a longer period, the data basis for issues like fauna, flora and biodiversity becomes obsolete considerably faster (Reinke et al. 2005 p.47).

13.5.7 Derivation of Impact Indicators for Monitoring the Effects of the Spatial Structure Plan on Individual Protected Items and the Evaluation of these Effects in Relation to Environmental Objectives

The central and most complex part of the SEA is evaluating the effects of spatial planning decisions of Impact factors attributable to planning decisions become linked to the relevance of the concerned area for the protected property. This link is established by impact indicators. Impact indicators are defined in keeping with assessment granularity (Heiland et al. 2006 p. 22). In the case of assessing the implications of a specific item for the environment, or in weighing up alternatives, one example of an impact indicator could be the spatial occupation (impact factor) of habitats or protected areas (protected property) caused by the plan measured in hectares (state indicator). If the focus is on the plan as a whole, a suitable impact indicator could be "the change of the number or size of habitats and protected areas". Impact indicators are usually evaluated in grades. The SEA is required to assess positive and negative implications on the environment caused by implementation of a plan, but also by the results of the plan alone or in interaction or cumulation with other actions. With regard to wind energy plants, impacts can be caused by the construction or removal phases (e.g. noise emissions, temporary occupation of areas), the operational phase or by the mere existence of an installation (Janssen et al. 2007 6.2.2.7).

13.6 Criteria for Avoiding Duplication of SEA and EIA: the Example of Offshore Wind Energy Use (Tiering)

The introduction of SEA for plans and programmes gives rise to concerns that planning and authorization proceedings are becoming more complex and time-consuming, thwarting efforts to accelerate and simplify these procedures. Duplicated, redundant assessment in the context of other planning or approval procedures must therefore be avoided. For this purpose, tiering assessment content is needed. It has to be recognized that subsequent proceedings form a vertical planning hierarchy (Bunzel 2003 p.29). Proper assignment of individual steps in evaluating environmental impacts allows steps to be planned specifically at the level with the most appropriate scope and degree of detail, and only at that level. Just as overloading high-level planning with detailed evaluations must be avoided, the consideration of non-local environmental impacts at a low-level phase of planning needs to be obviated (Bundestag printed paper 15/3449, p.74). In the case of the EEZ, the spatial structure plan is the lowest-level spatial planning phase. Criteria for tiering in relations between the spatial structure plan and the approval level of projects therefore need to be investigated.

13.6.1. Overlap between SEA and EIA

The problem of duplicated assessment arises from the fact that plans are subject to multiple proceedings on identical topics (Bunzel 2003 p.31). In the case of the SEA, a considerable overlap and thus the highest potential for tiering exists with environmental impact assessment (EIA) for approval of projects under Sections 2 seq. UVPG (Janssen et al. 2007 6.2.3.1). This is because of the similarities regarding content and structure between EIA and SEA, most noticeable in the identical protected properties under Section 2 (1) and 4 UVPG (cf. Schomerus and Busse 2005, p.48). Furthermore, the EIA like the SEA, is to be integrated into existing procedures (Kloepfer 2004 p.350). Following amendment of the SeeAnlV as from 5 April 2002, an EIA under Section 2a SeeAnlV is now mandatory for most projected facilities in the EEZ (BSH 2003 p.5). For instance, offshore windfarm projects comprising more than 20 turbines require an EIA based on the UVPG (BSH 2006). The UVPG requires applicants to investigate the marine environment in the project area and predict the impact of the projected windfarm. The Federal Maritime and Hydrographic Agency (BSH) has issued regulations specifying the required scope of the investigations to be carried out by the applicants with respect to each of the features to be protected, the so called "Standards for the Environmental Impact Assessment" (BSH 2003).

13.6.2. Legal Instructions Concerning Tiering

The starting point for tiering is Section 14f (3) UVPG which contains the following rule for scoping within the SEA: Where plans and programmes form part of a hierarchy, in view of avoiding duplication of assessment, it shall be determined at which level of the hierarchy certain environmental impacts are to be assessed. The nature and magnitude of environmental impacts, technical requirements as well as the content and the aim of the plan or programme have to be taken into account. In subsequent plans or programmes, as well as for subsequent approval procedures, environmental assessment shall be limited to the additional or other relevant environmental impacts and to the required updates and shall increase in depth of analysis. From this it follows that aspects already assessed do not need to be reassessed in subsequent planning steps or later planning efforts if they are sufficient detailed and up-to-date (Janssen et al. 2007 6.2.3.1). The granularity of evaluation is to correspond to the depth of planning. It must, however, be ensured that the combination of all planning steps at all hierarchical levels form a complete environmental assessment, and the participation of the general public and the authorities is properly implemented (Bundestag printed paper 15/3449 p.74; Bunge 2003 p.126). Tiering is limited by the requirements of UVPG on the assessment reports of the SEA or the EIA, respectively. Evaluations to be incorporated into the assessment report of the SEA cannot be shifted to the EIA for subsequent evaluation of an installation (Schomerus et al. 2006 p.206).

13.6.3. The Example of Offshore Wind Energy Use

More precise guidelines for tiering than the ones mentioned above cannot be deduced from legal regulations, nor can this be possibly expected (Bunge 2003 p.126). Detailed tiering has to be carried out in each case, taking into account the relevant aspects of environmental assessment and the quality of the information available (Bunzel 2003 p.31). Tiering assessment content at the scoping stage in keeping with Section 14f (3) UVPG appears appropriate since the focus of subsequent evaluations is addressed by the provision anyway. Certain limitations have to be observed, however, in examining tiering options. They are described below, taking the example of offshore wind energy use.

Table 13.1. Criteria for tiering assessment contents in SEA and EIA: the example of offshore windfarms

	Level of spatial planning (SEA)	Level of approval of individual windfarms (EIA)
Analysis of impacts on protected properties	Rough assessment of impacts of windfarms on all protected properties	Detailed analysis of the various protected properties
Assessment of cumulative effects	Integrated assessment of various multicausal impacts of windfarms (the totality of each individual, still tolerable impact can require another site for the windfarm)	Assessment of cumulative effects of individual wind energy plants within one windfarm
Alternatives of projects	Alternative concepts for sites, alternative site within larger areas, temporal alternatives for a site	Alternative positions of the individual plants within the windfarm, technical alternatives
Measures to offset significant adverse effects on the environment[11]	Inter-site balancing and compensating concepts	Compensating measures related to the concrete site of the windfarm

[11] Since the Federal Nature Conservation Act (Bundesnaturschutzgesetz, BNatSchG) does not apply to the EEZ (Risch 2006 p.131; Gellermann 2004 p.81; Klinski 2001 p.21), compensating measures according to Sections 18 seq. BNatSchG do not need to be considered. However, compensating measures according to Section 2 (2) no. 8 ROG (principle of comprehensive spatial planning) are part of the assessment (Köppel et al. 2006 1.4.2.2.4).

Table 13.2. Criteria for tiering assessment contents in SEA and EIA in relation to the impact of offshore windfarms on individual protected properties

Issue	Level of spatial planning (SEA)	Level of approval of individual windfarms (EIA)
Landscape	Visual impacts by selected site of windfarm in a spatial comparison	Visual impacts in relation to the concrete size and height of individual wind energy plants
Fauna, Flora, Biodiversity	Large-area impact by selection of the site of the windfarm (e.g. barrier effects on birds during migration, blocking of paths between different resting and/or feeding areas, use of special protected areas for macrozoobenthos)	Impacts of special technical features of the individual wind energy plants (e.g. impacts of the size and colour of rotor blades on collisions with birds, impacts of the intensity of noise emissions and electric fields on the disturbance of mammals)
Soil	Impacts on morphologically special protected structures (e.g. sandbanks)	Impact on sediments at the specific site during the construction and removal phase
Water	(-), since no criteria for differentiation in large areas known	Impact by mobilisation of sediments during the construction phase, discharge of pollutants (oils, greases) during the operation phase
Cultural goods	(-), since no large-area cultural goods are known on sea bottom	Damage to cultural goods (e.g. wrecks) during the construction and removal phase

13.7 Conclusions and Recommendations

The establishment of the spatial structure plan for the EEZ of the Baltic and the North Seas is a new field of spatial planning in Germany. It aims for the environmentally compatible coordination of traditional marine uses like shipping, fishing and the extraction of mineral resources as well as new kinds of marine uses like the generation of offshore wind energy. SEA, which was implemented in Germany in 2004, is a promising instrument promoting this aim. The success of the SEA, however, depends on how to assess impacts on the environment efficiently. The methods presented in this chapter show the importance of standards and thresholds at the various stages of the SEA process and their degree of aggregation in relation to the EIA. In contrast to the importance of standards and thresholds for the SEA in the German EEZ there is still a lack of objectives, standards and data concerning the marine environment. Further research on marine ecological cause-effect-relationships and the derivation of appropriate environmental indicators is accordingly needed.

References

Albrecht J (2007) Umweltqualitätsziele im Gewässerschutzrecht. Eine europa-, verwaltungs- und verwaltungsrechtliche Untersuchung zur Umsetzung der Wasserrahmenrichtlinie am Beispiel des Freistaates Sachsen, Berlin 2007 (forthcoming)

BMU – Federal Ministry for Environment, Nature Conservation and Nuclear Safety (2002) Strategie der Bundesregierung zur Windenergienutzung auf See, January 2002. Internet address: http://www.bmu.de/files/pdfs/allgemein/application/pdf/windenergie_stra tegie_br_020100.pdf, last accessed on 02.02.2007

BSH – Federal Maritime and Hydrographic Agency (2003) Standards for Environmental Impact Assessments of Offshore Wind Turbines in the Marine Environment. Hamburg and Rostock 2003. Internet: http://www.bsh.de/en/Marine%20uses/Industry/Wind% 20farms/standard_environmental.pdf, last accessed on 02.02.2007

BSH (2006) Wind farms, 2006. Internet: http://www.bsh.de/en/Marine%20uses/Industry/ Wind%20farms/index.jsp, last accessed on 02.02.2007

Bunge T (2003) Möglichkeiten und Grenzen der „Abschichtung" bei der strategischen Umweltprüfung, in: Eberle, Dieter/Jacoby, Christian (eds), Umweltprüfung für Regionalpläne, Hannover 2003, pp. 20-26

Bunzel A (2003) Abschichtung der Umweltprüfung zwischen Regional- und Bauleitplanung. In: Eberle D, Jacoby C (eds) Umweltprüfung für Regionalpläne. pp. 27-37

Dahlke C (2002) Genehmigungsverfahren von Offshore-Windenergieanlagen nach der Seeanlagenverordnung. In: Natur und Recht 2002, pp. 472-479

Gellermann M (2004) Recht der natürlichen Lebensgrundlagen in der Ausschließlichen Wirtschaftszone (AWZ). In: Natur und Recht 2004, pp. 75-81

Heiland S, Moorfeld M, Regener M (2006) Entwicklung eines anwendungsbezogenen Ziel- und Indikatorenkatalogs für Umweltprüfung und Monitoring im Rahmen der Fortschreibung des Regionalplanes der Region Stuttgart. Final report. Leibniz Institute of Ecological and Regional Development Dresden. Dresden, February 2006. Internet: http://www.ioer.de/ioer_projekte/p_186.htm, last accessed on 02.02.2007

Jacoby C (2005) SUP in der Raumordnung: Positionen und Praxishinweise von ARL und MKRO. UVP-Report 19(1):26-30

Janssen G, Sordyl H, Albrecht J, Konieczny B, Wolff F, Schabelon H (2007) Anforderungen des Umweltschutzes an die Raumordnung in der deutschen Ausschließlichen Wirtschaftszone (AWZ) – unter besonderer Berücksichtigung des Nutzungsanspruchs Windenergie (forthcoming)

Klinski S (2001) Rechtliche Probleme der Zulassung von Windkraftanlagen in der ausschließlichen Wirtschaftszone (AWZ), Berlin 2001 (UBA-Texte 62/01)

Kloepfer M (2004) Umweltrecht, 3rd edn, Munich 2004

Köppel J, Peters W, Steinhauer I (2004) Entwicklung von naturschutzfachlichen Kriterien zur Abgrenzung von besonderen Eignungsgebieten für Offshore-Windparks in der Ausschließlichen Wirtschaftszone (AWZ) von Nord- und Ostsee. Bonn 2004 (BfN-Skripten 114)

Köppel J, Wende W, Herberg A, Wolf R, Nebelsiek R, Runge K (2006) Naturschutzfachliche und naturschutzrechtliche Anforderungen im Gefolge der Ausdehnung des Raumordnungsregimes auf die deutsche Ausschließliche Wirtschaftszone. Gutachten im Auftrag des Bundesamtes für Naturschutz. Umweltforschungsplan 2004, Forschungskennziffer 804 85 017 K2, Schriften des Bundesamtes für Naturschutz, Bonn

MKRO (Conference of Ministers Responsible of Spatial Planning) (2004) Umweltprüfung von Raumordnungsplänen (Plan-UP). Erste Hinweise zur Umsetzung der RL 2001/42/EG, May 2004, Internet: http://www.vm.mv-regierung.de/raumordnung/doku/AG_Plan_UVP_030401.pdf, last accessed on 02.02.2007

Molitor L (2005) Raumplanung in der AWZ: Eine die verschiedenen Nutzungs- und Schutzinteressen im Bereich des Meeres koordinierende Gesamtplanung mit dem Ziel einer nachhaltigen Raumplanung. In: Bundesministerium für Verkehr, Bau und Wohnungswesen/Bundesamt für Bauwesen und Raumordnung (eds.), Nationale IKZM-Strategien – Europäische Perspektiven und Entwicklungstrends, Bonn 2005

Reinke M, Bölitz D, Helbron H (2005) Strategische Umweltprüfung für die Regionalplanung – Entwicklung eines transnationalen Prüf- und Verfahrenskonzeptes für Sachsen, Polen und Tschechien. 1st interim report on the Interreg IIIA project. Dresden, February 2005. Internet: http://www.ioer.de/ioer_projekte/p_165.htm, last accessed on 02.02.2007

Risch J (2006) Windenergieanlagen in der Ausschließlichen Wirtschaftszone. Verfassungsrechtliche Anforderungen an die Zulassung von Windenergieanlagen in der Ausschließlichen Wirtschaftszone (AWZ), Berlin 2006

Schomerus T, Busse J (2005) Strategische Umweltprüfung bei planerischen Ausweisungen für Offshore-Windparks in der deutschen ausschließlichen Wirtschaftszone (AWZ). In: Zeitschrift für öffentliches Recht in Norddeutschland 2005, pp. 45-51

Schomerus T, Runge K, Nehls G, Busse J, Nommel J, Poszig D (2006) Strategische Umweltprüfung für die Offshore-Windenergienutzung. Grundlagen ökologischer Planung beim Ausbau der Offshore-Windenergie in der deutschen ausschließlichen Wirtschaftszone, Hamburg 2006

Stratmann L, Helbron H, Heiland S, Schmidt M (2007) Prüfmethodik und Bewertungsmaßstäbe für die SUP in der Regionalplanung. UVP-report 20 (5): 229-235

14 Standards of Implementing Renewable Energy Technologies in Cameroon

Ernestine A. Tangang Yuntenwi[1], Victor Ngu Cheo [2] and Jürgen Ertel[1]

1 Department of Industrial Sustainability, Brandenburg University of Technology (BTU) Cottbus, Germany
2 Department of Journalism and Mass Communication, University of Buea, Cameroon

14.1 Introduction

Renewable Energy (RE) sometimes referred to as 'clean energy' or 'green power' can be defined as any energy source that is naturally occurring and that cannot, in theory, be exhausted. RE sources include solar energy, biomass, tidal, wind or wave power, geothermal energy (English Dictionary Online 2003). Renewable energy is obtained from sources that are essentially inexhaustible. The reverse is true for non renewable such as fossil fuels, of which there is a finite supply (NATSOURCE 2005).

Meanwhile, Renewable Energy Technology (RET) can be defined as any technology that exclusively relies on an energy source that is naturally regenerated over a short time and derived directly or indirectly from the sun, or from moving water or other natural movements and mechanisms of the environment. Renewable energy technologies include those that rely on energy derived directly from the sun, wind, geothermal, hydroelectric, wave, or tidal energy, or on biomass or biomass-based waste products, including landfill gas (Evolution Markets LLC 2005).

Nowadays, there is a shift from the use of fossil fuel to renewable energy, and this can be attributed to the diminishing and unsustainable use of fossil fuel. This is particularly true for developing countries as shown in Table 14.1. Furthermore, conventional fuels are associated with a high emission of greenhouse gases (GHGs) which are considered environmentally unfriendly and greatly contribute to climate change crisis thereby causing health, social and environmental hazards. A shift towards the use of sustainable technology can alleviate the crisis.

Standards and Thresholds for Impact Assessment. Edited by Michael Schmidt, John Glasson, Lars Emmelin and Hendrike Helbron. © 2008 Springer-Verlag

Table 14.1. Energy and Resources (adapted from Earth Trends 2003)

Energy Consumption by Source 1999 (in thousand metric tons oil equivalent)	Cameroon	World
Total Fossil Fuels	939	7,689,047
Coal and coal products	0	2,278,524
Crude oil and natural gas liquids	1,454	3,563,084
Natural gas	0	2,012,559
Nuclear	0	661,901
Hydroelectric	287	222,223
Renewables, excluding hydroelectric	4,877	1,097,889
Primary solid biomass (includes fuel wood)	4,877	1,035,139
Biogas and liquid biomass	0	14,931
Geothermal	0	43,802
Solar	0	2,217
Wind	0	1,748
Tide, wave, and ocean	0	53

The poorest household can save emissions of GHGs by using sustainable energy technologies (Goldemberg 1996).

In line with this, the 1997 Kyoto protocol in its article 12 elucidated the Clean Development Mechanism (CDM) to promote worldwide cooperation and for climate protection and sustainable development. The Clean Development Mechanism (CDM) projects have as main criteria to fulfil the transfer of sustainable technology, additionality and the use of modalities and procedures with the objective of ensuring transparency, efficiency and accountability through independent auditing and verification of project activities. Furthermore, the 2005 World Summit from September 14 – 16, came up with an action plan for promoting international security and for achieving the 8 UN Millennium Development Goals by 2015 (Seifert et al. 2006). However, it is worthwhile stating that RET have their own drawbacks, reason for which there is the need for policies, standards and thresholds to be set which act as control tools and basis for their implementation. These standards and thresholds can perform many roles in Environmental Impact Assessment (EIA) (Murray and McGranahan 2003).

14.1.1 Statement of the Problem

The threat of energy insecurity in the future, particularly, non-renewable energy based sources is a cause for concern preoccupying many nations. Besides, there are also certain environmental consequences emanating from the use of non-renewable energy sources such as fossil fuels. While the gradual adoption, worldwide, of renewable energy resources is a potential guarantee of energy security, it also has negative consequences, mainly health wise due, partly, to the nature of technologies adapted for use with these energy resources.

In Africa, and Cameroon in particular, some of the most commonly used renewable energy technologies are the three-stoned-fire side, saw dust cookers etc. In spite of their economic and environmental advantages, they too have serious health repercussions for users except where certain thresholds and standards of implementation are clearly defined. Even with these, a hundred percent compliance is impossible. Hence, what we perceive is a crisis-mitigated circumstance rather than total avoidance. The situation, obviously, would be worse in countries where such standards or their equivalents are not implemented.

Apparently, despite the widespread use of these technologies in Cameroon, there seems to exist no proper elaborated standards of implementation of renewable energy technologies compared to what is obtained in other countries. To address the issues raised in this paper, the following *hypotheses* were adopted;

- Cameroon has no standards or equivalent parameters which serve as reference for the implementation and regulation of renewable energy technologies.
- because of the absence of such standards, the consequences on the environment and human health are enormous.
- there is an urgent need, of some form, of standards to mitigate the debilitating effects of the use of these technologies on human health.

The advantages emanating from the use of renewable energy technologies in Africa in general, and in Cameroon in particular cannot be overemphasized. The *overall objective* of this paper is to investigate the status of Cameroon with regard to standards of implementing appropriate renewable energy technologies and the potential effects of these on their users and the community as a whole.

This study deals with standards and thresholds for implementing RET in general in Cameroon, but more specifically those related to indoor air pollution sources, predominantly through the use of biomass technologies. The importance of a study of this nature, which focuses on guidelines for the implementation of RET in a developing country, therefore, is vital. Cameroon government's cosmetic guidelines for the implementation of RET projects in general and the failure to address basic causal agents of indoor air pollution such as saw dust cookers and the three stoned firesides is a serious flaw. This study is therefore a step forward in attempting to make the government realise the need for a more comprehensive guidelines on the effects on human health and the environment, especially air quality. It is intended to act as a catalyst in the enhancement of RET standards in Cameroon.

The *methodology* employed in this study comprises mainly secondary data collection. Information from books, journals, internet as well as a textual analysis of ministerial decrees N° 2005/0577/PM of 23rd Feb 2005 laying down modes of carrying out environmental impact assessment from the Prime Minister's Office, head of Government, decree N° 0069/MINEP of 8th March 2005 fixing the different categories of operation on which environmental impact studies are based from the then Ministry of Environment and Nature Protection, decree N° 2005/087 of 29th March 2005 laying down modes of the organisation of the Ministry of Energy and Water from the Ministry of Energy and Water, decree N° 2004/320 of 8th December 2004 from the Prime Minister's Office, and a Compendium of Environmental Laws of African Countries Volume 1: Framework Laws on EIA Regulations (1997 Supplement to Volume 1, 1996 edition).

14.2 Institutional Framework of the Ministry of Energy and Water and EIA Procedures in Cameroon

14.2.1 Structural Organisation of the Ministry of Energy and Water

Following decree no 2005/087 of 29th May 2005; the Ministry of Energy and Water (MINEE) was created (see figure 14.1). The Renewable Energy department, notably Biomass Department and Solar Energy department constitute the Energy Control Department of the Ministry of Energy and Water. The other Renewable Energy sources are more often not represented. This creates a big lapse as to the substantial management of energy issues in the country. Cameroon mostly relies on Fossil fuels and water for its energy (electricity) and despite the huge potentials the country has in Renewable Energies, very little has been done by the Government regarding the implementation of such technologies.

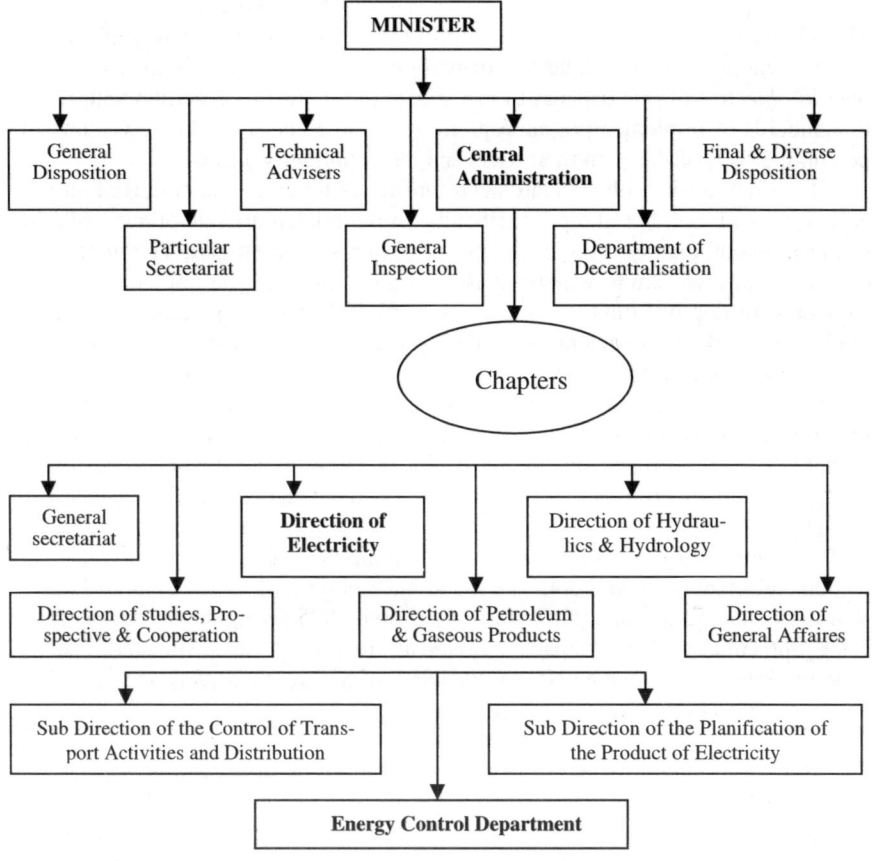

Fig. 14.1. Organisational Chart of the Ministry of Energy and Water (adapted from the Organisational chart of MINEE 2004)

14.2.2 A Synopsis of the Process and Procedural Framework Governing EIA in Cameroon

As per Ministerial Order No. 0069/MINEP of 08th March 2005 that fixes the different categories of projects necessitating an EIA, all Renewable Energy Projects are required to undergo a brief EIA before these projects are authorised. Following this Order, under Art. 2, sections 2 and 3, certain guidelines have been laid down for brief EIS before the implementation of and exploitation of Renewable Energy. The execution of the project cannot take place without the approval of a relevant Environmental Impact Statement (EIS). These guidelines as in the Order include:

- a summary of the study in English and French languages.
- a description of the environment surrounding the project site and region
- a description of the project
- an environmental impact statement
- an inventory and a description of the project impacts on the environment as well as the envisaged mitigating measures and the estimated corresponding expenditure
- approved terms of reference
- related bibliographic studies

An example of a project which has been executed following this Order is the Kai Mini Hydropower project, Momo Division, North West Province Cameroon. This program was executed by a non governmental organisation (NGO), the Centre for Appropriate Technology (CAT) Bamenda, after undergoing the EIA process and getting approval by the government for its implementation. Today, the people of the Kai Village do not only have portable water, but they enjoy living in a protected environment. The project could be Inferior (produces less than 2 megawatts (MW) of energy), micro (produce less than 50MW of energy) and macro (produce 50MW and more of energy).

Figure 14.2 is a sketch of Cameroon's EIA procedure which is laid out by Decree No 2005/0577/PM of 23rd Feb. 2005. Following Decree No 2005/0577/PM of 23rd Feb. 2005 laying down modes of carrying out EIA, the Administration in charge of the environment which, in this case, is the Ministry of Environment and Nature Protection has a leading role to play. Before a Proponent executes a project there are certain standards to be met.

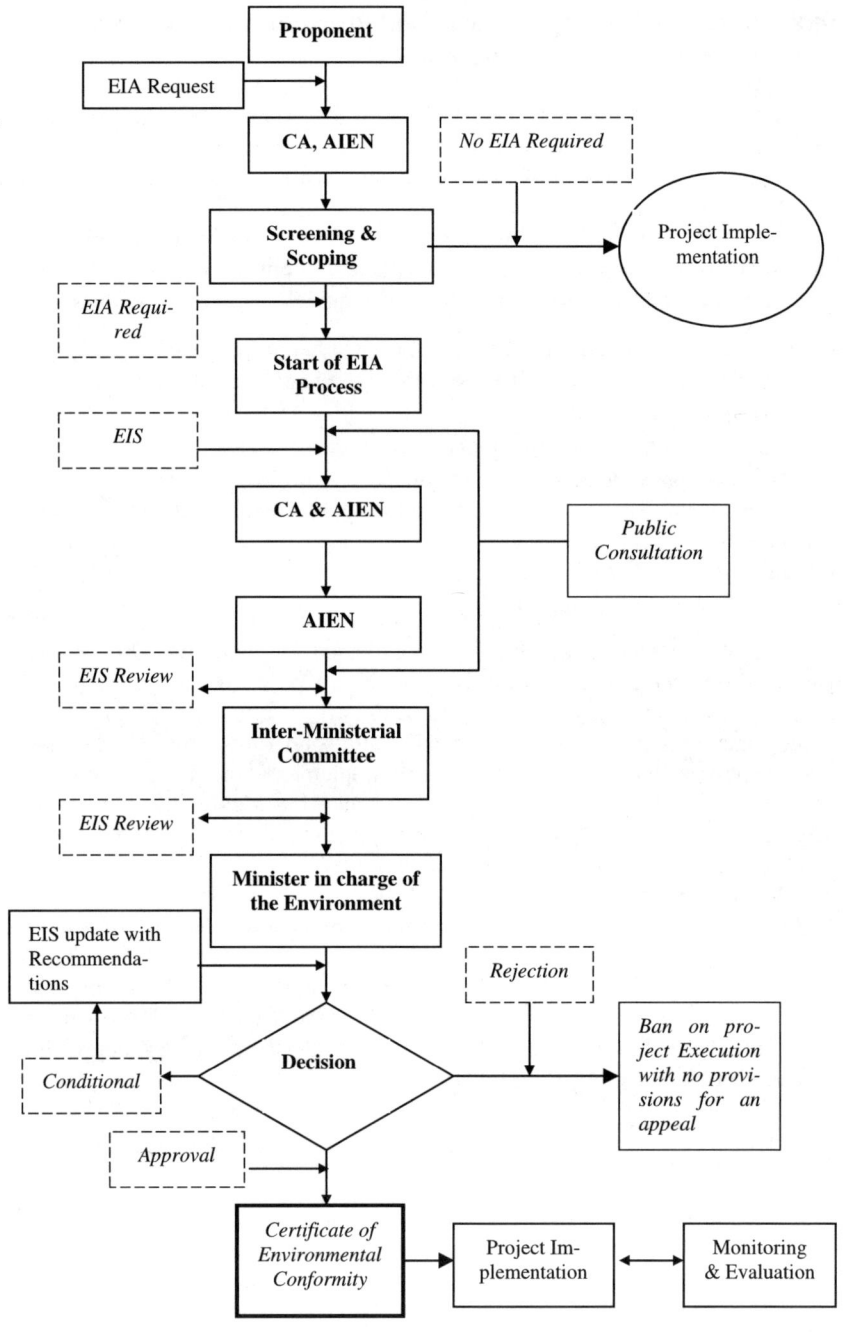

Fig. 14.2. Flow chart of Cameroon's EIA procedure (from Alemagi et al. 2006)

Following the afore-mentioned legal disposition, the Proponent is expected to initiate an EIA by submitting an EIA project file to the Competent Administrator (CA) who evaluates and sends this file to the Administration in charge of the Environment (AIEN). The AIEN then carries out screening and scoping, and if no EIA is required the Proponent is given the authorization to execute the project. Otherwise, review comments are sent to the Proponent for a full scale environmental impact study to be initiated. The report that emanates from the environmental impact study is termed the Environmental Impact Statement (EIS). The EIS is formulated by the proponent and the current legal disposition requires that this be done in consultation with the public. When the EIS is formulated it is forwarded to the CA and AIEN (who at this level work together as a body) for these authorities to opine. The CA and AIEN now go to the field and consult the public and verify if the handed information is in conformity with their findings. Thereafter, their submissions are sent to the AIEN for review. The AIEN then forwards its final submission to the Inter-ministerial Committee for the Environment which is the final opinion-making body. After reviewing the file, this committee forwards its opinion with recommendations to the Minister in Charge of the Environment. The Minister in charge of the Environment is the final decision maker. Therefore, based on the recommendations of the Inter-ministerial committee, a final decision is made within 20 days by the Minister. At this level, three possible decisions could be taken: an approval, a conditional decision, and a rejection. In the case of an approval, a Certificate of Environmental Conformity (CEC) is rewarded to the proponent and the project is implemented with some monitoring and evaluation by the competent authority. It is important to underscore the fact that the CEC is valid for a period of three years. Thus, if the project is not commissioned within this requisite time frame an updated EIS is mandatory for revalidation. If the decision is conditional, some recommendations would be made to the Proponent to help update the EIS. Finally, if the decision is a rejection, the project execution is banned with no provision for an appeal. Despite the presence of an EIA procedure in Cameroon, its effect is not very much felt.

14.2.3 Limitations to Effective Implementation of RET in Cameroon

The law itself is a hindrance – it is not all encompassing. Emphasis is placed mainly on electricity and water. Following decree No 0069/MINEP of 8th of March 2005, the EIA, for project implementation, could be a brief or a detailed study. However, basis for carrying out such a study for renewable energy projects (mainly hydro and biomass) have been well stipulated under the brief EIA. The other renewable energy sources are not well exploited in Cameroon reason for which they are not mentioned in the draft of the guiding document for their implementation. This, however, creates a gap which in the near future has to be filled as there is a shift from the use of biomass technology to the exploitation of solar energy, for example.

In this same light, the synopsis of the process and procedural framework governing EIA in Cameroon is plagued with a flaw which nevertheless could be redressed if the document is well studied and subjected to corrections. A proponent, whose project proposal after a thorough EIA study has been rejected, should be given an opportunity to upgrade the project and file in an appeal. The total denial of an appeal after a project has been rejected should not be promulgated.

Despite the presence of an EIA procedure in Cameroon, its effect is not very much felt. The issue of bribery and corruption overrides the positive effects of such a program because many NGOs and Companies easily buy their way through. Reason for which unsustainable projects are being executed which lead to persistent environmental degradation. A more encompassing and well defined EIA procedure will, very much, improve sustainability provided appropriate technologies are used.

14.2.4 Cameroon's Energy Action Plan

Based on the Global Village Energy Programme's (GVEP) experience in Cameroon, the country team is confident that Cameroon, like other countries, have adopted the six points outline process as a methodology for the development of National Energy Action Plans in their respective countries. These include:

1. Government – UNDP – GVEP consultations leading to the signing of the project document to get the government involved, and leading from the onset,
2. The creation of an inter-ministerial multi-Actors/multi-sector based work group to supervise the country level GVEP process and create ownership of the process,
3. Collecting and analyzing information and data (from the different authorities or institutions involved) on the use of energy services in the different key sectors in the fight against poverty,
4. Regional workshops to facilitate the participation of grassroots actors within a participative framework (local authorities, NGO, private sector actors, rural women's groups, etc.),
5. National multi-sector and multi-actors workshop to discuss and validate the methodology for the integration of all aspects linked to modern energy services in the PRSP and to identify all priorities and actions necessary for its implementation,
6. The finalization of the National Energy Action Plan (NEAP) on the basis of the recommendations and outputs drawn from the data collected and analyzed, the discussions at the regional workshops, taking into consideration the priorities of the RSP to infuse a multi-sector/multi dimensional approach to energy (Nstama 2005).

14.3 Global Efforts to Mitigate Problems Caused By RET

Cameroon like many other developing countries should involve more in International activities for a better implementation of Renewable Energy Technologies and meeting up with set standards. In his opening remark at the fourteenth session of the Commission on Sustainable Development (CSD -14) on 10th of May 2006, the former UN Secretary General Kofi Annan called for new approaches to increase access to improved energy in developing countries. He called for a revolution in energy efficiency and encouraged the use of the Clean Development Mechanism (CDM) to support projects in developing countries (Annan 2006). In this wise, in September 2000, Cameroon took part at the millennium summit with the presence of the head of state, his Excellency, President Paul Biya, and Cameroon fully agreed on the MDGs (Millennium Development Goals) and the associated targets. Based on this, the associated MDGs targets were put into the national context within the framework of the Poverty Reduction Strategy Paper (PRSP) in April 2003.

The eight *Millennium Development Goals (MDGs) of the United Nations* ranging from halving extreme poverty to stopping the spread of HIV/AIDS and providing universal primary education, all by the target date of 2015 – form a detailed plan of action agreed on by all the world's countries and all the world's leading development institutions. They have stimulated unprecedented efforts to meet the needs of the world's poorest countries.

The involvement of Cameroon in the MDG program is a milestone in solving the existing energy crisis plaguing the country as it will now embark on

- avoiding unsustainable logging, erosion and desertification
- avoiding GHG-emissions.
- reducing the 'push' factor in rural-urban migration by improving the energy services in rural areas.
- taking advantage of new technologies to avoid energy-intensive, environmentally unsound development paths.

Although by September 2000, Cameroon was not a signatory of the Kyoto Protocol, the Clean Development Mechanism (CDM) which has, as one of its targets, to support climate-friendly sustainable development projects in developing countries (UNFCCC-United Nations Framework Convention on Climate Change 2005) is very important for the country's development. The set targets of the protocol could be implement in the country and could act as standards used to reduce or control emissions from energy technologies.

14.4 Standards and Thresholds for Mitigating Indoor Air Pollution from Biomass Use

Biomass smoke contains a large number of pollutants such as particulate matter (PM), carbon monoxide (CO), nitrogen dioxide, formaldehyde, and polycyclic organic matter, including carcinogens such as benzo[a]pyrene and benzene which greatly affect human health. Exposure to indoor air pollution from the burning of solid fuels has been attributed, with varying degrees of evidence, as a causal agent of several diseases in developing countries, including acute respiratory infection

(ARI) and *otitis media* (middle ear infection), chronic obstructive pulmonary disease (COPD), lung cancer (for coal smoke), asthma, nasopharyngeal and laryngeal cancer, tuberculosis, perinatal conditions and low birth weight, and diseases of the eye, such as cataracts and blindness.

An estimate of global mortality due to indoor air pollution from solid fuels shows that in 2000, between 1.5 million and 2 million people died from exposure to this risk factor. This accounts for approximately 4% to 5% of total mortality worldwide. Approximately 1 million of the deaths were due to childhood Acute Lung Respiratory Illnesses (ALRI) while the remainder resulted from other causes, dominated by COPD and lung cancer among adult women.

Due to the increase health loss associated with exposure to indoor smoke as well as its concentration among the marginalized socio-economic and demographic groups (women and children in poorer households and the rural population), international development and public health organisations have recently prioritized and seen the need to set standards such that exposure to indoor air pollution is mitigated (Ezzati and Kammen 2002).

Emissions in the kitchen vary from day to day and from season to season, due to the moisture content and density of the fuel, the amount of airflow, the type of food being cooked, the fuel type and stove technology. Therefore, the concentration levels especially of inhalable particulates ($< PM_{10}$) and CO levels will greatly vary. The commonly used measures for particulates are PM_{10} (inhalable) and $PM_{2.5}$ (respirable). These refer to particle sizes of less than 10 micrometres (Âμm) diameter and less than 2.5Âμm respectively. The small size of these particles are easily carried deep into the lungs, with $PM_{2.5}$ being the most penetrating and appearing to have the greatest potential for damaging health (Practical Action 2006). *The European Union "Council Directive 1999/30/EC states that a PM_{10} 24-hour limit value of 50 Âμg/m³ should not be exceeded more than 35 times per year by 1ˢᵗ of January 2005 and no more than seven times per year by 1ˢᵗ of January 2010 in the member states. Also, a PM_{10} annual limit value should not exceed 40Âμg/m³ by 1ˢᵗ of January 2005 and 20 Âμg/m³ by 1ˢᵗ of January 2010 (Practical Action 2006). "* World Health Report of 2002 states that Indoor Air Pollution caused about 36 % of ALRI, 22 % of COPD and 1.5 % of cancers on the Trachea, lungs and bronchus.

Table 14.2. WHO statement on indoor air pollution allowable standards (source: practical action (2004) Based on the UNDP/DESA/WEC World Energy Assessment)

Pollutant(mg/m3)	Emission (mg/m3)	Allowable standard
Carbon Monoxide	150	10
Particles	3.3	0.1
Benzene	0.8	0.002
1, 3-Butadiene	0.002	0.0003
Formaldehyde	0.7	0.1

Note: applicable for burning 1 kg of wood in a traditional stove in a 40 m³ kitchen with 15 air changes per hour (ACH).

World Health Organisation (WHO) statement on Indoor Air Pollution allowable standards (Table 14. 2), the European Union Standards and the comparison of typical levels of PM10 and CO in less developed countries (LDC) with WHO and USEPA guidelines (Table 14.3) could be a baseline for controlling emissions and promoting the use of sustainable technologies. The lack of such standards in the country creates a void in the organizational and managerial schemes of renewable energy issues. Implementing such standards will safeguard the proper use of technology and resources thus optimizing efficiency and ensuring environmental sustainability.

Table 14.3. Comparison of typical levels of PM10 and CO in less developed countries (LDC) with WHO and USEPA guidelines (adapted from Bruce et al. 2000)

Pollutant	Range of ambient levels in LDC studies, for simple stoves		WHO guidelines (WHO 1999)		US EPA guidelines (USEPA 1997)	
	Period	Level	Period	Level	Period	Level
Particles PM$_{10}$ ($\mu g/m^3$)	Annual	Not available but expected similar to 24hr	Annual	Guidelines presented as exposure-response relationships. Levels as low as 10$\mu g/m^3$ associated with excess risk	Annual	50
	24 hour	300 - 3,000+	24 hour		24 hour	150 (99th pecentile)
	During stove use	300 – 20,000+ some 30,000+				
Carbon Monoxide (ppm)	24 hour	2 - 50+	8 hour	10	8 hour	9
	During stove use	10 - 500+	1 hour	30	1 hour	35
			15 mins	100		
	COHB (%)	1.5 -13%	COHB (%)	Critical level < 2.5%, Typical non smoker 0.5-1.5%, Typical smoker 10%		
	During stove use	1000+				

ppm = parts per meter, $\mu g/m3$ = microgram per cubic meter.
COHb = Carboxyhemoglobin.

14.5 Conclusions and Recommendations

Within the framework of emerging trends to curb indoor air pollution which is a major cause of health and environmental hazards, a good number of countries have begun adopting or adapting internationally recommended standards e. g. from WHO, to mitigate the effects of these technologies as well as re-orientate their use in their respective countries. Cameroon which has also recognised the need for the regulatory standards to curb the effects of these technologies has, however, not put in place a well elaborated guideline to this effect. The so-called guidelines implementing the use of these technologies in Cameroon, is still cosmetic and not all encompassing to say the least. More emphasis is rather oriented towards hydroelectricity technologies to the near total oblivion of non-negligible widespread indoor air pollution-causing technologies like saw-dust cooker and three-stoned fire side.

On the basis of the above mentioned lapses, this paper recommends an urgent need for Cameroon to re-think it's RET policy standards and to be more domain-specific rather than a "shallow" generalization. The government should initiate a study that will compile the available RET usage in Cameroon, their potential effects and come up with standards which adequately address the problems or provide mollifying solutions. In this direction there is the need to examine carefully, some prototype documents of international organisations like the WHO and countries which remain a success story like China and India. The need for experts who will develop such a document should not be overlooked.

References

Alemagi D, Sondo VA, Ertel J (2006) Constraint to environmental impact assessment practice: a case study of Cameroon. Journal of Environmental impact assessment review (April 2006)

Annan k (2006) New approaches to increasing access to improved energy in developing world. Speech delivered on the 14th session of the commission on sustainable development (CSD-14) on the 10th of May 2006 "New York, USA." Internet address http://www.un.org/News/Press/docs/2006/sgsm10455.doc.htm, last accessed on 26/6/2006

Bruce N, Perez-Padilla R, Albalak R (2000) The health effect of indoor air pollution exposure in developing countries. WHO/SHE/OEH/02.05. Geneva, Switzerland

Compendium of environmental laws of African countries Volume 1: framework laws on EIA regulations (1997 Supplement to Volume 1, 1996 edition). UNEP/UNDP joint project on environmental laws and institutions in Africa. Nairobi, Kenya

EarthTrends (2003) Energy and Resources in Cameroon "Washington DC" Internet address: http://earthtrends.wri.org, last accessed on 15/6/2006

Evolution Market LLC (2005) "New York, USA." Internet address: www.evomarkets.com/mk_em_rec_gloss.html, last accessed on 4/6/2006

Ezzati M, Kammen D M (2002) The health impacts of exposure to indoor air pollution from solid fuels in developing countries: knowledge, gaps, and data needs. Resources for the Future, Washington DC

Global village energy partnership (2005) "Douala, Cameroon" Internet address: http://www.gvep.org/section/actionplans/africa/cameroon/, last accessed on 27/6/2006

Goldemberg J (1996) Energy, Environment and Development. Earthscan, London, http://www.allwords.com/word-renewable%20energy.html, last accessed on 07/06/2006

Murray F, McGranahan G (2003) Air pollution and health in developing countries–the context. ed in Air Pollution and Health in rapidly developing Countries. Earthscan London, UK pp 1-9

NATSOURCE (2005) "New York, USA" Internet address:http://www.natsource.com/markets/index.asp?s=104, last accessed on 12/6/2006

Nstama J (2005) Report on the global village energy partnership (GVEP) in Cameroon. Brasilia, Brazil

Practical Action (2004) Indoor Air Pollution - the Killer in the Kitchen "Geneva, Switzerland" Internet address: http://www.itdg.org/?id=iap_who. Accessed 10/6/2006

Practical Action (2006) Smoke's increasing cloud across the globe "Geneva, Switzerland" http://www.itdg.org/?id=smoke_report_2. last accessed on 10/6/2006

Seifert D (2006) Clean Development Mechanism (CDM)–a powerful instrument to fulfil the UN millennium goals-experiences, visions and suggestions". Paper presented at 2006 solar cooking and food processing international conference, Granada/Spain.

UNFCCC-United Nations Framework Convention on Climate Change (2005) A summary of the kyoto protocol "Bonn, Germany." Internet address:http://unfccc.int/essential_background/feeling_the_heat/items/2879.php, last accessed on 8/6/2006

United Nations (2005) United nation millennium development goals "New York, USA" http://www.un.org/millenniumgoals/, last accessed on 30/5/2006

15 Standards for Mining and Quarrying

Stanislaw Gruszczyński

AGH Krakow, Poland

15.1 Introduction

Mining includes many different technologies used in searching and exploiting mineral resources in the form useful for their further use. Apart from obviously associated with mining underground mining enterprises (hard coal, lignite, metal ores, chemical resources), the objects running mining activities also include: open cast mines (hard coal, lignite, metal ores, sulphur, chemical resources), quarries and bore-hole mines (water, mineral waters, natural gas, oil, salt, sulphur). Mining technologies are also applied in searching for and identifying deposits and in tank-free storing of resources and wastes in the rock mass.

Each applied mining technology to smaller or greater degree affects the environment. The scale and kind of this impact depend on many circumstances: geological properties of the deposit, technical solutions, scale of exploitation, presence of factors transferring or limiting the influence and properties of environmental components - receptors of such impact. Basic condition of mining: strict dependence on the situation and character of the exploited deposit, demands that the decisions referring to the scale, technique and level of the accepted environmental impact are made considering the impact on natural and technical objects and its economic and social effects. It has to be stated that apart from strictly mining technologies, in also techniques not connected with mining exploitation are used in mining industry: related to the enrichment of the offered resources, their purification, transport and storing. Also certain amount of facilities is necessary. All these technologies, apart from pure mining make multifaceted impact on environment.

Standards and Thresholds for Impact Assessment. Edited by Michael Schmidt, John Glasson, Lars Emmelin and Hendrike Helbron. © 2008 Springer-Verlag

15.2 Factors of Impact

The factors of the impact of mining, causing numerous direct and in direct kinds of impact are: taking out the deposit and actions uncovering the deposit, dealing with the overburden, treatment (enrichment) of the resources and disposal of wastes (including sterile rocks), functioning of the infrastructure and industrial buildings and providing facilities (water, electricity etc.) necessary for the work. Potentially harmful influence of respective factors can make changes in: air and water quality, hydrogeological system, soils, vegetation and fauna - affecting bio-diversity and ecological stability, human health, state of natural resources, social conditions, cultural objects and technical infrastructure (Fig. 15.1).

15.2.1 Exploitation of Deposits

Obtaining valuable minerals is the purpose of mining exploitation of the deposits. It is preceded by geological survey of the resources to assess economical aspects of the project and make the plan of exploitation. Direct effects of the mining ex-ploitation include making exploitation voids or excavations on the surface. Indi-rect effects include the changes in hydrogeological conditions by the deformation of the rock mass and drainage of the exploited deposit. Environmental implica-tions of such influence can include: geomechanic deformations of the terrain sur-face with further consequences and the disturbance of the hydrological regime.

Geomechanic deformations, in case of open-cast exploitation mean drastic changes in the morphology of the area, which exclude a smaller or larger surface from the carried out so far way of management. A new special form requires rec-lamation after the end of mining. The old working is also a serious deformation of the landscape.

More complicated are the consequences of geomechanic transformations in case of underground mines. Arising of post-exploitation voids, causes deforma-tions on the surface of the area. Their scale depends on the size of voids and the thickness and properties of the rock making overburden. These deformations are observed as subsidence of the ordinates of terrain points, with further geometric consequences (changes of curves and slope angles). The biggest economic costs involve such impact on heavily invested areas, with dense water supply and sew-erage networks, road networks and communication objects, railways and residen-tial buildings. The effects of deformations include: breaking the continuity of the facility networks or problems with their functioning, damage to communication routs, excluding them from traffic or making traffic more difficult, cracks, disloca-tions or destruction of buildings. This involves social effects, which are difficult to assess. High costs of geomechanic transformations of areas of developed infra-structure encourage exploitation under areas of much smaller level of develop-ment: agricultural and forest areas. However, geomechanic transformations of these areas can bring dangerous economic and ecological effects. In case of deep underground exploitation, when the continuity of the terrain surface is kept, the implications connected with the changes of the erosion conditions (changes in surface slope angle) can be relatively less important. They can locally increase or decrease the intensity of erosion, compared to the initial state. Changes in the slope angle of the area cause the disturbance in the flow of streams (increase or decrease of flow speed), leading into flooding or the intensification of bottom erosion. In case of very shallow exploitation the continuity of the terrain surface can be broken and faults or hollows can be made. Safety reasons require to stop the agriculture in such an area. In some regions strains occurring in the rock mass as the result of post-exploitation voids, tend to express in a form that is very dangerous for buildings and constructions, namely – as underground tremors.

Morphology is responsible numerous environmental deformations of the area. Particularly complicated can be the results of geomechanic transformations in the areas covered by hydrogenic or semi-hydrogenic soils (the following groups of soils: histosols, gleysols, some fluvisols and phaozems). Soils of these groups usually occur in hollows and are characterized by a shallowly situated table of ground water. Deformations of surface, especially subsidence of the terrain ordinates, can lead into very serious consequences related to land use, namely – flooding of grounds (qualitative changes in the usefulness of soils for the existing way of use) and ecological changes (in the developmental trends, habitat changes, decrease of the resistance of trees in forest areas, disturbance to natural processes responsible for biodiversity). Changes manifested in too high amount of water (average humidity of soils) are the result of one of three mechanisms: the formation of local hollows in the area in the regions of a shallowly situated table of ground water, formation of troughs in the regions where grounds have low permeability for water (surface flow) and formation of new subsidence troughs covering river beds and streams causing floodings.

The disturbance of hydrological regime can be also the result of the mining-caused dehydration. It is necessary in those cases when the deposit is so humid that its exploitation is impossible. The possibility of its exploration is conditioned by efficient drainage meaning hydrologic depression in the deposit. Hydraulic continuity between dehydrated deposit and situated higher water-bearing horizons carrying the depression also into different levels, including water-bearing horizons pumping water into soil layers (in the region of hydrogenic soils). Drying phenomena are observed, leading into changes in the usability of soils and transformation of ecosystems.

Particular threat is connected with less popular technologies of exploitation using boreholes in the process of exploitation. This type of technology is applied in the exploitation of gas, oil, salt and sulphur. In every such case, except of the exploitation of gas deposit, the main problem is the preservation of sealing mining and transferring installations. Great catastrophic leakages or even small tolerated leakages of oil and liquid sulphur are always a serious factor of environmental threat. Specific threats are connected with the exploitation of aggregates with the use of explosives (Batko et al. 2001). We deal with the dispersion of explosives. They are thrown in certain radius. Explosions cause tremors.

15.2.2 Overburden and Sterile Rocks

Dealing with the overburden is the only significant factor of environmental impact of mining on environment. In open-cast mining reaching the deposit usually requires the removal of certain volume of the overburden. The necessity to achieve a sufficiently long advance in exploitation in the relation to the overburden, causing additional landfill (external heap) of the overburden, outside the exploitation area. Together with the landfill situated within the heap (internal heap) it makes main objects subdued to reclamation. Apart from significant formation of the landscape (in large open-cast mines external heaps achieve even over 200 metres of relative height and surface exceeding 25 square kilometres), they are for the long time raw grounds, without soils (lack of micro-organisms, humus substances and other features ensuring the stability of soil environment), additionally built of the material that does not occur on the surface. In some circumstances these undoubted disadvantages of depositing the overburden could be diminished by a thought-out formation of the landscape on the occasion of the necessity to implement large scale soil works connected with open-cast mining.

In case of underground mines the factor moderating the course of the rock mass deformations over the exploitation area is dealing with post-exploitation voids. Possible solutions in this area are: exploitation combined with systematic filling post-exploitation voids with a material of low compression (exploitation with stowing), and exploitation without such preventive measures (exploitation with caving). Filling post-exploitation voids can diminish the volume of exploitation troughs observed in the surface, even up to 90 % compared to exploitation without such measures. Particularly valuable objects, threatened even with smallest surface deformations have to be protected by the use of a proper quality of stowing, or leaving not mined parts of the deposit making a so-called protective pillar.

15.2.3 Treatment of the Mining Product

Very often, in mining making a product that can be sold requires making additional actions increasing the concentration of components desired in the product (by the separation from undesired components or valuable minerals, offered separately, e.g. silver in copper ores). Such processes are carried out in the hard coal mining, metal ores mining (zinc, lead, copper), chemical substances and rock mining. The processes of separating components are technologies applying physical and chemical methods. As industrial procedures they are the source of environmental impact, specific to a given technique. They use energy and media (mainly water), and cause noise emission, emission of dust and gases and release of wastewater making nuisance and threatening the state of soils and waters.

15.2.4 Deposition of Wastes Coming from Treatment

Unfavourable components are separated with methods using physico-chemical processes. Their separation makes the problem of deposition and utilization of large quantities of wastes. The scale of the problem depends on the range of the mining exploitation, concentration of the useful component in the deposit and the properties of the waste product. Depending on the circumstances, they can be utilized or deposited, and the objects where they are stored are subdued to reclamation. The list of possible impact (Adamczyk et al. 2000) of the landfills includes (among others):

1) covering large surfaces, usually excluded from ecological use for many years (the landfill of the wastes after the exploitation of copper ores in Poland cover above 14 square kilometres, landfills of the wastes after the coal mining cover up to several square kilometres, landfills of the wastes after zinc and lead mining are slightly smaller),
2) dispersion of the deposited wastes during their transport and the result of eolic erosion, leading to the load of soils with substances contaminating them (landfills with ore-bearing dolomites in the region of Olkusz caused pollution of several square kilometres of land with the compounds of Zn, Pb and Cd, in the degree significantly exceeding the established standards for soil),
3) washing out by precipitation waters from the deposited wastes polluting ground waters,
4) prolonged lack of soil-forming activities connected with unfavourable properties of the deposited too coarse-grain wastes (from hard coal mining, aggregate mining, mining of zinc and lead), or too fine-grain wastes (copper ores, sulphur and other chemical deposits); also unfavourable chemical properties, e.g. too acid reaction (wastes of copper ores) or alkaline (wastes of zinc and lead ores).

The way to reduce the deposited wastes is their utilization. Unfortunately it is not possible. There are also attempts (bringing good effects) to use wastes as material filling post-exploitation voids in underground mines.

15.2.5 Infrastructure, Supply of Facilities and Energy

Infrastructure is a source of direct impact on environment. It is practically indispensable component of the mining process, including: office objects, social objects, workshops, car-parks, power supply equipment and transport equipment. Its use causes specific types of environmental impact. Transport lines (pipelines, assembly lines, road transport or rail) make potential source of hazard connected with the dispersion of the transferred materials (product, overburden, wastes), especially in emergency situations. The 24 hours cycle of work in mining enterprises, particularly open-cast, can have a specific impact on environment caused by the night artificial light on the workings, heaps and fields of technical services. This light can influence the direction of migration and gathering of fauna (including insects), which makes problems, particularly in the regions of the occurrence of valuable species. It is hard not to notice the fact that the mining enterprise is often a huge receive of different facilities necessary for its functioning. In the places where they are made influence specific to the product is seen. Some of these media (water, fuel) are precious and difficultly renewable or not renewable products.

15.3 Symptoms of Significant Environmental Threat

A general look at the factors of impact allow the conclusion that it is difficult to make one model explaining the functioning of different types of mining. Figure 15.1 and Table 15.1 show the scheme of potential influences of mining.

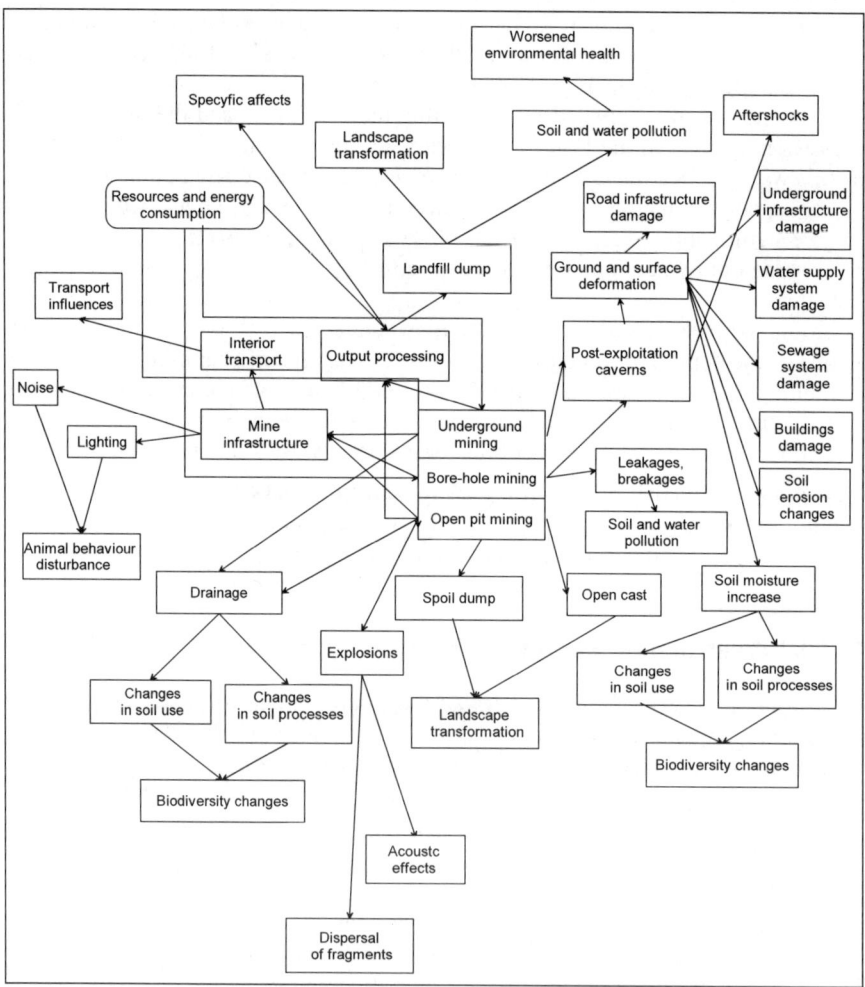

Fig. 15.1. Scheme of potential influence of mining exploitation to the environment

Some of them are caused by technological reasons, which means that their occurrence or degree of influence depends on the technical solutions or applied preventive measures (for instance stowing of underground mines to much degree reduces the deformations of the rock mass and surface). Many consequences depend on environmental circumstances: financially costly damage in the infrastructure is avoided by - if possible – transferring the exploitation under poorly invested areas. Obvious conditions for drying or flooding soils is the occurrence of adequate circumstances: muddy soils or soils characterized by strong sensitivity to the disturbance of relief and its hydrography. Like in other cases of the investments affecting in many directions – the prediction and assessment of their environmental impact requires confronting the probable scale of the impact of respective factors with the sensitivity of the receptors. Naturally, things determining the scale and importance of the transformations are spatial conditions resulting from the situation of the deposit and layout of the objects necessary for exploitation. The difference between mining and other investments is the fact that mining directly interferes in geological structures deciding on the direction of many significant processes affecting the environment.

The key problem in the assessment of the impact of mining investment is practical possibility of quite accurate prediction of the influences and their results. In every case we have two unknowns: uncertainty about the time, place and scale of a factor destructive for the environment, and uncertainty about the reaction of the receptors to the factor. Here also impacts can be forecasted. The degree of the credibility of the forecast in terms of their place, scale, significance and certainty of the reaction of receptors can also vary.

15.3.1 Geo-Mechanical Transformations

Among two categories of geo-mechanical transformations: (1) connected with open-cast exploitation, and (2) resulting form underground or borehole exploitation, the first can be considered to be controlled in compliance with geological conditions and determined by the decisions made at the stage of planning (Kasztelewicz 2004). In this sense, the design decisions can be a subject of optimisation analyses, including environmental criteria. Factors subdued to optimisation can be the following:

- location of overburden heaps, regarding criteria of soil protection and preserving the continuity of the hydrographical network,
- geometric shape of the heaps (horizontal projection, slope of scarps, relative height), regarding border conditions of the predicted use of the object and landscape context,
- the kind of the formation of overburden that, regarding the conditions of predicted use of the object, should make the surface layer of the heap.

A particular kind of problems results from the exploitation of small deposits of aggregates (gravel, sand, stone quarries). Especially quarries, where the properties of rock give the possibility of shaping the scarps with a very big slope angle often make the objects that are very difficult in further management. Reasonable landscape planning should regard all the aspects of forming the final shape of the quarry, adjusting it to the planned in advance function.

Geo-mechanical transformations associating underground and borehole exploitation have to much degree random character (Dżegnuik et al. 1997; Knothe 1984, Kowalski 2001), thus the indicators of the risk of making mistakes in the assessment of the situation are here much greater than in open-cast mining. In various countries different models of rock mass and surface deformation are applied: in Poland the Budryk-Knothe's model dominates, in Germany – Ehrhardt-Sauer's (Ruhrkohle) model, in other countries iteration models based on the theory of finite elements are used (Pielok et al. 1996; Popiołek 1983; Popiołek et al. 1997). According to these models it can be accepted that the size of the deformation observed on the surface and their course depend on:

- size and shape of the post-exploitation void in the rock mass (the void of a greater surface causes the formation of a large subsidence trough of the maximal depth depending on the height of the excavation),
- the way of filling the post-exploitation void (exploitation with caving causes much deeper subsidence trough than the exploitation combined with stowing),
- speed of exploitation (increase of exploitation speed accelerates the revealing of the influences and increases their dynamics harmful for technical objects on the surface),
- depth of exploitation (greater depth of exploitation makes a subsidence trough of a large surface, milder slopes, increasing the range of the influences exploitation in this way),
- rigidity of the rock mass (more rigid rock mass decreases the degree of subsidence, increasing the risk of more radical discharge of stress in the form of seismic tremors).

To assess the risk of occurring significant environmental consequences of the surface deformation quantitative indicators characterising subsidence trough should be compared with the indicators of the level of the allowed impact of a given object. In many cases this comparison comes across difficulties caused by the lack of proper assessment of the sensitivity of the objects (Kwiatek 2002; Ostrowski 1998).

The main indicators of the threat to soils and habitats are cases of the subsidence of the area surface and the angle of the slopes of the trough. The subsidence should be compared to local depths of the oscillation of the water-table. The comparability of subsidence and depth of the situation of the table of ground water is a signal of a significant risk of flooding the grounds, leading to the creation of swamps. The presence of soil types made in the conditions of high humidity makes a significant risk of soil degradation. Angles of the slopes in the trough make the indicator of the threat from floods. The comparability of the angle of the slopes with the fall of streams means the possibility of making swamps. Both types of threat can be successfully predicted with the analyses of the morphology of the area and the distribution of subsidence. Environmental effects of these phenomena, typological changes and qualitative changes in soils, require proper modelling based on analogics, for instance by the application of adaptation algorithms.

The main indicators of the threat to building objects and infrastructure are (Kwiatek 2002; Sroka 1993; Wodyński et al. 2000): slope angle, radius of the trough curvature and deformations of the surface. Slope angle causes the loss of the verticality of constructions, making threat to their durability and causing different nuisance. Appearance of curvatures and deformations (local mutual dislocations of points in horizontal situation) inflicts damage to elements of buildings, roads and other objects. Respective kinds of technical objects are characterized with different resistance to deformations, additionally depending on the construction and materials. The issues of the assessment of the resistance of objects are regulated by relevant national instructions. Dangerous consequences for technical objects can be caused by mining tremors. The basic indicator of the assessment is in this case acceleration of horizontal waves. One can assume that the acceleration of waves exceeding 50 mm/s^2 causes a serious problem.

15.3.2 Drainage

The scale of mining drainage is modelled by empirical formulae and with the application of specialist computer algorithms. The disturbance of natural regime of underground and surface waters can cause significant environmental effects (Wohlrab et al. 1970). On the surface of the area, hydrologic depression in the conditions of sufficiently permeable rock mass can be observed in the form of drying out of certain soils. The basic indicator of the assessment of the threat to soils is the size (depth) of the depression. Its spatial distribution should be compared to the spatial distribution of the areas with shallow table of ground water. It can be assumed that in the soils where the water table, at least temporarily oscillates within the soil profile (i.e. shallower than 1.5 m below ground surface) unfavourable consequences of depression can be revealed. The results of soils dehydration, typological and bonitation effects can be modelled by adaptation methods (Gruszczyński 2000). In general they depend on the size of the changes of average situation of ground water, water management before drainage and retention abilities of soils (grain size, structure, thickness of humus level).

15.3.3 Deposition of Wastes

Effects of the deposition of post-mining wastes depend on properties of the material, way of its transport and the construction of the heap. Apart from direct effects of the deposition of wastes after the enrichment of minerals (taking new areas, influence of the landscape), one can assume that the most significant hazard is their impact as a source of the emission of environmentally harmful substances. Coarse grain wastes (for instance wastes from hard coal mining) are resistant to wind erosion, but chemical compounds are relatively intensively washed out from them. Particularly arduous are repositories of fine grained wastes from the enrichment of ores (e.g. wastes Zn, Pb, Cu). In this case their high fraction compared to the total product is characteristic (85-97 %). They still contain large quantities of harmful for soils and waters components (Trafas et al. 2003; Wójcik et al. 2002) – mainly compounds of the exploited (e.g. Zn, Pb, Cu, Cd). The ways of their impact are eolic erosion and permeation of precipitation waters. The basic procedure of the assessment of hazard is analogy to the existing objects. A model called SESOIL is known. It is used for the modelling of the propagation of these pollutants in soils and underground waters. Alternative modelling tools can show similar usefulness. The level of the reference to the existing or predicted hazard is set by proper standards of water and soil cleanliness. The approach to these standards is differentiated: from the establishment of single threshold values for individual kinds of pollutants and determining the land use by indicating a few thresholds conditioning the undertaking of proper actions (Dutch list), till complex classification rules regarding the level and kind of the contamination, and properties of soils determining the impact to vegetation. For instance, it is assumed that the level of risk connected with the accumulation of heavy metals in soils depends on the texture of soils and their reaction: consistent soils can safely accumulate a higher dose of pollutants, while medium and light soils - lower dose. Acid reaction of soils increases the risk of the accumulation of pollutants in plants. Numerous difficulties in the issue of reliable assessment of the spatial distribution of present pollution with metals and organic substances must be emphasised, as well as their forecasting.

15.3.4 Infrastructure

Infrastructure is a factor that is often neglected as the source of the potential influence on the environment. More thorough analysis, however, indicates that also this factor can be analysed in terms of the hazard to ecological stability. Large masses of the product are connected with the necessity of its transport, at least within the mining enterprise (treatment, enrichment, disposal of wastes). Regardless the way of transport, it is hard to exclude the break-down of the loading or transporting equipment. In this aspect characteristic are for instance bore-hole mines: despite the apparent construction simplicity and easiness to undertake security measures, emergency leakages (gas, oil, salted water, sulphur) very often make threat for soils and vegetation. Communication lines, noise and permanent light can make a very important factor making free migrations of fauna more difficult, indirectly influencing the safety and sustainability of animal populations. Serious arguments against starting the exploitation can be predicted changes in the behaviour of ecologically valuable or rare insect species that would be attracted by necessary (for safety reasons) illumination of open-cast workings, squares and roads.

15.3.5 Seismic Tremors

Mining tremors have a character of seismic phenomena. They arise as a result of rapid translocation, cracking or breaking of the overburden rocks. Strong phenomena of this type include tremors of energy exceeding 10^9J. The distribution of the intensity of tremors is irregular. The main factors conditioning the occurrence of tremors are considered to be: exploitation depth, presence of thick and resistant layers in the overburden, tectonics of the deposit, local strains made by the mining.

In the period of very intensive exploitation carried out in the region of the Upper Silesia, over 16 years nearly 52 000 tremors have been recorded, among them 20 % could be considered strong. Also in the region of the of exploitation copper ores, mining-caused seismic tremors occur: over eight years over 15000 tremors of different strength have been recorded. Over recent years also single tremors in the region of the open-cast mine „Bełchatów" have been recorded. The moment of the occurrence of tremors is not possible to be predicted. The tremors are strongly dependent on local mining and geological conditions and properties of the rocks in the rock mass. It has been observed that the damage to buildings situated on not very condense grounds are more common than in the buildings put on rock basis.

15.3.6 Reclamation

Reclamation is carried out in different kind of post-mining wastelands: final open-cast workings, heaps of the overburden and sterile rocks, repositories of wastes after the enrichment and treatment of minerals, mining areas polluted with chemical substances, agricultural and forest areas changed by the surface deformations or with disturbed hydrology. The liquidation of mines requires the dismantling of equipment, demolishing unnecessary buildings, liquidation of repositories, roads and squares. This means that finishing of mining activities includes different actions tiding up the area and preparing it for new management. The ways of reclamation cannot be generalized: they have to be adjusted to the kind of the transformations and local conditions, including the access to the materials isolating from toxic soils etc. The principle of reclamation include: planning of future management of all the objects, still at the stage of designing, forming the objects taking into account the rules of future management, consecutive reclamation immediately following the primary reclamation. It can be stated that the range of reclamation activities should be the object of planning and the optimisation of the decisions before the start of the exploitation (Chwastek et al. 2000; Gołda et al. 2001; Olschowy 1993, Trafas 2000).

15.4 Mining and EIA

Mining activities (understood in a broad context) are listed among the projects demanding compulsory EIA (Environmental Impact Assessment) procedures. This is the way of assessing the level of potential threat caused by mining of mineral resources, geological surveying of the deposits and deposition of wastes in the rock mass. The first stage where the level, range and possibly the intensity of exploitation-related threat are defined is the application for the licence. The licence to exploit deposits is issued by a governmental body responsible for licensing. The licence can include the obligation of securing all the claims, including the ones connected with the deterioration of the state of environment, if there is an important interest of the state or society. The licence-giving organ, based on geologic documentation and the results of environmental impact assessment, determines the borders of the mining area and mining terrain. The mining area would be the area of carrying out the activities covered by the licence, then the mining terrain is the area extended by the range of harmful influence, unavoidable with the application of the accepted techniques of exploitation. Indirectly it means accepting the influence of exploitation within the borders of the mining area.

Mining exploitation, by its multidirectional influence makes numerous limitations in the management of the terrain. Polish mining regulations impose the duty of making the plan of the management of the area for the mining terrain. Following the intention of the law-maker, the investor has to provide the integration of all the activities undertaken within the mining terrain, including: obtaining licences, providing general security, environmental protection, including the protection of buildings. A traditional way of providing general security and the protection of the elements of environment requiring the preservation of their present state is making protective pillars, i.e. areas excluded from exploitation. Such protective pillars, designed with a proper safety margin, are made to protect particularly precious buildings (cultural monuments, objects of high requirements in terms of ground stability, objects of high reconstruction or repairing costs) and natural objects (nature reserves, nature monuments). Making pillars does not dismiss the licence-holder from the duty to repair the damage connected with exploitation, caused despite the pillars. This also refers to other terrains where there is a decrease of value, loss of profits or decrease of comfort or safety resulting from the carried out exploitation. For decades the problem of Polish mining was the lack of financial resources for the liquidation of mines terminating their activities. Present regulations make a duty for the enterprise holding a licence to have a special fund for the liquidation of the mine. The collected financial resources are used to finance the liquidation of objects, securing the deposit, reclamation of the area, devastated by the exploitation.

15.4.1 Mining and Environmental Standards

Almost all the kinds of mining exploitation make environmental standards likely to be transgressed. Thus limiting the negative impact is one of the most important tasks in planning and carrying out exploitation.

Mining and the Standards of the Cleanness of Environmental Components

The main factors of the impact of mining on the chemical components of environment is production, deposition and disposal of mining wastes. Waste management, including the deposition of wastes, is probably the most important environmental problem in mining. In particular the disasters with landfills cause serious threat. Among them, ecological catastrophes in Aznalcollar (Spain) and Baia Mare (Romania) at the end of 20th century made a stimulus for the document „Safe Operation of Mining Activities" (COM(2000)664). Among numerous recommendations the document emphasizes the necessity to make a reference document on the best available techniques (BAT) of waste management in mining. A proper document (BREF: Reference Document of Best Available Techniques for Management of Tailings and Waste-Rock in Mining Activities) was published in July 2004. It includes multifaceted approach the analysis of all the problems connected with waste management in mining.

The limitation of harmful impact of the operations connected with the management of mining wastes includes decisions regarding: properties of deposited materials, optimisation of the localization of the landfill, securing from short- and long-term influence and the monitoring of the object. The starting point to many subsequent decisions is the establishment of the properties of deposited materials, including: mineral composition, chemical properties, mechanical properties, acidifying or alkalising potential, hydrological features, density, humidity, plasticity etc. These features should also be considered in the context of using the material outside the place of deposition (filling in the excavations, ground construction work) or re-using waters applied in their transport.

Depending on the properties of the deposited materials and local opportunities, a proper method of the protection from spreading pollution should be chosen. The following methods can be applied: keeping the layer of water over the surface of the wastes, covering the wastes with the layer of formations poorly permeable for water and air, constructed wetland, elevation of the level of ground waters to decrease the oxidation of wastes, separation of acidifying factors (sulphides) or segregation of wastes into fractions to be treated separately. Protecting the environment from the emission of harmful substances depends on local conditions, including the conditions of water areas, availability of a good-quality isolating material, level of the ground water.

The problem of leakage from the landfills depends on the localization of the object and properties of deposited materials. An efficient solution is locating the landfill in the area where grounds have possibly low permeability. It is reasonable to cover the bottom of the landfill with loam that is poorly permeable by water;

satisfying permeability is lower than 10^{-8} m/s. Also membranes of sufficiently high resistance are acceptable. In some cases, the limitation of leakage is possible when water used for hydro-transport is sent back. The diminishing of emission to surface and underground waters is the first protective task. The solutions can be: recycling of transport waters to use them in industry, adsorption of water pollutants on specially selected reagents, dilution of polluted waters for recycling, construction of decantation ponds to capture residues, neutralization of waters.

A key question connected with the safety of the surroundings of the repository of mining wastes is related to designing, construction and building of embankments and their monitoring. In designing, it is recommended to take a wide safety margin for securing from natural threats to the durability of the construction. It is suggested, for instance, that 100 years old water should be used as relevant indicator of threat to the embankment of the repositories of low level of threat, and 5000-10000 years old water - for objects of high risk. The monitoring of the repository is a basic duty of the owner. This refers to the stability of the embankment and the movements of grounds in the surrounding of the repository, the efficiency of draining, observation of the symptoms of damage to the object to prevent a catastrophe.

To decrease the impact of the repository it is reasonable to secure it from the inflow of waters from outside, by locating it, if possible, in the excavations after open-cast mining, regarding a proper coefficient safety in designing embankments and slopes, systematic, continuous reclamation of the area excluded from deposition. The termination of the use of waste repositories should be predicted earlier, to make a safety margin for the durability of the elements of the objects (dams, embankments) and planning the reclamation work.

In some cases it is necessary to consider filling the open-cast and underground mining excavations with wastes. It is justifiable for different reasons, one of the most important is „wiping out" the traits of industrial exploitation. Considering the problem of EIA in terms of the necessity of the disposal and utilization of mining wastes, one can give the following ranking of actions, preferred as leading to the minimization of environmental impact:

- minimization of the mass of wastes,
- complete utilization of wastes,
- use of wastes in filling excavations,
- deposition of wastes in places that naturally create conditions for safe deposition, without the emission of components into the environment,
- security measures to eliminate the emission to soils and ground waters.

It should be mentioned that correct construction of a repository does not dismiss from the duty of an on-line monitoring of the emission, imission and safety of technical elements. Observations of the existing objects indicate great difficulties in the remediation of soils and waters changed as a result of neglecting the issues of the protection from excessive emission from wastes.

Mining and Ecosystem Stability

Stable balanced development of ecosystems is conditioned by the stabilization of the conditions shaping the relations between the components of environment. In this case, however, unlike the threats of chemical provenience, it is difficult to establish the level of influence providing the conservation of ecosystem stability. Determination of accepted border levels of the changes in the morphology and hydrological conditions depends on the present state of the management of the area. The observed trend involves the resignation from the exploitation under the urbanized areas, in favour of the areas such as fields, meadows, forests. It is caused by quite a big difference between the costs of the reclamation of the two kinds of the area. The assessment of losses in natural areas is extremely complicated, while formal individual costs referring to the prices of the reconstruction of buildings or infrastructure are incomparable. All this makes the question of the stability of ecosystem possible to be analysed only in terms of the quality (bonitation) of the habitat. Possibly explicit defining the duty to preserve the stability of ecosystem, rather than repair the damage (which is obvious) would rise the requirements to technological rules of exploitation (filling the excavations, protective pillars).

15.5 Conclusions and Recommendations

Mining, as a part of economy, as a matter of fact, has multifaceted impact on environment. There is a big list of its impacts treated as unavoidable, which means their acceptation already on the stage of early planning of mining enterprises. Thus the comparison of environmental impact (see Table 15.1) with the accepted environmental standards, particularly within the EIA procedure should be made.

Table 15.1. Factors of impacts of mining with the assessment of the possibilities of their reliable forecasting

REASON	I-level effect	Condition of attendance /Prediction	Countermeasure	Further effects	Condition of attendance/Prediction	Countermeasure
Post exploitation caverns (under ground and bore-hole mining)	After-shocks	Uncertainty; unpredictable	Fundamentally not allowed, alternatively: seismographic observations	Infrastructure damage	Poor technical quality of infrastructure; unpredictable	Reinforcement of infrastructure
	Ground and surface deformation	Always; predictable: many models (Budryk-Knothe, Bals)	Exploitation (direction, depth, speed) control; cavern infill, protective pillars	Infrastructure damage	Poor technical quality of infrastructure; predictable	Infrastructure accommodation
				Soil moisture increase	Dense hydrographic network; swampy soils; clay soils; predictable: simulation models, deductive models, GIS	Levees; drainage; modification in soil use
				Changes in soil processes	Material alteration of air-water relation; poor predictable: assumption	Drainage

Table 15.1. (cont.)

REASON	I-level effect	Condition of attendance /Prediction	Countermeasure	Further effects	Condition of attendance/Prediction	Countermeasure
				Biodiversity changes	Substantial habitats changes	Exploitation abandonment, security pillars
Leakages, breakages (bore-hole mining)	Soils and water pollution	Uncertainty; unpredictable; eventually: monitoring of infrastructure	Reinforcement of transport infrastructure	Changes in soil processes: salinization, acidification, alkalisation, PAH and heavy metals overload	Weak buffered soils (acidification, alkalisation), overloading of natural cleaning capacity (salinization, PAH and metals overloading)/ Predicted by analogy; GIS models: Sesoil, AT123D, Waterflow	Soil maintenance in good culture, remediation
				Changes in hydrological processes: salinization, acidification, alkalisation, PAH and heavy metals overload	Good soil permeability. Predictable: Sesoil, AT123D, Waterflow	Soil remediation
Spoil dump (open pit mining)	Landscape transformation	Thick spoil layer; predictable: geological investigations	Interior spoil dump, earth building	Changes in water circulation, changes in soil use	Always. Construction design	Good environmental planning, land reclamation
Open cast	Landscape transformation	Always. Predictable: geological investigations	Interior spoil dump.	Water reservoir in open cast after the exploitation	Bottom of open cast below water table. Predictable: hydrological investigations	Drainage
Drainage of deposit (open pit mining and underground mining)	Draw-down	Deposit below of the water table. Predictable: hydrogeological investigations	Impermeable barrier around open pit: possibly in some cases	Soil drainage, changes in soil processes. Soil destruction. Biodiversity changes	Permeable spoil, shallow water table in soils: some soil classes. Predictable: deductive and adaptive methods	Land reclamation, soil hydratation
Explosions (open pit mining and underground mining)	Shocks	Hard spoil or/and deposit. Big pyrotechnics	Pyrotechnic decreasing, another exploitation method	Building damages, lifestyle deterioration	Weak building construction. Predictable: physical models	Reinforcement of buildings.
	Fragments scattering (dispersion)	Hard spoil or/and deposit. Big pyrotechnics	Pyrotechnic decreasing, another exploitation method	Danger for health and life	Bad computation of pyrotechnic. Predictable: physical models	Protecting zone.
	Acoustic effects	Hard spoil or/and deposit. Big pyrotechnics	Pyrotechnic decreasing, another exploitation method. Anti-acoustic barrier	Lifestyle deterioration, changes in animal behaviour	Bad computation of pyrotechnic. Predictable: physical models	Protecting zone.
Output processing	Resources an energy consumption	Always. Technical project.	Best available technical apply	Specific for technology	Specific for technology	Specific for technology
	Landfill dump	Big participation of wastes in deposit. Predictable: geological investigations.	Richest deposits exploitation	Destruction (coverage) of soils	Always. Technical project	Landfill dump optimisation

Table 15.1. (cont.)

				Soil pollution (wind erosion). Risk for health	Fine-grained wastes (include pollutant), absence of plant coverage. Predictable: chemical and physical investigations of wastes	Land reclamation, artificial coverage of landfill dump surface
				Water pollution (pollutants leaching). Risk for health	Absence of barriers, permeable ground. Predictable: wastes investigations	Barriers. Land reclamation
				Landscape transformation	Always. Technical project	Good project
Mine infrastructure	Interior transport	Technical requirements (spoil and output transport)	Mine project optimisation	Noise	Transport methods. Predictable	Shielding of routes
				Scattering of material	Transport methods. Predictable	Shielding of routes
	Noise	Many technical devices. Predictable	Best available technical apply			
	Lighting of places, routes and open casts	Conditions of safety	Absence	Changing in animal behaviour	Rare species attendance. Predictable: biological investigations and methods	Absence

Probably the most serious threat for environmental standards are factors connected with collecting, utilization and deposition of wastes. Strict procedures and regulations of the management of mining wastes were made. The main task in this field is the isolation of wastes from potential pathways of the propagation of the pollutants contained in these wastes. The morphology of the terrain and hydrological conditions make threat for the stability of ecosystems and the direction of the development of existing ecosystems. On the other hand, the far from perfect system of the valuation of costs of such phenomena, encourages moving exploitation from urbanized areas, where the damage is easy to asses and relatively high, to the areas such as forest, meadows, fields, the economic value of which is officially much lower.

References

Adamczyk Z, Motyka J, Witkowski AJ (2000) Impact of Zn–Pb ore mining on groundwater quality in the Olkusz region. 7th International Mine Water Association Congress : on Mine Water and the Environment': Katowice–Ustroń, Poland 11–15 September 2000: proceedings / eds. Andrzej Różkowski, Marek Rogoż. [Katowice : The International Mine Waters Association, Uniwersytet Śląski, 2000]. Earth Sciences pp. 27–37

Batko P, Modrzejewski S (2001) Modern blasting agents and getting technique in open-cut mining (in polish) Przegląd Górniczy. t. 57 nr 4 s. 14–21, Katowice, Poland

Chwastek J, Gruszczyński S, Trafas M (2000) The use of post-mining areas and objects for recreation and leisure 11-th international congress of the International Society for Mine Surveying : Cracow, September 2000. Vol. 2, Katowice: Zarząd Główny SITG

Dżegniuk B., R.Hejmanowski, A. Sroka, 1997: Evaluation of the Damage Hazard to Building Objects on the Mining Areas Considering the Deformation Course and Time. Xth International Congress of the International Society for Mine Surveying, Fremantle, Australia

Gołda T, Uberman R, Kozioł W, Naworyta W (2001) The problems connected to utilization of mine infrastructure and reclamation of terrain after liquidation of hard coal mines (in polish) Warsztaty 2001 nt. Zagrożeń naturalnych w górnictwie : materiały sympozjum : sesja okolicznościowa: Przywracanie wartości użytkowych terenom górniczym : sesja specjalna: Stare kopalnie – nowe perspektywy. Polish Academy of Sciences - Sympozja i Konferencje, Wieliczka

Gruszczyński S (2000) Soils Transformations Consequences Simulation on Mining Terrains with Application of Neural Classifiers (in polish), AGH-University of Science and Technology, Kraków

Kasztelewicz Z (2004) Polish brown coal mining (in polish). Górnictwo Odkrywkowe, Wrocław

Knothe S (1984) The forecast of the affects of the mining exploitation (in polish). Wydawnictwo "Śląsk", Katowice

Kowalski A (2001) Surface subsidence and rate of its increments based on measurements and theory. Archiwum Górnictwa No 46, Katowice

Kwiatek J (2002) Buildings on mining terrains (in polish). Główny Instytut Górnictwa, Katowice

Olschowy G (1993) Bergbau und Landschaft. Verlag Paul Parey, Hamburg-Berlin

Ostrowski J (ed) (1998) Protection of environment on mining terrains (in polish). Szkoła Eksploatacji Podziemnej, Kraków

Pielok J, Piwowarski W (1996) Deforamtions of the rock mass – formal definitions and physical references. Proceedings of the Symposium "Investigations using Geodetic Techniques", Szakesfeherwar (Hungary)

Popiołek E (1983) Probabilistic Method of Verification od Prognosis of Post-Mining Deformations of Terain Surface, Polish Academy of Sciences, Geodesy 30, Kraków

Popiołek E, Ostrowski J (1997) Estimation of Mining Area Hazard in the Light of Accuracy of Terrain Surface Deformation Caused by Underground Exploitation. X-th International Congress of the International Society for Mine Surveying, Fremantle, Australia

Sroka A (1993) Zum Problem der Abbaugeschwindigkeit aus bergscheidenkundlicher Sicht. Polish Academy od Sciences, CPPGSME, Kraków

Trafas M (2000) Documentation of the state of environmentally valuable areas in case of mine liquidation summary (in polish). Górnictwo zrównoważonego rozwoju. Konferencja 5: Ochrona środowiska naturalnego Wydawnictwo Politechniki Śląskiej, nr 1480. Seria Górnictwo ; z. 246), Gliwice

Trafas M, Eckes T (2003) The studies on a vertical distribution of the contents of heavy metals in soil profiles (in polish) Inżynieria Środowiska t. 8 z. 2 s. 207–216. Kraków

Wodyński A, Kocot W (2000) Analysis of technical wear of traditionally structured buildings on mining areas. Proceedings 11th International Congress of the International Society of Mine Surveying, Cracow

Wójcik J, Kowalik S (2002) The evaluation of the efficiency of a forest-type land reclamation of the waste rocks heaps in the Copper Mining Enterprise Inżynieria Środowiska. t. 7 z. 1 s. 51–65, Kraków

Wohlrab B, Bahr R (1970) Zur Entstehung von Bergsenkungsgebieten un ueber ihre Flaechentwaessurung. Schriftenreihe der Landesanstalt fuer Imisions- und Bodennutzungchutz des Landes Landes Nordrhein-Westfalen, Essen

Part II – Thresholds and Standards for Different Types of Projects

Part IIb – Thresholds and Standards for Spatially Dispersed Projects

Part IIb presents three cases of spatially dispersed developments: tourism, urban and industrial settlements and forest management. These distinguish to site-specific developments of Part IIa by their large spread land use, which can be found world-wide.

Chap. 16 by Buckley gives an insight into conflicts and interrelations between commercial tourism developments, housing and protected and recreational areas in Australia. Overlaps in these land uses may lead to some uncertainty in triggering thresholds for tourism EIA and may influence assessment of cumulative impacts. In Australia, World Heritage listing can be a significant triggering factor for tourism EIA.

Storch and Schmidt Chap. 17 concern the proceeding land consumption in Germany despite the national 30 ha target to be achieved by 2020. They suggest improved control of the efficiency of land-use at the regional planning level by using the potential of strategic environmental assessment. At the higher planning level definitions on the proposed type and scale of residential areas should be more specific and detailed than at present including for instance residential density, area size and designations on former built-up sites.

Chap. 18 by Nunoo and Schmidt delivers standards and thresholds for sustainable use of forest resources with the case study of Ghana. The country adopted a forest product certification scheme with a computerised log tracking system to monitor timber harvesting, processing and export. In the future better information and communication is needed to compromise between interests among stakeholders with regards to resource ownership, management responsibilities and disbursement. The author therefore recommends intensive environmental education and campaigns and the active involvement of forest communities in forest resource management.

All three chapters show that environmental objectives and standards are essential for spatially dispersed projects in combination with public participation and awareness raising, if the overall goal of preventive and sustainable development shall be met and a long-term high living quality of humans shall be guaranteed in the future. SEA has got the potential to detect cumulative and indirect impacts of spatially dispersed projects, which will need more attention in the future of EA.

16 Thresholds and Standards for Tourism Environmental Impact Assessment

Ralf Buckley

International Centre for Ecotourism Research, Griffith University, Australia

16.1 Introduction

Tourism is a highly heterogeneous industry sector, and different environmental planning tools are applied at different scales and in different jurisdictions. In most countries only certain components of the tourism industry, and particular types of tourism development, are subject to project-scale environmental impact assessment. Precisely because of its diffuse distribution and variable scale, tourism can provide a useful tool to test the effectiveness of EIA systems. Tourism can also illustrate the dilemmas involved in designing EIA systems which are both effective and efficient, in the sense that they require just enough environmental information, commensurate with the scale of each individual proposal, to make well-considered development control decisions. Currently, there are many cases where identical tourism development proposals in adjacent legal jurisdictions would yield very different EIA requirements (Warnken and Buckley 1995, 1996). This is perhaps an indication of how difficult it can be to set thresholds and standards for tourism EIAs.

Tourism development is often characterized as geographically diffuse and this can indeed be the case (Warnken and Buckley 2000). This, however, is not necessarily a major distinction from other industry sectors. The extractive industries sector, for example, includes scattered sand and gravel quarries as well as billion-dollar mega-mines and oil fields; and the manufacturing sector includes small-scale cabinetmakers and surfboard shapers as well as global car, clothing and appliance suppliers.

Perhaps more significant is that commercial tourism development is closely linked with residential housing, transport infrastructure, and often also with public utilities such as power and water supply, and sewage and other waste treatment systems. In most regions, these are subject to specialized development control regimes, which operate in parallel with project-scale EIA for primary and secondary industries.

Standards and Thresholds for Impact Assessment. Edited by Michael Schmidt, John Glasson, Lars Emmelin and Hendrike Helbron. © 2008 Springer-Verlag

Developments in those sectors also require both worker housing and utilities, but they are not linked to large-scale residential development, and indeed permanent population migration, in the same way as tourism.

In addition, a significant proportion of the tourism industry is concentrated in areas of high conservation value, in and around public national parks and wilderness areas, private nature reserves and other protected areas. Since these areas are allocated principally for conservation, thresholds and criteria for EIA should be more stringent than elsewhere. In some countries, certain other types of industrial development may also be permitted within protected areas. The main examples are mining, oil and gas production, and public infrastructure such as power lines and sometimes transport corridors. These, however, are built despite rather than because of the area's conservation value; and then only if the economic returns are sufficient to fund major mitigation measures.

The issues considered in this chapter apply worldwide but the examples are taken principally from Australia. Australia is a federated nation with three tiers of government. Planning and development control at local scale is carried out principally by local government authorities (LGAs), which have elected Councilors but whose legal powers are delegated from State governments. The State Governments are responsible for regional-scale planning, pollution control, protected areas, and most large-scale public infrastructure. Each State has its own project-scale EIA legislation. The federal government has its own environmental laws, including EIA law, but their application is limited by the Australian Constitution, which grants the Australian Government specific heads of power. The federal government is generally only involved in tourism EIA when one of Australia's 14 World Heritage areas could be affected.

16.2 EIA for Different Types of Tourism

Table 16.1 summarizes some of the issues associated with environmental assessment for tourism development and activities at various scales. At the largest scale, namely tourist towns, there may be no EIA because tourist accommodation, facilities and infrastructure are so completely integrated with those of local residents. On the Gold Coast in Australia, for example, high-rise residential apartment buildings are largely indistinguishable from adjacent high-rise hotels (Buckley and Araujo 1997; Warnken et al. 2003). In addition, there are many building and apartment complexes, both high-rise and low-rise, which are held under strata title and where individual accommodation units may be occupied by their owners, leased by their owners as holiday accommodation, or operated as hotel accommodation by the manager of the complex. In tourist towns, therefore, project-scale EIA is commonly required only for very large-scale development proposals which are not catered for under current urban development control plans. This occurred recently on Australia's Gold Coast, for example, when a large-scale cruise ship terminal was proposed.

Table 16.1. Types of tourism development in an EIA context

Tourism Types	Categories	Impact Issues	Planning Approach	Triggers and Thresholds	Approval Criteria
Tourist town accommodation	Mountain Rivers Coastal Arid	Tourist accommodation similar to residential, often mixed, load on urban utilities, local ratepayers	Urban planning, LEPs, large-project EIA	Scale of development triggering project EIA; cumulative loss of environment; boundary effects, fragmentation	Criterion = fit with DCP; project EIA rare; low data requirements
Greenfields resort residential	Mountain (eg ski ± golf; coastal (eg marina-based)	Roadworks, sewage treatment, infrastructure corridors, water pollution, habitat clearance, noise, weeds, feral animals	Standard project EIA as for any high impact industrial development	Zoning changes, prior land-use, baseline for EIA	Criterion = balance between economic gains and environmental costs
Cumulative holiday development	Ski dormitory towns, amenity migration areas, lakeshore and residential, coastal strip	Cumulative impacts, groundwater pollution, vegetation clearance, weeds, pets, ferals	Planning, zoning, REA/LEP, LGA DAs	Cumulative-impact trigger for project EIA, incorporation in planning/DA procedure	Criterion = cumulative impact < regional threshold, but poorly defined standards, processes
Single-activity infrastructure, private	Lodges, boat docks, cableways, equestrian and off-road areas, airstrips	Access, impacts (as above) on neighbouring land areas	Project EIA or none	Triggers depend on zoning; synergistic impacts	Criterion = little or no impact on neighbours. Small-scale low-tier EA as part of DA.
Infrastructure in parks, publicly-owned	Roads, tracks, lookouts, boardwalks, shelters, toilets, visitor centres	Secondary impacts, visitor management	Part of parks agency operations	Less scrutiny than for private developments	Criterion = reduce net impact on conservation: EIA tests if impact of infrastructure < impact of visitors without infrastructure
Private or concessionaire infrastructure in or near parks	Lodges, huts, camps, buses, snowcats, ski areas, boat moorings, pontoons, climbing anchors, shops	Impacts on park and buffer zones; enclaves; management; tenure	Project EIA with more stringent evaluation	Special conditions, responsibilities; lower thresholds for EIA in parks	Criterion = no net impact on park. Difficulties in legal application of criterion outside park boundary
Private or concessionaire activity in public parks	ORVs, bikes, horses, snowmobiles boats; guided hike, ride, climb, abseil	Increased visitor numbers, high-impact activities, net increase in impact on parks	Permitting by parks agency, may involve limited EA	EIA rare, but permit conditions can cover activity, site management	Criterion = activity permitted has no greater impact than aggregate impacts of existing individual visitors

DA development approval ;
EA environmental assessment (general process)
DCP development control plan

LEP local environmental plan
LGA local government authority
REA regional environmental assessment

EIA environmental impact assessment (legislated procedure)

At the other extreme, many types of tourism infrastructure and activity are at too small a scale to trigger EIA unless they happen to be in or adjacent to a designated protected area; and sometimes not even then. Such activities may still have significant conservation impacts, and depending on land tenure and legislation they may still be subject to some form of environmental assessment and to environmental conditions in operating permits, but they are rarely subject to project-scale EIA.

Large-scale greenfields tourism developments, commonly involving a mixture of activities, infrastructure and residential accommodation, are indeed generally subject to project-scale EIA which is directly analogous to that for any large industrial development such as a mine or manufacturing facility. EIA is triggered under planning law because for these greenfield sites, a large tourism or resort-residential development represents a major and material change in land use. As with major-project EIA for any sector, the purpose of the procedure is to provide adequate and accurate information on the probable environmental costs of the proposed development. It is then up to government decision makers, generally at national or subsidiary-state rather than local level, to consider these costs in conjunction with the likely economic benefits in order to decide whether or not the project should proceed.

In practice, there appears to be only one systematic study of the technical quality of tourism EIA, carried out for Australia a little over a decade ago (Warnken and Buckley 1998). That study showed that the standard of scientific information provided in tourism EIA at that time was rather uniformly poor. A subsequent comparison of EIA documents specifically for ski resorts in Australia and the USA (Buckley et al. 2000) found that ski-resort EIA documents from the USA, principally Colorado, were more detailed than those from Australia. The US documents, however, were all for more recent developments or extensions, whereas the Australian ones were older. Hence it was not possible to determine whether the difference is due to the date or the country. EIA legislation in the two countries is very similar; and in both countries, the ski resorts are in or adjacent to areas otherwise set aside for conservation.

In some cases and some countries, major greenfields tourism developments have escaped EIA entirely through special enabling legislation which exempts them from some or all of the state or country's planning and pollution control laws. This approach was adopted some decades ago for large coastal tourism developments in both east-coast and west-coast Australia, for example (Buckley, 1979). Currently it is much less likely, because of potential political repercussions; but by no means impossible.

The remaining categories of tourism development listed in Table 16.1 illustrate two issues which are particular to the tourism industry. These are considered in turn in the following sections.

16.3 Holiday Housing Clusters

When a single proponent seeks planning approval for a major resort-residential development, the thresholds and standards for EIA in any given jurisdiction are relatively straightforward. Certainly, there may be significant differences between different jurisdictions. Warnken and Buckley (1996), for example, comparing two neighboring Australian States, found that tourism developments were much more likely to trigger State-level EIA in New South Wales than in Queensland. Because of a third-party right of appeal in the NSW legislation, and a specialized Land and Environment Court, issues such as, e.g., the scale of expansion of an existing development which would trigger a new EIA had been established through litigation.

If a residential area of corresponding overall scale is constructed piecemeal, however, through a series of independent applications for individual holiday houses, then State-level EIA is very unlikely to be triggered. Roads, power, water supplies, sewage and garbage collection services are built and operated by government agencies, whether local or State, and brought to the boundary of each individual block. Such areas, whether used for holiday homes or amenity migration, commonly start out rather sparsely settled, with individual generators or gas supplies, rainwater tanks or dams, and septic tanks for sewage treatment. Only the areas immediately surrounding each individual house are likely to be cleared.

As such areas become more popular, however, and residential density increases whether through subdivision, multiple occupancy or other mechanisms, cumulative environmental impacts can become highly significant. Habitats once of significant conservation value become heavily fragmented through access roads; largely cleared for houses, yards, fire breaks and/or gardens; broken up by fences; and increasingly invaded by weeds brought in on vehicles or as garden escapes. Native wildlife are decimated by road kill and/or attacks from pet dogs or cats, disturbed by noise and light, and indirectly affected by changes in food supplies, competitive interactions, etc. In areas with skeletal soils overlying shallow bedrock, a proliferation of septic tanks can soon lead to groundwater pollution, with associated impacts on downslope watercourses. This issue has been raised, for example, in relation to residential development on a plateau in the hinterland and watershed of Australia's Gold Coast.

Project-scale EIA would not be an efficient tool to address such cumulative impacts (Buckley 1997). The marginal impacts of each new house are limited, and it would probably not be either feasible or equitable for each new development application to consider the cumulative impacts of all the houses which have already been built. Environmental planning in such circumstances is the province of the relevant local, or occasionally State government authority. Such planning processes can often be highly contentious. Existing residents who have bought property for its amenity value are likely to oppose continuing development, whereas those who have bought property as an investment are likely to be in favour of it.

Attempts to distinguish in legal terms between permanent residential housing and tourist accommodation are fraught with difficulty. There are so many different designs to which houses can be built; so many different combinations of housemates either related or not; and so many different reasons and patterns by which people may need or want to move away temporarily from their principal residence, and allow or invite others to live there during the periods concerned. There has recently been extensive controversy in Byron Bay in northern NSW, Australia for example, over the practice of holiday letting, where residents move out of their houses during peak holiday periods and rent them out at high prices to visiting holidaymakers. Neighbors complain that this leads to excessive noise disturbance from late-night partying by holiday revellers. This could well be true; but trying to ban holiday letting will not be a workable solution.

At a larger scale, the social divisions and conflicts which can occur when a rural farming area becomes popular for holiday homes and later for amenity migrants have been explored in some detail for areas such as Greater Yellowstone in the USA (Johnson et al. 2003, 2004) and elsewhere (Moss 2006). Amenity migration is not tourism, but it is closely related. Indeed, in many areas tourism can be seen as a transitional economy between the production of agricultural commodities, and the production of creative knowledge. To apply project-scale EIA to such a gradual social and land-use change is clearly problematical.

16.4 Tourism and Protected Areas

One particularly critical and contentious issue for tourism development in many countries is its relationship with the conservation estate. Different protected areas have been established historically with different goals, even in the same jurisdiction. For the majority of parks and similar areas worldwide, however, recreation is seen either as core function equal to conservation; as a legitimate additional use, or as an unavoidable necessity to maintain political support for the protected area estate. Conflicts between recreation and conservation in public protected areas, and indeed between different forms of recreation, have a very long history and are heavily researched (Buckley 2003, 2006; Hendee and Dawson 2002; Eagles and McCool 2002; Pigram and Jenkins 2006).

Commercial tourism in and around protected areas ranges greatly in scale. At one extreme are small-scale commercial tours, where people pay guides to equip and assist them in carrying out the same kinds of recreational activities, at similar scales, as are carried out by individual independent visitors to the park. Such small-scale tours are different in a legal sense from independent visitors, and commonly require permits from the protected area management agency (PAMA), but from an environmental perspective they can be managed in the same way as other visitor activities. These are the types of tourism listed in the lowest row of Table 16.1.

Commercial tourism in and around protected areas, however, is by no means limited to these low-key guided activities. The PAMAs themselves build a wide range of visitor management infrastructure, and since a large proportion of park visitors are tourists rather than local residents, this infrastructure is itself part of the tourism industry. Many PAMAs also have a routine system where they grant concessions to private individuals or corporations to operate particular activities or facilities within public protected areas. These facilities may be built and owned by the PAMA, but leased to a concessionaire to operate, as in the case of campgrounds in many US national parks. Alternatively, they may be funded and built in part or in whole by the concessionaire, as in the case of the visitors' centre and glacier bus terminal in Jasper National Park, Canada (Buckley 2004).

These are large-scale tourism facilities which are built inside a protected area because it is the protected area which provides the primary tourism attraction. Environmental impact assessment for such facilities therefore needs to be triggered at a lower threshold, and evaluated with a different criterion, than EIA for corresponding projects on private land outside the public protected area estate. Tourist infrastructure inside parks is there to help the parks agencies manage visitors (Buckley 2002a). From a commercial perspective, clearly the concessionaire will not be interested in such an opportunity unless it is profitable. From a public policy perspective, however, the key function of the public protected area estate is to contribute to global conservation of biodiversity, ecosystem function, and clean air and water. This is a far more significant role than contributing to regional economies, for example.

For project EIA on private land, the usual criterion for governments to assess the proponent's proposal is to balance predicted environmental costs against potential economic gains at a local or regional level. This is also the criterion used for large-scale developments which are unavoidably located within protected areas, such as mining or oil production. Whether or not such developments are permitted within parks differs from country to country, and indeed from government to government within the same country, but some overriding national interest must generally be invoked before permission for any such development is ultimately granted. Only major precious-metal mines or oil and gas facilities, for example, would commonly be considered, not quarries or coalmines. In Australia's Great Barrier Reef Marine Park, oil exploration has not been permitted by governments of any political colour. In the Arctic National Wildlife Reserve in Alaska, USA, proposals by the Bush administration to permit oil exploration proved extremely controversial.

Similar considerations apply for infrastructure corridors. Electricity corporations, both public and private, do quite often propose high-voltage transmission lines across national parks, because for them it is much cheaper than going through private land. Only rarely, however, are such developments permitted, and then under much more stringent environmental conditions than would be required in other land tenures. In addition, development proposals within protected areas are generally subject to much more intensive EIA than for those on other public or private land, with strong public involvement at all stages, and a much higher likelihood of top-tier approaches such as judicial or parliamentary commissions of inquiry, yielding information of much higher technical quality (Buckley 1979).

All of the above applies equally for large-scale tourism development proposals within or immediately adjacent to public protected areas. That is, they would generally be subject to more stringent evaluation both in terms of triggering threshold and technical quality, than if they were proposed on private land. There are two additional considerations, however, which are specific to tourism. The first is that the tourism sectors, and particularly government tourism agencies, often argue that tourism deserves special privileges in relation to the use of public protected areas. This argument is illogical and incorrect, but nonetheless reappears frequently. It takes various forms (Buckley 2003). At the crudest level, it is sometimes suggested that since historically, governments allocated large areas of land for other industry sectors such as forestry and agriculture, tourism deserves a corresponding allocation and the only land left is the national parks. Slightly more sophisticated is the argument that many public protected areas, including World Heritage sites, were established specifically for recreation as well as conservation. This is indeed correct, but it refers to private individual recreation, and does not confer any particular rights for commercial tourism development. Most recently this argument

Table 16.2. Queensland tourism development in or near protected areas and subject to EIA

Development	Features	Protected Area	Relative Location	EIS?	Built?	Special Circumstances
Hinchinbrook Harbour	Resort-residential and marina	Great Barrier Reef WHA, Queensland Wet Tropics WHA	Adjacent to both	Yes	Yes	In narrow corridor between two World Heritage Areas
Kingfisher Bay	Resort-residential and ferry	Fraser I./ Great Sandy WHA	Enclave	Yes	Yes	Stringent environmental conditions
Couran Cove	Resort-residential, ferry and	Stradbroke I. NP	Enclave	Yes	Yes	Harbour already excavated decades ago
Skyrail Cableway	Cableway and three visitor cen-	Queensland Wet Tropics WHA	Within	Yes	Yes	Part of package to convert logging to conservation
Naturelink Cableway	Proposed cableway	Springbrook NP, CERRA* WHA	Within	Yes	No	Critical differences from Skyrail, higher social and environmental costs
O'Reilly's Mountain Bowers	Residential addition to existing lodge	Lamington NP, CERRA* WHA	Enclave	Yes	Yes+	Enclave pre-dates park; WHA triggered EIA
SE Qld Great Walk	Walking track extensions and upgrade	Lamington and Border Ranges NP, CERRA* WHA	Within	Yes	Yes+	WHA triggered EIA; several endangered frog species at risk

*Central Eastern Rainforest Reserves Australia.   NP National Park.
+Approved and under construction 2006.   WHA World Heritage Area.

has reappeared in a third form, using the terminology of partnerships (Buckley 2002b, 2004). The critical issue is that protected areas are there for public good, not private profit. From the policy perspective, commercial tourism in protected areas should be managed for public good, through contributing either to recreational opportunities, to visitor management, or to financial support for conservation (Buckley 2002a). That is, the critical criterion for evaluating development proposals in protected areas is quite distinct from the criterion used to evaluate a corresponding proposal on other land tenures.

In addition, it is enormously more difficult and expensive to restore pristine native ecosystems than it is to restore anthropogenic agricultural or other primary-industries landscapes. If impacts in protected areas reduce populations of particular species below minimum viable population size, restoration may be impossible irrespective of expenditure. Therefore, the threshold minimum scale of development for which EIA is required within a protected area should be significantly smaller than for other land tenures, and in practice this is indeed the case.

Some examples of formal EIA carried out for tourism development projects within or immediately adjacent to public protected areas in Queensland, Australia, are summarized in Table 16.2. For the larger resort-residential developments at the upper end of the scale, project EIA would probably have been required irrespective of land tenure. The same probably applies for the two cableways. The relatively small-scale residential development at O'Reilly's, however, would not have required project EIA except that it is in an enclave within a World Heritage area. Similarly, the walking track construction works required under the South-East Queensland Great Walks Project would not have required EIA if they were outside a protected area. Indeed, if they were wholly within a Queensland National Park which was not also World Heritage, they would have been treated as part of routine park maintenance. The Great Walks Project was subject to formal EIA for two reasons. Firstly, because it traversed a World Heritage area for which the federal government also had responsibility and requirements; and secondly, because part of the track, accessed from Queensland, is in fact within a designated wilderness area of New South Wales, and there were concerns about liability as well as conservation.

16.5 Conclusions and Recommendations

Environmental impact assessment for tourism is broadly similar to EIA for any other major industry sector. There are two principal differences. Firstly, there is considerable overlap between tourism and residential development. Residential development has a rather different planning and development-control regime from industrial and major-infrastructure development. The overlap between tourism development and residential development can hence lead to some uncertainty in triggering thresholds for tourism EIA. This issue is particularly critical where cumulative impacts accrue from a series of small residential-style developments, especially where these may be used interchangeably for commercial tourism, holiday homes, or long-term residents. This may apply both for dispersed development in areas of high recreational amenity, such as coastlines and mountains, lakes and rivers; and secondly, in tourist towns where the attractions include casinos, nightclubs and theme parks rather than natural features.

A second distinguishing characteristic of tourism development is that a significant proportion of tourism developments are focused in and around national parks and other protected areas, which provide attractions for scenic and nature-based tourism. The primary purpose of these protected areas is to provide the public good of biodiversity conservation. Since the natural environment in these areas is relatively undisturbed, a relatively small anthropogenic stress can create a relatively large environmental response. In addition, they are specifically selected for their high conservation value, often including rare and endangered plant and animal species; and it may be difficult or impossible, and commonly very expensive, to reverse any unanticipated impacts. The threshold for triggering EIA is hence lower in protected areas than in other land tenures; the degree of scientific detail required is higher; and the criterion for evaluating development applications is different. The extent to which these distinctions are followed in practice depends on applicable legislation in the country or jurisdiction concerned. In particular, it appears that at least in Australia, World Heritage listing can be a significant triggering factor for tourism EIA.

References

Buckley RC (1979) Precision in Environmental Impact Prediction. Australian National University, Canberra

Buckley RC (2000) Strategic environmental assessment. Impact Assessment & Policy Appraisal 18:209-215

Buckley RC (2002a) Draft principles for tourism in protected areas. Journal of Ecotourism 1(1):75-80

Buckley RC (2002b) Public and private partnerships between tourism and protected areas. Journal of Tourism Studies 13(1):26-38

Buckley RC (2003) The practice and politics of tourism and land management. In: Buckley, R.C., Pickering, D. and Weaver, D.B. (eds) Nature-Based Tourism, Environment and Land Management. CAB International, Wallingford UK, pp. 1-6

Buckley RC (2004) A Natural Partnership, vol 2. Innovative Funding Mechanisms for Visitor Infrastructure in Protected Areas. TTF Australia, Sydney

Buckley RC, Araujo G (1997) Environmental management performance in tourism accommodation. Annals of Tourism Research 24:465-469

Buckley RC, Pickering C, Warnken J (2000) Environmental management for alpine tourism and resorts in Australia. In: Godde, P.M., Price, M.F. and F.M. Zimmerman, FM (eds.) Tourism and Development in Mountain Regions, 27-45. CABI, Wallingford

Eagles PFJ, McCool SF (2002) Tourism in National Parks ands Protected Areas. CABI, Wallingford UK

Hendee JC, Dawson CP (eds) (2002) Wilderness Management (3rd eds.) Fulcrum, Golden CO

Johnson J, Maxwell B, Aspinall R (2003) Moving nearer to heaven: growth and change in the Greater Yellowstone Region, USA. In Buckley RC, Pickering CM, Weaver DB (eds) Nature-Based Tourism, Environment and Land Management, 77-88. CABI, Wallingford, UK

Johnson J (2004) Impacts of tourism-related in-migration: the Greater Yellowstone Region. In Buckley RC (ed) Environmental Impacts of Ecotourism, 25-40. CABI, Wallingford, UK

Moss L (ed) (2005) The Amenity Migrants. CABI, Wallingford UK

Pigram JJ, Jenkins JM (2006) Outdoor Recreation Management (2nd edn). Routledge, Oxford

Warnken J, Buckley RC (1995) Triggering EIA in Queensland: a decade of tourism development. Environmental Policy and Law 25:340-347

Warnken J, Buckley RC (1996) Coastal tourism development as a testbed for EIA triggers: outcomes under mandatory and discretionary EIA frameworks. Environmental Planning and Law Journal 13:239-245

Warnken J, Buckley RC (1998) Scientific quality of tourism EIA. Journal of Applied Ecology 35:1-8

Warnken J, Buckley RC (2000) Monitoring diffuse impacts: Australian tourism developments. Environmental Management. 25:453-461

Warnken J, Russell R, Faulkner B (2003) Condominium developments in maturing destinations: potentials and problems of long-term sustainability. Tourism Management, 24:155-168

17 Spatial Planning: Indicators to Assess the Efficiency of Land Consumption and Land-Use

Harry Storch and Michael Schmidt

Department of Environmental Planning, Brandenburg University of Technology (BTU), Cottbus, Germany

17.1 Introduction

The aim of this chapter is to discuss the importance of socio-environmental efficiency indicators for Strategic Environmental Assessment (SEA) in spatial development planning. Although environmentally inefficient settlement development structures are resulting in an ongoing unsustainable use of land-resources, even the new instrument of SEA has its limitations to promote the necessary structural changes in spatial development planning. The chapter starts with a description of the spatial development trends in Germany and introduces the national sustainability target concerning efficiency of land-use for settlements and traffic. Based on current urban growth research, Section 17.3 offers a detailed overview of available indicators that can describe the efficiency of regional and urban spatial structures in relation to land use and land consumption. Section 17.4 verifies the relevance of these indicators in relation to contrasting urban development models. Section 17.5 gives conclusions and recommendations for the German spatial planning framework, promoting the use and integration of indicator-related socio-environmental data to assess the efficiency of zoning of new developments for residential areas and traffic infrastructures in SEA procedures in regional planning.

17.2 Land Consumption: Actual Trends in Germany

Spatial development in Germany, like in most other countries in Europe, is characterised by the continuing use of mainly agricultural land, a finite resource, for settlement and traffic purposes. In many European countries, most urbanised regions have been subject to population and employment decentralisation over the past decades resulting in an ongoing spread of urban peripheries, a growth in motorised transport modes and thus by pollution of the air, changes in the global climate and

Standards and Thresholds for Impact Assessment. Edited by Michael Schmidt, John Glasson, Lars Emmelin and Hendrike Helbron. © 2008 Springer-Verlag

the loss and fragmentation of natural and cultural landscapes. These irreversible land use changes are an important issue for sustainability. In contrast to similar problems in urbanised areas in Asia, land-use changes in Europe are minimally affected by employment and population growth. Driving forces are mainly changing patterns of lifestyle, production and retail structures, and transportation patterns.

Land consumption in Germany is characterised by a continuous and high rate of more than 100 hectares per day over the last decades, mainly caused by the construction of new housing projects and transportation infrastructure. Moreover, there are no signs that this trend will change in the next two decades (table 2.1). Therefore, land consumption is, because of its persistence, one of the main unsolved environmental problem areas. This problem is highlighted by the fact that existing environmental policy instruments have had in the past only limited long-term effects. Consequently, because a reduction of the rates of land consumption in Germany could not be achieved by the existing spatial and environmental planning instruments (Runkel 2004), it is important to analyse the possible strategies, instruments and measures of new policy instruments, like Strategic Environmental Assessment (SEA), to optimise the policies and instruments in urban and regional planning and traffic management to reduce and manage the demand for land use in Germany in a more sustainable way (UBA 2004).

17.2.1 Target: Reduction of Daily Land Consumption to 30 ha

The achievement of sustainable regional development is dependent on a policy direction that integrates social, economic and environmental dimensions in a manner which reduces the ecological footprints of urbanised regions while offering quality of life for citizens and vital economic conditions.

The sustainable use of land resources is an important indicator for the evaluation of settlement structures and transport infrastructures from the point of view of an efficient spatial development. Therefore the reduction of the actual high rate of land consumption for settlements and transport issues is one part of seven priority policy areas of Germany's sustainability strategy. The major goal for a sustainable spatial development is therefore a significant reduction in the annual increase of land consumption. In order to establish more sustainable land-use management, the federal government has announced that the rate of land consumed for residential areas and transport must be reduced, from the current average of 120 hectares per day during the past decade, to 30 hectares per day by 2020 (GFG 2002).

Table 17.1. Average daily land consumption in Germany for settlement and transportation

Time Period	Average Daily Land Consumption	Remarks
1993 - 1996	120 hectares	4-year period[1]
1997 - 2000	129 hectares	4-year period[2]
2001 - 2004	115 hectares	4-year-period[2]
2005 - 2020	104 hectares	Development Trend[3]
2020	30 hectares	National Sustainability Target

Compiled from: [1]Dosch (2002 p.33), [2]Statistisches Bundesamt (2005 p.6), [3]BBR (2005)

The main reason for the establishment of this so called '30 ha target' was that despite the broad range of available spatial planning instruments and the integrated environmental assessment procedures currently in use, land consumption constantly shows growth rates that are positively related to economic development rates (Jakubowski and Zarth 2003). Thus, the recently observable decline of the daily rate of land consumption under 100 hectares is caused only by the current state of the German economy. In the longer term, land consumption is expected to have to average 104 hectares per day until 2020 (BBR 2005). There is no doubt that this future direction of spatial development is not environmentally and economically (infrastructure and transport costs) desirable. Finally, this ongoing development shows that the available instruments, spatial planning assessment methods, and many measures on regional and urban planning levels used to establish a more efficient land-use management in Germany are highly ineffective in the final evaluation (Apel et al. 2001). This is observed in the nearly constant rates of land consumption for residential areas and the resulting urban sprawl into surrounding areas.

To close the gap between the current trend and the formulated sustainability target for land consumption, it is important to determine what regional development models and strategies and what spatial planning instruments and related environmental assessment tools can be used or should be changed to reach the ambitious sustainability '30 ha' target (Bachmann 2005).

Any economic activity involves the use of environmental resources. The principle of sustainable development requires using natural resources as efficiently as possible. In Germany, Environmental Economic Accounting (EEA) productivity indicators are used to measure eco-efficiency at the national level. The environmental-economic accounting report of the Federal Statistical Office of Germany (table 17.2) documents important environmental-economic performance indicators, such as

- Energy productivity: Energy productivity is the measure of efficiency in the handling of energy resources. It is expressed as a GDP (gross domestic product) to primary energy use ratio.
- Raw materials productivity: Indicator of how efficiently a national economy uses raw materials. It is expressed as a ratio of GDP to the consumption of raw materials.

In contrast to these two performance indicators, the use of land resource for residential and transportation purposes is one important environmental performance indicator that could not be decoupled from the economic growth rates (table 2.2). To reach the proposed national sustainability target, the efficiency of land use must dramatically be increased. In this policy area of central importance are measures to decouple land consumption from economic growth (Hülsmann 2001). In the next section, it will be explained that the significant reduction in land consumption in the direction of the required decrease from 129 hectares (1997–2001) to 30 hectares (2020), as the National Sustainability Strategy of the German Government formulates, requires a strategy that does not further ignore the spatial structure, dimension and the socio-demographic dependencies of indicators of land consumption and land-use changes.

Table 17.2. Use of environmental resources for economic purposes in Germany. Long-term examination of productivities (modified from Höh et al. 2002 Fig. 5)

	Productivity (GDP per unit) - Average annual change in %		
	Former territory of the Federal Republic		Germany
	1960 - 1980	1981 - 1990	1991 - 2000
Energy	0.4	2.2	1.6
Raw materials	1.2	2.8	1.8
Built-up land and land used for traffic purposes	1.6	0.7	0.6
CO_2	1.7	3.4	3.0
Acidification gases	2.9	10.5	13.9
Hours worked	4.5	2.4	1.9
GDP (real)	3.6	2.2	1.4

17.3 Efficiency Indicators for Land Use and their Impacts

In the spatial planning debate regarding sustainability impacts and spatial consequences of poorly managed expansion of residential areas in the outskirts of urban agglomerations, there is a need to have an agreed upon method to measure and evaluate the dominant structural changes in the urban landscape on a regional level. Beyond that, it is important for environmental and spatial planners to be able to demonstrate how the monitored sprawl of residential areas has real implications for an efficient land-use management and real impacts on the environment.

In the field of spatial and urban planning research, the main impacts of an inefficient use of land for settlement development are described as a spatial development in which the spread of residential development across the rural landscape far outpaces population growth (Nechyba and Walsh 2004). The efficiency of the resulting regional and urban spatial structure that this spatial development process creates can be measured and analysed (Apel et al. 2000, Ewing et al. 2002, Flacke 2003) by the use of the following spatial and structural indicators:

1. Residential density and density of use,
2. Variety of uses and mixed urban land use: Neighbourhood mix of homes, jobs, and services,
3. Strength of agglomeration centres: Concentration and of polycentric structure settlements,
4. Accessibility of public transportation infrastructures: Non car-based transport systems compatible with the city network.
5. Recycling of land: Focusing new development in already built-up areas: Reactivation of brownfield sites, utilisation of conversion and changed use potential in existing build-up areas, mobilisation of abandoned areas in urban contexts.

In the following sections, these core indicators are explained in more detail with their relevance for and impacts on efficiency of use of land resources.

17.3.1 Residential Density

The spatial structure of a metropolitan region or agglomeration area is not always visible from the ground but it can be detected when analysing demographic and land-use related data. In spatial planning, the first step to analyse the basic spatial organisation is to describe the spatial pattern of population distribution within the built-up area. In general, average population density is the most common spatial indicator of land consumption; the higher the density, the lower the consumption of land per person. The focus can be separated into two main indicators: land consumption per resident (density) and the spatial distribution of these densities (density profile). To be meaningful, density should be equal to the city population divided by the built-up area, because densities measured by administrative area do not allow cross-regional comparisons. Residential density is the most important indicator to evaluate the efficiency of residential development and land-use management (Apel et al. 2000; Ewing et al. 2002). Dispersed suburban residential areas are the spatial manifestation of urban sprawl. In general, these low density areas make it difficult to provide adequate and accessible public utilities, services and infrastructures, to establish activity centres, and to offer access to public transportation options. As most of the older European cities are showing, higher residential density does not necessarily mean high-rise buildings. Residential density is an attempt to measure the efficiency of land use in a metropolitan area. It quantifies the amount of land used per person and measures the degree to which housing is spread out or compact.

For several decades, the residential density in the European towns and cities has been declining, mainly due to a reduction in the size of households and an increase in the living space per person (Couch et al. 2005). The necessary technical and social urban infrastructure and other public service facilities must be funded by an ever lower number of residents, and simultaneously new public infrastructure and service facilities can only be built on the urban periphery if, at the very least, a certain minimum residential density is established. The preservation and restoration of a high, but socially acceptable and qualified, population density in the core urban areas is a centrally important environmental and economic goal.

17.3.2 Variety of Uses and Mixed Urban Land Use

The dominant structural element of the traditional European city can be described with its building structure along streets with the variety of uses. These traditional urban structural patterns tend to mix different land uses, placing housing near business centres and shops, or offices above storefronts. Measuring the degree of mix is therefore an important descriptor of urban compactness.

In contrast, the important spatial structure of urban sprawl is the strict segregation of different land uses. In sprawling regions, settlement areas are typically separated from shopping, business districts and cultural activity centres (Ewing et al. 2002). At the level of agglomeration areas the allocation of mixed land use with a variety of uses could lead to a spatial urban structure that generates less traffic. Securing dense structures with mixed land uses in settlement areas is therefore a priority for spatial planning. The mixed use of urban land implies a balance of housing, employment and facilities in each district of a city. This offers the potential for the more efficient use of urban land, while facilitating the use and maintenance of public transport, thereby contributing to energy efficiency and pollution reduction.

17.3.3 Strength of Agglomeration Centres

This indicator is representing the degree of the structural spatial concentration of population and employment in agglomerations. Metropolitan centres are concentrations of activity. This centeredness can be generated by concentrations of either population or employment. It can be distinguished between two types of spatial concentration. At the urban planning level, the strength of activity centres describes how centred it is near a central activity district. At the regional planning level it describes how activities are clustered in a more polycentric pattern. Urban compactness, as an efficiency parameter, is associated with centres of all types (business, shopping, leisure, culture). Urban sprawl, as inefficient land-use structure (Janssen-Jansen 2005), is defined with the absence of activity centres of any type.

17.3.4 Accessibility of Transport Infrastructure and Public Services

The dispersion of settlements isolates different land uses causing increased reliance on car-based private transport and limits the efficient use of public transport modes. In general, residents living in widespread settlement structures own more cars per household and commute greater distances to work, shopping or travel. The strong dominance of car transport in turn leads to a further unsustainable expansion of residential land into the countryside. Existing land-use patterns and the resulting spatial structures limit the choice of transportation alternatives. Spatial structures with a high residential density cannot rely on private cars as the dominant transport system; low density use patterns cannot maintain an effective public transport system. The amount of land taken for traffic thus depends primarily on the accessibility and use of public transport systems. The area taken up for traffic in large German urban agglomeration centres is approximately 25 per cent, but in car-based North American urban sprawl regions, it is up to 50 per cent of the total settlement area (Apel et al. 2000). Giving priority to public transport systems is an important element of the economical use of land.

Concerning transportation and other network based infrastructures, two strong empirical regularities are related to regional spatial structures (Steinocher and Tötzer 2001). In densely populated regions, more surface area is used for the transportation infrastructure, but the surface area per capita used for net-based infrastructures in general is much lower compared to requirements of large and scarcely populated regions.

Urban sprawl encourages populations to move outside of established densely populated urban areas, but standing expenses for infrastructure (e.g. transportation, water supply) remain the same. The infrastructure costs must be borne by ever fewer residents or requiring a reduction of services to the remaining population (Apel et al. 2000; Ewing et al. 2002). Therefore the efficient spatial development of settlements in agglomeration regions requires a minimum residential density defined per specific housing area (urban districts, suburbs). The resulting compactness of settlement forms facilitates the provision of the following public infrastructures and services:

- public transport system which is sufficiently accessible and used,
- sufficient local supply facilities (private and public services).

The primary goal of spatial development planning must be to concentrate urban development in the centre of urbanised areas and to reduce the ongoing process of dispersion of settlements to rural areas (Janssen-Jansen 2005).

17.3.5 Focusing New Development in Already Built Up Areas

Efficient spatial planning is not about restricting growth or even limiting growth, rather it is about focusing new residential developments in places where it can be best accommodated (Bergmann et al. 2006). Most important for an efficiency strategy are those areas that are currently available within the urban footprint. Most urbanised regions contain huge redevelopment opportunities such as brownfields (old industrial sites), empty shopping areas (greyfields), and vacant housing areas (Bergmann and Dosch 2005). These properties in generally have existing public infrastructure and access to public serves (public transportation, roads, water and other utilities) and can accommodate new urban developments with a variety and mix of uses (homes, shops, office etc.) linked together by existing streets and public transport infrastructure. These urban development investments within existing build-up areas are much more efficient than investments in new urban development in the outskirt of agglomeration areas. Due to the lower residential densities in the urban periphery and the new land needed for traffic and public infrastructures, one hectare of recycled land in the urban core area helps to save approximately three hectares of settlement and traffic land at locations on the urban periphery (Apel et al. 2001). In the next decade spatial development and urban planning in Germany must therefore change its focus from urban-growth planning to urban reconstruction (Preuß and Ferber 2005).

17.4 Linking Indicators to Urban Development Models

Urban development planning of the last decades and the current discussion on regional planning are characterised by two contrasting and conflicting urban planning models (Apel et al. 2000):

- Network city - this widespread city is signified by the gradual dissolution of the
  traditional compact European urban structures. The network city represents a
  car-based urban planning model and is in line with the previously described
  trends in urban development: less residential density, fewer mixed land uses,
  decentralisation, dispersed structures, growth of private car-based transport.
- Compact city - this urban model is based on European urban culture and can be
  adapted to urban districts in polycentric, public transport-based regions. The
  compact city as an urban model represents an efficient use of resources such as
  land, energy, materials and time, and at the same time enables, through the concentration of human activities, the preservation of large greenfield areas in the
  countryside.

Because efficiency indicators for residential land-use can be easily used to contrast and separate the two competing urban development models of the current
spatial planning discussion (table 4.1), the efficiency of regional and urban development structures is a real, measurable phenomenon with real implications for
Strategic Environmental Assessment (SEA) procedures in regional planning.

Table 17.3. Efficiency of land-use for urban development strategies (compiled from: Apel
et al. 2000, Ewing et al. 2002)

Indicator	Network City	Compact City
Residential density	The population is dispersed in low density development	High density of use, high residential density
Neighbourhood mix of uses	Rigidly separated uses (homes, shops, and workplaces)	Variety and mix of uses
Centeredness	Lack of well-defined activity centres (business, shopping)	Concentration of settlements, well-defined activity centres
Transportation choices	Poor access to public transpor-tation choices: higher commut-ing rates and car ownership	Environmental-friendly public transportation choices, areas suitable for walking and cycling
Recycling of land	New developments mainly on greenfield sites	Redevelopment of brownfield sites and already built-up areas

17.5 Conclusions and Recommendations

In general, spatial planning involves the setting of frameworks and principles to guide the location of residential development and physical infrastructure. Therefore spatial planning coordinates land-use related public and private investment decisions across space. As mentioned above, the trend of land-use changes and the resulting land consumption can be monitored in terms of spatial structural distribution, a system of activity centres and use densities and the patterns of interaction (commuting).

For the implementation of measures to reduce land consumption and to optimise land-use in Germany, the level of regional planning is most important (Von Haaren and Nadin 2003). But as the ongoing degree of land consumption shows, the effectiveness of regional planning in this policy area is actually very low. Major deficits in the current regional planning framework to limit the environmental pressures associated with sprawl are a lack of spatially detailed data required to create indicators related to sprawl and land consumption and the resulting inappropriate zoning (Runkel 1999). Strengthening the regional planning competence requires an appropriate use of available data and a more precise and transparent zoning of future settlement areas. How the current lack of usage and availability of socio-environmental efficiency indicators (Steinocher and Tötzer 2001) is limiting an appropriate zoning of land uses and the spatial assessment of environmental impacts of land-use changes resulting from spatial planning policies at the urban and regional levels will be discussed in the next sections.

Information Requirements for the SEA of Spatial Development Plans

As a prerequisite for the necessary quantitative reduction of land consumption in Germany, the control of the territorial setting of new residential areas must be acknowledged as the key task of SEA of spatial development plans. Therefore any assessment procedures against urban sprawl require detailed demographic information and land-use data to evaluate the spatial patterns of urbanisation and residential areas (Siedentop and Kausch 2004). The core data required are focused on the spatial concentration of land consumption as well as the distribution of newly created settlement areas. This information characterises driving forces and pressures related to demographic developments in agglomeration areas, their manifestation in the resulting land consumption and impacts on the compactness of urban structures. Special attention must be paid to the importance of understanding the spatial structure of regions, agglomeration centres and cities in order to develop standards and thresholds for indicators which are compatible with the observed spatial structures of the assessed planning region.

The question of what type of residential development can be regarded as efficient in the use of land resources, and therefore limiting land consumption and protecting the countryside, must be more precisely defined by reference values on regional-level. A pragmatic approach to the assessment of settlement developments involving core indicators should be used (Apel et al. 2000; Wrbka et al.

2001; Flacke 2003), because they can largely be derived from the above-mentioned available land-use und socio-demographic base data. A main component used to describe the spatial pattern of agglomeration areas is population distribution within the built-up area:

- Land area taken up for settlement development and transportation infrastructure.
- Degree of soil sealing by construction and paving.

Landscape-specific indicators that describe impacts on the environment resulting from land consumption are mainly represented by land-use change patterns and rural landscape fragmentation by urban development and transportation infrastructure:

- Degree of fragmentation and segmentation of open space by settlement areas.
- Length of traffic routes and the number of cars as an indicator to estimate the environmental impacts of the road network.

These indicators form the basis for spatial typologies that are based on intersections of land-use related environmental data and statistical socio-demographic information. Among these key indicators, land consumption as the specific area used for settlement and traffic per capita is the most important factor. This significant core indicator is determining the values of the whole set of indicators.

Deficits in Zoning on Regional Planning Level

Urban land-use plans in Germany must be adjusted to spatial development targets that are specified in regional planning regulations for a larger territory. In principle, development restrictions in the suburbs and urban periphery (e.g. green belts) can restrict the amount of land zoned for residential development. Land use policy based on strict zoning drives up the value of residential land in the already urbanised area inducing high-density settlement areas on the outskirts of cities and towns. An analysis of the differences in land consumption in England showed that England is only consuming about on third of land-resources per resident compared to the German situation (Von Haaren and Nadin 2003). One important factor of this success lies in the stricter preservation and zoning of green belts in England.

The relatively high strength of community planning interests in the spatial planning system of Germany weakens the capability of development restrictions for residential areas on regional level (Runkel 1999). Even the instruments of SEA have limited force to control the development of settlement areas in order to make efficient use of land or to define minimum residential densities. SEA is currently concentrated on assessment procedures which have only limited and indirect intervention possibilities. The assessment of regional plans is more or less centred on the protection of areas according to environmental standards and ecological defined standards for landscape conservation. This form of indirect control has lead to a greater preservation of areas that can be zoned based on legally defined environmental standards.

In order to provide improved control of the efficiency of land-use at the regional planning level, the use of SEA requires the primary definition of the basic parameters of urban development based on efficiency indicators (Flacke 2003, Wrhka et al. 2001). This means that definitions on the proposed type and scale of residential areas should be more specific and detailed than at present. This includes, for example, residential density and area size and basic information about the development potential in already build up areas or brownfields (Preuß and Ferber 2005). Only if the actual regional planning conditions for the use of land resources for settlements and traffic are reformed, can SEA directly promote the primary use of available spatial development reserves within already urbanised areas, which is the basic requirement to fulfil the national 30 ha target in 2020.

References

Apel D, Böhme C, Meyer U, Preisler-Holl L (2000) Szenarien und Potenziale einer nachhaltig flächensparenden und landschaftsschonenden Siedlungsentwicklung. UBA-Berichte 1/00, Erich Schmidt Verlag, Berlin

Bachmann G (2005) Das «Ziel-30-ha» in der Nachhaltigkeitsstrategie Deutschlands: Ein Schritt zur modernen Urbanität. DISP, Bd. 41 (2005), H. 160, pp. 102-103

BBR (Bundesamt für Bauwesen und Raumordnung - Federal Office for Building and Regional Planning) (2005) Raumordnungsbericht 2005. BBR-Berichte, Band 21, Bon

Bergmann E, Dosch F (2005) Auf dem Weg zur Flächenkreislaufwirtschaft - Visionen und strategische Anknüpfungspunkte. In: Böhm H R (ed) Unendliches Wachstum auf endlicher Fläche? Darmstadt, pp. 75-86

Bergmann E, Dosch F, Einig K, Jakubowski P (2006) Flächenkreislaufwirtschaft - eine bestandsorientierte Perspektive des städtischen und stadtregionalen Flächenmanagements. In: Sinning H (ed): Stadtmanagement – Strategien zur Modernisierung der Stadt(-Region). Dortmunder Vertrieb für Bau- und Planungsliteratur, Dortmund, pp. 214–230

Couch C, Karecha J, Nuissl H, Rink D (2005) Decline and sprawl: an evolving type of urban development – observed in Liverpool and Leipzig. European Planning Studies, vol. 13 (2005), no. 1, pp. 117-136

Dosch F (2002) Auf dem Weg zu einer nachhaltigeren Flächennutzung? In: Informationen zur Raumentwicklung Heft 1/2 2002, pp. 31-45

Ewing R, Pendall R, Chen D (2002) Measuring Sprawl and Its Impact. Smart Growth America. Available online at: http://www.smartgrowthamerica.org/sprawlindex/ MeasuringSprawl.pdf

Flacke J (2003) Nachhaltigkeit und GIS. Räumlich differenzierende Nachhaltigkeitsindikatoren in kommunalen Informationsinstrumenten zur Förderung einer nachhaltigen Siedlungsentwicklung. In: Raumforschung und Raumordnung 3/2003 (61), pp 150-159

GFG - German Federal Government (2002) Federal Government's National Sustainability Strategy. Perspectives for Germany,Our Strategy for Sustainable Development. Available online at: http://www.nachhaltigkeitsrat.de/service/download_e/pdf/ Perspectives_for_Germany.pdf

Jakubowski P, Zarth M (2003) Nur noch 30 Hektar Flächenverbrauch pro Tag. Vor welchen Anforderungen stehen die Regionen? In: Raumforschung und Raumordnung 3/2003 (61), pp. 175-197

Janssen-Jansen L B (2005) Beyond Sprawl: Principles for achieving more Qualitative Spatial Development. In: DISP 41, pp. 36-41

Höh H, Schoer K, Seibel S (2002) Eco-efficiency indicators in German. Environmental Economic Accounting. In: Statistical Journal of the United Nations ECE 19, pp 41–52.

Hülsmann W (2001) Potenziale und Strategien einer flächensparenden Siedlungsentwicklung. In Umweltbundesamt Wien (ed) Versiegelt Österreich? Der Flächenverbrauch und seine Eignung als Indikator für Umweltbeeinträchtigungen. UBA Conference Papers, CP-030, Wien, pp. 134–139

Nechyba T J, Walsh R P (2004) Urban Sprawl. Journal of Economic Perspectives 17(4): 177-200

Preuß T, Ferber U (2005) Flächenkreislaufwirtschaft: Neue strategische, planerische und instrumentelle Ansätze zur Mobilisierung von Brachflächen. In: Besecke,A, Hänsch R, Pinetzki M (eds) Das Flächensparbuch. Diskussion zu Flächenverbrauch und lokalem Bodenbewusstsein, TU Berlin, ISR-Diskussionsbeiträge, H. 56, Berlin, pp. 177-175

Runkel P (1999) Zur Zukunftstauglichkeit des planungsrechtlichen Instrumentariums für eine nachhaltige Siedlungsentwicklung. In: Raumforschung und Raumordnung 57 (4):255-258

Siedentop S, Kausch S (2004) Die räumliche Struktur des Flächenverbrauchs in Deutschland. Eine auf Gemeindedaten basierende Analyse für den Zeitraum 1997 bis 2001. Raumforschung und Raumordnung, 1/2004 (62):26-49

Statistisches Bundesamt (2005) Bodenfläche nach Art der tatsächlichen Nutzung. Land- und Forstwirtschaft, Fischerei, Fachserie 3/ Reihe 5.1, Wiesbaden

Steinnocher K, Tötzer T (2001) Analyse von Siedlungsdynamik durch Verknüpfung von Fernerkundungs- und demographischen Daten. In Umweltbundesamt Wien (ed) Versiegelt Österreich? Der Flächenverbrauch und seine Eignung als Indikator für Umweltbeeinträchtigungen. UBA Conference Papers, CP-030, Wien, pp. 39-47

UBA – Umweltbundesamt (2004) Verringerung der Flächeninanspruchnahme durch Siedlungen und Verkehr - Strategiepapier des Umweltbundesamtes. Erich Schmidt Verlag, Berlin

Von Haaren C, Nadin V (2003) Die Flächeninanspruchnahme in Deutschland im Vergleich mit der Situation in England. In: Raumforschung und Raumordnung, 5/2003 (61):345-356

Wrbka T, Peterseil J, Szerencsits E (2001) Versiegelung, Zersiedelung, Zerschneidung und Fragmentierung – „Neue" Indikatoren für die Belastung Österreichischer Landschaften? In: Umweltbundesamt Wien (ed) Versiegelt Österreich? Der Flächenverbrauch und seine Eignung als Indikator für Umweltbeeinträchtigungen. UBA Conference Papers, CP-030, Wien, pp. 79-96

18 EIA Performance Standards and Thresholds for Sustainable Forest Management in Ghana

Edward K. Nunoo

Brandenburg University of Technology (BTU), Cottbus

18.1 Introduction

Forests and its ecosystem dynamics constitute nature's most bountiful and versatile natural resource. Tropical forests epitomises its diversity. These are vital assets that provide a wide range of environmental, economic and socio-cultural benefits and services to local communities, national economies and the global environment at large. However, unsustainable use of the resources over the decades has now been a major cause of global concern. Areas under tropical forest continue to dwindle at alarming rates to the detriment of its productive and protective functions.

The annual rate of forest cover destruction worldwide is in the domain of 40-50 million acres (ITTO 1998). Ghana's share of 8.2 million hectares, a century ago, reduced to only 1.6 million hectare in 1998 (Kotey 1998). With the shift in paradigm from unsustainable harvesting practices to *not harvesting the capital of the forest stock*, a clear signal is being sent to stakeholders that any forest project likely to have significant impact on the environment needs to be assessed.

Deliberations on how to maximize utilization of forest resources and at the same time safeguard its protective and productive functions have illicit various scientific technologies to aid in its protection, and environmental impact assessment (EIA) is one of such advances. EIA of forest projects, where performance standards and thresholds are established, is thus seen as a hallmark in the measure of successes towards sustainable forest management (SFM).

This chapter examines implementation of the EIA process in establishing thresholds and standards for sustainable forest management (SFM) with emphasis on forest projects in Ghana. Although a relatively young field in Ghana, application of the EIA process has produced major environmental breakthroughs paramount to the sustainable development objectives of the country.

Standards and Thresholds for Impact Assessment. Edited by Michael Schmidt, John Glasson, Lars Emmelin and Hendrike Helbron. © 2008 Springer-Verlag

18.2 Country Background

Ghana is located on the West Coast of Africa with a total area of 239 460 sqkm (CIA 1994). In the world economic order the country is classified into the 'developing statuses with a real growth rate of 5.9 % and a per capita income of $ 2000. Agricultural activities dominate the economy with the sector employing 70 % of the rural labour force. It supplies 90 % of the country's staple food, contributes 45 % to the gross domestic Product and accounts for 55 % of exports. The population is estimated at 20 million based on year 2000 population census figure of 18.9 m with an inter-censual growth rate of 2.7 %. Forest resource harvesting is a way of life among forest communities although the degree of association varies across the ecological zones. Timber exploitation for commercial purposes has been an integral part of the economy since the colonial era. However extraction reached its climax (Senamede 1995) during the Economic Recovery Programme in the 1980s as logging was perceived to be a panacea for resuscitating a virtually battered economy. Although it raked in some immediate needed assistance in terms of giving a facelift to the national coffers, cost to environmental damage is yet to be fully assessed.

Following after cocoa and mineral export proceeds, the sector still accounts for 6 % of GDP and employs over 70 000 people (Asabere 1987) annually. It also meets all domestic timber requirements and supplement 70 % of domestic energy needs in terms of fuelwood consumption.

18.2.1 Resource Utilisation and the Environmental Problem

Forest resources are mainly classified into savannah woodlands in the north and the tropical high forest zone (HFZ) to the south. The HFZ are all found within reserves with only half of the estate in favourable conditions (FAO 1995). Off reserves they are seen as small patches of forest or trees on farms. They contain most trees of economic value and rich mineral deposits. The region also happens to be the most densely populated (Ghartey 1990).

Interest in the use of forests in Ghana reflects differences in the people's way of life. Whilst scientist and environmentalist argue for its carbon sequestering and conservation, farmers are waiting to carve off a piece of it into ploughing. Forests are also catching up very fast as alternative resources for satisfying recreational needs within a booming wood industry.

The quest for a progressive socio-economic growth and development has brought with it remarkable damage to the country's landscape and untold hardship to forest communities. The most affected are people who happen to live below the poverty line (Ghartey 1990) in environmentally sensitive areas where their economic activities often compromises environmental vitality. The magnitude of pollution is seen in annual field productivity losses (0.5-1.5 % of GNP), sustainable logging potential, erosion prevention, watershed stability (IUCN 1988b), carbon sequestration and loss of potential new drugs as a result of endangered genetic resources.

18.2.2 Sustainable Forest Management in Ghana

Development of SFM takes its root from the Stockholm UN Conference on the Environment in 1972. The aftermath institutionalized an agency (Environmental Protection Agency-EPA) to regulate activities within the environment using Environmental Assessment Administration procedures for achieving its aims (EPA 2005).

Working towards SFM has underpinnings of the country's vision of sustainable development. It means utilizing resources to cater for the needs of today while ensuring that they continue to exist at acceptable levels for the benefit of the future generation. Interventions identified so far include conservation and preservation of forest resources which constitutes the major natural resource management problem, enhancing resource vitality and restoration of degraded land. New development areas are participatory management and setting threshold limits. As a result logging rates which exceeded sustainable annual allowable cuts (AAC) have been worked out and step-down to acceptable standards (ITTO 1996).

18.3 EIA and Institutional Framework in Ghana

EIA is one of the scientific advances towards environmental protection. This is a systematic process by which the likely significant effects of a project on the environment are identified, assessed and taken into account by a competent authority in the decision-making process. It is a useful tool for forecasting environmental implications of proposed projects (FAO 1995).

EIA is applied to development projects including the forestry sector as well as other under-takings as an environmental permitting pre-requisite (EPP) and a major environmental management tool (EPA 2005). EPA is responsible for enforcing EIA regulations and administration as stipulated by law (EPA Act 490).

EPA operates as a district, municipal and metropolitan assemblage responsible for implementing the EIA process. Carrying out an EIA is manifested in three main parts. First, a preliminary review of projects with potential environmental impacts is done followed by a full EIS describing the proposed project and predictable environmental effects (Gilpin 1995).

The EIA lays out alternatives for decision-makers and calculates the costs and benefits of each alternative. The proponent is requested by law to register the project and carry out the EIA study. Final decision on the project is made by EPA if the EIS is approved by issuing an "Environmental Permit".

The EIA Procedural Standards

EPA Act 490 mandates the agency to "ensure compliance with laid down EIA procedures in the planning and execution of projects, including forest developments". The procedures provides a step-by-step guidelines stating who does what relevant to operations of the Agency staffs, developers, consultants and give guidance on the expected content of the EIS.

It make provisions for the *registration, screening, scoping and terms of reference*, the *EIA study, review and public hearing, appeals, timelines for decision-making and public participation* at all levels of the process.

Undertakings requiring EIA are registered with EPA. After receiving notice, EPA, in consultations with stakeholders, makes a decision by placing the project at the appropriate level of assessment (screening). At this stage the minimum standards to satisfy include taking a closer look at *the project location, size and output levels; proposed technology to be involved; concerns to the general public; land use considerations;* and *other factors relevant to the project.*

From the screening report one of the following decisions (Figure 18.1) may be reached. In the event that decision point *iv* is the case, EPA subjects the reconnaissance information to environmental scoping which identifies key issues of concerns to be addressed. A scoping report is submitted from which the draft Terms of Reference (TOR) is set. Provisions are also made to consider views of state agencies, the public and other relevant bodies.

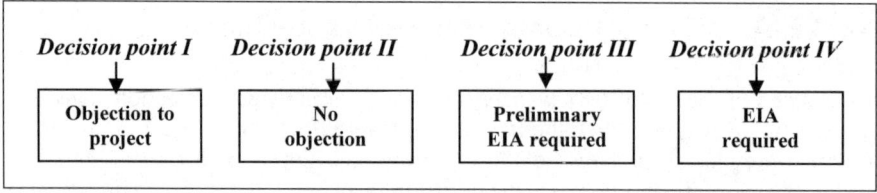

Fig. 18.1. EIA decision point

A technical committee determines whether issues referred to in the TOR have been addressed and notify the proponent of any inadequacies for the necessary amendments and re-submission. Once the minimum standards are met, in the concurrence of the sector ministry, EPA grants approval for implementation of the proposed project subject to specified conditions or refused approval for implementation with reasons. In the case of logging a timber utilization contract need to be obtained as well.

18.4 Standards and Thresholds of Significance

Standards are "benchmarks" or statements indicating expectations, movement towards some larger goal, claims about improvement in, or accommodation to a norm or value (Taylor 1961). In the case of forestry it relates to a set of principles, criteria and indicators that serve as reference for assessment of SFM (Sayer 1996).

"Threshold of Significance," on the other hand, is a quantitative or qualitative standard, or set of criteria, pursuant to which the significance of a given environmental effect may be determined (Miner 1994). In other words, it is that level at which the lead agency responsible for environmental decision making finds the effect of a forest project to be significant.

Setting standards for SFM in Ghana covers a broad spectrum of stakeholder's views and concerns. They include socio-economic well-being and environmental vitality indicators, and rights and responsibilities of stakeholders. Issues bordering biodiversity conservation, workers rights and responsibilities, compliance with regulations, land tenure, and resource rights as well as other relevant issues have been captured and categorized under seven principles (Box 18.1). These form the basis of C&I development within the hierarchical framework for measures of successes towards SFM. Thresholds for biodiversity conservation in Ghana take the form of classifying tree species with star values (black star to green star) based on spotless samples to show priority species distribution and protection (Musah 1990) (see Table 18.1).

Box 18.1. Underlying principles for SFM in Ghana (NCFC 1996)

Principle One	Compliance with laws and regulations
Principle Two	Land tenure, stakeholder and resource rights
Principle Three	Conservation and maintenance of biological diversity
Principle Four	Rights and responsibilities of workers
Principle Five	Benefits from the forest, revenue generation and the equitable distribution of costs and benefits
Principle Six	Maintenance and enhancement of ecosystem productivity
Principle Seven	Forest management system

A black star for example connotes scarcity of the resource in and outside Ghana. Such species call for high protection management (Table 18.1). The tabulation implies that SFM in Ghana is committed to meeting sustainable supplies of forest products. The threshold is intended to ensure that genetic diversity of the forest, its productivity and environmental protection functions are not jeopardized.

The conservation strategy gave rise to the Genetic Heat Index (GHI) which exhibits the genetic value of forest in terms of the composition of tree species by computer programming (Box 18.2). GHI is a weighted index that draws its strength on tree species distribution with the multiplier 3 (Swine 1981).

Values express scarcity or availability of the resource. Higher figures denotes endangered or near extinct species whilst lower numbers show species with no particular or very little conservation concerns. A similar calibration, using the same weights as GHI, give rise to the economic index which is an indication of the number of economic tree species, with values ranging between 0 and 120 (Hawthorne 1995b).

Table 18.1. Standards for Biodiversity Conservation in Ghana (FC 2006)

Star Label	Implication for Species	Implication for SFM
Black	Rare Internationally, Uncommon in Ghana	Highly protected
oldG	Fairly rare internationally, Fairly rare in Ghana	High Protection
Blue	Common internationally, Rare in Ghana (and vice versa)	Require Protection
Scarlet	Common internationally, Common in Ghana, High pressure from over exploitation	Strict control on AAC if to remain commercially viable Level of cut > 200 % of ACC
Red	Common internationally, Common in Ghana, Tend to be over exploited	Restriction needed Level of cut (100-200) % of ACC
Pink	Utilizable but not popular to the international trade, Utilizable in Ghana	Harvest levels set below ACC
Green	No particular conservation concern	Judicious resource utilization

The thresholds, combined with Hall and Swain's system of fine-grain protection measures (Hall 1981) have committed 4.4 % of the total forest reserve area of rare species, ecosystems and economic trees to biological genetic conservation, 19.2 % to environmental protection (Vulnerable reserves like hills and swamps), 5.2 % to fire protection in the dry forest types and 21.1 % of degraded areas to forest rehabilitation.

Box 18.2. Genetic Heat Index applicable in Ghana (FC 2006)

$$GHI = \frac{\left| [BK \cdot 27] + [GD \cdot 9] + [BU \cdot 3] + [RD \cdot 1] \right| \cdot 100}{[BK+GD+BU+RD+GN]}$$

Values range from 0 – 533, GN = 0, Where

BK = tree species labelled as black
GD = tree species labelled as gold
BU = tree species labelled as blue
RD = tree species labelled as red
GN = tree species labelled as green

18.4.1 Applying Thresholds

Threshold of significance for forest projects are identifiable during EIA screening with the minimum threshold limits being compliance to environmental laws of the country. 'Cut-off points' under which projects are considered unlikely to have *significant effects*, are allowed to proceed without being caught by screening (Sheate 1994). Once thresholds are set certain projects could proceed without screening if they are found below the minimum requirement sizes, meet certain criteria or if they are located in certain areas.

Threshold limits apply to a wide range of developments in Ghana split into two categories. EPA schedule I list 30 major of project types and schedule II have 16 undertakings to which threshold are administered. Areas of interest include developments related to forestry, agriculture, drainage and irrigation, land reclamation and fisheries (Box 18.3).

Box 18.3. Selected Projects requiring EIA and Threshold (EPA 2005)

1. Forestry
   a. conversion of hill forest land to other land use.
   b. logging or conversion of forest land to other land use within the catchment's area of reservoirs used for water supply, irrigation or hydro-power generation or in areas adjacent to forest, wildlife reserves.
   c. Conversion of wetlands for industrial, housing or agricultural use.
2. Agriculture
   a. land development for agriculture purposes not less than 40 hectares;
   b. Agricultural programmes necessitating the resettlement of 20 families or more.
3. Drainage and Irrigation
   a. construction of dams and man-made lakes.
   b. drainage of wetland.
   c. Irrigation schemes.
4. Land Reclamation
   a. Coastal land reclamation.
   b. Dredging or bars, estuaries.
5. Fisheries
   a. Construction of fishing harbors.
   b. Harbour expansion;
   c. Land based aquaculture undertaking.

Thresholds simplify the EIA process. They also enhance a case-by-case project assessment (Donelly 1998), guard against potential negative effects and minimise the chance that acceptable and beneficial projects may be caught inadvertently.

Environmental sensitive zones (Box 18.4) under schedule 5 of the EIA regulation explicitly mapped 12 areas of interest (EPA 2005). They include sites of special scientific interest, national parks, historical sites, religious or sacred sites, archaeological sites, water bodies and mangrove sites.

Box 18.4. Definition of Environmental Sensitive Areas in Ghana (EPA 2005)

1. All areas declared by law as national parks, watershed reserves, wildlife reserves and sanctuaries including sacred groves
2. Areas with potential tourist value.
3. Areas which constitute the habitat of any endangered or threatened species of indigenous wildlife (flora and fauna)
4. Areas of unique historic, archaeological or scientific interests.
5. Areas which are traditionally occupied by cultural communities.
6. Areas prone to natural disasters (geological hazards, floods, rainstorms, earthquakes, landslides, volcanic activity etc
7. Areas prone to bushfires.
8. Hilly areas with critical slopes.
9. Areas classified as prime agricultural lands.
10. Recharge areas of aquifers
11. Water bodies characterized by one or any combination of the following conditions
    a) water tapped for domestic purposes;
    b) water within the controlled and/or protected areas;
    c) Water which support wildlife and fishery activities.
12. Mangrove areas characterized by one or any combination of the following conditions)
    areas with primary pristine and dense growth;
    b) areas adjoining mouth of major river system;
    c) areas near or adjacent to traditional fishing grounds;
    d) areas which act as natural buffers against shore erosion, strong winds or storm floods.

Developments and Threshold Limits

A field assessment in Ghana identified five forestry projects where threshold limits are being tested. The test areas cover logging, forest roads, forest quarry, agriculture, deforestation and afforestation (Table 18.2).

Table 18.2. Forest projects and area thresholds (Field data analysis 2006)

Type of projects	Area thresholds	
	Sensitive Areas	Non-sensitive areas
Agriculture (Commercial)	0 hectares. Opinion always needed	> 40 hectares
Logging	Hill sanctuary slope $\geq 30\%$ Close to water course > 25m Other sensitive areas, 0 hectares. Opinion always needed	Logging intensity in all contract areas >2 trees/hectare
Forest Roads	0 hectares, in all sensitive areas. Opinion always needed	> 5 hectares

In all cases EIA is mandatory where the project area is greater than the area threshold. Projects identified are defined as follows:

i *Agriculture:* Developing forestlands for commercial agricultural purposes.

ii *Logging:* Harvesting timber for economic gains.

iii *Forest Roads:* The construction, alteration or maintenance of ways on land use (or to be used) for forestry purposes.

iv *Forest Quarries:* Prospecting for materials required for forest road works on land that is used or will be used for forestry intended purposes.

v *Deforestation:* Converting forest or woodland area into development infrastructure or for other purposes.

vi *Afforestation:* Planting new trees or forest from either natural regeneration or direct seedlings.

Enforcing Thresholds and Standards in Ghana

Threshold and standards are strictly enforced by EPA with support from the law enforcement agencies. As a result logging by the clear-cut-felling system is considered illigetimate. Timber concessions (Froggie 1957) are protected from activities likely to destroy the forest canopy (Froggie 1962).

They are divided into compartments according to harvesting schedules (Fig 18.2) within a forty-year management cycle (Hall 1987). Harvesting is actually carried out within the confines of these and monitored by the forestry commission. Limited number of trees per hectare (< 2) are allowed (Hall 1987) and trees are individually selected before logging. The commission ensures that access route to the compartments are carried out through topographic survey to plan trucking roads and major hauling routes (Box 18.3).

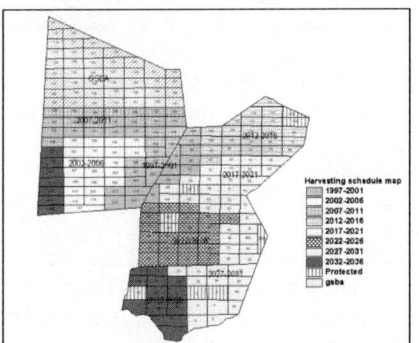
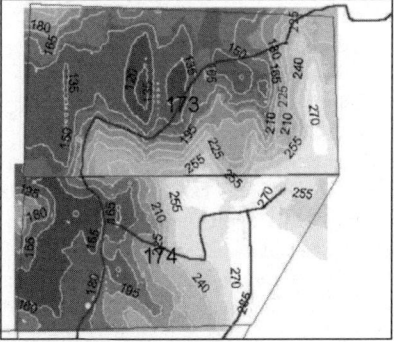

Fig. 18.2. Timber harvesting schedule routes   **Fig. 18.3.** T-survey of access routes

18.5 Conclusions and Recommendation

In conclusion, setting standards for sustainable use of forest resources in Ghana has shown some signs of positive impact on forest vitality and protection.

Although a relatively new field, the process has succeeded in enacting threshold limits for major development projects. The process promotes consistency, efficiency and predictability. Adopting threshold on impact assessment criteria to the locale has enabled a forest product certification scheme that has rolled out a computerized log tracking system to improve on monitoring of timber harvesting, processing and export.

The initiative is however challenged by conflict of interest among stakeholders with regards to resource ownership, management responsibilities and disbursement of proceeds due to poor information on the new forest management paradigm.

For an efficient monitoring system and strict adherence to the standards, the author recommends for a periodic intensive environmental education and campaigns for stakeholders to become abreast with current issues and especially to get forest communities more involved in forest resource management. This could improve resource use maximization and its sustainability concurrently.

References

Abu-Juam Musah (1995) biological inventory and monitoring. Ghana Forestry Department. Kumasi

Barbier E B (1995) raising revenue for sustainable forest management and biodiversity conservation: should the tropical timber trade pay? Workshop Paper to IUCN/ CSERGE

Barbier EB (1994) the economics of the tropical timber trade. Earthscan, London

Byron N, Ruiz Perez M (1996) What future for the tropical moist forests 25 years hence? Commonw. For. Rev.75(2):124-129

CIA World Fact Book (1994) Ghana Country Profile and Maps. Electronic Version. Posted on the internet May, 2004. http://www.cia.gov.cia/publications/factbook, last accessed on 12.08.2006

Dickinson MB, Dickinson J C, Putz F E (1996) Natural forest management as a conservation tool in the tropics: divergent views on possibilities and alternatives. Commonw. For. Rev. 75(4):309-315

Donelly A, Dalal Clayton B, Hughes R (1998) A Directory of EIA guidelines. International Institute for environment and development. Russell press, Nottingham UK

EC – European Commission (2002) Third Annual Survey on the Implementation and Enforcement of Community Environmental Law: Office for Official Publications of the European Communities. Official Journal (1997), L73 14.3.97, Council Directive 97/11/EC on the Assessment of the Effects of Certain Public and Private Projects on the Environment

Environmental Protection Agency (2005) Ghana Environmental Impact Assessment Procedures. EPA, Accra

FAO (1993) Conservation of Genetic Resources in Tropical Forest Management:Principles and Concepts. FAO For. Pap. 127, Rome

FAO (1995) Model Code of Forest Harvesting Practice. Rome

FAO (1998) Forest plantation areas 1995. November 1997, revised July 1998. En Report to the FAO project GCP/INT/628/UK, Rome

FAO (2005) National report on the forestry policy of Ghana. FAO Forestry Paper no 132. Rome

Foggie A (1957) Forestry problems in the closed forest zone of Ghana. J. W. African Science Assoc. 3:141-147

Foggie A (1962) The role of forestry in the agricultural economy. In: Wills (ed), pp. 229-235

Foggie A, Piasecki B (1962) Timber, fuel and minor forest produce. In Wills (ed), pp. 236-251

Ghartey KF (1990) The Evolution of Forest Management in the Tropical High Forest of Ghana. African Development Bank. Abidjan

Gilpin A (1995) Environmental Impact Assessment: cutting edge for the twenty first century. Cambridge, UK: Cambridge University Press

Gregersen HM (1995) Valuing Forests: context, issues and guidelines. FAO For. Pap. 127, Rome

Hall JB (1981) Distribution and ecology of vascular plants in a tropical rain forest. Forest vegetation in Ghana. Geobotany 1. Junk, The Hague.

Hall JB (1987) Conservation of Forests in Ghana. Universities 8:33-42 (University Of Ghana, Legon, Ghana

Hawthorne WD (1992) Froggie: Forest Reserves Of Ghana: Graphical Information Exhibitor. Part 1: Manual for the program. Now published as Hawthorne 1995a

Hawthorne WD (1995a) Froggie (Forest Reserves Of Ghana: Graphical Information Exhibitor). Programme and database. Distributed with Hawthorne & Abu-Juam

Hawthorne WD (1995b) Forest Reserves of Ghana: Graphical Information Exhibition Manual. Froggie, Accra

Hawthorne WD, Musah Abu-Juam (1995) Forest Protection in Ghana. IUCN. Gland. (Circulated since 1993 as unpublished ODA report)

ITTO – International Tropical Timber Organization (1996) Tropical Forest Update vol 6, No.3 (includes several articles on reduced impact logging). ITTO, Yokohama

ITTO (1998) Annual review and assessment of the world timber situation 1997. Yokohama, Japan

ITTO (1999) Tropical forest update. Yokohama, Japan

IUCN Tropical Forestry Programme (1988b) Ghana: Conservation of Biological Diversity. Conservation Monitoring Centre, Cambridge, UK

Keeling W (1991) Ghana aims for a fitter forestry sector, Financial Times, 6 June

Kemp RH Whitmore TC (1978) International cooperation for the conservation of tropical and subtropical forest genetic resources exemplified by South East Asia, in: Eighth World Forestry Congress, Djakarta, 16-28 October 1978. FQL/26-11

King KFS (1974/75) Forest policies and national development. Unasylva, 9:27(107)

Klemperer W (1979) On the theory of optimal forest harvesting regulations. Journal of Environmental Management 1

Kotey (1998) Falling into place, Ghana, policy that works for forest and people. IIAD, United Kingdom. IIAD

Laing E (1991) Ghana Environmental Action Plan (Volume 1), Ghana

May E (1978) Canada's moth war. 4 New Ecologist 115

Miner PF (1994) Thresholds of Significance: Criteria for Defining Environmental Significance. CEQA Technical Advice Series (916) 445-0613), California

Pardo RD (1978) A review of forestry legislation in Indonesia, the Philippines, Papua New Guinea, Malaysia and Thailand. FAO

Poore D (1976) The values of the tropical moist forest ecosystems and the environmental consequences of their removal. FAO. FO: FDT/76/8(a). Rome

Prieur M, Lambrechts CL (1979) Modèle cadre relatif à l'impact sur l'environnement dans l'optique d'un aménagement ou d'une planification intégrée du milieu nature. Council of Europe, SN-PM (79) 2. Strasbourg

Ranjitsinh MK (1979) Forest destruction in Asia and the South Pacific VII (5) Ambio 192

Rees J (1985) Natural Resources, Routledge UK

Sayer J (1996) A Hierarchical Framework of Criteria and Indicators. Center for International Forestry Research. Boor, Indonesia 1996

Schmithusen F (1977) An annotated bibliography on forest legislation in developing countries. FAO Background Paper No. 12, Legislation Branch. Rome

Sheate WR (1994) Making an Impact: A Guide to EIA Law and Policy, Cameron May, London, UK. p. 23

Siegel WC (1974) State forest practice laws today. J. For., 208:72(4)

Swine MD (1981) Distribution and ecology of vascular plants in a tropical rain forest. Forest vegetation in Ghana. Geobotany 1. Junk, The Hague.

Taylor PW (1961) Normative discourse. Englewood Cliffs, NJ: Prentice Hall.

World Bank (1978) Forestry sector policy paper. Symposium on Federal Lands Forest Policy. Special number of the revue Environmental Law, vol. 8. IBRD. Washington, D.C

Part III – Thresholds and Standards for Environmental Media

Part III focuses on the use of standards and thresholds for the protection of the environmental media defined in the EIA and SEA Directives of the European Commission: human health, fauna, flora, biodiversity, soil, water, climate and air, landscape, cultural heritage and material assets and their environmental components and interrelations. The assessment of impacts is looked at from the perspective of the environment and ecosystems within each country's legal and planning context. The emphasis here lies on EU and German regulations, other countries looked at are Poland and the United Kingdom. Strict standards and assessment thresholds for especially sensitive and valuable states of environment (including its carrying capacity) are deemed to be crucial with the aim to further strengthen EIA's efficiency and quality in the future and to prevent from further damage of human's natural living basis. The question is addressed, which quantitative and qualitative values do exist and which additional thresholds will be important in the future for the assessment, prevention and mitigation in EIA of adverse effects such as pollution and consumption of natural resources.

Part III covers the essential environmental media of the EC EIA and SEA Directives. In the same way as the environmental media are very much interrelated with each, the chapters of Part III have dependencies. The Part starts with Chapter 26 by Therivel and Bennett on the application of noise standards in the UK. Noise will be one of the essential cumulative impacts in the future and demands special attention, if airports are extended.

Chapter 20 by Chorus deals with impacts on human health through pollution of drinking water and respective preventive measures followed by Chap. 21 by Socher, which delivers an excellent case study of flood prevention at the River Elbe in Saxony. The management of dams in transnational river basins with the help of sustainability impact assessment is one important measure to deal more intensively with flooding in the frame of global climate change.

Chapter 22 by Antwi and Wiegleb then discusses the assessment and monitoring of disturbances on biodiversity. It analyses the potential of large-scale mining areas for mitigating adverse effects on fauna and flora. Recultivation is closely linked to specific thresholds applied in these anthropogenic devastated areas.

The following chapters 19 and 20 by Mayer and 21 by Ritschel address impacts – especially chemical pollution – of soil resources. The emphasis in Chap. 19 is on the critical loads and level concepts which determines limit values for the acidification of soils. Chap. 20 gives a German overview on standards and thresholds for the essential environmental media soil, which is the basis for agricultural food production and a healthy human living quality. Chapter 21 recommends specific background and reference values for polycyclic aromatic hydrocarbons (PAH) and polychlorinated biphenyls (PCB).

Chapter 24 by Nixdorf explores the guidelines and standards of the EU Water Framework Directive and their contribution towards sustainable management of water resources, which are directly linked to groundwater (Chap. 27 by Chorus).

Air and climate on particulate matter evaluates how best to prevent the bioclimate with relevance for urban areas from harmful levels of dust particles (Chap. 23 by Johansson). The chapter, in contents closely connected to human health, proposes for instance to implement guideline levels of particulate matter for specific monitoring sites in relation to population exposures.

The last Chap. 25 by Chmielewski of Part III deals with the important issue of landscape protection. The chapter suggests at the example of Polish protected areas to consequently shift form a top-down to a bottom-up systems of nature conservation policy and enhance public involvement in decision-making.

19 Critical Loads and Levels Concept for Ecosystems

Robert Mayer

Department of Architecture, Urban and Landscape Planning, University of Kassel, Germany

19.1 Introduction

The concept of critical loads and levels, a method to estimate sensitivity towards stress factors and environmental risks to ecosystems, is introduced here with Environmental Impact Assessment (EIA) as its legal frame. The concept has been very much stimulated by environmental issues developing simultaneously in North America and in Europe: vegetation damage and acidification of lakes and rivers on the Canadian Shield and in various regions of northern and central Europe became known to a broader public. Dramatically reduced salmon population in Scandinavian rivers and regional forest dieback in remote European mountain regions were suspected to be caused by far reaching emission of acidifying components of industries and coal firing.

Environmental Impact Assessment (EIA) is an instrument for the prevention of environmental damage resulting from human activities in their widest sense. Formerly, negative effects from industry and manufacturing on human health, on goods and objects of economic or cultural value, were prevented or reduced mainly by technical means, applied during the production process. Enforcement for use of such techniques, imposed upon those who caused the damage or disturbance, was put into a legal frame, and norms and standards were established which allowed to control or limit negative effects at a tolerable level.

This situation characterizes the earliest stages of environmental protection in times where the term *environment* was just about to emerge. It ended towards the early 50ties of the last century with the rise of an environmental consciousness within the scientific community and the public in general. It was recognised more and more that the approach described above is not adequate to prevent a low level impact which at the beginning does not obviously exceed the no-effect level.

Standards and Thresholds for Impact Assessment. Edited by Michael Schmidt, John Glasson, Lars Emmelin and Hendrike Helbron. © 2008 Springer-Verlag

In such cases the impact is absorbed by the environmental system (as e.g. soil, atmosphere, water or vegetation, organisms, humans) without showing any visible effects, though slowly leading to a degradation of the system. In system theory the term *elasticity* is used to describe this property of a system. Many environmental systems show such an absorption or buffering capability in respect to external impacts: soils may neutralize or buffer acids from incoming rain or from plant roots and micro-organisms, while at the same time pH remains constant. Also the aqueous system of rivers and lakes can display an acid neutralizing capacity which allows many water organisms to survive in spite of a steady acid input.

In consideration of these facts, EIA was developed, beginning in the US and Canada, followed by the European Countries, as an instrument to avoid and to counteract these type of impacts. The objectives expressed in Article 3 of the EIA Directive states that environmental impact assessment will

> *identify, describe and assess (…) the direct and indirect effects of a project on the following factors: human beings, fauna and flora; soil, water, air, climate and the landscape; the inter-action between these factors (…); material assets and the cultural heritage* (EIA Directive 85/337/EEC).

The scope of questions raised in context of Strategic Environmental Assessment (SEA) is very similar to this (Schmidt et al. 2004). In both procedures, EIA as well as SEA, the systems to be examined can be comprehensively characterized as follows:

- low or moderate intensity of impact by a multitude of factors
- prediction of the system's reaction upon impact is difficult, with absence of easily visible or detectable effects in multi-component systems (biological systems, ecosystems)
- source of impacts or emissions often hidden or unknown
- multitude of sources
- source-effect-relations often hidden, unknown or consisting in a chain or sequence of processes, or knowledge insufficient

The situation thus described calls for a scientific base which must reach beyond a uni-dimensional dose-effect-relationship, as it is the traditional approach in toxicology, and in ecotoxicology, as well, a rapidly developing field in the last decades. It is necessary here to point out that the judgment whether an effect is significant or not completely depends upon the *sensitivity* of the receptor system to which a pollutant is transferred. It is of no practical use to claim that all human activities shall be reduced to a level at which the most sensitive systems remain undisturbed and can survive. In contrary to this extreme position, it is much more useful to evaluate the size, magnitude and regional distribution of the systems (for instance forests, rivers, lakes, alpine meadows, bogs etc.) likely to be affected by a project and then to open up this matter for a political and public discussion on the cost, public appreciation and boundary conditions under which the systems can be maintained in future.

19.2 Research in Dynamic Environmental Systems

A principle of life in all levels of organisation from cells to ecosystems is its non-static nature. Biological systems are characterised by their change in time, either in the form of structural or material fluctuations or cycles, or unidirectional to a definite end or death. Ecology teaches us that consequently the environment of living systems, through steady interaction with the organisms living in it, is concurrently subject to temporal changes. Simultaneously, the anorganic world is steadily subject to geological alterations by numerous exogen (sun, atmospheric) and endogen (geological, tectonic) forces which find their expression in the formation of the land surface (geobiosphere).

The scientific treatment of the biosphere which includes living organisms in their natural environment is closely linked with ecology as an emerging science in the late 19th century. The further development of ecology saw the incorporation and adaptation of findings from system sciences, mainly from physics and chemistry, and computer sciences. But it was not until post Wold War II that the study of natural systems was done in experimental ecology. One of the motors of development was the public interest and political commitment in environmental matters. Society and political representatives tried to cope with the problems, putting funds into research development, and science responded to these necessities. The International Biological Programme (IBP) was the first to initiate comprehensive studies of complex ecosystems in an international context (for Germany see Ellenberg et al. 1986; for Belgium: Duvigneaud, Denaeyer-De Smet 1964). Although not having in their focus environmental aspects at the beginning, the studies were the first in which a complete mass balance for all main chemical components was established (Ulrich et al. 1978; 1979). The mass balance clearly revealed element fluxes through the ecosystem, starting with atmospheric input, followed partly by storage in soil and/or biomass, and losses of materials to groundwater from the soil pool.

It is not possible here to give a comprehensive view upon todays progress in experimental ecology. The matter is complicated by the fact that the terminology has developed very differently in different countries, wich is reflected in national legislations. I will restrict myself to the German and English speaking area. An excellent overview of this matter is presented by Wulf (2001).

After what was mentioned in the introduction, the goal of EIA is defined very much in analogy to the definition of 'stress' by Berrett 1981 as

> *a perturbation that is applied to a system by a stressor which is foreign to that system or which may be natural to it, but in the instance concerned, is applied at an excessive level (e.g. phosphorus or water)"*.

Ten years earlier, Ellenberg (1972) had coined the term *"Belastung"* (engl. „*load*") in almost the same words as

> *an impact of factors, or a complex of factors, mostly caused by man, exerted upon a system.*

Both definitions have in common the view that stress, or load, is only to be stated if the magnitude or intensity of the initiating factor (or complex of factors) is in excess, or goes beyond, the scope of *normal* or *natural* conditions of the system. For example, If we consider the influence of the factor water, exerted upon an ecosystem, then only an excess or a deficit of water, compared with a long-term average, would be qualified as stress which may eventually destabilize or alter the system.

The example shows that the concepts of stress or load are closely related to stability of biological systems. When a system is not able to regenerate after the stress factor has come to an end, the impact is often qualified as *critical load.*

19.3 New Instruments in Environmental Policy

Increasing knowledge on how biological systems and ecosystems react upon anthropogenic impact stimulated the development of a method by which the sensitivity of ecosystems could be quantified. Such method should be suitable for making decisisions in environmental management and politics and, simultaneously, bring together the competing national based environmental programmes of neighbouring countries. A starting point for the concept of critical loads and levels was the Geneva convention for the reduction of long-range, transboundary air pollution transport in 1979 between UN-ECE countries (Economic Commission for Europe Convention on Long-range Transboundary Air Pollution) (Jahresgutachten Wiss. Beirat 1994). The agreement between the governments was put down in several protocols for the reduction of SO_2 emissions (Helsinki 1985), NO_x (Sofia 1988) and volatile organic compounds (VOCs, Geneva 1991). The following steps already included the critical loads concept, an approach capable to reveal and quantify various contributions to the same effect. The most intensive research was lanced in the fields of air pollution, water pollution and eutrophication of soils and aquatic systems (groundwater, surface water). Since soil plays a decisive role in each of these fields, a special chapter related to soil standards and thresholds is included in this volume.

The increasing awareness of a broader public to environmental changes and damages mentioned in the introductory chapter, and the political response to it, opened a long-lasting dispute over the cause of such damage. Often it led along the question what is just *natural* and what is man-made and, if so, what or who is responsible for the impact. The answer to these questions was very often given according to the interest of the participant in the discussion, without sufficient evidence based on scientifically sound observations for either of the answers. Simultaneously with the upcoming pressure in public opinion towards a sound explication for the environmental damages, the first results of comprehensive ecosystems studies mentioned above supported the view that long-range transport of air pollutants have a considerable impact upon the element cycles even in environments which had, so far, been considered as *natural* or *almost*-natural. Further studies and measuring networks were initiated on international as well as on various national levels.

19.4 The Concept of Critical Loads and Levels

Primary task of environmental policies is to distinguish *natural* changes of the environment from *man-made* alterations. This distinction is a prerequisite and a starting point for all action to be taken in order to prevent unwanted impacts. Considerations of this kind form the base for a concept and strategy which is known as that *concept of critical loads and levels*. It can be described as follows: the dynamics of undisturbed (or quasi-natural) environments within distinct limits (areas, patches, space elements) are observed as far as possible with all relevant aspects (structural parameters, chemical/mineralogical and biological components) in its variability in space and time. The set of all parameters characterizing this environment represents the a reference system, i.e. the natural state which all man-man impacts have to be related to.

Applied to any case of environmental assessment this means that the impact from a project under consideration is to be judged in the light of *further* man-made impacts, if there are any, *plus* all *natural* alterations of this very patch of environment in which the project is located.

In this concept, the behaviour of a system (i.e. a patch of environment including its biotic contents) - as characterized by a specific set of parameters - will constantly remain in this state within a certain corridor of fluctuation (also termed steady state), but it will leave this corridor and change its characteristics under the influence of impacts or stress factors. The sum of all impacts or stresses, man-made or natural, is considered as the total *load* acting upon the system. The point from which on the system leaves the corridor of steady state is called the *critical load*.

When the impact from various sources, man-made as well as natural, is defined as a *critical load* imposed upon the environment, then the focus lies predominantly upon the input of matter as, e.g. the deposition of atmospheric dust particles, chemicals, from volcanic action etc. Impacts of this type are usually expressed in the form of capacity parameters like mass-transfer per area and time (or grams per m^2 and days). The mass inputs are additive and depend on the capacity of the soil to transform/ mitigate harmful inputs, the effect may be absorbed or buffered over a long time period. The gradual build up of harmful substances will sooner or later exceed the buffer capacity and effects will be seen (i.e. damage to plants or animals living in the system, raise of toxic substances in seepage water etc.). The *mass balance* is the best means to estimate future effects. It provides the only suitable starting point to counteract unwanted effecs

Besides this, there is another group of environmental impacts which can be characterized by *intensity* parameters rather than by mass transfers. These may be concentrations of chemicals or any other substances in water or in air/atmosphere in units mass per volume, but also temperatures (expressed in °C or °F), chemical potential like pH, or radiation units. In this case it is adequate to talk about *critical levels* as the limits at which steady state of a system becomes transitory towards a new state.

If we look upon environmental problems, it is often easy to distinguish the types of impacts that lead to a disturbance or deterioration. Typically soils have a capacity to buffer or absorb effects from all sorts of pollutants upon secondary, dependant compartments such as groundwater, plants growing in and animals/microorganisms living in the soil. Therefore the *critical loads* concept is the adequate approach to deal with these impacts. The *critical level concept* is more often used to describe and counteract direct impacts of atmospheric pollutants upon plants, and of water pollutants upon aquatic organisms or microorganisms.

The pH in precipitation water (rain, fog, snowmelt), by its nature a chemical potential, is an intensity parameter highly relevant in the environment. It plays a role in direct biotic effects, if plants and animals are in direct contact, and it is important for soil mediated processes as well by claiming the soil acid neutralising capacity.

19.5 Advantages of Using the Critical Loads and Levels Concept

What then is the advantage of using the concept of *critical loads and levels,* compared to other strategies of environmental protection? It offers the opportunity to take into account the stresses and impacting factors on any environmental system, irrespective of their source, thus making it possible to weight the importance of the project under consideration, in relation to other sources which are correspondingly in effect. It allows estimating the *sensitivity* of a specific environment with respect to specific stress, which is a primary task of environmental policy. The term sensitivity/sensibility is often used when the capacity of a system with respect to stress input is filled up, or exhausted, as e.g.:

- adsorption capacity of a soil for the storage of chemicals in an inert or slowly reactive form
- the buffer capacity of soils, sediments or water bodies for acids/bases, also termed acid/base neutralizing capacity (ANC, BNC)
- transformation capacity of humic soil horizons, responsible for the metabolization and mineralization of organic chemicals
- water holding capacity

Exhaustion capacities as a result of environmental impact leads to a change in the steady state conditions, or system destabilisation. One way to describe such chains of processes is the use of *mass balances*, i.e. by balancing all inputs and outputs of a substance to each compartment of the system within a definite period of time, taking into account all forms and speciations this substance may take (see e.g. Baccini 1996).

Table 19.1. Critical Loads and Levels

Critical loads		
Environmental Compartments		
Soils, sediment and rock layers	Water	Plants
Capacity to absorb, store and/or transform, metabolize substances (chemicals, organisms) that have the potential to disturb or damage the system or related systems. The value of this parameter limits the load that can be neutralized or bufferd by the system		
adsorption capacity transformation capacity acid neutralizing capacity	Alkalinity (acid neutralization capacity)	leaf surface and/ or mass
Pathways by which an external impact may take place (load)		
- atmospheric input of particles, aerosols and gases (rain, snow, fog, dustfall) - liquid input from seepage water	- atmospheric input of particles, aerosols and gases (rain, snow, fog, dustfall), input from erosive processes - liquid inflow from point or diffuse sources	- atmospheric input of particles, aerosols and gases (rain, snow, fog, dustfall) - radiation
Sources for natural and anthropogenic impacts (contributing to the load)		
industrial emissions (anthropogenic) wind erosion (natural; anthropogenic) volcanic (natural)	industrial emissions (anthropogenic) agriculture (anthropogenic) soil and groundwater flows (natural, anthropogenic) wind and water erosion (natural; anthropogenic) volcanic (natural)	industrial emissions (anthropogenic) agriculture (anthropogenic) wind erosion (natural; anthropogenic) volcanic (natural)
Critical levels		
effect of a disturbance upon.....		
Human Health		Plants & Animals
Parameters indicating the level of disturbance		
element concentration and pH in drinking water element concentration in ambient air ambient radiation intensity physiological parameters		element concentration and pH in rain snow, fog and in ambient air of organisms ambient radiation intensity physiological parameters

Table 19.1 shows some examples for different environmental compartments to which the concept was applied. The critical load must specify three aspects of the environmental problem to be solved:

1. the type of substance which provokes an effect, or *stressor* (e.g. pollutant; chemical substance; physical, chemical or biological stress)
2. the path through which the stressor comes into effect (e.g. atmospheric transport, land use practices, flooding, irrigation, drainage etc.)

3. the target of the stress where an impact is observed, i.e. the system to be protected (e.g. soil, atmosphere, health of humans, animals, plants)

It is important to notice that several stressors (1) may come into effect on different paths (2) to act upon the same target to provoke the same symptoms (3).

It has been, so far, not much being said with regard to the substances or type of pollutants appropriate to be handled by the critical load concept. In fact, there is no principal restriction to this, as long as a substance can be individually distinguished from other substances in its emission, transport and reaction in any target system. There is one exception to this: the concept seems not very suitable to substances which are ubiquitous emitted from diffuse sources, though unevenly distributed, which have long life cycles and which are globally distributed in the atmosphere by transboundary transfer. Best example is the carbon dioxide (CO_2), which is emitted from numerous anthropogenic sources (burning of fossil fuels), but also from countless, globally distributed, biologically active surfaces (decay of organic matter by micoroorganisms). The result is a slow increase in the CO_2 concentration in an atmospheric layer where concentration differences are rapidly levelled out be rapid mixing of the layer. The contribution of single sources can not easily be distinguished in this mixed layer, and any effect or impact in environmental systems develop only gradually. It can not be attributed to a definite source neither can it be avoided by local or regional counteraction. A similar situation is found for other climatically relevant gases like methane. Consequently, the solution of these problems is discussed on a supranational level, as e.g. UNEP and WHO, as part of global climatic change.

19.6 The Deduction of Critical Loads and Legislation in Germany

According to Nagel and Gregor (1999 p. 10), a successful implementation of the critical loads concept into international policy has to proceed step by step, beginning with (1) cause-effect field experiments and environmental monitoring, followed by (2) an iterative process of finding a state-of-the art agreement on best guess thresholds for sensitive ecosystems. As a next step there must follow (3) a mapping programme by which the ad-hoc situation of site-specific environmental stress in European regions is represented, depicting the areas with highest loads, where critical loads are reached or exceeded. The final step, and ideally the objective of the whole procedure, must be (4) an intergovernmental agreement to reduce the environmental stress to the extent that critical loads are not exceeded on the whole territory.

For the time being we can say that, in 2006, the most of the European countries have arrived at step (3), while step (4) is still a hope for the future. Mapping of critical loads, based upon a large number of existing data sets imported into a Geographical Information System (GIS), is available for the territory of Germany

- for acids from atmospheric deposition to terrestrial ecosystems

- for nitrogen from atmospheric deposition to terrestrial ecosystems
- for heavy metals and persistent organic compounds.

The critical loads are mapped on a topographic base in the scale of 1:100 000. Critical loads for aquatic ecosystems with respect to acidification and eutrophication inputs are in progress.

As soils take a mayor role in site specific critical loads, the following Chapter 20 on soil standards will give an example for the calculation of critical loads for acid deposition. For a more detailed explanation and discussion of the concept of critical loads & levels see Nagel and Gregor (1999).

19.7 Conclusions and Recommendations

In summary it must be stated that the critical loads and levels concept has, until now, not been implemented in any legal frame. This does not mean that mapping and monitoring of environmental loads had no practical results so far. At the European level it has been recognized as a strategic instrument in environmental policies, useful to evaluate regional pattern of loads. In Germany, planning authorities make use on the national level, state and community level of the information provided by critical load maps to explain to the public and justify necessary political decisions in the environmental sector. For public awareness is, as we know, often a better impellent for environmental protection than standards and thresholds which exist on paper only, but are not legally enforced against an uniformed public. The critical loads and levels concept offers an appropriate way to give necessary information to the public. With the mass balance it has a sound scientific base which is easy to understand by a non-scientific community.

With the knowledge gathered so far, and with the methods developed by international effort over the last 20 years, the critical loads concept can turn into a powerful tool for environmental protection. The most urgent step forward seems to be improvement of the legal frame in EU legislation, and adjustment of national legislations with appropriate inclusion of the concept. Splitting into different acts for interdependent and interrelated environmental compartments must be replaced by a comprehensive environmental code. The critical loads and levels concept is an appropriate base for such a code, for it is comprehensive with regard to sources and target systems.

The concept itself certainly can still be improved and, provided its inclusion in a legal frame, it merits some further efforts. In view of the wealth of environmental data available in all member states of the EU it seems promising to link the concept with a Geographic Information System. As an example, we can look at soil as an environmental compartment decisive for the bearing capacity with regard to acidic pollutants. Today the geological and soil surveys of the federal states of Germany offer complete maps and data sets on a scale of at least 1:50 000 for all relevant soil chemical and physical parameters, offering a perfect base for critical loads map.

Finally it should be pointed out that the concept of critical loads and levels should more and more be introduced into the discussion of environmental issues and on political priorities for planning. The concept reveals the limitations to land use and industrial as well as infrastructural development imposed by naturally fixed site factors. It shows the values, monetary as well as emotional, which we may loose from environmental impact when the bearing capacity is exceeded on the long term. The cost of this loss – it may be a certain type of vegetation, a historical landscape, buildings as part of the cultural heritage, or a groundwater body with drinking water quality – can than be evaluated und be balanced with cost for avoidance or technical reduction of the impact.

References

Baccini P (1996) Regionaler Stoffhaushalt, Erfassung, Bewertung und Steuerung, Spektrum Akademischer Verlag Heidelberg; Berlin; Oxford

Barrett GW (1981) Stress ecology: an integrative approach. In: Barrett GW, Rosenberg R (eds) Stress effects on natural ecosystems. J Wiley and Sons, Chichester, pp. 3–12

Duvigneaud P, Denaeyer-De Smet S (1964) Le cycle des éléments biogènes dans l'écosystème forêt. Lejeunia (Bruxelles) NS 28:1-148

Ellenberg H (1972) Belastung und Belastbarkeit von Ökosystemen. Ges. f. Ökologie, Tagungsbericht 1972, pp. 19–26

Ellenberg H, Mayer R, Schauermann J (eds) (1986) Ökosystemforschung. Ergebnisse des Solling-Projekts 1966-1986. Ulmer-Verlag, Stuttgart, 507 p.

Jahresgutachten Wiss Beirat (1994) Welt im Wandel: Die Gefährdung der Böden, Jahresgutachten 1994 des Wissenschaftlichen Beirats der Bundesregierung für Globale Umweltveränderungen, Economica Verlag, Bonn, 263 p.

Nagel H-D, Gregor H-D (eds) Ökologische Belastungsgrenzen - Critical Loads and Levels. Berlin, Heidelberg usw. (1999), Springer Verlag, 259 p.

Schmidt, M., E. Joao, E. Albrecht (2004): Implementing Strategic Environmenal Assessment, Springer Verlag Heidelberg, 742 p.

Ulrich B, Mayer R, Khanna PK, Prenzel J (1978) Ausfilterung von Schwefelverbindungen aus der Luft durch einen Buchenbestand. Z. Pflanzenern. Bodenk. 141:329-335

Ulrich B, R Mayer, PK Khanna (1979) Deposition von Luftverunreinigungen und ihre Auswirkungen in Waldökosystemen im Solling. Schrift. Forstl. Fak. Univ. Göttingen und Nieders. Forstl. Versuchsanstalt, 58 (J. D. Sauerländers Verlag, Frankfurt am Main), 291 p.

Wulf AJ (2001) Die Eignung landschaftsökologischer Bewertungskriterien für die raumbezogene Umweltplanung, im Selbstverlag, Augustusdorf, 560 p.

20 Soil Standards and Thresholds

Robert Mayer

Department of Architecture, Urban and Landscape Planning, University of Kassel, Germany

20.1 Introduction

There are two completely different perspectives under which standards and threshold values for soil related parameters are defined in the context of environmental protection. The different views arise, due to the fact that soil is a dynamic system, containing by definition living organisms as part of an ecosystem. A soil can be degraded, or even destroyed, gradually or rapidly changing its functionality in the ecosystem. It is therefore considered as an object of environmental protection *per se*. In Germany the Federal Soil Protection Act forms a legal frame for this protection.

A quite different perspective results from the interrelationship between soil and the other compartments of ecosystems like vegetation, crops, ground and surface water and the atmosphere close to the earth's surface. Hereby soil comes into the focus of the environmental protection of water, human health, health of crops and livestock etc. As a result, there are standards and threshold values aiming at the protection *of* soils (their functionality and integrity), but also protection *from* soil-mediated impacts to other targets and media (health, vegetation, water and atmosphere). In the discussion of soil standards these differences are not always clearly distinguished.

20.2 Standards for Soil Protection

Primary intention of the German Federal Soil Protection Act *(Bundes Bodenschutzgesetz- BBodSchG)* is the safeguard and restoration of soil functions. The soil functions, specified as *natural* functions, are:

- offering space of living for man, animals, plants; maintaining the cycles of water, minerals and organic matter;

Standards and Thresholds for Impact Assessment. Edited by Michael Schmidt, John Glasson, Lars Emmelin and Hendrike Helbron. © 2008 Springer-Verlag

- medium for the decomposition, and buildup of organic and inorganic substances by its buffering, filtering and transformation capacities;
- offering raw materials and growth medium for agriculture and forestry uses.

Standards and threshold values under the Soil Protection Act are set primarily with the intention to secure and maintain this functionality, and to avoid anthropogenic impact from land use practiques, waste deposition, immissions etc. Restoration of hazardous waste sites is included expicitly in the act as a special form of soil degradation in the past.

The aptitude or function of a soil to be used for anthropogenic purposes such as crop or timber production, and the capability to maintain the cycles of water, energy and biomass in the human environment is linked, like other functionalities, to physical, chemical, and biological soil properties which define the different site-specific aptitudes of soils. Specific functionalities are, e.g. storage capacity for water and chemicals, buffer capacity, nutrient supply for plant growth. These functionalities are also at the base of the critical loads concept explained in the previous chapter.

Many of these parameters are fairly constant in time under natural conditions, given a defined area (*pedon*), and irrespective of minor seasonal fluctuations. This is true for most soil contents of chemicals in their various forms or associations, and also for most physical and chemical parameters with their characteristic pattern of distribution in the soil profile.

Yet, under anthropogenic impact the parameters may leave their *natural* scope of values, as a consequence of, for instance atmospheric deposition of chemicals, agricultural use or change in the temperature or water regime. Change in one of the soil parameters may imply changes in related parameters, but also in the functionality of the soil. The limiting value taken by any parameter beyond which reduced functionality or disfunctionality is observed when exceeded, can be seen as a thereshold value. Expressed in the terminology of environmental impact assessment (EIA), the exceedance of a threshold value is to be apprehended as a significant environmental impact. In contrast to this situation, the *standard* or *background* values represent the natural scope that a parameter may take under the absence of any anthropogenic impact, or its non-relevance with regard to given quality criteria.

The establishment of standards and the definition of thresholds must strictly be based upon scientific experiences. We will concentrate here on the most important type of soil parameters under a practical view. These are the soil contents of potentially toxic substances which are found in low concentrations under natural conditions, or which are even non-existent in natural environments (xenobiotic). In literarture we find extensive lists of background values for element concentrations in natural soils (see e.g. Scheffer/Schachatschabel 2002, p.329, tab. 6.3-1) . Such values are somtimes included in official lists of standards and thresholds as, e.g., in the so-called *Berliner Liste (1996)* as part of the environmental regulations in the federal state and city of Berlin (Hoffmann-Hoeppel et al. 2005, pp. 3510). Such values have no legal relevance, but rather serve as indicators for the strength of anthropogenic impact. Historically, the scope of background and threshold values has been developed over decades by close cooperation between analytical chemists, scientists and biologists, very often in a step by step approach treating upcoming environmental problems and deriving benefits from methodological and instrumental developments. Two strains of development can be seen, one in the field of heavy metals (Hg, Pb, and Cd as the most prominent elements), the other in the field of persistent organic compounds, both of which show a tendency for accumulation in surface soils. These compounds tend to associate with clay minerals and humic substances to such a degree that considerable accumulation levels are reached (Mayer 1991, 2003).

Accumulation of toxics substances in soils may have a long-term effect on soil chemical or biological processes, thus affecting soil functions. But this must not necessarily go along with a reduction in the productivity function of the soil (for crops, timber), or with pollution of seepage and ground water. The impact and deterioration which soil has undergone can be hidden for a long period of time, because it could only be manifested by changes in microorganisms, followed by a decrease in the decomposition potential for organic matter which may have been overlooked, or not examined.

20.3 Standards for the Protection from Soil-Mediated Hazards

Standards and thresholds of this type are set in order to avoid hazards to environmental media other than soil (water, atmosphere, vegetation) and to human health. The values are not in principle different from the standards for soil protection but their scaling is determined rather by the quality requirements of the target system (e.g. human health, water quality) than by structure, dynamics and integrity of natural soils. The hazards of this type have in common that they are soil-mediated, although the paths along which the target system may be affected can be very diffent. Usually the standards and thresholds are specificly corresponding with this path, as shown in the following selective list of frequently applied values (choice of values in the following chapter):

- limiting the application of sewedge sludge to agricultural soils: the target system is the crop produced on the soil to which the sewedge sludge is applied. The eventual transfer of toxic substances (path) from sludge - to soil - to plants - to man/animal determines the scaling of threshold values, together with the quality requirements in respect to health of animals and/or humans.
- limiting the atmospheric input of acids: acid air pollutants from various sources may have impact upon soils. Acid input is followed by the loss of basicity and drop of pH. The critical load of acids marks the threshold beyond which such impact takes place. Again the target system which determines threshold values in actual legislation is not the soil but vegetation, here primarily forest vegetation, since agricultural crops are usually protected by measures (liming,

fertilization) which counteract acidification. Soil protection is here a by-product of protection of vegetation, and as it will be shown in the following chapter 22.4, comes into effect under the Federal Immissions Control Act, the EIA Act and the Building Code.

- limiting the atmospheric input of potentially toxic chemicals such as heavy metals, nitrogen compounds, and organics. As in the case of acidification, the target system is primarily the vegetation cover, but next to it soil organisms and roots of higher plants invading soils and taking up nutrients and water. Consequently, soil protecting standards are again set primarily with regard to vegetation and microorganisms, and soil protection is a function thereof. Legislation is covered by Immission Control, EIA and Soil Protection Acts[1].
- restricting polluted soils for specific land uses: the concentrations of toxic elements in polluted soils may be unsuitable for specific uses because of health risks for sensitive target systems (human health, drinking water, crops and animals). For instance, governmental and administrative planning authorities use chemical soil standards for land use and building zone planning.
- the restoration of degraded or damaged soils (e.g. former industrial sites; brownfields; polluted soils).

20.4 Soil Standards and Threshold Values under Various Acts in Germany

The current situation in Germany with respect to soil standards and thresholds is characterized by great inhomogeneity, complexity and insufficiency in the sets of values which often can be explained only as a result from different historical roots and very different developments in environmental protection. It is for this reason that we have soil standards developed within very different legal frames, as mentioned above, and on many administrative levels under state, federal and european legislation. In most libraries the literature is out-of-date, and the newest regulations in vigour are often found via Internet only.

Another difficulty rises from the fact that soils are a very difficult and complex matter to handle because of its variability and dynamics. For this reason each soil parameter, when being measured or monitored as a single value representing a definite surface, must be accompanied with a complicated, very detailed description of sampling procedure, replication, digestion and chemical or physical analyses. This situation requires the observation of complicated technical specifications in the assessment of soil parameters. Without going into further detail, it is tried in the following to give a choice of relevant soil standards and

---

[1] In German legislation the term emission denotes the mass of a pollutant - usually expressed by a capacity term (freight, flux) – emitted into the environment by any source. The term immission is indicating the result of one or several such emissions in the environment, usually express by an intensity parameter (concentration of the pollutant in ambient air).

thresholds which are actually in use in different fields in Germany for instance under the Building Code, the Spatial Planning Act, EIA Act, Federal Immissions Control Act, Water Management Act and Federal Soil Protection Act. The choice is far from complete, but it shows at least the different approaches and strategies under which soil-related environmental protection is working today. Table 20.1 gives an excerpt from the regulatory values under the German Federal Soil Protection Act, Annex 2 (Hoffmann-Hoeppel et al. 1999 -2005).

Table 20.1. Soil standards and thresholds under the Soil Conservation Act (BBodSchV), Annex 2 (Hoffmann-Hoeppel et al. (eds) (1999-2005) vol 1)

	path of impact: soil/humans (direct intake) test medium: dry soil (solids) dimension of values: mg per kg dry soil							
	childrens playgrounds	residential areas	parks and recreation	industrial and bussiness	children's playgrounds	residential areas	parks and recreation	and bussiness
limit of examination	Pb				PCB$_6$			
	200	400	1.000	2.000	0,4	0,8	2	40

path of impact: soil/plant (uptake via roots) test medium: dry soil (solids) dimension of values: mg per kg dry soil

land use type: agriculture

examination	As: 0,4			Cu: 1	Ni: 1,5	Zn: 2,0		

land use type: grassland

action	As: 50	Pb: 1.200	Cd: 20	Cu: 1.300	Ni: 1.900		Hg: 2	PCB$_6$: 0,2

land use type: residential areas presumably polluted from unknown source

precaution	Pb in clay: 100 loam: 70 sand: 40	Cd in clay: 1,5 loam: 1 sand: 0,4	Cu in clay: 60 loam: 40 sand: 20
acceptable maximum input (g per hectare and year)	Pb: 400	Cd: 6	Cu: 360

path: soil/groundwater test medium: soil water at the border between saturated and unsaturated zone dimension of values: µg per L in soil water

examination	As: 10	Pb: 25	Cd: 5	Cu: 50	Ni: 50	Zn: 500	Hg: 1	PCB$_{tot}$: 0,05

The complete list is more extensive, especially the organic compounds are quite numerous and only the polychlorinated biphenyls (PCBs) are presented here as an example to show how threshold values are organized in the regulatory annex to the Soil Protection Act. The system of standards and threshold values are applicable relative to the path through which an impact may come into effect.

1. from soil to man (direct contact or intake, for instance intake of sand from playgrounds by children)
2. from soil to plants (agricultural crops, eaten by man and/or fed to animals)
3. from soil to groundwater (soil water transport)

It is organized in 3 steps with increasing levels of hazard.

1. examination level (Prüfwert): if this value is exceeded by any impact, the authorities are obliged to examine whether a noxious or detrimental soil change can be observed/recognized
2. precautionary level (Vorsorgewert): if this value is reached or exceeded, authorities shall take it as a fact, without further examination, that a noxious or detrimental soil change has occurred
3. action level (Maßnahmenwert): if this value is reached or exceed, authorities shall take it as a fact that a noxious or detrimental soil change has taken place, prompting measures for restoration of the soil and/or protection against health hazards.

The list is relevant for governmental or administrative authorities in specific situations:

1. planning of land use and rezoning according to the Building Act
2. in decision-making in EIA and Strategic Environmental Assessment (SEA)
3. in the treatment and sanitation of contaminated former industrial sites

Table 20.2. Regulation for sewage sludge applications under the German Waste Management Act - Sewage Sludge Directive (AbfKlärV)

	sewage sludge[1]	soil[2]
	values in mg per kg dry matter	
Pb	900	100
Cd	10 (sandy and acid soils: 5)	1,5 (sandy and acid soils: 1,0)
Cr	900	100
Cu	800	60
Ni	200	50
Hg	8	1
Zn	2500 (sandy and acid soils: 2000)	200 (sandy and acid soils: 150)
PCB_{tot}		0,2
PCDD/PCDF		100 ng per kg dry matter

[1] Sewage sludge application prohibited on arable soils if at least one of the listed values in the sludge are exceeded.

[2] Sewage sludge application prohibited on arable soils if at least one of the listed values in soils (plough layer) are exceeded.

Table 20.2 shows the threshold values in agricultural soils to which sewage sludge is applied. Acceptable sludge loads (freight) depend upon the elemental content in potentially toxic heavy metals and organic compounds of the soil *before* application as well as from the content of the sludge to be applied. The values are to be observed by facilities for waste water treatment producing and selling the sludge and by the farmers who distribute it over their fields. The whole matter is regulated in Germany by the Sewage Sludge Directive (AbfKlärV). It is the intention of this Directive to avoid negative impacts on soils by accumulation of potentially toxic substances, and related hazards for plant production and groundwater quality. The Directive also specifies the number of consecutive applications over the years, the application of sewage sludge mixtures with composted organic matter, and the monitoring of soil contents in the listed elements or compounds.

Soil protective regulations are also found under the Federal Immissions Control Act (BImSchG), TA Luft Directive. If an industrial project requires administrative authorisation, or if it is subject to environmental assessment, or if air quality in an area has led to a designation as immission protection zone, the immission concentrations as well as the deposition rates of pollutants can be regulated by prescription of emission control measures.

An example for maximum deposition rates based upon this legislation is given in Table 20.3. Values in the list are determined as annual averages. The regulation requires that the atmospheric deposition of dust and particulate matter shall at no point of the protected area exceed the limit. Implicitly, this regulation considers a deposition rate below that limit as non-relevant in respect to soil functionality. From a scientific point of view, this assumption is rather weakly justified. The behaviour of the chemicals in the list is certainly very much dependent upon specific site conditions such as texture and pH of soil, soil minerals and hydric conditions. For this reason, a universally applied threshold will be too high for very sensitive soils and unnecessarily high in other cases.

The concept of critical loads dealt with in Chapter 20 tries to counter this shortcomings in the methodology by taking into account the sensitivity of different soils in respect to environmental impacts, but also the influence of major land use types. It is based on an complete consideration of the mass balance of the chemical in question, including inputs, outputs and storage changes at each specific site. Different methodologies have been developed in a number of *Inter-*

Table 20.3. Allowable input of arsen and heavy metals with atmospheric deposition of dust particles under the German Federal Immissions Control Act (BImSchG), TA Luft (TA Luft, Annex A, 2006)

Elements and their inorganic compounds	Deposition rate in µg per m² and day averaged over a period of 1 year
As	4
Pb	100
Cd	2
Ni	15
Hg	1
Tl	2

national Coopertive Programmes (ICPs) on the assessment and monitoring of air pollution effects on forestes, on acidification of surface water bodies, on crops, and on materials including historical and cultural monuments. An integrated monitoring programme has been launched in 1989 under the UN/ECE Convention on long-range transboundary air pollution with the goal to the assess the actual state of the environment in its regionalized pattern. Compared to the use of fixed environmental standards, the concept of critical loads and levels is much more demanding when it comes to availability of site specific data. But in countries with a dense network of soil information systems, as in many EU Member States, this is deemed to be possible. The methods of how to derive critical loads has been well demonstrated and documented by Nagel and Gregor (1999).

In Table 20.4 as an example, the critical loads for acids applied to the surface of Germany mapped on topographic base in the scale of 1:100.000, is represented from this work. Similar presentations exists for Nitrogen.

The critical load for acids must be understood as an expression of threshold with regard to atmospheric deposition of acids and acidifying substances (see Sect. 20.3). When the actual input is in excess of this threshold, a chain of subsequent processes comes into effect. Together with its vegetation cover, the acid neutralizing capacity of the soil is consumed, followed by a pH drop and leaching of basic cations, eventually of aluminium ions and other metal cations. The sum of all this is to be stated as a negative environmental impact, soil degradation and a hazard for water quality. Table 20.4 shows that critical loads for acids are quite variable and the distribution of site sensitivity over the area of the country is very uneven. One single threshold value for the whole of the country would certainly not be adequate. On the other hand it must be recognized that there is, until now, no legal instrument by which these standards could be implemented.

Table 20.4. Critical loads for acid deposition to terrestrial ecosystems in Germany according to Nagel and Gregor (999) p. 246f., Table 2)

Tolerable annual input of acids to soil (eq per hectare and year)	Area percentage of the respective land class (% of total surface of Germany)
< 200	2,8
200 - 500	9,0
500 - 1000	33,0
1000 - 2000	37,8
> 2000	17,4

20.5   Conclusions and Recommendations

The situation in Germany is characterized by the fact that soil standards and thresholds have been established over a long period of time under different legal frames, mostly independent from each other, for different protection goals: human health, quality of groundwater and surface water, vegetation. The Federal Soil Protection Act, coming into effect in 1998, was the last one in environmental legislation. It is therefore a subsidiary act which comes only into effect if no other law takes the part.

In a future perspective the situation in soil protection must be improved by a comprehensive environmental law which overcomes the splitting of environmental legislation and strengthens the link with planning and building law. At present, European law is not developed to a degree that it could compensate for the deficits in national legislation.

What can scientific research contribute to secure and improve soil conservation and protection in future? It is no question that setting of standards and thresholds for soils is a matter of scientific research. It will be necessary in future, to widen our knowledge of mechanisms how soils respond upon impacts and on the interdependence of environmental compartments, where often soil is at a crosspoint. A new and promising aspect in the methodology is opened by the concept of critial loads and levels which is treated seperately in chapter 19.

References

Bachmann G, Thoenes H-W (eds) (2000) Wege zum vorsorgenden Bodenschutz. Erich Schmidt Verlag, Berlin

Blume H-B (ed) (1992) Handbuch des Bodenschutzes, 2. ed., ecomed Verlag, Landsberg/Lech

EIA Directive 85/337/EEC, 27.06.1985, Internet address: http://eur- lex.europa.eu/ LexUriServ/LexUriServ.do?uri=CELEX:31985L0337:EN:HTML last accessed on 11.09.2006

Ellenberg H, Mayer R, Schauermann J (eds) (1986) Ökosystemforschung. Ergebnisse des Solling-Projekts 1966-1986. Ulmer-Verlag, Stuttgart

Fiedler H-J, Rösler H-J (1993) Spurenelemente in der Umwelt. 2 edn, Fischer Stuttgart

Hoffmann-Hoeppel J, Schumacher J, Wagner J (eds) (1999-2005) Bodenschutzrecht-Praxis

Mayer R (1991) The impact of atmospheric acid deposotion on soil and vegetation. In: J-P Vernet (ed.): Heavy Metals in the environment (Trace Metals in the Environment 1), Elsevier, Amsterdam, London, 1:1-36

Mayer R (2003) Schwermetallhaushalt, (ecomed Verlag) Landsberg/Lech 2003. (= Kap. IV - 2.2.7 In: Fränzle, Müller, Schröder, Handbuch der Umweltwissenschaften)

Mayer R (2003) Schwermetallhaushalt, Kap. IV - 2.2.7 In: Fränzle, Müller Schröder, Handbuch der Umweltwissenschaften, ecomed Verlag, Landsberg/Lech, pp. 1–13

Nagel H-D, Gregor H-D (eds) (1999) Ökologische Belastungsgrenzen - Critical Loads & Levels. Springer Verlag Berlin, Heidelberg

Scheffer F, Schachtschabel P (2002) Lehrbuch der Bodenkunde, 15 edition, Spektrum Akademischer Verlag

Sewage Sludge Directive of Germany (2006) http://www.gesetze-im-internet.de/ bundesrecht/abfkl_rv_1992/gesamt.pdf/, last accessed on: 11.09.2006

TA Luft (Germany), Annex A, 2006. Internet address: http://www.bmu.de/files/pdfs/ allgemein/application/pdf/taluft.pdf, last accessed on 14.09.2006

Ulrich B, Mayer R, Khanna P K (1979) Deposition von Luftverunreinigungen und ihre Auswirkungen in Waldökosystemen im Solling. Schriftenreihe Forstwiss. Fak. Univ. Göttingen vol 58:291

Ulrich B, R Mayer, Khanna P K, Prenzel J (1978) Ausfilterung von Schwefelverbindungen aus der Luft durch einen Buchenbestand. Z. Pflanzenern. Bodenk. 141:329-335

21 Soil Background and Reference Values for PAH and PCB

Jürgen Ritschel

Brandenburg State Office for Environment, Germany

21.1 Introduction

Registering the actual condition of soil as regards pollutants plays an important role within the framework of implementing the legislation on soil protection. It is decisive for the characterisation of the soil functions and forms the basis for deriving specific background values. In this regard, the assessment of the condition of the soil focuses on dealing with existing substance-related soil pollutions and restricting new harmful entries of materials into soil.

Data on soil pollution has to allow representative statements; it has to be comparable and to be available with the required differentiations. The establishment of a uniform register on soil condition as well as the derivation of specific reference values in dependence on soil properties, the geological origin of the initial material for soil formation, the type of land use and the regional type are therefore the basis for assessing the soil condition.

A space-covering stock taking that marks the soils comprehensively includes statements on pollutants (heavy metals and organic pollutants), nutrients, properties of location and environment as well as the structure of the soil profile.

The indivisibility of data on substance-related pollution and site data within the soil condition register form the basis for deriving reliable and representative data which is to be generalised within the framework of an Information System on Soil Protection.

Whereas there were partially comprehensive data on natural soil properties, nutrients and heavy metals in the soil in the Brandenburg state, the existing data on the content of organic pollutions in the soil was, however, insufficient. Until 1994 only a few examinations were carried out which did not allow an extensive evaluation. The Brandenburg Ministry for the Environment therefore implemented projects to determine the content of pollutants of topsoils ("Soil Standard Values: Organic Pollutants and Heavy Metals (MUNR 1998) and "Transfer Factors for Pollutants Soil – Plant for Typical Brandenburg Soils and Crops" (MUNR 1997)).

Standards and Thresholds for Impact Assessment. Edited by Michael Schmidt, John Glasson, Lars Emmelin and Hendrike Helbron. © 2008 Springer-Verlag

Based on these projects the following was performed for the groups of polycyclic aromatic hydrocarbons (PAH_{16}) and polychlorinated biphenyls (PCB):

- analysis of soil pollution in dependence on soil use
- derivation of background and reference values in soils of different types of land use
- Characterization of the transfer of PAH and PCB from soil to plant for typical Brandenburg soils and crops.

The transfer of organic substances from soil into plant mainly depends on the kind of the organic combination and is therefore also determined by organic combinations of the respective pollutant in the soil, as well as the plant species and, as the case may be, the concrete place of accumulation (part of the plant). The transfer factors deliver the fundamentals to assess the entry of pollutants into the food-chain of human being and animal.

21.2 Determination of Terms and Fundamentals of Assessment

Background and reference values are derived dependently on the substrate (soil properties, origin of the initial material for the formation of soil), type of land use and regional type. Whereas the geological origin of the initial material the formation of soil essentially influences its content in the soil as regards heavy metals, it is the diffuse entry into soil for the organic pollutants.

The publication "Background values for inorganic and organic substances in soils" of the Joint Federal State Working Group for Soil Protection (LABO 2003) and the Federal Soil Protection Act (BMU 1998) describes and legally fixes once again, among other things, background values, reference and soil precaution values.

The *background content* of a soil is made up of its geogenic basic content and the ubiquitous distribution of substances as a consequence of diffuse entry into the soil.

The "ubiquitous/diffuse" formulation distinguishes between the background content and the actual contents which are decisively higher compared to the background values due to high penetration levels of substance at specific sites (influence of selected emitters, contaminated sites). The background content of organic substances is mainly identical with the ubiquitous entries into soil that are redistributed by pedogenic processes and effects of land use in the soil.

Background values are representative values for generally spread background contents of a substance or a group of substances in soils.

Background values for soils are based on the ascertained background contents and denote the representative substance contents in soils by indicating statistic parameters and the differentiation with regard to soil properties and local conditions, as well as reference parameters of land use and regional type.

Reference values are used, under certain framework conditions, to compare different concentrations regarding the spatial reference and the features of soil science. They are not ascertained on an effect-related basis.

Reference values can be derived from background values by the latter being evaluated for a specific issue and concrete individual cases in a specific form; those are used as reference values.

In the following, the 90th percentile is used as a reference value. It serves to exclude soils with increased contents of organic pollutants without considering the phyto-, zoo-, human- and eco-toxicological aspects.

Soil precaution values are those values from which it has generally to be assumed, that, if exceeded, there is the risk of a dangerous change of the soil, in considering geogenic or extensive settlement-related contents of pollutants. Soil precaution values must ensure long-term protection of the soil from future impacts. Long-term protection aims at using soils in any different way also in future (Hüttl 1999).

When assessing "precaution" or, as the case may be, "concern" the natural functions of the soil have particularly to be considered as:

- basis of life and living space for human beings, animals, plant and soil organisms,
- part of the natural household in particular with its hydrological and nutrient cycles,
- reducing, balancing and constructing media for the impact of substances due to the filter, buffer and substance change properties, particularly also for the protection of groundwater.

Based on the above-mentioned definitions it can be derived that background values are only useful for the types of site/land use "arable land", "grassland", "woodland topsoil" and "woodland organic layer", because in these cases it is not proceeded from a special or selected emission situation. Concerning the types of site/ land use of "recultivation spaces", "settlement area", "river pastureland", "allotment gardens", "roadside area" and "industry" an immense local anthropogenic component has to be reckoned with, besides the ubiquitous entries into soil. Concerning these anthropogenic characterised types of site/land use it seems to be reasonable to indicate tolerance ranges (x_{min}-x_{max}, preferably 10th percentile, 90th percentile).

21.3 Taking Samples and Status Report on Soil Site Parameters

To derive the background values for organic pollutants about 1 000 topsoil samples were taken in Brandenburg and examined for their content of organic pollutants. The determination of the points for taking samples based on two goals. On the one hand, a space-related representative registration was to realise (space representation) and, on the other hand, as much samples as possible had to be taken from problematic, but less occurring types of site / land use compared to the size of the area to allow assessment (problem relevance).

The square 8 x 8 km grid-network of the "Soil Status Report Woodland" (BMELF 1994) formed the basis for the extensive collection that was extended to areas without woodland. This led to altogether 479 sampling points for Brandenburg. For the purpose of capturing data on spatially low representative uses further 299 samples were taken at certain points. The precise determination of the points for taking samples in the grid-network was realised by the application of a "dynamic grid" to show the typical type of land use in the surrounding of the sampling point. Problem relevance was ensured by considering the respective settlement-structural type of area. Furthermore, some 200 tested soil samples of other projects were taken into consideration.

The status report was performed on the basis of the "Instructions on Soil Science Mapping" (BGR 1994). Apart from the pollutant parameters the soil condition parameters ph-value, humus and clay content as well as exchanging capacity were also determined. The site conditions (location, geology, emitters, etc.) and the use were documented respectively.

21.4 Implementation of Pot Trials and Taking Samples in the Field

The pot trials were carried out in Mitscherlich vegetation pots. One sample of sandy soil and another mixed soil with a high content of fine particles were used as trail soils. PAH_{16} were introduced in the test soil in the form of finely cut, highly polluted brickwork. PCB_6 were added to this brickwork dissolved in hexane. Despite different exchange capacities of both soils the recovery of the added PAH_{16} and PCB_6 was satisfactory in different concentrations.

The examination of the acceptance of PAH_{16} and PCB_6 was performed with the help of the crops maize, potatoes, carrots, lettuce and spinach. Two tests were carried out. Within the framework of test 1 four different concentration levels of pollutants were examined for the two mentioned kinds of soil and five crops each, in test 2 only the highest concentration level of pollutants was used apart from the zero variant (without accumulation of pollutants). three repetitions per variant of test resulted in 96 Mitscherlich pots for test 1, in 64 Mitscherlich pots for test 2 at four repetitions.

Within the framework of these examinations in the field 30 soil samples, as well as the respective wheat samples (straw, grain) were taken on various fields, as well as 20 soil samples and the respective grassland samples.

It was the target of the tests to make statements on the transfer of pollutants from soil into plant. Furthermore, during the pot trials the transfer factors were determined as well. The results of the field tests were used to check the pot test results.

21.5 Results

21.5.1 Basic Parameters

The medians of the pH-values of most types of site/land use are in the weakly acid range (pH 5.4–6.6). The wood soils are clearly more acid (median for organic layer and topsoil 3.4 each). The content of clay of the soils is relatively homogeneous (about 3–5 %, range of variation 0–35 %). River pastureland shows a content of clay of 8 % (median) due to the increased portion of fine grained substances. The content of organic substance (OS) ranges between 0.3–63 %. The median value for arable land of 1.4 % is decisively below the average of all soils (about 3 %), and that one in the examined river pasturelands is decisively above average with a value of 4.3 %. Higher OS-portions were also found in grassland (median 6.2 %). The highest OS-content was noticed for woodland topsoil horizon (median 3.7 %).

21.5.2 PAH_{16}-Content in Brandenburg Soils

The PAH_{16}-content (according to EPA) of the tested samples ranged independent of use between 22 µg/kg and 111 mg/kg DS[1]. The median was 350 µg/kg DS. Because of the relatively low settlement and industrial density of Brandenburg and the predominantly sorption-weak soils the PAH16-content is lower compared to that one in other federal states (e.g. Thuringia, Baden-Württemberg and Bavaria).

It is generally known that PAH_{16} are strongly bound by organic substances. The Federal Soil Protection Provision (BMU 1999) therefore defines soil precaution values in dependence on the content of organic substance to protect the soil from material changes. This differentiation shall take the different exchange capacities of the soils into account and thus the changed availability. Difference is made between soils with a content of organic substance (OS) > and ≤ than 8 %. The soil precaution value for soils with a content of OS ≤ is 3 mg PAH_{16}/kg DS and 10 mg PAH_{16}/kg DS for soils with > 8 % OS. These soil precaution values orientate by the effective levels of the carcinogen combination benzo[a]pyrene (BaP) on the condition that the proportion BaP:PAH_{16} is relatively uniform 1:10. A transfer into the plant does not take place at such concentrations or only to an extent that can be ignored.

The soil precaution values for PAH_{16} of the Federal Soil Protection Provision are decisively exceeded for the types of site/use "arable land use" and "use of grassland", as well as for "woodland" in the median.

The BaP content of the examined samples varies between 0.5 µg/kg DS and 9.1 mg/kg DS. The median is 22 µg/kg DS. The portion of BaP in the sum of PAH_{16} is on average 7 % for the examined topsoil, and it is thus below the proportion of BaP:PAH_{16} of 1:10 as specified by LABO. At locations marked by a special emission features (types of site/land use, "roadside area", "settlement area", "allotment

[1] DS – dry substance

gardens", "river pasturelands") the portion of BaP in the sum of PAH_{16} is above average for all samples. Concerning BaP the 90th percentile that depends on use is also below the soil precaution values of the Federal Soil Protection Provision here.

The increasing anthropogenic influence (proximity of emitters) has led to increasing PAH-contents for the types of site/land use of "allotment garden", "industry" and "roadside area" (median 0.9–2 mg/kg DS), (Tab. 21.1). Apart from the extensive diffuse entry into the soil there was a selective and, as the case may be, small scale PAH_{16}-pollution that led to an increase of the median PAH-concentrations and a high variation range of the analysis values.

Table 21.1. PAH16-contents of topsoil in dependence on land use /sites in Brandenburg

Type of site/land use		Benzo[a]- pyrene	PAH_{16} acc. to EPA
		µg/kg DS	
Arable land	Results above DL[2]	162	162
	Median	10	149
	90th percentile	36	434
Grassland	Results above DL	93	93
	Median	17	208
	90th percentile	76	804
Woodland topsoil	Results above DL	121	121
	Median	19	345
	90th percentile	88	1,207
Woodland organic layer	Results above DL	56	56
	Median	50	1,170
	90th percentile	155	2,892
Recultivation space	Results above DL	5	7
	Median	7.0	190
	90th percentile	3,742	45,584
Settlement area	Results above DL	42	43
	Median	130	1,397
	90th percentile	660	5,322
River pastureland	Results above DL	25	25
	Median	67	1,658
	90th percentile	448	5,048
Allotment garden	Results above DL	22	22
	Median	120	1,466
	90th percentile	658	6,946
Roadside area	Results above DL	38	38
	Median	215	2,010
	90th percentile	1,150	13,473
Industry	Results above DL	15	15
	Median	130	1,733
	90th percentile	920	12,619

[2] DL = Detection Limit.

The sites under agricultural and forest use without specific emission influence reflect, however, a large scale PAH$_{16}$-pollution in the sense of a ubiquitous distribution. A differentiation of the agriculturally uses of topsoils of arable land and grassland is useful due to the results of the performed examinations, because grassland shows higher PAH$_{16}$-concentrations than arable land. This can be explained by the fact that arable soils generally show a higher horizon thickness than grassland soils, they have got a lower OS-content; and thinning and subsurface shifting effects have to be expected due to thorough mixing during the ploughing process.

In order to be able to determine the PAH$_{16}$-contents of the soils more in detail to derive background values further influencing factors were focused on. Concerning the settlement structure differentiation was made between rurally characterised areas and conurbations. It was, however, not possible to derive a reliable differentiation from the existing data. The relatively equal settlement structure of Brandenburg might be a reason for this. A definition of background values for PAH$_{16}$ in dependence on the settlement structure seems thus not to be useful.

The exchange capacity of the soils in dependence on the content of OS and clay, etc. also influences decisively the content of pollutants. About 200 samples taken from topsoil and humus layer horizons showed OS-contents of >8 % whereas those were mainly sites with humus layers and woodland topsoil (Tab. 21.2). The median PAH$_{16}$-content of the group >8 % OS is - with a value of 0.8 mg/kg DS - about three times higher than that one of the group <8 % OS (types of site/land use "arable land" and "grassland"). It was not possible to make reliable statistic statements on the correlation between PAH$_{16}$ and the OS-content for soils without any reference to land use, but also for sites with arable land and grassland. A reliable connection could only be established for "woodland soils" (organic layer and topsoil), that is on the one hand determined by the higher sorption potential, but, on the other hand, it has to be considered also in connection with the different immission-relevant leaf surface. The PAH$_{16}$-contents of Brandenburg arable land and grassland are calculated based on the clay content for light (≤ 12 % of clay), medium (>12 % up to 25 % of clay) and heavy (>25 % of clay) soils.

Table 21.2. PAH16- und BaP- contents (µg/kg DS) of topsoil in dependence on OS-content according to the Federal Soil Protection Provision in the Brandenburg state

Type of site/land use		≤ 8 % OS		> 8 % OS	
		PAH$_{16}$	BaP	PAH$_{16}$	BaP
Without reference to use	Number	443	443	139	139
	Median	250	18	785	39
	90th percentile	2,529	250	3,344	190
Arable land	Number	160	160	2	2
	Median	149	10	205	22
	90th percentile	438	34	-	-
Grassland	Number	58	58	35	35
	Median	171	9.5	436	22
	90th percentile	713	65	966	80

In principle, the medium and heavy soils are proven only by few sample figures. The median PAH_{16}-content of the arable land of the group > 12 % up to 25 % of clay is only 1.5 times higher with a value of 223 µg PAH_{16}/kg than arable land with ≤ 12 % of clay. Grassland soils are represented by only two samples in the group >12 % up to 25 % of clay so that a comparison seems not to be useful. It was not possible to determine a correlative connection between PAH_{16}-content and that one of clay for the types of site/land use of "arable land" and "grassland". A weak correlative connection only exists for topsoils of woodland. Two statistically not proven tendencies can be derived from the results of the examinations: there is both a tendency of slightly increasing PAH_{16}-concentrations from sand via loam to clay sites, as well as of moderately higher PAH_{16}-contents by grassland use compared to the use of arable land. To establish correlations between site and land use towards the use of arable land makes obviously more sense than a differentiated preventive evaluation according to the use of the soil.

The results prove that a derivation of background values is only meaningful for types of land use without a special emission situation, whereas difference should be made between "arable land", "grassland" and forestry use ("woodland organic layer" and "woodland topsoil"). As regards the types of site/land use of "recultivation space" "settlement area", river pastureland", "allotment garden", "roadside area" and "industry" it is not justified to determine background values for PAH on the basis of the median values because of possible local components of pollution.

21.5.3 PCB$_7$-Content in Brandenburg Soils

PCB_7- sum contents (congeners: C28, C 52, C 101, C 181, C 138, C 153, C 180) were ascertained for 35 % of the analysed soil samples above the detection limit. The low chlorinated PCB C 28 and C 52 could be detected extremely seldom because, on the one hand, mainly medium and highly chlorinated PCB were produced industrially and found their way into the environment, and on the other hand, the low chlorinated PCB show comparably low persistence in the soil. The PCB_7 sum contents of the soils range within a low detection limit (1 µg/kg) up to 1 mg/kg DS (Tab. 21.3).

The soil precaution values for PCB_6 of 0.05 or 0.1 mg/kg DS (OS-content <8 % or >8 %) based on the Federal Soil Protection Provision are exceeded only in a few cases. Low median PCB-contents (<1 µg/kg DS) were determined for the types of site / land use "arable land", grassland", "woodland topsoil", "woodland organic layer", "allotment garden" and "settlement area". PCB_7-concentrations >5 µg/kg DS were determined for the types of site/land use "river pastureland", "roadside area" and "industry". The measured maximum value of PCB_7 was 1.64 mg/kg DS for the type of site/land use "industry". The PCB_7 sum contents of up to 0.2 mg/kg DS have to be assessed to be mainly unproblematic in phyto-, zoo-, human- and eco-toxicological terms. A significant relation was shown between the PCB_7- content and the OS-content of the soils which is mainly determined by the organic layer horizons of woodland. The types of site/land use of "river pasture land" and "roadside area" showed less significant relations regarding OS-content.

Table 21.3. Content of PCB7 of topsoil in dependence on the types of site/land use (µg/kg DS) in the Brandenburg state

Type of site/ land use		C28	C52	C101	C118	C153	C138	C180	PCB$_7$
Ara-ble land	Number	162	162	162	162	162	162	162	162
	Results above DL	0	0	1	0	2	3	1	-
	Median	0.50	0.50	0.50	0.50	0.50	0.50	0.50	3.5
	90th per-centile	0.50	0.50	0.50	0.50	0.50	0.50	0.50	3.5
Grass-land	Number	93	93	93	93	93	93	93	93
	Results above DL	0	1	3	0	6	9	4	-
	Median	0.50	0.50	0.50	0.50	0.50	0.50	0.50	3.5
	90th per-centile	0.50	0.50	0.50	0.50	0.50	0.50	0.50	3.5
Wood-land topsoil	Number	120	120	120	120	120	120	120	120
	Results above DL	1	1	8	3	34	40	25	-
	Median	0.50	0.50	0.50	0.50	0.50	0.50	0.50	3.5
	90th per-centile	0.50	0.50	0.50	0.50	2.0	2.0	1.0	7.0
Wood-land orga-nic layer	Number	56	56	56	56	56	56	56	56
	Results above DL	4	4	14	6	24	24	20	-
	Median	0.50	0.50	0.50	0.50	0.50	0.50	0.50	3.5
	90th per-centile	0.50	0.50	1.5	0.75	3.0	4.0	2.0	13
Recul-tiva-tion area	Number	7	7	7	7	7	7	7	7
	Results above DL	0	1	1	0	3	3	2	-
	Median	0.50	0.50	0.50	0.50	0.50	0.50	0.50	3.5
	90th per-centile	0.50	0.70	0.70	0.50	2.8	3.4	1.4	10
Settle-ment area	Number	43	43	43	43	43	43	43	43
	Results above DL	1	0	9	0	21	25	21	-
	Median	0.50	0.50	0.50	0.50	0.50	1.0	0.50	4.5
	90th per-centile	0.50	0.50	1.0	0.50	2.8	4.0	2.0	11

Table 21.3. (cont.)

Type of site/ land use		C28	C52	C101	C118	C153	C138	C180	PCB$_7$
River pasture land	Number	25	25	25	25	25	25	25	25
	Results above DL	5	5	7	5	11	11	10	-
	Median	0.50	0.50	0.50	0.50	0.50	0.50	0.50	3.5
	90th percentile	1.0	1.6	4.0	2.6	12	13	6.2	42
Allotment garden	Number	22	22	22	22	22	22	22	22
	Results above DL	1	2	2	2	9	10	8	-
	Median	0.50	0.50	0.50	0.50	0.50	0.50	0.50	3.5
	90th percentile	0.50	0.50	0.50	0.50	2.0	2.9	1.9	8.8
Road side area	Number	38	38	38	38	38	38	38	38
	Results above DL	1	2	10	5	19	19	17	-
	Median	0.50	0.50	0.50	0.50	0.75	0.75	0.50	4.3
	90th percentile	0.50	0.50	3.0	1.3	20	30	8.1	72
Industry	Number	15	15	15	15	15	15	15	15
	Results above DL	0	1	5	1	6	7	7	-
	Median	0.50	0.50	0.50	0.50	0.50	0.50	0.50	3.5
	90th percentile	0.50	0.50	1.6	0.50	5.6	10	4.0	24

The types of site/land use "settlement area", "river pastureland", "roadside area" and "industry" showed numerous soil samples with an OS-content of > 8% PCB$_7$-concentrations that partially exceeded the soil precaution value. In most cases, local emitters were the sources of pollutants. A significant relation between PCB$_7$ and OS-content does not exist for the use of arable land and grassland. Due to the mentioned correlation between PCB$_7$ and OS-content statements on the influence of content of clay and settlement structure on the PCB$_7$-content of the coils focused on the types of site /land use "arable land", "grassland", "woodland organic layer" and "woodland topsoil". The influence of the clay content of arable land, grassland and woodland on the PCB$_7$-content could not be proven. The settlement structure was also not in a plausible connection with the PCB$_7$-concentration of the soils of the mentioned types of site/land use.

21.5.4 Derived Background and Reference Values

Table 21.4 shows the background values (rounded median values) and reference values (rounded 90th percentile values) of selected organic pollutants and groups of pollutants in Brandenburg soils being under agricultural and forestry use derived from the examinations. The types of site/land use with large parts of space only show very low pollutions as regards the examined pollutants. No harm to the environment or disturbance of the natural soil function is therefore to be feared.

Table 21.4. Background and reference values for PAH16, BaP and PCB7 in topsoil

Types of sites/land use		PAH_{16} (acc. to EPA) µg/kg DS	BaP µg/kg DS	PCB_7 µg/kg DS
Arable land	Median	150	10	< 7
	90th percentile	450	40	< 7
Grassland	Median	200	20	< 7
	90th percentile	800	80	< 7
Woodland –	Median	350	20	< 7
topsoil	90th percentile	1,200	90	10
Woodland –	Median	1,200	50	< 7
organic layer	90th percentile	3,000	150	15

Table 21.5 shows the ranges of variation of organic pollutants and groups of pollutants (rounded 10th and 90th percentile values) in Brandenburg topsoils for the types of site/land use 'settlement area', 'river pastureland', 'allotment garden', 'roadside area' and 'industry'. Due to possible local sources of pollution it is not suitable to indicate background values for these types of site/land use on the basis of the median values. The condition of pollution of these anthropogenic characterised types of site/land use is better characterised by indicating the 10th and 90th percentile values. The 90th percentile exceeds partially decisively or often the soil precaution values of the Federal Soil Protection Provision at certain types of site/land use.

Table 21.5. Ranges of variation for PAH16, BaP and PCB7 in topsoil

Types of site/land use		PAH_{16} (EPA)	BaP in µg/kg DS	PCB_7
Recultivation area	10th percentile	150	< 1	< 7
	90th percentile	50,000	4,000	10
Settlement area	10th percentile	300	20	< 7
	90th percentile	5,000	700	10
River pastureland	10th percentile	250	15	< 7
	90th percentile	5,000	450	50
Allotment garden	10th percentile	200	15	< 7
	90th percentile	7,000	700	10
Roadside area	10th percentile	250	20	< 7
	90th percentile	13,000	1,200	70
Industry	10th percentile	50	15	< 7
	90th percentile	13,000	1,000	30

21.5.5 Pot and Status Report Examinations

In the following the transfer factors derived from the results of examination are assessed compared to the PAH_{16}-contents and PCB_6-contents found in Brandenburg soils. The harvest products showed PAH_{16}-contents between 0.04 (potatoes, hulled) and 0.84 mg/kg DS (lettuce) in the arithmetical mean. The transfer factors are between 0.002 and 1. High transfer factors result for low soil contents; they decrease with increasing soil contents. An influence of the soil properties on the volume of the PAH_{16}-tansfer could not be determined. The reason for this might be the stronger influence by ambient air. Any connection between the degree of condensation of the PAH-single compounds and their re-sorption by the plant could not be detected.

All PAH_{16}-contents found out during the examinations in the field were below 1 mg /kg DS. Thus, they do not reach the concentration checked in the pot trials, but they are typical of Brandenburg arable land and grassland. The existing data was insufficient for validating the transfer factors of soil-plant detected during the pot trials. It is generally to state that numerous findings are below the detection limit. In such cases it was not possible to evaluate the transfer factors. The determined transfer factors were extraordinarily low and confirm the low PAH_{16}-transfer from soil to plant. As far as the dimensions are concerned the evaluated transfer factors correspond well with those known from literature.

The medium PCB_6-contents of the harvest products were low and ranged between the detection limit of the procedure (0.001 mg/kg DS per congener) and 0.5 mg PCB_6/kg DS. Due to the low contents it was not easy to evaluate the transfer factors. It is, however, to state that the PCB_6-plant contents do not increase to the same extent as the PCB_6-soil contents change. This means, the transfer rate into the plant decreases at an increasing PCB_6-pollution of the soil, mostly that one of the substance transport in the plant parts above ground.

The PCB_6-contents found during the status report examinations were below 0.02 mg/kg DS. Thus, they do not reach the concentration generated during the pot trials, but correspond to those contents being typical of Brandenburg arable land and grassland. A transfer into the plant was not noticed. All results of the examinations for wheat grain, straw and grass were below the detection limit of the applied procedure. It was therefore not possible to evaluate the transfer factors.

The parts above ground of crop plants with a cuticle having a high lipid content (reference plants: lettuce, spinach and maize) proved to be accumulation places for atmospheric PAH_{16} and PCB_6. The thus caused accumulation of pollutants falsifies the statement on the soil-plant-transfer. This applies to the tested crops maize and lettuce. When assessing the pot trials and status report examinations it can generally be stated that the PAH_{16} and PCB_6-transfer in the plant is on a low level under the conditions of the relatively light Brandenburg soils that are badly provided with organic substances as it has been published for most of the federal states. A decisive influence of soil properties on the transfer behaviour could not be noticed. It is not necessary to grade the PAH and PCB soil standard values as regards the content of clay and humus in the soil.

21.6 Conclusions and Recommendations

Within the framework of the examinations background values were derived for organic pollutants (PAH, BaP and PCB) in Brandenburg topsoils. The background value of most organic pollutants corresponds to those concentrations reached by ubiquitous entries into soil. These are partially redistributed in the soil by pedogenic processes and different types of land use. Based on the background contents statistic procedures were used to evaluate background values. Here, the focus was put on the median value and the 90th percentile. The median value is used as a background value. It characterises the median concentration of organic pollutants of Brandenburg soils and by this the "normal state" in this federal state. It is mainly determined by the anthropogenicly caused, ubiquitous spreading of the organic pollutants. The 90th percentile is a concentration of pollutants exceeding the "normal state" without considering phyto-, zoo-, human- and eco-toxicological aspects. This value is therefore used as a reference value for Brandenburg soils. But as far as we know, the current 90th percentile values of organic pollutants do not endanger the natural soil function in Brandenburg soils.

Based on the definitions and terms it was concluded that background values can only be defined for the types of site/land use of "arable land", "grassland", "woodland topsoil" and "woodland organic layer". As regards the types of site/land use of "recultivation space", "settlement area", "river pastureland", "allotment garden", "roadside area" and "industry" a considerably local anthropogenic component has to be reckoned with apart from the ubiquitous entry into soil, so that it would not make any sense to derive background values. It would therefore be more suitable to indicate ranges of variation ($x_{min} - x_{max}$, preferably 10th percentile and 90th percentile) for these anthropogenic characterised land uses and sites.

In order to be able to further validate the ascertained background and reference values it has to be concentrated on extending the database to derive values for continuous data capturing. The same applies to the registration of value pairs "soil-plant" to ensure the transfer statements in qualitative and quantitative terms. To increase the meaningfulness of the results it is to check whether the reference values can be further differentiated (e.g. type of site/land use, settlement-structural regional type, and substrate).

According to preventive soil protection it should be focused on avoiding an increase of the concentration of organic pollutants in Brandenburg soils, characterised by the median values. This means to get a balance between entry and drag out of organic pollutants on a level as low as possible. The 90th percentile values should be used at present to exclude soils with increased contents of organic pollutants, i.e. of soils that are polluted beyond the "normal" ubiquitous pollution.

Pot trials were carried out for soils and plants typical of Brandenburg, as well as status surveys on agricultural spaces of the Brandenburg state to be able to describe the transfer of PAH_{16} and PCB_6 from soil into plant under specific Brandenburg conditions and to draw conclusions regarding a transfer risks for the currently found contents. The transfer of the organic pollutants into the crops maize, potatoes, carrot, lettuce, spinach, grass, wheat grain and wheat straw was checked.

By assessing the pot trials and status surveys it can generally be stated that the PAH_{16}- and PCB_7—transfer into the plant is on low level under the conditions of Brandenburg soils with relatively light soils being provided badly with organic substances. The results of these examinations correspond well with those published from North Rhine Westphalia, Lower Saxony, Baden-Württemberg and Bavaria as well as the European foreign countries.

The food chain is sufficiently protected also under Brandenburg conditions based on the soil precaution values for PAH_{16}, BaP and PCB_6 in the soil provided for in the Federal Soil Protection Provision.

References

BGR – Bundesanstalt für Geowissenschaften und Rohstoffe (Federal Institute for Geosciences and Natural Resources) (1994) Bodenkundliche Kartieranleitung (Instructions on Soil Science Mapping), 4th edition, E. Schweizerbart'sche Verlagsbuchhandlung, Stuttgart

BMELF – Bundesministerium für Ernährung, Landwirtschaft und Forsten (Federal Ministry for Food, Agriculture and Forestry) (1994) Nation-wide survey of soil status woodland (BZE-working instruction)

BMU – Bundesministerium für Umwelt, Naturschutz und Reaktorsicherheit (Federal Ministry for the Environment, Nature Conservation and Nuclear Safety) (1998) Federal Soil Protection and Contaminated Site Act

BMU – Bundesministerium für Umwelt, Naturschutz und Reaktorsicherheit (Federal Ministry for the Environment, Nature Conservation and Nuclear Safety) (1999) Provision on the implementation of the Federal Soil Protection and Contaminated Site Ordinance (Federal Soil Protection and Contaminated Site Ordinance)

HÜTTL RF (1999) Bodenschutz muss Zukunftsaufgabe bleiben (Soil protection must remain a task for the future). Berlin: E. Schmidt, Soil Protection 3, pp.8-9

LABO (2003) Joint Federal State Working group for Soil Protection. Background values for inorganic and organic substances in soils. In: ROSENKRANZ/ EINSELE/HARREß [pub.]: Soil Protection 28; III/04

MUNR – Ministerium für Umweltschutz, Naturschutz und Raumordnung des Landes Brandenburg (Brandenburg Ministry for the Environment, Nature Conservation and Regional Development) (1998) Soil standard values for the Brandenburg state: organic pollutants and heavy metals, project part 1995, final report

MUNR – Ministerium für Umweltschutz, Naturschutz und Raumordnung des Landes Brandenburg (Brandenburg Ministry for the Environment, Nature Conservation and Regional Development) (1997) Transfer factors of pollutants from soil to plant for typical Brandenburg soils and crops, final report

22 Standards and Indicators for Monitoring Impact of Disturbance on Biodiversity in a Post-Mining Area Using GIS

Effah Kwabena Antwi and Gerhard Wiegleb

Department of Ecology, Brandenburg University of Technology (BTU) Cottbus

22.1 Introduction

The Convention on Biological Diversity (CBD), the Ramsar Convention, and the Convention on Migratory Species (CMS) recognize Impact Assessment as a significant decision-support tool that assist planning and implementation of development with biodiversity "in mind." Article 8 of the Biodiversity Convention gives justification for the application of impact assessment toward protection of ecosystems and natural habitat IAIA (2005). In order words, CBD require impact assessment to identify possibilities to quantify land cover characteristics or indicators that influence habitat transformations in an ecosystem. Ecosystem structure, a considered level of our assessment, concerns the organisation of various biological units in space and time.

The Institute of Ecology and Environmental Management (IEEM) developed a guideline that promotes Ecological Impact Assessment relating to terrestrial environments in the UK. Ecological Impact Assessment being a process of identifying, quantifying and evaluating the potential impacts of defined actions on ecosystems or their components Treweek (1999); may be carried out as part of a formal environmental impact assessment (EIA) or to support other forms of environmental assessment or appraisal (1).

Biodiversity concerns and EIA are often poorly integrated in local planning frameworks and even where Biodiversity information is available; it is usually limited, descriptive and cannot be used as a basis for numerical predictions. There is therefore the need to develop or compile biodiversity criteria for impact evaluation and to have measurable standards or objectives that can evaluate significance of individual impacts UNEP (2001). Spatial statistics of an ecosystem provide such a measurable standard that reflect the state of diversity before and after an EIA project. Though EIA should address biodiversity at all levels, (gene, species, and ecosystem), it is necessary to start at ecosystem-level to allocate resources efficiently to other levels. Failure to consider biodiversity at ecosystem level results in, risk of negative impacts on important life-support functions, risk of overlooking ecosystem services, failure to understand variation in time and space TBAG (2000). Therefore our assessment on the impact of disturbance on regional biodiversity (habitat or ecosystem) contributes to a needed assessment that helps to predict present and future impact of disturbance on an ecosystem.

Biodiversity of terrestrial ecosystem is expected to be mainly affected by land use changes within the next 100 years (Sala et al. 2000). Here in the Lower Lusatia regions of Germany, the Federal Office for Building and Regional Planning in March 2001 reported that, extensive mining of coal has permanently changed the ecosystem, land use potential and the attractiveness of the landscape. Furthermore, mining in this area has left behind damage stretching over large areas. It is therefore an important task of spatial planning policies to restore the damage to ecosystem and to open up new land use possibilities in this region.

Landscape ecology considers vegetation as a mosaic of patches with unique landform, species composition and disturbance gradient (Ravan 1998). Johst and Huth, in 2005, ascertained that, succession after disturbance generates a mosaic of patches in different successional stages. The effects of disturbance on patches vary with distance from patches and connectivity to other patches. To research and manage these characteristics of patches, spatial and temporal understanding of landscape is required. Ravan and Roy (1998) have proved the potential of GIS in landscape ecology by mapping disturbance zones in natural ecosystem and quantifying its impacts on biodiversity.

In the context of this research, habitat fragmentation shall be defined as the breaking up of habitat, ecosystem or land cover types into smaller parcels (Forman 1995). These smaller fragments of land are referred to as habitat patches. Habitat fragmentation has a major impact on the regional survival of plant species and is one of the most important causes of worldwide loss of biodiversity (Vitousek et al. 1997). The threat posed by forest loss and fragmentations to local biodiversity have been popularised for nearly two decades (Harris and Miller 1984). Although spatial heterogeneity is a natural phenomenon, human activity are altering natural landscape by changing the abundance and spatial pattern of habitat. The two most significant effect of forest fragmentation are a decrease in population size and reduction of species diversity (Zuidema et al. 1996).

Plant species richness is relatively simple to measure for a small area, e.g. a sampling quadrate of a few square meters. Several methods exist for estimating the species richness of an assumedly homogeneous larger area (Bunge and Fitzpatrick 1993, Colwell and Coddington 1994, Palmer 1995). However, assessing and predicting diversity of a complex landscape remains a problem (Stohlgren et al. 1997a). Dale and Pearson (1997) suggested four general types of fragmentation indices to describe spatial pattern in habitat maps. They include patch or habitat area, frequency distribution of patch size, measure of patch shape and length of edge between different habitat types. In landscape analysis, indices of shape, richness and diversity provide additional evaluation of land cover spatial distribution within a particular landscape. Landscape analysis also provides an outline of the degree of disturbance and biodiversity change within a period of time (Roy and Joshi 2002). In disturbance dominated landscapes such as post-mining areas, patterns may fluctuate widely over time in response to the interplay between disturbance and succession processes that leads to change in biodiversity.

Felinks et al. (1998) ascertained the exact location of each site in the study area by satellite imagery (LANDSAT TM) and high-resolution scanning data. Vegetation types as defined by Wiegleb and Felinks (2000) best describe land cover types. The loss in biodiversity in Schlabendorf Süd can be attributed to habitat loss, fragmentation and disturbance of the natural landscapes or the interaction of the three (Wiegleb 2001).

Box 22.1. Integrated checklist of EIA process and biodiversity requirements for legislation and guidelines (adapted and revised from Bagri et al. 1998)

Screening
Group of activities likely to impact biological diversity, e.g. activities that would damage the natural ecosystem (structure and function)
Durations, severity and threshold of biodiversity measure leading to overexploitation of ecosystem composition (plants and animals)

Preliminary Assessment
Measurements of landscape characteristics that reflects present ecosystem health

Scoping
Temporal and spatial parameters reflecting biodiversity considerations
Involving technologies and tools with capabilities for monitoring biodiversity trend at various levels
Cumulative effects on biodiversity are taken into account
Public participation is used to minimise bias in defining impacts
List of impact shall consider impacts on ecosystems and habitats

Identification
Methodologies include direct and indirect impacts on biodiversity such as habitat loss, fragmentation, land use changes, introduced species, pollution of soil, climate change
Landscape characteristics metrics shall be used as criteria

Alternative
Alternatives are assessed for their potential impacts on biodiversity and for the distribution of their costs and benefits

Prediction
Baseline biodiversity information is obtained from information provided from sources such as the CBD's clearinghouse mechanism
Existing baseline data is supplemented by further studies such as measurement of landscape characterisation metrics, which give indication of present biodiversity
Data produced through studies and predictions is available to the clearinghouse mechanism and BCIS thereby furthering the exchange of information (Art 17)

Evaluation of Significance
Stakeholders are involved in the process of attaching significance to impacts thereby furthering the equitable sharing objective of the CBD.
Biodiversity trend assessment should be done at relevant level with relevant baseline data

This present study first of all, aids the investigation of spatial and temporal fluctuations that have occurred in the post-mining area. Secondly, Patch Analyst, an Arc View GIS extension, would be used to compute landscape metrics needed to assess biodiversity of the area, degree of disturbance and nature of fragmentation processes. Finally we aim at developing measurable biodiversity indicators based on the landscape or vegetation structure.

Since landscape ecology regards landscape as a mosaic of patches (Ravan 1998), patch would be used interchangeably with habitat. The biodiversity requirements for legislation and guidelines by Bagri et al. (1998) shown below shall be adapted with revisions. The checklist (see Box 22.1) includes the involvement of measurable diversity indcators in the EIA process.

22.2 Materials and Methods

Area of Study

The post-mining area of Schlabendorf Süd, regarded as an extremely disturbed area in Lower Lusatia region, Germany, was selected as the study area. The area is 13 years of age since dumping with a total land cover area of 3,035 hectares. Mining activity in this area started in 1975 with the last mining activity taking place in 1991. Wiegleb (1996), Blumrich et al. (1998), and Wiegleb and Felinks (2000) gave detailed account on the ecological and socioeconomic problems in this area. Land cover types found in this area are shown in Table 22.1, figures 22.1 and 22.2. The two figures 22.1 and 22.2 are part of satellite images digitized by Monika Pilartki. This area is selected based on the following reasons (Wiegleb 2001a).

- The landscape has the potential of forming extremely large man made lakes with high concentration of acid (down to pH 2.1) that affects water bodies, other plants and animal lives (e.g. seasonal migratory birds) in the ecosystem.
- The need to restore the damage to the ecosystem and open up new land use possibilities in the region (Federal Office for Building and Regional Planning 2001).

Human impacts on biodiversity operate differently at different spatial scales and resulting in changes over time within-habitat (local), habitat-mosaic (landscape) and macro-scale ('regional') diversity (Weber et al. 2004). The scale of our assessment is based on diversity within habitat types in the landscape ('habitat-mosaic').

22.3 Data Acquisition and Spatial Database

Satellite imagery (LANDSAT TM) of the post-mining area was used to carry out a survey of the changing landscape. The basic data required were land cover maps of 1995 and 2000. The land cover types were classified based on the dominant vegetation type present at the area in 1995 and year 2000. Land cover types were divided into 13 vegetation classes (land cover types) in the year 2000, represented by aggregation code (Table 22.1 Figure 22.1). The number of vegetation classes in Schlabendorf Süd 1995, were 9 (Table 22.1 Figure 22.1), thus, 4 classes fewer than that of year 2000 due to absences of classes such as deciduous trees, meadow and seeded grassland in 1995. The presence or absence of a particular class is indicated with a yes (Y) and no (N) key.

Figures 22.1 and 22.2 present different classes showing the vegetation structure of the landscape in 1995 and 2000. The patch edges in 1995 appear to be more linear compared to edges in 2000. Patches in year 2000 are smaller and more homogeneous in area than patches in 1995. The land cover maps in 1995 have larger patches and straight patch edges whiles patches of 2000 become smaller with curvy patch edges.

Table 22.1. Land cover types with different aggregation codes (adapted from Felinks et al. 2000)

| Agg Code | Vegetation classes | Present | |
		In 1995	In 2000
1	Pine Forest	Y	Y
2	Mixed Pine forest and agricultural land	N	Y
3	Deciduous trees	Y	Y
4	Afforestation of mixed forest	Y	Y
5	Afforestation of deciduous tress	N	Y
6	Sparse vegetation	Y	Y
7	Dry vegetation	Y	Y
8	Bare sand	Y	Y
9	Seeded grass land	N	Y
10	Meadow	N	Y
11	Agricultural land	Y	Y
12	Lake	Y	Y
13	Wetland	Y	Y

Y = Yes; N = No.

22.3.1 Calculating Measures of Landscape Pattern

Quantifying the land cover structure, its change over time and associated diversity change requires the use of a variability of statistical measures or metrics. Area, Patch Density & Size, Edge, Shape and Diversity, and Interspersion Metrics have been used. These metrics describe landscape composition and configuration. The metrics were used as indicators. Hargis and Bissonette, (1994) examined the behaviour of landscape metrics by generating artificial landscapes that mimicked fragmentation processes while controlling the size, shape, and placement of disturbance patches.

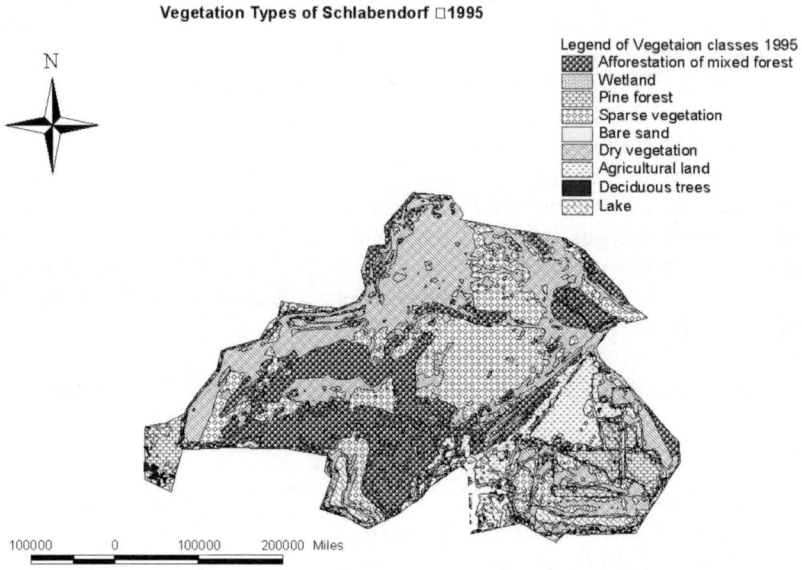

Fig. 22.1. Land cover map of Schlabendorf Süd 1995

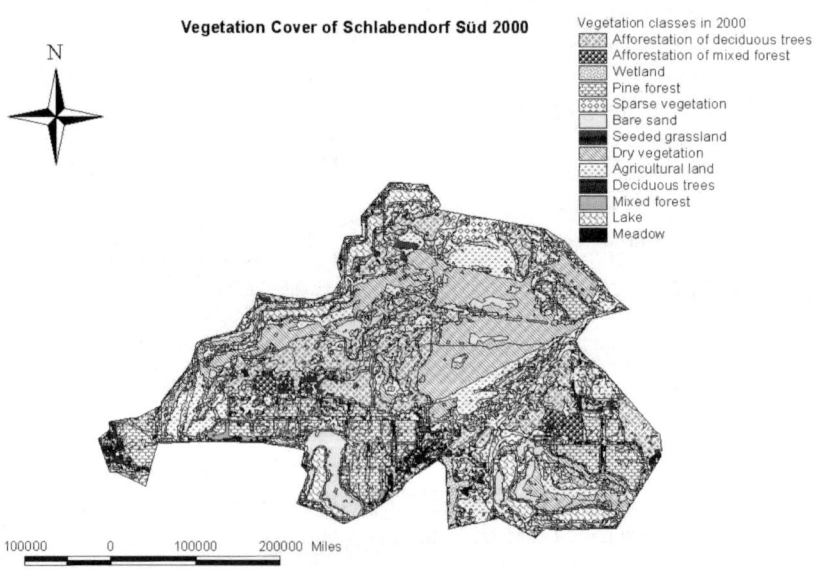

Fig. 22.2. Land cover map of Schlabendorf Süd 2000

They created nine series of increasingly fragmented landscapes that differed in the patch size or shape used to represent disturbance and used these landscapes to investigate patch edge density, perimeter-area and diversity. For each map year, land cover statistics were calculated and landscape metrics were computed using Patch Analyst, for the entire study area including individual vegetation classes. In Table 22.2, a complete overview of the landscape metrics used is given.

The Patch Analyst offers a comprehensive choice of landscape metrics at the patch, class and landscape levels. It calculates spatial statistics on both polygon files (vector format such as shape files) and raster files (e.g. Arc/Info grids). Selection of metrics in the Table 22.3 was based on their applicability to vegetation and landscape analysis in question. In order to overcome an artificially high number of patches, the boundaries between adjacent polygons with the same meaning or attribute were dissolved before all calculations were made. Table 22.3 shows metrics analysis at landscape level.

22.3.2 Landscape Metrics as Indicators of Biodiversity

Considering the focus of this paper, a standard would be defined as a biological condition or function needed for the normal functioning of an ecosystem. Table 22.2 shows diversity change standards adapted to study the impact of disturbance on biodiversity.

Table 22.2. Standard, Indicators and Indicator Ranges adapted from the landscape metrics for assessing the impact of disturbance on biodiversity

Standard	Indicators	Description	Range
Habitat Richness/ Number of Patches, Fragmentation	Number of Patches NP	It is a measure of the extent of subdivision or fragmentation of the habitat type NP = 1 when the landscape or class consists of a single patch	$NP \geq 1$, without limit
	Edge Density ED	It measures habitat length in a landscape. ED = 0 when the entire landscape and landscape border, if present, consists of the corresponding patch type.	$ED \geq 0$, without limit
Patch /Habitat Size	Mean Patch Size (MPS)	The range in MPS is limited by the grain and extent of the image and the minimum patch size in the same manner as patch area.	$MPS > 0$, without limit.
	Mean Shape Index (MSI)	It measures the average patch shape or perimeter-to-area ratio, for a patch type or patches in the landscape. MSI = 1 when all patches of the corresponding patch type are circular (vector). It increases without limit as the patch shapes become more irregular.	$MSI \geq 1$ without limit.

Table 22.2. (cont.)

Evenness	Shannon	It measures distribution of area among	$0 \leq SHEI$
	Evenness	patch types	≤ 1
Habitat Het-	Index	SHDI = 0 when the landscape contains	
erogeneity	(SHEI)	only 1 patch (i.e., no diversity) and ap-	
		proaches 0 as the distribution of area	
		among the different patch types becomes	
		increasingly uneven. SHDI = 1 when dis-	
		tribution of area among patch types is	
		perfectly even (i.e., proportional abun-	
		dances are the same).	
Habitat Diver-	Shannon	It is a measure of diversity in community	$SHDI \geq 0$,
sity	Diversity	ecology	without
	Index	SHDI = 0 when the landscape contains	limit
	(SHDI)	only 1 patch (i.e., no diversity). SHDI in-	
		creases as the number of different patch	
		types (patch richness, PR) increases.	

22.4 Results

Output of landscape metrics computation using the Patch Analyst revealed significant transformation in the biodiversity and nature of fragmentation in the Post-Mining Landscape. The results are as shown in the table below.

Table 22.3. A summary of relevant landscape metrics used for the study

Metrics/Indicator	Indicator	in 1995	in 2000	Change
Area Metrics				
Total Landscape Area	TLA	3 014.2	3 035.3	21.1
Patch Density & Size Metrics				
No. of Patches	NumP	1 030.0	3 106.0	2 076.0
Mean Patch Size	MPS	2.9	1.2	-1.9
Patch Size Coefficient of Variance	PSCV	893.6	1 149.2	255.6
Patch Size Standard Deviation	PSSD	26.2	11.2	-15.0
Edge Metrics				
Edge Density	ED	201.5	337.2	135.7
Shape Metrics				
Mean Shape Index	MSI	1.5	1.3	-0.2
Mean Patch Fractal Dimension	MPFD	1.5	1.5	0.0
Diversity & Interspersion Metrics				
Shannon's Diversity Index *	SHDI	1.7	2.1	0.4
Shannon's Evenness Index *	SHEI	0.7	0.8	0.1

*Applicable only at the landscape level (FRAGSTATS*ARC, Manual). Landscape metrics showed a general increase in the number of patches, edge density and diversity index from 1995 to 2000.

Patch Area Metrics

The number of patches increased significantly during the study period due to changes in the land cover distribution. At the landscape level, the number of patches increased from 1030 in 1995 to 3106 in 2000 (Table 22.3). This indicates Schlabendorf Süd has become more fragmented over the years. Fragmentation in this context means breaking up of habitat, ecosystem or land cover types into smaller parcels (Forman 1995). Commonly mentioned incidence of fragmentation is one resulting from transport project, which reduces the area of natural habitat into smaller patches. Similar to post-mining activities, fragmented landscape from transport network become surrounded by inhospitable landscape.

The mean patch size (MPS) is an indicator for the grain of landscape. It shows that, generally, patch sizes, relative to the total number of patches were getting smaller. MPS decreased more than three times (2.93 in 1995 to 0.98, in 2000) between the five-year periods (Table 22.4). This again supports the fact that Schlabendorf Süd has over the five-year period become more fragmented. A landscape with greater patch size standard deviation (PSSD) is an indicator for a more heterogeneous land cover while landscapes with lower PSSD is an indicator for a more uniform landscape. PSSD tended to decrease over time. Schlabendorf Süd that was relatively heterogeneous in 1995 with respect to variation in core area among patches (PSSD of 26.2) is becoming more homogeneous (PSSD of 11.2) Table 22.4.

Edge Density Metrics

Edge Density ED is 0 when there is no edge in the landscape; that is, when the entire landscape and landscape border, if present, consists of a single patch and the user specifies that none of the landscape boundary and background edge be treated as edge (Fragstats* Arc Manual). Edge density (ED) increased from 201.5 in 1995 to 337.2 in year 2000. Straight patch edges offer fewer niche habitat opportunities than curved edges (see http://www.class.uidaho.edu/italy2004/ecology2.htm, last accessed 01.01.05). The patch edges in 1995 appear to be more linear as compared to edges in 2000, i.e. the levels that an organism occupies however were fewer in 1995 than in Schlabendorf Süd 2000.

Shape Metrics

Mean shape index (MSI) is 1 when all patches (polygons) are circular (polygons) or square (grids). Decreasing MSI indicates that shape becomes simpler, and increasing, otherwise. Slight decrease in MSI recorded (1.5 in 1995 to 1.3 in 2000), favours simple patches formation (Table 22.4). The mean patch fractal dimension (MPFD) ranges from 1 to 2, approaching 1 indicates simple perimeters such as squares or circles, and increasing towards 2 for shapes with highly convoluted perimeters (Mandelbrot 1977, 1982). MPFD remained the same (1.45 in 1995 and 1.47 in 2000) over the years with slight decrease in 2000. Patch shapes have become simpler as indicated by the MSI (Table 22.4).

22.5 Diversity and Interspersion Metrics

Shannon's diversity and evenness index (SDI and SEI) are popular measures of diversity in community ecology. SDI is 0 when the landscape contains only 1 patch. SDI increases as the number of different patch types (i.e., patch or habitat richness) increases and/or the proportional distribution of area among patch types become more uneven.

Shannon's evenness index (SEI) is 0 when the landscape contains only 1 patch and approaches 0 as the distribution of area among the different patch types becomes increasingly uneven (i.e., dominated by 1 type). SEI is 1 when distribution of area among patch types is perfectly even (i.e., proportional abundances are the same). Increases in the SDI from 1.73 in 1995 to 2.10 in year 2000 are in agreement with higher diversity over a five-year duration (Table 22.4). SEI recorded showed an increase from 0.72 to 0.82 which means the distribution of the area among patch types is approaching perfect evenness, nevertheless, values of SEI less than 1 recorded, implies that the patch types are even hence proportional abundance are not the same (Table 22.4).

22.6 Discussions

Patch Area Metrics

Formation of more patches lead to high degree of fragmentation. This could be the result of the "afforestation" program at the post mining landscape. According to Pickett and Rogers (1995), patchiness occasionally results in higher species diversity. High patchiness recorded therefore supports the fact that species diversity in Schlabendorf Süd have increased. Habitat richness (patch richness) according to our results has also increased significantly due to high patchiness. This presupposed that the number of individual species in the area has as well increased.

On the contrary, following the values of NUMP and MPS at the landscape level, it is deduced that, the landscape has become more fragmented between the due to decreased mean patch size, increased number of patches and edge effect. High degree fragmentation due to decreased mean patch size, increased number of patches and edge effect reduced species richness in the area. According to Forman (1995), small patches often have high species richness, but they contain only common edge species, whereas larger patches contain more specialized interior species. With reference to Forman's findings in 1995, Schlabendorf Süd now exhibits high species richness in the year 2000 than five years previous, due to reduction in patch sizes. This is in agreement with high habitat richness record. Though the area now has high species richness but contains mainly common edge species, unlike in 1995 with larger patches containing more specialized interior species.

Edge Density Metrics

Straight patch edges offer fewer niche habitat opportunities than curved edges. The patch edges in 1995 appear to be more linear as compared to edges in 2000. The levels that organisms occupy however were fewer in 1995. The landscape transformation in effect has left relatively high niche opportunity in 2000. Edges of patches provide protection for the interior species; increased in Edge Density (130) therefore help maintain habitat structures in the landscape.

Shape Metrics

The slight decrease in MSI recorded indicates that most of the patches in Schlabendorf Süd are becoming simpler in form. Metrics such as mean shape index and mean patch fractal dimension both showed that patches in the area have become simpler and smaller. Many small patches or simple patches do not offer the same habitat opportunity as a single larger patch, especially for organisms that require interior habitat. Therefore formation of more fractal patches in year 2000 indicates variability in habitat opportunity leading to a more heterogeneous habitat conditions in year 2000.

Diversity and Interspersion Metrics

Though distribution of patch types was not perfectly even, values of SEI recorded (0.81 in1995 and 0.83 in 2000) support increase in diversity since it is more than zero. On the other hand, SEI was basically maintained (with an insignificant difference of 0.02). According to the above observation, it could be deduced that, there has been increase in habitat diversity at the post mining landscape; nevertheless, habitat richness is more in a state of equilibrium or steady state.

Increased of SDI (1.95 in 1995 to 21.3 in 2000) supports increase habitat diversity or biodiversity during the five-year duration. A higher SDI value indicates a higher diversity. Though increase diversity was recorded, a difference of 0.18 is not significant. Habitat richness has increased slightly but then such an insignificant increment also supports richness being almost in equilibrium or steady state.

Severe disturbance or even a prolonged absence of disturbance generally has depressing effect on biodiversity, but intermediate disturbance seems to enhance diversity in a system (Pickett and White 1985; Anon. 2001). The slight increase in diversity therefore implies disturbance at the post mining landscape was a severe and an intermediate disturbance. The intermediate disturbance hypothesis predicts that intermediate disturbance leads to the high diversity (Johst and Huth 2005).

Composite effects of high degree of fragmentation and high patch (habitat) richness support stable species richness. This is in agreement with the realisation that species abundance and composition at the early stages of disturbance did not decrease (Wiegleb et al. 2000). Generally however, the diversity results of both SDA and SEI are supported by the mosaic concept (Duelli 1992) that predicts increasing diversity with increasing habitat variability and with increasing habitat heterogeneity. Nevertheless, the area recorded decrease heterogeneity that may be due to the observation made by Wiegleb (2001b) due to the fact that, post-mining landscapes have some characteristics, which are often no longer found in cultural landscapes of Central Europe. These are in particular large undisturbed areas, low nutrient content of soils, dynamics of soil and landscape, and great habitat diversity on a small-scale.

Landscape metrics according to our observation, are reliable quantitative and measurable indicators for evaluating significance of individual standard for habitat diversity assessment. These standards and indicators are particularly very good during the baseline assessment of EIA in biodiversity projects where measurable standards are required for predictions. During monitoring stage of implemented post mining projects, the metrics as indicators have consistently aided in showing present land cover structure and predicting habitat condition (see Box 22.1). Major benefits integrating standards and indicators in biodiversity monitoring are:

- Evaluate present habitat condition
- Predict/forecast future habitat condition
- Provide early warning related to species extinction and restoration

22.7 Conclusions and Recommendations

According to Pickett and White (1985) severe disturbance or prolonged absences of disturbance have depressing effect on biodiversity but intermediate disturbance enhance diversity. Increase in diversity recorded in our analysis at Schlabendorf Süd confirms the presence of ecosystem disturbance and supports an intermediate level of disturbance. It is therefore explicit that disturbance and fragmentation due to land cover changes are related processes with strong relationships that affect habitat diversity in a post mining landscape.

Reduction of the post-mining landscape from large continues habitat into smaller fragmented habitats; accounts for the magnitude of impact on the biodiversity as required by Environmental Impact Assessment in project with biodiversity in mind as indicated by the CBD. Weber et al. (2004) states: landscape diversity is the most important information for conservationists and landscape planners; the observed landscape transformation therefore raises alarm concerning the urgent need for conservationist and landscape planners involve in the reclamation program to focus on measure that would not deteriorate the already deteriorating state of the landscape. Measurable biodiversity indicators based on the landscape structure offer supports dependable standards require at preliminary and post EIA projects with biodiversity considerations.

References

Bagri A, McNeely J, Vorhies F (1998) Biodiversity and Impact Assessment Paper presented at a workshop on Biodiversity and Impact Assessment, Christchurch, NZ

Blumrich H, Bröring U, Felinks B, Fromm H, Mrzljak J, Schulz F, Vorwald J, Wiegleb G (1998) Naturschutz in der Bergbaufolgelandschaft – Leitbildentwicklung. Studien und Tagungsberichte 17:1-44

Bunge J, Fitzpatrick M (1993) Estimating the number of species: a review. Á J. Am. Stat. Assoc. 88: 364Á 373

Colwell RK, Coddington JA (1994) Estimating terrestrial biodiversity through extrapolation. Á Philos. Trans. R. Soc. Lond. B 345: 101Á 118

Dale VH, Pearson SM (1997) Quantifying habitat fragmentation due to land use change in Amazonia. In: Laurance W, Bierregaard R (eds) Tropical Forest Remnants. The University of Chicago Press, Chicago, pp 400-414

Denkinger P, Mrzljak J, Wiegleb G (2002) The Importance of Small Scale Disturbance on Species Diversity of Vegetation in Post Mining Landscapes. Poster Presentation International Workshop of BIOLOG Europe Gießen

Federal Office for Building and Regional Planning (2001) Brochure, "Spatial Development and Spatial Planning in Germany", pp 35, FORUM GmbH, Bonn

Felinks B, Pilarski M, Wiggleb G (1997) A hierarchical classification of vegetation of the former brown coal mining areas of Eastern Germany (Lower Lusatia, Brandenburg). Conference Abstracts, IAVS Symposium, Ceske Budejovice, pp32-33

Felinks B, Pilarski M, Wiegleb G (1998) Vegetation survey in the former brown coal mining area of eastern Germany by integrating remote sensing and groundbased methods. Appl. Veget. Sci. 1: 233-240

Felinks B, Mrzljak J, Pilarski M, Wiegleb G (2000) Generalisierung biologisch/ ökologischer Daten der Bergbaufolgelandschaft vom Punkt in die Fläche. In: Wiegleb G, Bröring U, Mrzljak J, Schulz F (eds) Naturschutz in Bergbaufolgelandschaften - Landschaftsanalyse und Leitbildentwicklung. Physica, Heidelberg: 264-283

Forman RTT (1995) Land mosaics. The ecology of landscapes and regions. Cambridge University Press, Cambridge

Haridason K, Rao RR (1985) Forest Flora of Meghlaya, Volume 1. Bishen Singh Mahendra Pal Singh, 23-A Connaught Place, Dehra Dun

Harris LD, Miller KR (1984) The Fragmented Forest: Island Biogeograhpy Theory and the Preservation of Biotic Diversity. University of Chicago Press, Chicago

Hurd JD, Civco DL, LaBash C, August P (1992) Coastal wetland mapping and change detection in the northeast United States. in Proc. 1992 ASPRS/ACSM/RT'92 Convention, Washington, D.C. 1:130-139

Hurd JD, Wilson EH, Lammey S, Civco DL (2001) Characterization of Forest Fragmentation and Urban Sprawl using Time Sequential Landsat Imagery. Proc. 2001 ASPRS Annual Convention, St. Louis, MO

IAIA – International Association for Impact Assessment (2005) IAIA Special Publication Series No. 3, Biodiversity in Impact Assessment Fargo, USA. www.iaia.org

Johst K, Huth A (2005) Testing the intermediate Disturbance Hypothesis: when will there be two peaks of diversity, Diversity and Distribution, (Diversity Distrib). Biodiversity Research Leiptzig Germany. 111-120

Li H, Reynolds JF (1994) A Simulation Experiment To Quantify Spatial Heterogeneity In Categorical Maps. Ecology 75:2446–2455

McGarigal K, Barbara M (1995) FRAGSTATS. Spatial Pattern Analysis Program For Quantifying Landscape Structure. Gen. Tech Rep. PNW-GTR-351. Portland, Oregeon: USDA Forest Service, Pacific Northwest Research Station

Palmer MW, E van der Maarel (1995) Variance in species richness, species association, and nicher limitation. Oikos 73:203-213

Pickett STA, White (1985d) The Ecology of Natural Disturbance And Patch Dynamics. Academic Press, Inc., London, pp 5-9, 19-33, 161-170, 253-264

Ravan SA, Roy PS, Sharma CM (1998) Accuracy Evaluation of Digital Classification of Landsat TM data - An Approach to Include Phenological Stages of Tropical Dry Deciduous Forest, 33-43

Roy PS, Joshi PK (2002) Tropical Forest Cover Assessment in North East India- Potentials of Temporal Wide Swath Satellite Data (IRS 1C- WiFS). International Journal of Remote Sensing (In press)

Sala OE, Chapin FS, Armesto JJ, Berlow E, Bloomfield J, Dirzo R, Huber-Sanwald E, Huenneke LF, Jackson R. B, Kinzig A, Leemans R, Lodge DM, Mooney HA, Oesterheld M, Poff N. L, Sykes MT, Walker BH, Walker M, Wall DH (2000) Global Biodiversity Scenarios for the Year 2100. Science 287:1770-1774

Stohlgren TJ, Chong GW, Kalkhan MA, Schell LD (1997b) Rapid Assessment of Plant Diversity Patterns. A Methodology for Landscapes. Environ Monit Assessment 48:25–43

Stoms D, Estes J (1993) A Remote Sensing Research Agenda For Mapping And Monitoring Biodiversity. International Journal of Remote Sensing, 14:1839-1860

Szaro RC, Johnston DW (eds) (1996) Biodiversity in Managed Landscapes. Oxford Univ. Press, Oxford. (5) Gustafson EJ (1998) Quantifying landscape spatial pattern: What is the state of the art? Ecosystem

Treweek J (1999) Ecological Assessment. Blackwell Science, Oxford

Tropical Biodiversity Advisers on EIA. March 2000. Working Group Minutes, GTZ, Bonn

UNEP, Convention on Biological Diversity, Subsidiary Body on Scientific, Technical and Technological Advice, September 2001. Indicators and Environmental Impact Assessment. pp19, Montreal

Vitousek PM, Mooney HA, Lubchenco J, Melillo JM (1997) Human Domination of Earth's Ecosystems. Science 277:494-499

Weber D, Hintermann U, Zangger A (2004) Scale and trends in species richness:considerations for monitoring biological diversity for political purposes, Global Ecology and Biogeography, (Global Ecol. Biogeogr.). Switzerland. 13:97-104

Wiegleb G (2002) Management of Regional Environmental Problems with Special Respect to Post Mining Landscapes. Environment al and Resource Management Lecture Notes. pp 3,4 and 6

Wiegleb G, Felinks B (2001a) Primary Succession in Postmining Landscapes-Chance or Necessity? Ecological Engineering. 17:199-217

Wiegleb G, Felinks B (2001b) Predictability of Early Stages of Primary Succession in Post-Mining Landscapes of Lower Lusatia. Appl. Veget. Sci., 4,5-18

Zuidema PA, Sayer JA, Dijkman W (1996) Forest Fragmentation and Biodiversity: the Case for Intermediate-Sized Conservation Area. Environmental Conservation 23:4, 290-297

23 Air Pollution and Climate: Standards for Particulate Matter

Matti Johansson

United Nations Economic Commission for Europe, Environment, Housing and Land Management Division, Geneva, Switzerland

23.1 Introduction

Air pollutants have detrimental effects on the environment and human health. The 1979 Convention on Long-range Transboundary Air Pollution (LRTAP) is regarded as a pioneering instrument, which has brought about tangible results in reducing emissions and improving the environment by delivering legally binding protocols (UNECE 2004). There is increasing scientific evidence that current levels of particulate matter (PM) lead to a wide range of acute and chronic health problems and its long-range transport contributes significantly to these effects. PM also leads to other environmental problems such as corrosion and soiling of materials, heavy metals and climate change (figure 23.2). Guidelines to address human health risks exist at local and regional levels, but PM is currently not directly included in any of the Convention's protocols.

PM is an air pollutant consisting of a mixture of solid and liquid particles suspended in the air with both anthropogenic and non-anthropogenic sources. Particles can be directly emitted into the air as primary PM or be formed in the atmosphere as secondary PM from gaseous precursors, mainly sulphur dioxide (SO_2), nitrogen oxides (NO_x), ammonia (NH_3) and non-methane volatile organic compounds (VOC). PM is often characterized by the particle size (or aerodynamic diameter). Coarse particles (PM_{10} with a diameter <10 μm), fine particles ($PM_{2.5}$ with a diameter <2.5 μm) and even ultrafine particles ($PM_{0.1}$ with a diameter <0.1 μm) are used to describe the origin of the emissions, the transport in the atmosphere, the mass concentrations and the ability to be inhaled into respiratory system. The most important chemical constituents are sulfate, nitrate, ammonium, other inorganic ions, organic and elemental carbon, crustal material, particle-bound water and heavy metals. The coarse fraction is relatively easily deposited within tens of kilometres of their origin. Very fine particles can stay in the atmosphere for weeks and be transported over distances up to thousands of kilometres.

Standards and Thresholds for Impact Assessment. Edited by Michael Schmidt, John Glasson, Lars Emmelin and Hendrike Helbron. © 2008 Springer-Verlag

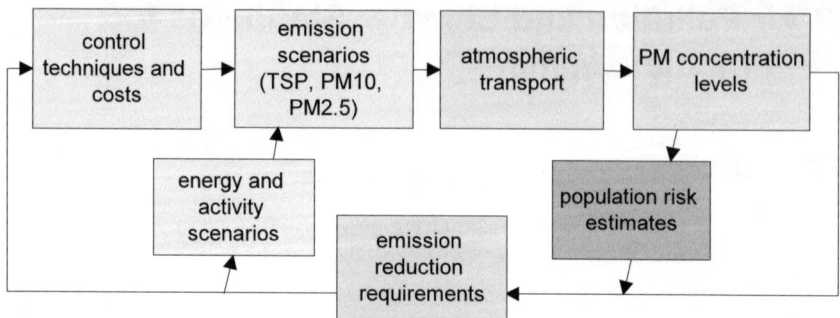

Fig. 23.1. Schematic assessment of particulate matter

This chapter aims to describe an integrated assessment approach to address the impacts of PM (figure 23.1). The emphasis is at the international level, in particular within the LRTAP Convention, and the complementary use of models and measurements. It assesses emissions, atmospheric transport, effects and abatement for air pollutants within the geographical area of the United Nations Economic Commission for Europe (UNECE), which comprises European countries extending to the Central Asia and two North-American countries. The Convention's eight protocols in force do not currently include explicit obligations for PM emissions to its 51 Parties (UNECE 2004). However, the 1998 Aarhus Protocol on Heavy Metals and the 1999 Gothenburg Protocol to Abate Acidification, Eutrophication and Ground-level Ozone indirectly address particles through controls for emission sources of heavy metals and precursor gases for secondary PM.

23.2 Integrated Assessment

The key integrated assessment model at the European level for the Convention is the RAINS model (Regional Air Pollution Information and Simulation), which describes the pathways of pollution from anthropogenic emissions to various environmental and health impacts. It comprises databases from all European countries with essential information on emissions, abatement options and costs, atmospheric transport coefficients and effects. It links these data to assess economic development scenarios and emission control strategies. The key environmental problems (acidification, eutrophication, ground-level ozone, health) are considered in a multi-pollutant context, quantifying the contributions of SO_2, NO_x, NH_3, VOC and primary fine and coarse (PM10–PM2.5) particles.

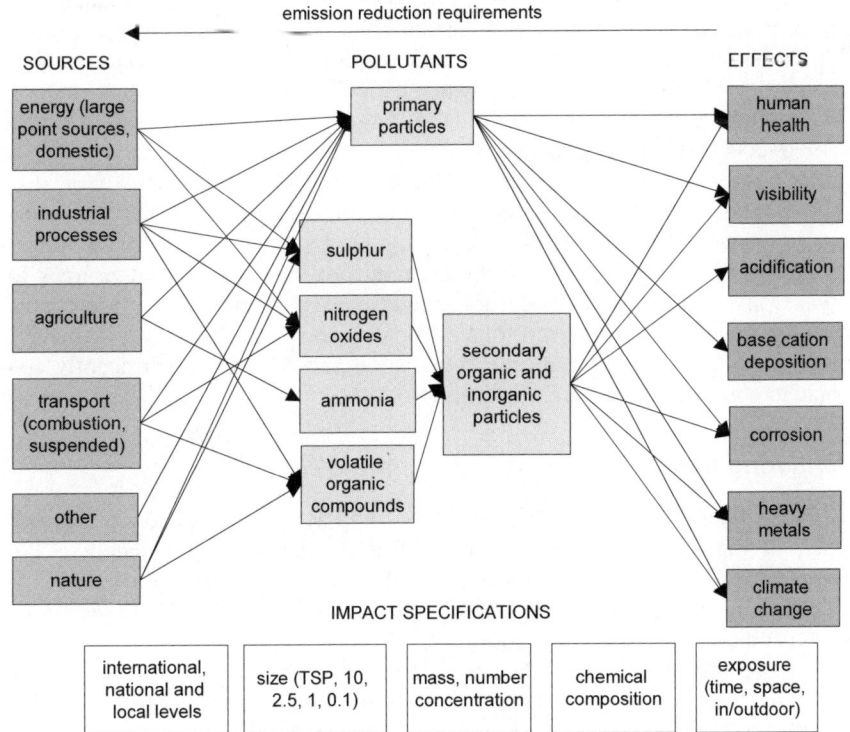

Fig. 23.2. Integrated assessment of particulate matter

Emissions

Official emissions are reported by the Parties of the LRTAP Convention every year on SO_2, NO_x, NH_3, VOC, carbon monoxide, PM, heavy metals and persistent organic pollutants (POP) for protocol base years and selected target years in the future. Every five years large point sources and the sectoral and spatial distribution (50 km × 50 km) are detailed. When official emission data are missing or clearly inconsistent, complementary emissions are estimated for modelling use. Although the Convention has required reporting on particle emissions and their sources since 2000, the quantitative knowledge is still incomplete (EMEP 2006). Only 28 Parties reported at least some data in 2004. Data were not always reported for sectors with likely PM emissions, and there is little information on the chemical composition. Emissions of the PM precursor gases SO_2, NO_x, NH_3 and VOC were reduced 30 %–70 % in 1990–2004 according to the inventories. Countries and municipalities have their own emission inventories, which may contribute to international reporting.

According to the RAINS model estimates, mobile sources, industry (including energy production) and domestic combustion contributed 25–34% each to primary $PM_{2.5}$ emissions in 2000 in Europe. These sectors are also major emitters of the precursor gases SO_2, NO_x and VOC, while agriculture is a dominant contributor to NH_3. Anthropogenic primary emissions of $PM_{2.5}$ and PM_{10} were halved in Europe in 1990–2000. Modelled projections suggest that existing legislation will lead to similar further reductions of primary PM emissions in the European Union (EU) and that transport and domestic sectors will be increasingly important. The reductions in the non-EU countries are not projected to be as large. In 2020 the highest PM emissions and the greatest technical reduction potential will at sources for which Convention's protocols do not specify emission limit values, in particular small non-industrial combustion sources such as wood and coal stoves. Low-level sources, especially urban domestic combustion and transport, significantly contribute to population exposure. (WHO 2006)

Monitoring Data

In 2004, 12 and 8 countries measured PM_{10} and $PM_{2.5}$ at 39 and 23 sites, respectively, of the monitoring network of EMEP (Convention's Co-operative Programme for Monitoring and Evaluation of the Long-range Transmission of Air Pollutants in Europe). The annual PM_{10} limit value 40 $\mu g\ m^{-3}$ of EU was not exceeded at these stations (EMEP 2006).

For 2002 the annual mean PM_{10} was 22 $\mu g\ m^{-3}$ in the rural areas, 26 $\mu g\ m^{-3}$ in the urban background and 32 $\mu g\ m^{-3}$ at traffic locations, calculated from extensive data of the "AirBase" database at the European Environment Agency. The limit values of EU were exceeded in several locations. Trend assessment did not indicate a clear relation to changes in emissions using available data since 1997. The rural background concentration for $PM_{2.5}$ was 11–13 $\mu g\ m^{-3}$ in 2001, but urban levels were considerably higher (15–20 $\mu g\ m^{-3}$ in urban background and 20–30 $\mu g\ m^{-3}$ at traffic sites). (EMEP 2006)

Atmospheric Transport

Transboundary concentrations of PM_{10} and $PM_{2.5}$ for 2004 were calculated with the eulerian EMEP model in a 50 km × 50 km grid in Europe (EMEP 2006). The chemical composition of primary PM was described with an extension to the EMEP model, developed from detailed aerosol modelling (Pirjola and Kulmala 2000). The modelled annual mean concentration for PM_{10} was 5–20 $\mu g\ m^{-3}$ and for $PM_{2.5}$ 5–15 $\mu g\ m^{-3}$ for most of Europe. The annual mean concentrations of primary $PM_{2.5}$ were 1–3 $\mu g\ m^{-3}$, reaching 3–5 $\mu g\ m^{-3}$ in several big cities.

The EMEP model generally underestimates the observed regional background levels of PM_{10} and $PM_{2.5}$ in Europe, a feature shared by other models. It can well reproduce spatial variability and observed levels of secondary inorganic aerosols across Europe, which contribute 20–30 % and 30–40 % of PM_{10} and $PM_{2.5}$ mass, respectively (EMEP 2006). The differences in organic aerosols are large. The regional background concentrations of anthropogenic $PM_{2.5}$ have a considerable transboundary contribution of about 60 %. Organic carbon and mineral dust seem to be major contributors to the differences between levels at traffic sites and the regional background. Regional modelling of individual PM components still bears uncertainties.

Effects

Acute and chronic exposure to PM in ambient air has been linked to many different health outcomes. They range from modest transient changes in the respiratory tract and impaired pulmonary function, through increased risk of symptoms requiring emergency room or hospital treatment, to increased risk of death from cardiovascular and respiratory diseases or lung cancer. Toxicological evidence supports the observations from epidemiological studies.

The effects of long-term PM exposure on mortality (life expectancy) seem to be attributable to $PM_{2.5}$ rather than to coarser particles. The latter may have more visible impacts on respiratory morbidity (illness). The primary carbon-centred and combustion-derived particles have been found to have considerable inflammatory potency. However, available evidence is still not sufficient to predict the health impacts of changing the composition of the PM mixture.

Health effects are observed at all levels of exposure. Within any large population there is a wide range of susceptibility and some people are at risk even at the lowest observed concentrations. People with pre-existing heart and lung disease, asthmatics, socially disadvantaged and poorly educated people and children belong to the more vulnerable groups.

In Europe the risks to health of PM have been assessed by WHO working groups, the work of the Joint WHO/Convention Task Force on the Health Aspects of Air Pollution (WHO 2006) and within the European Commission's Clean Air for Europe thematic strategy. Their shared analysis employed mortality as the main indicator of health impact. The concentration-response function indicated an increase in risk of all-cause mortality by 6 % per 10 μg m^{-3} of $PM_{2.5}$. Quantification of impacts on morbidity was deemed less precise than for mortality, since the available data on responses was insufficient, but selected estimates on morbidity were analyzed.

Reduction of the remaining substantial uncertainties in the assessment will require improved understanding of PM components crucial to observed impacts. Since the long-range transport of pollution contributes significantly to ambient PM levels and population exposure, international action must accompany local and national efforts to reduce both emissions and effects.

Long-range transboundary air pollution contributes to the background concentrations of heavy metals (e.g. cadmium and lead) attached to particles in the atmosphere, the deposition on crops or topsoil and the contamination of food. Although the inputs may be relatively small in comparison to stores already accumulated, their reduction is advised to minimize the health risks (UNECE 2006). PM also leads to increased corrosion and soiling of materials, whether used in constructions or in objects of cultural heritage, as demonstrated by the Convention's International Cooperative Programme on Materials. Recently updated concentration-response functions for corrosion of carbon steel, zinc, copper, bronze and limestone included PM (UNECE 2005). PM can include significant amounts of inorganic ions (such as calcium and magnesium) that can balance the acidity in precipitation and alleviate acidification of terrestrial and aquatic ecosystems, however, SO_2 emission controls also reduced dust emissions and have brought down this positive effect. In the atmosphere PM can affect the visibility, which is of concern in North America but has not been widely considered in Europe. Particles also influence the radiative forcing of the atmosphere, and the direction and magnitude largely depends on their chemical composition, for example the carbonaceous fraction.

Selected Examples

The results from the RAINS model analysis indicate that current exposure to anthropogenic $PM_{2.5}$ leads to an average loss of 8.6 months of life expectancy in Europe (Amann et al. 2005). The total number of premature deaths attributed to exposure was about 348 000 in 25 EU member states. Effects on morbidity included 100 000 hospital admissions per year; however, several expected impacts could not be reliably estimated due to weaknesses of existing data. Current existing legislation will reduce the impacts by one third. The RAINS model analysis included a method to address separately urban and rural exposure levels, which is currently being improved to cover 500 cities in Europe.

A strategic environmental impact assessment was made on the national energy and climate strategy in Finland (Hildén 2005). This study complemented an earlier environmental impact assessment on national greenhouse gas reductions, which may cause environmental impacts. The studied energy production alternatives had no significant effect on primary $PM_{2.5}$ emissions modelled with the Finnish regional emission model. Emissions decreased from 36 Gg in 2000 to 32–34 Gg and 27–28 Gg in 2010 and 2025, respectively. Large combustion plants are equipped with effective dust filters in Finland and the selection of primary energy source for electricity production, even if it were coal, did not significantly affect national total PM emissions. No regional method was operational at the time to make a health risk assessment. However, later studies in a 1 km × 1 km grid over Finland have indicated that as population density is high adjacent to roads, the $PM_{2.5}$ emissions from vehicular traffic (5.8 Gg) that are lower than from domestic wood combustion (7.6 Gg) lead to a higher total population exposure.

23.3 Air Quality Guidelines

The *Air quality guidelines* (AQG) of the World Health Organization (WHO) are designed to offer guidance in reducing the health impacts of air pollution based on expert evaluation of current scientific evidence. There is growing evidence of adverse health effects at low levels of exposure without a clear threshold below, which there are no adverse effects. The epidemiological evidence shows adverse effects of particles after both short-term and long-term exposures with a focus on outdoor sources, although in developing countries pollutants emitted indoors, in particular combustion of solid fuels, cause highest exposures. Risk coefficients in underdeveloped countries have been found to be similar to those in developed countries, suggesting transferability for policy making.

The most recent and extensive epidemiological evidence is largely based on PM_{10} as the exposure indicator. The majority of monitoring data is on PM_{10}. The composition of fine and coarse particles is likely to vary substantially across cities around the world depending upon local geography, meteorology and specific sources. Although few epidemiological studies exist comparing the relative toxicity of combustion from fossil fuels versus biomass, similar effects have been reported for cities in developed and developing countries. The numerical guideline value was chosen as an indicator based on studies using $PM_{2.5}$. The PM_{10} guide-

Table 23.1. Air quality guidelines (AQG) and interim targets (IT) for particulate matter as annual and 24-hour means (after WHO 2006). The values in square brackets indicate the 95% confidence interval. The use of the $PM_{2.5}$ guidelines is recommended

Annual PM_{10} $\mu g\ m^{-3}$	Annual $PM_{2.5}$ $\mu g\ m^{-3}$	Basis for the selected level	24-hour PM_{10} $\mu g\ m^{-3}$	24-hour $PM_{2.5}$ $\mu g\ m^{-3}$	Basis for the selected level
20 AQG	10 AQG	Lowest levels with shown total, cardiopulmonary and lung cancer mortality increase (>95% confidence) in response to PM2.5 (Pope et al. 2002)	50 AQG	25 AQG	Relation between 24-hour and annual PM levels
70 IT-1	35 IT-1	Levels estimated to be associated with about 15% higher long-term mortality than at AQG	150 IT-1	75 IT-1	Published risk coefficients from multi-centre studies and meta-analyses (about 5% increase of short-term mortality over AQG)
50 IT-2	25 IT-2	Additional lower risk of premature mortality by approximately 6% [2-11%] compared to IT-1	100 IT-2 *	50 IT-2 *	As above (about 2.5% increase of short-term mortality over AQG)
30 IT-3	15 IT-3	Additional reduced mortality risk by another approximately 6% [2-11%] compared to IT-2	75 IT-3 **	37.5 IT-3 **	As above (about 1.2% increase in short-term mortality over AQG)

* 99th percentile (3 days/year).

** for management purposes, based on annual average guideline values; precise number to be determined on basis of local frequency distribution of daily means.

line value is derived with a $PM_{2.5}$-PM_{10} ratio of 0.5. Both short-term (24-hour) and long-term (annual) guidelines are needed, based on known health effects. As air pollution levels in developing countries often far exceed the recommended levels, interim target levels were introduced. They are intended as incremental steps in a progressive reduction of air pollution in severely polluted areas to promote a shift from concentrations with serious acute health consequences to concentrations to significantly reduce chronic effects as well. The guideline values and the interim targets are shown in Table 23.1 with justifications for the chosen levels (WHO 2006). There is considerable toxicological evidence of potential detrimental effects of $PM_{0.1}$, however, the epidemiological evidence is insufficient to conclude a relationship between exposure and response.

In the European Union (EU), the framework directive on ambient air quality assessment and management was adopted in 1996. This directive introduced new air quality standards for previously unregulated air pollutants, setting the timetable for the development of daughter directives on a range of pollutants. The health limit values for PM_{10}, defined by the first daughter directive (EU 1996), were to be met by 2005. The annual limit was 40 µg m^{-3}, and an indicative stricter limit 20 µg m^{-3}, was proposed for 2010. The daily limit was 50 µg m^{-3}, not to be exceeded more than 35 times per year, and indicatively not more that 7 times in 2010. Current positions of the bodies of the EU may result in keeping the current PM_{10} limits.

23.4 Conclusions and Recommendations

The increasing importance of particulate matter has led to a wide body of methods to assess its impacts and abatement options. The integrated assessment approach leads to harmonized methods for emission estimates, quantification of control option efficiency and costs, incorporation of measured and modelled rural and urban background levels and consistent evaluation of impacts. General conclusions and recommendations can be made:

- An integrated assessment on the impacts of PM bring together all relevant aspects of the pollution cycle from emissions and atmospheric dispersion to impacts and abatement options; health impacts are important but not the only effects;
- The new *Air quality guidelines* of WHO provide a useful reference to aim at PM_{10} and $PM_{2.5}$ levels that indicate low health risks; interim targets may be used in severely polluted areas to progressively reduce impacts;
- Guideline levels refer to monitoring sites representative of population exposures, but they may be higher close to specific sources such as roadways and large stationary sources, and the $PM_{2.5}$-PM_{10} ratio of 0.5 may be changed with justified local data for setting local standards
- Health impacts can be quantified for mortality and morbidity endpoints;

- The use of observations and model results together is encouraged to estimate the contribution from regional background, urban background and local hot spots.

References

Amann M, Bertok I, Cofala J, Gyarfas F, Heyes C, Klimont Z, Schöpp W, Winiwarter W (2005) Baseline scenarios for the Clean Air for Europe (CAFE) programme. CAFE Scenario Analysis Report 1, International Institute for Applied Systems Analysis (IIASA), 79 p.

EMEP (2006) Transboundary particulate matter in Europe. Joint CCC & MSC-W status report 4, Norwegian Meteorological Institute, Oslo, 139 p.

EU (1999) Council Directive 1999/30/EC of 22 April 1999 relating to limit values for sulphur dioxide, nitrogen dioxide and oxides of nitrogen, particulate matter and lead in ambient air. Official Journal L163, European Communities, Brussels, pp. 41–60

Hildén M, Karvosenoja N, Kankaanpää S, Ratinen M, Liski J, Hämekoski K (2005) Environmental assessment of the national strategy for the implementation of the Kyoto protocol (in Finnish). The Finnish Environment 802, 81 p.

Pirjola L, Kulmala M (2000) Aerosol dynamical model MULTIMONO. Boreal Environmental Research 5:361–374

Pope CA et al. (2002) Lung cancer, cardiopulmonary mortality, and long-term exposure to fine particulate air pollution. J American Medical Association, 287: 1132–1141

UNECE (2004) Handbook for the 1979 Convention on Long-range Transboundary Air Pollution and its Protocols. ECE/EB.AIR/85, United Nations, New York and Geneva, 341 pp

UNECE (2005) Final results from the multi-pollutant programme including dose-response functions on effects on materials, Report ECE/EB.AIR/WG.1/2005/7, United Nations Economic Commission for Europe, Geneva

UNECE (2006) Health Risks of Heavy Metals from Long-range Transboundary Air Pollution. Report ECE/EB.AIR/WG.1/2006/12, United Nations Economic Commission for Europe, Geneva

WHO (2005) WHO air quality guidelines global update 2005. Report on a Working Group meeting, Bonn, Germany, 18-20 October 2005, 25 p.

WHO (2006) Health risks of particulate matter from long-range transboundary air pollution. Joint WHO / Convention Task Force on the Health Aspects of Air Pollution, WHO European Centre for Environment and Health, Bonn Office, 99 p.

24 Standards and Thresholds of the EU Water Framework Directive (WFD) – Phytoplankton and Lakes

Brigitte Nixdorf[1], Atis Rektins[1] and Ute Mischke[2]

1 Chair of Freshwater Conservation, Research Station Bad Saarow, Brandenburg University of Technology (BTU), Germany
2 Leibniz-Institute of Freshwater Ecology and Inland Fisheries, Department of Shallow Lakes and Lowland Rivers, Berlin, Germany

24.1 Introduction

Water protection policy of the European Union (EU) dates back to 1970s, especially to Paris Summit in 1972, where commitment was made to introduce common EU wide environmental policy that included protection of water resources (European Parliament 1972). Since then, several water quality acts have been passed, mostly regulating water pollution. As environmental issues took more important place in the political agenda and EU water policy was rather fragmented, consisting of many complex individual and fragmented regulations, the European Commission (EC) stressed out the need for an integrated approach to water resources covering both the quantitative and qualitative aspects of the policy, management of both surface water and groundwater, environmental protection and the links with other policies (EC 1994). The Water Framework Directive (WFD), proposed in 1994, was finally adopted in 2000. It applies to all water in the natural environment - rivers, lakes, estuaries and coastal waters as well as groundwater.

Currently WFD has been implemented in national legislation of 25 EU countries and in four candidate countries – Romania and Bulgaria (expected to join the EU in 2007), Croatia and Macedonia, as well as in Iceland, Norway and Liechtenstein, principles of the WFD are partly taken into consideration in other EU neighbouring countries. Therefore, WFD has truly become the most comprehensive normative act regarding management of European waters.

The WFD gradually (till 2013) repeals and replaces a number of older EC water directives - Exchange of Information Directive, Dangerous Substances Directive,

Standards and Thresholds for Impact Assessment. Edited by Michael Schmidt, John Glasson, Lars Emmelin and Hendrike Helbron. © 2008 Springer-Verlag

Fish Water, Shellfish Water and Groundwater Directives and incorporates the remaining existing water Directives - the Bathing Water Directive, Nitrates Directive and Urban Waste Water Treatment Directives, into its framework through its protected areas provisions. Also all Natura 2000 sites in which quality of water is concerned (The Directives on the Protection of Habitats and Birds) must be incorporated into river basin management plans. It is expected that during the next few years Daughter Directives, which are currently being discussed, will amend the WFD – the Groundwater Directive (EC 2004), Floods Directive (EC 2006) and Priority Substances Directive (EC 2001).

24.2 WFD – Aim and Present State of the Implementation

The aim of the WFD is to achieve the objective of at least good water status (when ecological functions of the water are not significantly altered) by defining and implementing the necessary measures; where good water status already exists, it should be maintained. Detailed definitions of good ecological status are provided in the WFD for different types of waters. They are given in terms of the quality of the biological community, the hydrological characteristics and the chemical characteristics. As there can not be fixed standards used throughout the EU for ecological quality, because of geographic variability, "good status" is defined as allowing a slight deviation from natural conditions due to anthropogenic impact.

Additionally, EU-wide minimum requirements will be developed for 33 priority substances, when Priority Substances Directive will be adopted. Water management is based on River Basin Districts (RBD), which consist of river basins and associated lake, groundwater, transitional (estuaries) and coastal (up to 1 nautical mile from the coastline) water bodies. Total number of water bodies and their area differ from country to country, not only because of geographic conditions but also national water management policies, e.g. France and Sweden, countries of comparable land area, have respectively 3 900 and 12 300 surface water bodies.

24.2.1 Artificial and Heavily Modified Water Bodies

Surface water bodies may also be designed as artificial or heavily modified, due to the fact that many watercourses, especially in heavily populated areas, have been modified from their natural conditions during the course of the last centuries and it might be impossible to restore them to natural conditions. Such deviations are applied to allow navigation, functioning of ports, recreation activities, activities for the purposes of which water is stored, such as drinking water supply, power generation or irrigation, as well as water regulation, flood protection, land drainage or other equally important sustainable human development activities. However, these exemptions are only possible when alternatives are technically unfeasible, they are disproportionately expensive or they produce worse environmental result. Artificial or heavily modified water bodies are required to achieve good ecologic potential, instead of good ecological status – it reflects the values of ecological status for the closest comparable water body type, but with conditions from the artificial or heavily modified characteristics of the water body taken into consideration.

24.2.2 River Basin Management Plans and Monitoring

For transboundary river basins, measures to achieve the environmental objectives set by the WFD, should be coordinated for the whole of the RBD, note separately for each country. Such approach is more natural and can also be more productive, there are already many successful examples in Europe, such as Danube or Rhine river basins.

To report how environmental objectives of the WFD will be achieved, countries are required to produce River Basin Management Plans, each for every RBD lying entirely within their territory. The plan must contain the following elements: the river basin's characteristics, a summary of significant pressures and impact of human activity, information on the status of water quality in the basin and detailed information on environmental objectives established in the WFD regarding the particular river basin, including a set of measures how these objectives will be fulfilled. Also economic analysis of water use within the river basin must be carried out.

In order to obtain information for preparation of River Basin Management Plans, data from water quality monitoring are necessary. The WFD defines three types of water quality monitoring:

- operational monitoring – for water bodies which are at risk not to achieve "good status" and to assess changes resulting from the programmes of measures;
- surveillance monitoring – to monitor long term trends and assessment of the overall surface water status;
- investigative monitoring – to investigate reasons of unknown exceedances and in cases of accidental pollution.

24.3 Main Objectives of WFD and Intercalibration

Main objectives of the WFD are defined in Article 4 and can be summarized as follows:

For surface waters:

- For all water bodies deterioration of the status should be prevented;
- All water bodies should be protected, enhanced and restored with the aim of achieving good surface water status till 2015;
- All artificial and heavily modified bodies of water, should be protected and enhanced with the aim of achieving good ecological potential and good surface water chemical status till 2015;

- All the necessary measures should be taken, with the aim of progressively reducing pollution from priority substances and ceasing or phasing out emissions,
  discharges and losses of priority hazardous substances.
  For groundwater:
- For all groundwater bodies the input of pollutants into groundwater and the deterioration of the status should be prevented or limited,
- All groundwater bodies should be protected, enhanced and restored, to ensure a
  balance between abstraction and recharge of groundwater, with the aim of
  achieving good groundwater status till 2015.

Additionally, the reference to "protected areas" is made, where compliance with
any relevant environmental standards and objectives should be achieved till 2015.

The WFD provides clear deadlines for carrying out the main activities, till
2006, RBDs had to be identified and elaborately characterized, also providing a
review of the impact of human activity and an economic analysis of water use.

Table 24.1. Timeline for WFD implementation

Date	Activity
22 December 2006	Monitoring networks are operational
22 December 2008	Drafts of River Basin Management Plans published
22 December 2009	First River Basin Management Plan published for each RBD, including programme of measures
2010	Water pricing policies are in place
22 December 2012	Programmes of measures for each RBD are operational, to achieve environmental objectives
22 December 2015	Main environmental objectives are achieved
Every sixth year after 2015	Review and update of River Basin Management Plans

Within this process every step of planning and design of River Basin Management
Plans is connected with public participation playing an important role. It is done
by providing that timetables, work programmes and draft River Basin Management Plans available for public to review and comment, up to three years before
the period to which the plan refers.

Intercalibration is essential for ensuring a comparable level of protection in
consistency with the Directive. It is in order to ensure that class boundaries are established consistent with the normative definitions and are comparable between
Member States. A number of additional research activities, provide support to the
intercalibration exercise and improve the quality of the results.

Intercalibration progress currently is slower than anticipated, as generally only
one element for each water body (phytoplankton for lake, benthic invertebrates for
river, macroalgae and angiosperms for coastal water bodies) is expected to sufficient intercalibration till the end of 2006 (ECOSTAT 2006). Therefore, it seems
unlikely that the exercise will be completed within the deadline (August 2007) and
its prolongation will be necessary. Main reason why intercalibration progress is
relatively slow and not all the quality the elements are covered yet is that it is very
difficult to find common metrics for comparatively large areas. There are signifi-

cant gaps of knowledge, also for phytoplankton in lakes, from which the most notable are:

- reference conditions in different lake types,
- threshold concentrations for high/good and good/moderate boundaries,
- taxonomic indicators for measuring impacts of nutrient pressures,
- establishment of supporting physicochemical conditions,
- effect of seasonal variability on classification schemes,
- ecological impact of nitrogen conditions (REBECCA 2005).

Usually different countries have different perceptions on issues which are not stated explicitly in the WFD, allowing certain flexibility – e.g. regarding design of monitoring networks, as only general principles are set. To deal with such situations, 14 non-binding guidelines (on intercalibration, monitoring, public participation, planning process etc) have been prepared and ongoing consultations and exchange of information between the actors involved are taking place.

24.4 Biological Quality Elements and Ecological Status

Overview about Biological Quality Elements

When assessing the ecological status of the water, biological quality elements are considered to be the most important, they are supported by hydromorphological and physico-chemical elements.
  Biological quality elements for the purposes of the WFD consist of:

- Composition, abundance and biomass of phytoplankton;
- Composition and abundance of other aquatic flora (macrophytes and microphytobenthos for lakes and rivers, macroalgae and angiosperms for coastal and transitional waters);
- Composition and abundance of benthic invertebrate fauna;
- Composition, abundance and age structure of fish fauna.

Although such approach is not new, it has never been introduced in Europe in so large scale before. Many countries have changed their monitoring programmes significantly to adapt to the WFD requirements, as previously many of them had relied rather heavily on physico-chemical parameters, when assessing the water quality. Therefore new assessment methods have to be developed and implemented. Assessment based primarily on biological parameters is more complicated, as effects that specific changes in environment have on living organisms are not always fully understood.
  Following the WFD each country has to divide the water quality status for each surface water category into five classes - high, good, moderate, poor and bad. „High status" for biological, physicochemical and morphological quality elements is similar to totally or nearly totally, undisturbed conditions, also called „reference conditions" and are associated with no or very limited anthropogenic pressures.

Further quality classes are based on the extent of deviation from the reference conditions. For surface waters, so-called „One out, all out" principle is used, meaning that the status of the water body is represented by the lowest of the values for the biological and physicochemical monitoring results for the relevant quality elements.

To ensure that class boundaries are comparable between different countries, intercalibration exercise is carried out. Each participating country in 2003 and 2004 had to select sites representing ecological status at the boundaries between the "high" and "good" and between the "good" and "moderate" classifications, on the basis of provisional classification of their current national assessment methods. These sites were entered into EC intercalibration register, that currently includes 1489 sites from all 25 EU Member States, Bulgaria, Norway and Romania (EC 2005) Member States are divided into Geographical Intercalibration Groups (GIG), such as Rivers Central/Baltic (RCE), Lakes Alpine (LAL) or Mediterranean Atlantic GIG[1], comprising Member States sharing particular surface water body types (EC 2005a).

The main tasks for countries involved in is to find quantitative expressions (criteria or metrics) for the response in abundance and taxonomic composition for the different biological quality elements along the gradient of main pressures. For example, eutrophication can be reflected by quantifying the increased algal/plant biomass and the impact on other organisms and water quality (EC 2005b), in order to establish boundaries for good, moderate and poor ecological quality classes.

24.4.1 Phytoplankton - Ecological Quality Element and Indicator for Eutrophication Status of Freshwater Bodies

Eutrophication is regarded as one of the most significant water protection issue in European waters; it is addressed in various international conventions, like Helsinki Convention, OSPAR Convention, Convention for the Protection of Rhine and others, as well as in other EU Directives – the Nitrates Directive and Urban Wastewater Treatment Directive (UWWT).

In the UWWT Directive (EC 2005b) eutrophication is defined as the enrichment of water by nutrients, especially compounds of nitrogen and/or phosphorus, causing an accelerated growth of algae and higher forms of plant life to produce an undesirable disturbance to the balance of organisms present in the water and to the quality of the water concerned (EC 1991). However, the assessment of eutrophication status is integrated in the classification of surface water bodies - the definition of good ecological status for the quality elements "phytoplankton" and "macrophytes and phytobenthos" uses similar wording as the definition of eutrophication used in the UWWT and Nitrates Directives and by OSPAR (EC 2005b).

---

[1] GIG acronyms: LAT – Lakes Atlantic; LCE – Lakes Central/Baltic; LME – Lakes Mediterranean; LNO – Lakes Northern; RAL – Rivers Alpine; REC – Rivers Eastern Continental; RME – Rivers Mediterranean; RNO – Rivers Northern.

The phytoplankton community is considered to be the first biological community to respond to eutrophication pressures especially in lakes (REBECCA 1005).

Phytoplankton is prescribed by the WFD as a quality element for lakes, but is not explicitly mentioned for rivers. Still, the appendix of the WFD provides a definition for high, good and moderate ecological status in rivers also for phytoplankton when it can be considered relevant for ecological status. Phytoplankton biomass is very low in low-order streams and increasing in high-order large rivers (Wetzel 2001). Germany (Mischke and Behrendt 2006) decided to restrict assessment on latter once.

The WFD defines high, good and moderate status for three phytoplankton quality elements that potentially can be used to assess the ecological status of lakes:

- Phytoplankton abundance and composition;
- Phytoplankton biomass;
- Planktonic bloom intensity and frequency.

According to intercalibration milestone reports from September 2006, phytoplankton will be the only element intercalibrated by all lake GIGs till the end of 2006 (ECOSTAT 2006). Common metrics will be developed and agreed only for *chlorophyll a* concentrations, partly also *phytoplankton biovolume*, but there is work in progress by GIGs to intercalibrate also *phytoplankton composition* and *proportion of cyanobacteria*. Results can not be expected sooner than in 2007. River GIGs have not considered intercalibration of phytoplankton in their plans.

For rivers the main uncertainties are relationships among nutrient concentrations and blooms, as nutrient concentration is not the only impact factor for triggering blooms (REBECCA 2005). Still, in slow flowing lowland rivers like river Elbe or Odra phytoplankton blooms can occur comparable to highly eutrophic lakes (Mischke and Behrendt 2006).

24.5 Phytoplankton Assessment System

24.5.1 Lake Types - Present State in Germany and Europe

In Germany the trophic status of lakes is assessed by a seven level classification that was developed in 1999 by German Working Group of the Federal States on water issues (LAWA). It takes chlorophyll *a* concentrations, as well as Secchi depth and two parameters of total phosphorus into account (Nixdorf et al. 2006).

However, there was an additional necessity to develop national ecological status assessment system for lakes, fully compliant with WFD requirements that would include composition, abundance and biomass of phytoplankton. Introduction of new parameters increased the sensitivity of evaluation. This work was completed in 2006 (Nixdorf et al. 2006) and is very lake type specific (Mischke et al. 2002).

Table 24.2. Lake types in Germany and in European intercalibration lakes in the Alpine and Central lake GIG

Characters for typing	Mixing type	VQ	Lake size	Residence time	Mean depth
Relevant in German type	yes	yes	yes	Yes in part	no
Relevant in IC lake type	Not in Central GIG	no	no	differs	yes

Mixing type: polymictic or stratified; VQ = area of catchment to lake volume = VQ large = >1.5 or VQ small <1.5; lake size (km^2) = lake surface area = in DE always >0.5; residence time (yr) = calculated from mean precipitation, catchment area and lake volume without exchange with groundwater; in DE only relevant for riverine lakes (type 12); mean depth (m) = mean depth of the lake).

The German lake types are generally compliant with lake types used in the inter-calibration exercise. Since the parameters defined for typing are different (see Table 24.2) the types are not completely compatible. The new assessment system was developed for 9 of the 14 lake types existing within Germany. Especially lakes and reservoirs of the eco-region Central Uplands cannot be assessed by now (see Table 24.3).

Table 24.3. Comparison of German (DE type) and European intercalibration lake types (IC type) and thresholds for lake types in Germany (Mathes et al. 2002) and in Al pine (AL) and Central (LCB) lake GIG. Legends see Table 24.2

DE - type number	DE - Type name	Residence time [yr]	Mean depth [m]	IC type name	Additional IC type character
1	Calcareous, polymictic pre-alpine lake				
2	Calcareous, stratified pre-alpine lake with large catchment area		IC = 3 - 15	AL4	Altitude 50 – 800m a.s.l.; Stratified
3	Calcareous, stratified pre-alpine lake with small catchment area		IC = 3 - 15	AL4	Altitude 50 – 800m a.s.l.; stratified
4	Calcareous, stratified alpine lake		IC > 15	AL3	Altitude 200–800m a.s.l. Alpine catchment area
5 - 9	Calcareous and siliceous reservoirs and few lakes in the altitude range 200 – 800m				
11.2*	Calcerous, polymictic very shallow lowland lake with large catchment area	>0.084 IC = 0.1 – 1	< and = 3m IC = < 3m	LCB2	Calcareous Alk. >1 meq/l Altitude < 200m a.s.l.
11.1	Calcareous, polymictic lowland lake with large catchment area	> 0.084	>3m	-	Most lakes excluded from LCB1 by residence time < 1

Table 24.3. (cont.)

DE - type number	DE - Type name	Resi- dence time [yr]	Mean depth [m]	IC type name	Additional IC type character
14	Calcareous, polymictic lowland lake with small catchment area			-	Excluded from LCB2 by residence time > 1
12	River lakes - Calcare- ous, polymictic low- land lake with large catchment area and low retention time	0.008 – 0.084		-	Excluded from LCB1 and LCB2 by short residence time
10	Calcareous, stratified lowland lakes with large catchment area	IC = 1 – 10	IC = 3 - 15	LCB1	Calcareous Alk. >1 meq/l Altitude < 200m a.s.l.
13	Calcareous, stratified lowland lakes with small catchment area	IC = 1 – 10	IC = 3 - 15	LCB1	Calcareous Alk. >1 meq/l Altitude < 200m a.s.l.
	No siliceous lakes in German with surface area >0.5km^2		IC = 3 - 15	LCB3	0.2-1 meq/l Siliceous moderate alk. Alti- tude < 200m a.s.l.

* Very shallow lakes with a mean depth smaller than 3m are a sub-type of DE type 11 for phytoplankton, only.

To carry out assessment, three to four metrics have to be used: total phytoplankton biovolume, algae class and phytoplankton taxa in a special Lake Index (PTSI). In specific cases the forth value is required – evaluation of composition of yearly self sedimented planktonic diatoms. The data obtained, using lake-specific weighting constants are recalculated to obtain a common Phytoplankton Index in Lakes (PSI), reflecting the quality status in the water body. Assessment requires type specific reference conditions which were developed also by paleolimnological in- vestigations (Nixdorf et al. 2006).

24.5.2 Class Boundaries for Common Phytoplankton Metrics

Currently the German assessment method both in Central/Baltic (CB 1-3) and Al- pine (AL3; AL4) GIGs is one of the most complete as most of the countries are still developing their metrics (EC, Joint Research Centre 2006 a, b). Further analy- sis are actually carried out to compare the results of national evaluation on a common European data set and to develop common metrics which are applicable in the whole eco-region. Preliminary results for thresholds of phytoplankton met- rics *Chl a- concentration* and *biovolume* are given in Table 24.5 for Central Baltic lake types and 24.6 for IC lake types in the alpine eco-region (AL 3 and AL 4).

Table 24.4. Overview of the German assessment method by means of phytoplankton

Biological indicator	Instructions for method
All	The sampling frequency is at least 6 sampling per year with 4 sampling within the period between May to September. Untreated, integrated samples are usually taken from the epilimnion, in clear lakes from the euphotic zone and preserved with Lugol solution. Details are summarized in the German instruction for sampling.
	The basic unit of all parameters is the biovolume (mm^3/l). It must be measured microscopically following the new CEN-norm for enumeration of phytoplankton by Utermöhl technique (EN 15204: 2006-12) in combination with the new national instructions, as to counting strategy, biovolume measuring and an operational German taxa list for phytoplankton.
	All biological parameters have a different indicative value in each lake type. Thus, the lake type of the investigated water body must be clear to select the type specific relevant parameters, thresholds and periods from the assessment tables, can not be shown here in detail.
Total bio-volume assess-ment	Total biovolume is the sum of all phototrophic taxa enumerated including benthic once found in the plankton.
	Value of this parameter is the average of all sampling sites of one water body and of the whole vegetation period following a given instruction how to make averaging.
	Assessment is made by comparing the resulting value with lake type specific thresholds defined for all five ecological status classes.
	Actually a new national biomass metric is under development including also chlorophyll a mean and maxima values to fulfil ECOSTAT results.
Algal class as-sessment	For each lake type and each algal class it is defined weather the specific biovolume of an algal class or its proportion on total biovolume is to use.
	To optimize the indicative value, the method focuses on special periods of the yearly natural succession. Sample values are averaged for specific periods, e.g. proportion of dinophytes in summer (June-July) in case of the lake type 4 (alpine lakes). Defined for each lake type up to four different algal classes are indicators. Assessment is made by comparing values with thresholds defined for all or some of the five ecological status classes. Results of all algal class indices are averaged to one value to algal class metric.
Indicator species assess-ment by PTSI	The level of required taxonomical detection is listed in the operational German taxa list. Each species biovolume (mm3/l) must be measured separately. The resulting species biovolume is attached to one of the 7 classes of abundance category defined for this method. Trophic values and weighting factors ("Stenökiefaktor") of all indicator species are given in three different lists for alpine & pre-alpine lakes, for stratified lowland lakes and for polymictic lowland lakes. Both values are multiplied with the measured abundance category for each indicator species to calculate the PTSI index by an integral calculus.
Phyto-plankton index as-sessment	The indicative value of the indices "total biovolume", "algal class metric" and "PTSI" is different in the lakes types, which is taken into account by incorporate weighting factors to calculate the final "ecological status".

Table 24.5. Reference values and H/G boundaries of chlorophyll-a concentration (μgL^{-1}) in Central Baltic lakes types

	L-CB1	min	max	L-CB2	min	max	L-CB3	min	max
Reference (median)*	3.2	2.6	3.8	6.8	6.7	7.4	3.1	2.5	3.7
H/G (75Th percentile)*	5.8	4.6	7.0	11	11	12	5.4	4.3	6.5
EQR H/G	0.55			0.63			0.57		

The EQR is calculated as reference value divided by the boundary value
*only for mid value.

Table 24.6. Class boundaries for the common metrics "total biovolume" (BV) and "chlorophyll-a concentration" Chl-a) for the IC lake types L-AL3 and L-AL4

	L-AL3				L-AL4			
	BV [mm^3 L^{-1}]		Chl-a [μg L^{-1}]		BV [mm^3 L^{-1}]		Chl-a [μg L^{-1}]	
	GIG	MS	GIG	MS	GIG	MS	GIG	MS
H/G (EQR = 0.8)	0.5	0.5	2.7	2.5	1.1	1.1	4.4	4.7
G/M (EQR = 0.6)	1.2	1.25	4.7	4.7	2.7	2.25	8.0	8.6
M/P (EQR = 0.4)	3.1	3.0	8.7	8.6	6.9	4.6	14.6	15.7
P/B (EQR = 0.2)	7.8	7.4	15.8	15.7	17.4	9.5	26.7	28.7
class width (In-scale)	0.9	0.9	0.6	0.6	0.9	0.7	0.6	0.6

GIG = boundaries using the common GIG dataset and the BSP described under section B, C and Annex C, MS = Boundaries using a MS dataset in the German classification method.

24.6 European Standards (CEN) for Alpine and Lowland Regions for Lake Assessment and Sampling Procedure

In order to ensure that in the process of assessment of ecological status sampling and analysis of parameters usually should be done according to international standards, accordingly verified and calibrated. The WFD (Annex V 1.3.6) requires for monitoring of quality elements standard methods to ensure scientific quality and comparability of these data throughout the EU (European Parliament 2000).

Since the WFD was adopted, CEN (European Committee for Standardization - CEN/TC (Technical Committee) 230/WG (Working Group) 2 is dealing with development of biological methods of water analysis has developed new standards, relevant for the WFD, which potentially can be considered for inclusion in the WFD (EC, Joint Research centre 2006).

A Harmonisation Activity of the WFD has been assigned the role of compiling information on national biological monitoring methods, evaluating their comparability, identifying standards to be included in the WFD and well as fostering the link between CEN, European Commission and implementation working groups. In 2005 a document was drafted by DG Environment of the European Commission, to establish and agree the process of developing standard methods for the assessment of water status (EC, Joint Research Centre 2006).

Currently there are 20 CEN Standards, which are suggested for inclusion in the WFD (EC, Joint Research Centre 2006), most of which are still under development or awaiting official approval, but it can be expected that most of them later will be included in Annex V of the WFD to supplement the existing information. These standards are mostly guidance standards, providing advice how to carry out certain activities, as it is not always possible to develop standards which would cover the wide variety of different climatic, geographic and other conditions in a large number of countries. To fully develop CEN standards there is a certain and comparatively lengthy procedure to be followed, to ensure that standards developed are fully functional.

The most recent standards are those on analysing phytoplankton, since up to now in Europe there are very different techniques and counting strategies in use. Thus, it is great step forward, that the enumeration standard, titled: "Water quality - Guidance standard on the enumeration of phytoplankton using inverted microscopy (Utermöhl technique)" (Draft CEN standard prEN15204), is already in the formal vote procedure of CEN Still this new standard lacks the instruction, how to evaluate the parameter "biovolume", which actually is suggested in a draft proposal by Germany for an additional guidance standard ""Phytoplankton biovolume determination using inverted microscopy (Utermöhl technique)" (German Draft Proposal for CEN guidance). Evaluation of biomass of phytoplankton instead of abundance allows more precise assessment and it can be considered more appropriate for food chain modelling than abundance. To calculate the biovolume of certain cells, for each phytoplankton taxa a fitting geometric shape is assigned and after measurements of cell dimensions have been carried out, the average cell volume is multiplied by the number of individuals (Draft CEN standard prEN15204).

24.7 Conclusions and Recommendations

It can be expected that once the standardisation of methods is complete, the wide variety of standards will ensure that ecological status assessments produced by Member States are of the best quality and easily and reliably comparable in within European Union ever been before.

On the other side, the ongoing process of European intercalibration within the GIG´s concerning defining of biological parameters and of common boundaries is not without risks: already finished national assessment methods come in conflict with common European boundaries, which could significantly differ from them. As it is stated by the ECOSTAT meeting in September 2006, the member states have to defend their national method, when deviations occur from a common GIG boundary.

One reason for deviation is defining the intercalibration water body types much more roughly than it had happened on national member state level. This fact leads to comparable narrow thresholds for the bio-components of the water bodies, which could differ widely in respect to triggering habitat factors such as residual time of water or geogenic background. So, the evaluation by common metrics could mislead in the special case. In the GIG's extreme strong criteria for selecting reference sites (EC REFCOND 2003) effect further deviations from the national methods. The latter consider regional aspects and included also water bodies, which were only stated to be "good" and not consequentially checked up for anthropogenic influences.

Thus, the biological standards recently developed still have to be improved and must be harmonized. Furthermore they are under significant pressure of political interests.

References

CEN – European Committee for Standardisation: standard EN 15204 "Water quality – Guidance standard on the enumeration of phytoplankton using inverted microscopy (Utermöhl technique)"

ECOSTAT – Working Group 2.A – Ecological Status (2006) Milestone 6 reports, WFD CIRCA, Internet: http://forum.europa.eu.int/Public/irc/jrc/jrc_eewai/home

EC – European Commission (1991) Council Directive of 21 May 1991 concerning urban waste treatment 91/271/EEC, OJ L135, 30.5.1991

EC (1994) Proposal for a Council Directive on the ecological quality of water - OJ C 222, 10.8.1994; COM(93) 680; Bull. 6-1994, point 1.2.179; Bull. 12-1994, point 1.2.206; http://europa.eu/bulletin/en/9601/p103149.htm

EC (2001) Bulletin EU 11-2001, 1.4.42. Decision No 2455/2001/EC of the European Parliament and of the Council establishing the list of priority substances in the field of water policy and amending Directive 2000/60/EC

EC (2004) Bulletin EU 1/2-2004, 1.4.56. Proposal for a European Parliament and Council directive on the protection of groundwater against pollution

EC (2004) Common implementation strategy for the Water Framework Directive (2000/60/EC). Moving to next stage in the Common Implementation Strategy for the Water Framework Directive, Progress and work programme for 2005 and 2006

EC (2005a) Commission decision of 17 August 2005 on the establishment of a register of sites to form the intercalibration network in accordance with Directive 2000/60/EC of the European Parliament and of the Council 2005/646/EC, Official Journal of the European Union 19.9.2005, L243/1

EC (2005b) Towards a guidance document on eutrophication assessment in the context of European water policies

EC (2006) Bulletin EU 1/2-2006, 1.20.19. Proposal for a directive of the European Parliament and of the Council on the assessment and management of floods

EC, Joint Research Centre (2005) Report on harmonisation of freshwater biology methods. Internet: http://ies.jrc.cec.eu.int/fileadmin/Documentation/Reports/Inland_and_Marine _Waters/EUR_2005/Harmonisation__EUR_21769_EN.pdf

EC, Joint Research Centre (2006a) Intercalibration Lakes GIG report Central/Baltic, Internet: http://forum.europa.eu.int/Public/irc/jrc/jrc_eewai/home

EC, Joint Research Centre (2006b) Intercalibration Lakes GIG report Alpine, Internet: http://forum.europa.eu.int/Public/irc/jrc/jrc_eewai/home

EC, Joint Research Centre (2006c) Proposal for revision of the CEN methods in Annex V of the WFD and identification of future standardisation items

EC, REFCOND (2003) Common implementation strategy for the water framework directive (2000/60/EC) - Guidance Document No 10- Rivers and Lakes – Typology, Reference Conditions and Classification Systems - Produced by Working Group 2.3 – REFCOND. Internet, Europa server: http://europa.eu.int

European Committee for Standardization, CEN TC/230 structure, Internet on: http://www.cenorm.be/CENORM/BusinessDomains/TechnicalCommitteesWorkshops/CENTechnicalCommittees/TCStruc.asp?param=6211&title=CEN%2FTC+230, last accessed on 24.03.2007

European Parliament (1972) Resolution of the European Parliament on the Results The First Summit Conference of the Enlarged Community, Paris 19-21 October, Bulletin of the European Communities, No 11, 1972

European Parliament (2000) Directive 2000/60/EC of the European Parliament and of the Council of establishing a framework for Community action in the field of water policy, Official Journal of the European Union (OJ L 327)

German Draft proposal for CEN guidance standard "Phytoplankton biovolume determination using inverted microscopy (Utermöhl technique)"

Mischke U, Behrendt H (2006) Handbook to describe the assessment method to evaluate rivers by phytoplankton [in German]. In press. In: Bundesweiter Praxistest eines Bewertungsverfahren für Phytoplankton in Fließgewässern Deutschlands zur Umsetzung der EU-Wasserrahmenrichtlinie. Report IGB, Berlin, p. 68

Mischke U, Nixdorf B, Behrendt H (2002) On typology and reference conditions for phytoplankton in rivers and lakes in Germany. In: Ruoppa M, Heinonen P, Pilke A, Rekolainen S, Toivonen H, Vuoristo H (eds) Nordic Council of Ministers, Copenhagen. TemaNord 566:44-49. Internet: http://www.norden.org/pub/webordering/sk/, last accessed on 24.03.2007

Nixdorf B, Mischke U, Hoehn E, Riedmüller U (2006) Leitbildorientierte Bewertung von Seen anhand der Teilkomponente Phytoplankton im Rahmen der Umsetzung der EU-Wasserrahmenrichtlinie, Bad Saarow, Internet: http://www.tu-cottbus.de/BTU/Fak4/Gewschu/, last accessed on 24.03.2007

REBECCA (2005) Relationships between pressures, chemical status, and biological quality elements. Analysis of the current knowledge gaps for the implementation of the Water Framework Directive

Response to the proposal of Dr. Georg Wolfram 13.01.2006 for the Alpine GIG-boundary setting draft for phytoplankton in lakes in the Alpine region

Water Framework Directive, Directive 2000/60/EC of the European Parliament and of the Council of 23 October 2000 establishing a framework for Community action in the field of water policy, OJ L 327 of 22.12.2000, p. 1, corr. OJ L 017 of 19.01.2001, p. 39

Wetzel RG (2001) Limnology. Elsevier, Amsterdam, p. 390

25 Landscape and Protected Areas –
Polish Experiences

Tadeusz J. Chmielewski

Department of Landscape Ecology and Nature Conservation, Lublin University of
Agriculture, Lublin, Poland

25.1 Introduction

The objective of this chapter is to discuss the problems related to methods of landscape ecological systems impact assessment, with special emphasis on the issue of
protected areas. The problems in question will be presented for the example of Poland, where the system of protected areas is particularly well developed and covers 32.5 % of the total territory of the country (Grzesiak and Domanska 2005).

Section 25.2 presents the issue of landscape quality objectives (LQO) which
should be developed and implemented in all member countries of the EU, in accordance with the European Landscape Convention (ETS No 176). It also indicates those EU countries in which work in this field is the most advanced. The
first part of Section 25.3 provides a general characterisation of the landscapes of
Poland and a description of the system of protected areas created in the country. It
also provides a set of the fundamental legal thresholds for land-use and management, pertaining to the primary elements of the system. The second part of Section
25.3 presents a discussion of the main problems related to the observance of prohibitions and restrictions imposed on the inhabitants and users of protected areas
in Poland. The third part of the section presents the results of the first Polish project devoted to the identification of LQO for a region of particular natural and cultural values. Section 25.4 presents standards for environment impact assessment
(EIA) reports that should be developed for investment projects to be realized
within protected areas. The section also presents a discussion of problems related
to the merit aspect of such reports in Poland and indicates the difficulties involved
in the practical enforcement of the observance of report recommendations in the
process of investment implementation. Section 25.5 contains conclusions concerning the directions of subsequent studies on landscape quality objectives for protected areas, as well as recommendations for the improvement of standards and
thresholds for protected areas management.

Standards and Thresholds for Impact Assessment. Edited by Michael Schmidt, John Glasson, Lars Emmelin and Hendrike Helbron. © 2008 Springer-Verlag

25.2 Landscape Quality Objectives: Problems of Identification and Conservation in Europe

The second half of the 20th century and the beginning of the 21st have been a period of notable intensity in the transformation of the landscapes of Europe, resulting in a decrease in biological diversity, weakening of cultural identity and unification of the physiognomy of regions (Zonneveld and Forman 1990; Paneuropean Strategy 1995; Walker at al. 1999). At the same time, landscape– due to its immense complexity – is an area of interaction of numerous, frequently mutually conflicting interests and processes. These features cause that landscape is a system that tends to evade precision control. Lasting conservation of characteristic quality and quantity features of the landscape of individual regions of Europe has become one of the greatest contemporary challenges for environmental scientists, nature conservation staff and spatial management planners.

The need for setting standards of landscape quality (landscape quality objective) was formulated in the "European Landscape Convention", adopted by the member states of the Council of Europe on 20th October, 2000 (ETS No. 176). In Poland the document was not published until the year 2006 (Europejska Konwencja 2006). Art. 1 letter b of the Convention states that "Landscape quality objective" means, for a specific landscape, the formulation by the competent public authorities of the aspirations of the public with regard to the landscape features of their surroundings" (ETS No. 176 p. 3).

Research concerned with standards of landscape quality has been conducted in Europe for only several years and is still at the stage of preliminary studies. The highest activity in this field can be observed at research centres of Spain, Portugal, Holand, Scandinavian countries, and Italy, and in Central-Eastern Europe – in Slovakia (Kuiper 2000; Yusim et al. 2002; Bulcão et al. 2004; Soren et al. 2004; Ribeiro and Barão 2006).

It should be emphasized that development and observance of such standards is currently of particular importance in areas with the highest natural, cultural and aesthetic values.

In Poland the first works aimed at the determination of landscape quality standards were performed for the proposed „Roztocze–Solska Forest" Biosphere Reserve (Chmielewski and Sowinska 2006), and in a somewhat different approach (as so-called „landscape attributes") – for the Wigry National Park (Solon 2006).

25.3 Natural Landscapes of Poland and System of their Protection

25.3.1 Zonal System of Natural Landscapes and Network System of their Protection

Poland is located in Central Europe, within the basin of the Baltic Sea. In the eastern part of the country it is traversed – north to south – by a physiographic border between Eastern and Western Europe. It is also a zone that the Atlantic, the boreal, and the eastern-continental climates influence. To the north-east of the country, there is a biome of boreal coniferous forest, to the west – a biome of deciduous forest of the moderate zone, and to the south-east – a biome of steppe and forest-steppe (Starkel 1991; Weiner 1999).

Characteristic of the Polish landscape is the strip-like, latitude-like system of the main types of natural landscape. In the north area there is the lowland landscape (late-glacial and early-glacial), in the central regions – the highland landscape (loess, aluminosilicate and carbonate-lime), and to the south – the medium and high mountain landscape (Richling and Solon 1996).

That parallel layout of landscape zones is criss-crossed by a branched system of river valleys, with predominantly meridian orientation. Some of the river valleys have still retained semi-natural character, with numerous meanders, flood plain lakes and wetlands.

In the land-use structure, the dominant form is agrocenosis – 59.3 %. Forests constitute 28.7 %; water 2.7 %; peat, bog, steppe, and sand dune areas – 1.5 %; and urban areas – 7.8 % (Grzesiak and Domanska 2005).

The average population concentration is 120.8 inhabitants per sqkm. Almost 40 % of the population are farmers, which causes the spatial pattern of arable fields and meadows to be fine-mosaic and differentiated.

All the above features make Poland a country of great biological and landscape diversity. In many regions extreme ecological values have been preserved right up until the beginning of the 21 century.

Table 25.1. Main elements of the ecological system of protected areas in Poland (status of 31st December, 2004)

Category of protected area	Number of objects	Percentage of country area (%)[*]
National park	23	1.0
Nature reserve	1385	0.5
Natura 2000 habitat sites	184	3.8
Natura 2000 bird sites	72	9.9
Landscape park	120	8.1
Landscape protecting areas	445	22.5

[*]Sum of all values given in the column exceeds 32.5 %, since certain types of protected areas overlap fully or partially (e.g. national parks, Natura 2000 habitat sites and Natura 2000 bird sites).

Poland has high achievements in nature conservation network design. In 1976 the first draft of the project of 'Ecological System of Protected Areas' [Polish abbreviation: ESOCh] was prepared (Gacka-Grzesikiewicz 1976). It was a prototype of ECONET network, introduced in European Union approximately 25 years later (Bennet 1991; Liro 1995). The ESOCh concept is still consistently developed in Poland. At the end of 2005 all the elements of the system covered a total of 32.5 % of the area of the country (Table 25.1).

The structure and functionality of this system were completed by small scale forms of nature protection known as: ecological grounds, natural landscape complexes, documentary plots and nature monuments (Nature Conservation Act of Poland 2004). They covered in total c.a. 0.4 % of the country area, but most of them were situated inside the territory of other above mentioned larger forms. Different types of protected areas are cross linked, forming a spatially comprehensive network (Fig. 25.1).

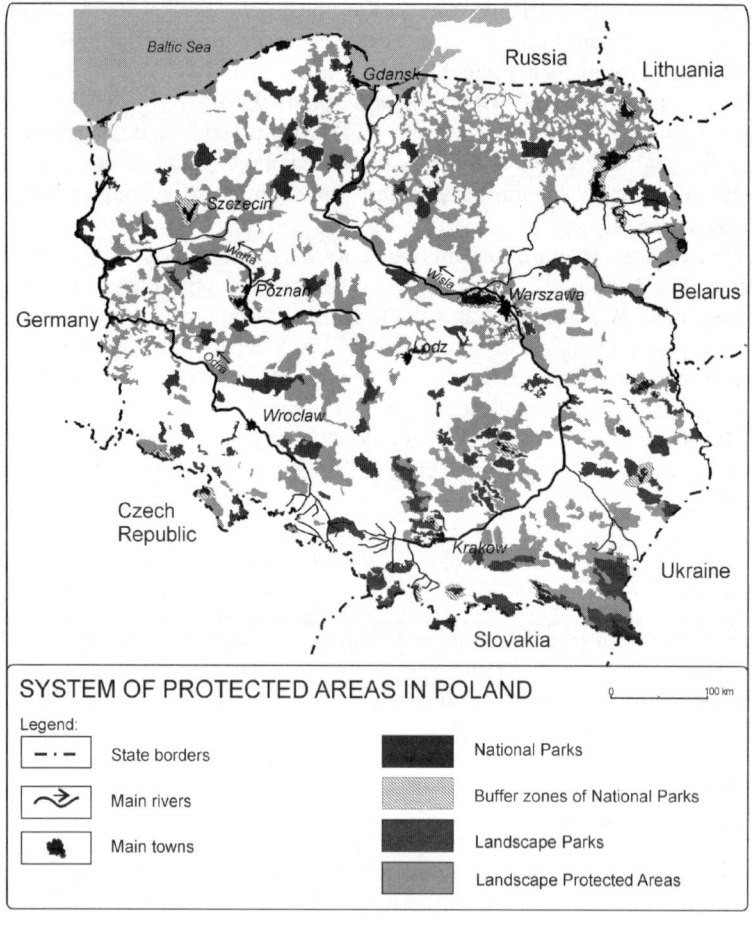

Fig. 25.1. System of protected areas in Poland (changed from Wojcik et al. 2001)

Each of these legal nature conservation forms fulfils different tasks and shows different limits in land use and different purposes. All of these forms together create a very wide group of steps to realize nature conservation aims in practice.

25.3.2 Legal Thresholds for the Land-Use and Management of Protected Areas in Poland

The creation of particular types of protected areas in Poland involves the introduction, within their territories, of a specific set of prohibitions and restrictions concerning the possibility of implementation of investment programs and land-use. A catalogue of various restrictions is contained in the law on nature conservation (Nature Conservation Act 2004). For example, the list of restrictions that may be introduced in the area of national parks and nature reserves comprises 27 items that include, among other things, prohibition of:

- construction or extension of buildings and technical installations that are not designed to serve nature conservation,
- fishing,
- game hunting,
- acquisition, destruction or damaging plants and fungi,
- regulation of rivers and changing hydrological conditions, except for measures aimed at nature conservation,
- acquisition of stone, peat, and fossils,
- application of chemical and biological herbicides and pesticides,
- introduction of genetically modified organisms,
- disturbing natural silence, etc. (Nature Conservation Act 2004, Art. 15 para 1).

The Act, however, provides for a number of situations when the restrictions do not apply, or may be waived. For example, securing a permit from the Minister for the Environment, performance of a rescue operation, performance of justified farming or husbandry measures on private grounds, etc. (Nature Conservation Act 2004, Art. 15 para 2).

The list of prohibitions that may be introduced in the area of landscape parks comprises 14 items that include, among other things:

- localisation of investment projects with high environmental impact,
- killing wild animals and destroying their breeding places,
- elimination and destruction of river bank, midfield and roadside trees (except for situations when they create a hazard to the safety of people and property),
- exploitation of mineral raw materials, peat and fossils (including amber),
- deformation of terrain surface relief (except for situations of flood or landslide hazard),
- introducing changes to the water relations (though this prohibition does not apply to works related to agriculture, fishery and forestry),

- elimination and transformation of water reservoirs, flood plain lakes and wet-lands,
- localisation of buildings within a distance of 100 m from river and lake banks (except for structures serving the needs of agriculture, fishery or forestry) etc. (Nature Conservation Act 2004, Art.17 para 1).

However, the Act provides for a number of situations when the prohibitions do not apply or may be waived. For example, an investment project with high environmental impact may be realized in a landscape park if the relevant environmental impact assessment report shows that its effects in a specific case will not be harmful. The list of the above prohibitions does not apply also in a situation when a particular investment program realizes important public functions or purposes (Nature Conservation Act 2004, Art. 17 para 2).

Moreover, the regulations of the Act do not provide for any restrictions or sanctions related to the degradation of the aesthetic values of the landscape, not only in landscape parks but even in national parks and nature reserves. Only Art. 16 para 1 contains a statement that a landscape park should possess high physiognomic values. However, no regulation is provided on how these should be protected.

Such formulations cause that in practice it is very difficult to protect natural and semi-natural areas against anthropogenic transformations of their environment and against degradation of their aesthetic value. Studies show that many areas in Poland, even though under legal protection, irreversibly lose their natural and aesthetic value (Chmielewski 2004a, 2004b). This means low effectiveness of top-down nature conservation management and induces more and more numerous social groups to undertake local bottom-up nature conservation initiatives.

A good example of such forms of activity is provided by certain Polish biosphere reserves, i.e. the West Polesie Biosphere Reserve (Chmielewski 2005), or the Roztocze-Solska Forest Biosphere Reserve, currently at project preparation stage.

25.3.3 Effects of the First Polish Project Involving the Identification of Landscape Quality Objectives

The Roztocze-Solska Forest Biosphere Reserve was designed in central-eastern Poland. It covers a range of gentle lime-stone hills covered with loess and sandy soils, with exceptionally attractive visually mosaic of fields and forests (Fig. 25.1). At the foot of those hills lies an expansive valley with pine forests, with groups of sand dunes interspersed with peatlands. The total area of the two natural complexes is 263 363 ha. In order to create conditions for effective protection of the unique values of the area, an attempt was undertaken in 2006, to identify landscape quality objectives in accordance with the requirements of the European Landscape Convention (ETS No. 176). Expectations of the society as to the features that the landscape of the Roztocze–Solska Forest Biosphere Reserve should retain or acquire were tested by the public survey method. The survey comprised seven social-professional groups that play a key role in the protection and shaping

Fig. 25.2. Characteristic landscape of Roztocze Region (photo of W Lipiec)

of the landscape of the region and in its management The respondents of the survey were representatives of the following social-professional groups:

- scientists
- nature conservation staff and Public Forests staff in this region
- local governments
- tourists who took their rest in the Roztocze and the Solska Forest
- teachers
- students
- members of NGO`s and members of photographic societies.

The survey was made for 30 respondents from each of the groups specified above, i.e. for a total of 210 people. The participants of the survey identified those features of landscape that, in their opinion, were the most important for preserving high natural, cultural and recreational values of the region, and graded those features in a system of points. Similarities and differences in opinions of various social groups were analysed. The greatest differences of opinion occurred between scientists and tourists, and between students and representatives of local governments. The maximum difference in the valuation of a single landscape feature by various social-professional groups reached the level of 29.9 %, so the opinions were not extremely divergent. Summing up the points given by the various groups to particular landscape features the resultant ranking list was formulated for parameters that, in the opinions of the local communities and experts, should be preserved as the regional landscape quality objectives. The same method was applied to determine the ranking list of hazards to the environment and the aesthetic values of the regional landscape. The results of the survey showed that the landscape protection policy followed so far, based primarily on a top-down system of regulations and restrictions, proved to be ineffective, and the scale of hazards to the values of the region – greater and greater. Therefore, making use of the newly identified landscape quality objectives, a set of local rules of management were proposed, the observance of which is necessary for effective protection of the unique qualities of the landscape of the region. The essence of the proposed approach was joint development of a bottom-up landscape conservation policy (Chmielewski and Sowinska 2006).

25.4 Polish Standards for Environment Impact Assessment in Protected Areas

Requirements concerning the application of the Environment Impact Assessment (EIA) procedure and rules for the preparation of EIA reports in Poland are laid down in the act on "Environmental Law" (Environmental Law 2001, Part VI). Art. 40 para. 1 of this document provides that:

The Environment Impact Assessment procedure is required for the following:

1.) draft concept of national development plan, draft regional development plans and regional development strategies;
2.) drafts of policies, strategies, plans and/or programs pertaining to industry, electric power systems, transport, telecommunications, water management and conservation, wastes management, forestry, agriculture, fisheries, tourism, and land use (...), within the scope of their impact on the Natura 2000 sites;
3.) drafts of policies, strategies, plans and/or programs other than those enumerated in item No. 2, that are not related directly with the conservation of Natura 2000 sites (...), if the realization of the stipulations of such policies, strategies, plans or programs may have a significant impact on the sites.

No other categories of protected areas – apart from Natura 2000 sites – have been specified in the regulations of Art. 40 para. 1 No. 2 of the Act.

The EIA procedure identifies, analyses, and evaluates the following:

1.) direct and indirect impact of a given undertaking on:
    a) the environment and human health and living conditions,
    b) material property,
    c) historical monuments,
    d) mutual interactions between the factors specified under letters a)-c),
    e) accessibility to mineral deposits;
2.) possibility of negative impact on the environment and methods of its prevention and limitation;
3.) required scope of monitoring (Art. 47).

The regulations do not mention the evaluation of the impact of investment programs on the spatial arrangement, harmony, or the aesthetic values of landscape, even in respect to areas of national parks, landscape parks, or Natura 2000 sites.

With respect to investment projects that may have a significant impact on the environment, preparation of "EIA Reports" is required (Art. 51 para. 1).

An EIA Report should include, among other things:

1.) description of the planned undertaking, and especially:
   a) characterization of the whole project and conditions of land use at the stage of implementation and subsequent operation,
   b) main characteristic features of production processes involved,
   c) projected levels of the emission of pollutants, resulting from the operation of the planned undertaking,
2.) description of natural components of the environment affected by the scope of impact of the planned undertaking (...),
3.) description of analysed variants, including the following:
   a) variant consisting in resignation from implementation of the undertaking,
   b) variant that is the most favourable for the environment, giving the justification for the choice made,
4.) definition of the projected impact of analysed variants on the environment, including the possibility of occurrence of a major industrial disaster and the possibility of trans-frontier impact on the environment,
5.) justification of the choice of variant made by the applicant, indicating its impact on the environment, and in particular on the following:
   a) human population, animals, vegetation, water and air,
   b) earth surface, with the inclusion of mass movements of earth, the climate and landscape,
   c) material property,
   d) historical monuments and cultural landscape, covered by existing documentation, and in particular listed in the register of historical monuments,
   e) mutual interactions among the elements listed under letters a)-d), (...)
7.) description of projected actions aimed at the prevention, limitation, or compensation of negative impact on the environment, (...)
11.) analysis of potential social conflicts related to the planned undertaking,
12.) presentation of proposals for monitoring the impact of the planned undertaking at the stage of its implementation and operation, (...)
16.) sources of information constituting the basis for the preparation of the report (Art. 52 para 1).

The word "landscape" appears in that list only twice – in No. 5 b), d).

Investment programs that may have a significant impact on the environment of Natura 2000 sites can take place only on the basis of a special administrative document known as the "Decision on environmental conditions for the realization of an undertaking" (Art. 46 para 1). Issuing of such a decision should be based on careful consideration of the results of the EIA Report. In the case of investment programs realized in other protected areas such a decision is not required.

At the same time, Polish regulations do not introduce more restrictive standards of environment quality for the protected areas. Pollution standards for the air, waters and soils, as well as regulations pertaining to wastes management, are uniform for the whole area of the country (Environmental Law 2001). Only the standards for permissible levels of noise are more stringent for health-resort and recreation areas (Decree of the Minister for the Environment on permissible noise levels in the environment 2004).

25.5 Conclusions and Recommendations

Poland has a varied and very well developed system of protected areas. The Polish ecological network is made up of ten types of protected areas and objects that cover a total of 32.5 % of the country area. However, ensuring high standard of environment quality in those areas is not easy. Regulations concerning the rules of economic activity and management in the protected areas are based mainly on a long list of prohibitions and restrictions. At the same time, however, there numerous methods of obtaining waivers or limitations to the prohibitions. The best legal protection is granted to the Natura 2000 sites and to the national parks. But the total area of those categories of nature conservation in Poland is relatively small. Nature resources management is the least effective with respect to the landscape protecting areas, and that is the form of conservation that has the dominant share in the total area of the ecological system in the country.

There is no detailed Polish legislation pertaining to the standards of landscape quality. The protection of spatial arrangement order and visual values of landscape are given marginal treatment, like in the case of the regulations concerning the EIA.

Local communities are searching for a way out of this situation, undertaking more and more numerous bottom-up nature conservation initiatives. In Poland such initiatives have the best chances of success within the areas of the UNESCO Biosphere Reserves.

In order to improve nature resources management in protected areas for human health and the environment it is necessary to:

- reduce the scope of waivers (exemptions) from laws and regulations valid in the protected areas,
- successively move away form the top-down to the bottom-up systems of nature conservation policy,
- develop research and methodological studies concerned with the identification and permanent implementation of landscape quality objectives,
- devote greater attention to the problems of estimation of the impact of investment programs on the landscape ecological systems function and landscape visual values,
- promote local initiatives and achievements in the range of ecological space conservation and landscape design.

References

Bennet G (ed) (1991) Towards a European Ecological Network. Institute of Environmental Policy. Arnhem, The Netherlands:1–80

Bulcão L, Ribeiro L, Arsénio P, Abreu M M (2004) The protection of landscape as a resource: Case study – Monte da Guia protected area (Faial-Azores), Agriculture, Ecosystems and Environment Vol. 77, Issue 1-2:143–156

Chmielewski TJ (2004a) Processes taking place in the ecological structure and spatial development of areas with natural value in the central and central-eastern Poland. In: Markowski T (ed) Space in regional and local development management. Bulletin of the Committee for Spatial Development of Poland, Polish Academy of Sciences, vol. 211. Warsaw: 267-292 (in Polish, English summary)

Chmielewski TJ (ed) (2004b) New quality of landscape: ecology – culture – technology. Bulletin of the Committee „Man and Environment", Polish Academy of Sciences, vol. 36. Warsaw – Lublin: 1-158 (in Polish, English summary)

Chmielewski TJ (ed) (2005) The Roztocze – Solska Forest Biosphere Reserve. MaB-UNESCO Nominating Form. Lublin Province. Lublin:1–155

Chmielewski T J, Sowinska B (2006) Landscape quality standards of the Roztocze–Puszcza Solska Biosphere Reserve: estimation and protection problems. In: Wojciechowski K H ed. Cultural landscape: features, values, protection. Polish Association for Landscape Ecology. Kazimierz Dolny (in Polish, English summary) [in press]

Gacka-Grzesikiewicz E (1976) Ecological problems of the creation of new types of protected areas as a form of environment conservation. Wiadomości Ekologiczne. Warszawa 1:3–25

Grzesiak M, Domanska W (eds) (2005) Environment Protection 2005. Information and statistical analyses. Główny Urząd Statystyczny. Warszawa: 1-539 (in Polish, English foreword)

Kuiper J (2000) A checklist approach to evaluate the contribution of organic farms to landscape quality, Management of Environmental Quality: An International Journal Vol. 15, Issue 1:48–54

Liro A (ed) (1995) Concept of the national ecological network ECONET – Poland. Fundacja IUCN-Poland. Warszawa: 1-202 (in Polish)

Paneuropean Strategy for Biological and Landscape Diversity (1995) Council of Europe, Brussels: 1-48

Ribeiro L, Baräo T (2006) Greenways for recreation and maintenance of landscape quality: five case studies in Portugal, Landscape and Urban Planning vol 76, Issue 1-4:79–97

Richling A, Solon J (1996) Landscape ecology. Państwowe Wydawnictwo Naukowe. Warszawa: 1–319 (in Polish, English summary)

Solon J (2006) In search of indicators for landscape sustainability in rural areas – the STAMEX approach. Agriculture, Ecosystem and Environment. Elsevier Editorial System (in press)

Søren R, Silja S, Bòra E, Haveraan E, Eriksson C, Adolfsson E (2004) Nordic landscape, The Nordic Council of Ministers, Kopenhaga: 1-76

Starkel L (ed) (1991) Geography of Poland. Natural environment. Państwowe Wydawnictwo Naukowe. Warszawa: 1-670 (in Polish)

Walker B, Steffen W, Canadell J, Ingram J (1999) The Terrestrial Biosphere and Global Change: implications for natural and managed ecosystems. Int. Geosphere – Biosphere Programme Book Series, vol. 4. Cambridge Univ. Press, UK, pp. 1–439

Weiner J (1999) The life and evolution of the biosphere. Państwowe Wydawnictwo Naukowe. Warszawa. pp. 1–591 (in Polish)

Wojcik J, Walczak M, Smogorzewska M (2001) Protected Areas in Poland. Institute of Environment Conservation. Warszawa: 1-281 and map 1: 25000 (in Polish, English summary)

Yusim O, Papadema N, de Pundert L, Berghuysen A, Pols D (2002) EPCEM project report: Landscape Quality in Europe Wood elements as illustrative features for landscape quality in the Netherlands and Poland. Internet address: http://www.leidenuniv.nl/cml/sem/projects/epcem/2002-2.pdf, last accessed 14.06.2006

Zonneveld IS, Forman RTT (eds) (1990) Changing Landscapes: an Ecological Perspective. Springer-Verlag. New York: 1–291

26 The Use and Misuse of Noise Standards

Riki Therivel[1] and Chris Bennett[2]

1 Levett-Therivel sustainability consultants and Department of Planning, Oxford
  Brookes University, Oxford, England
2 Stop Stansted Expansion (SSE) Noise Group, Takely, England

26.1 Introduction

Noise standards aim to protect people from being annoyed or physically harmed. They thus try to translate an essentially subjective response by individuals ("I can't concentrate because of the traffic noise") into quantitative measures that can be applied to either the noise source or the ambient environment.

Sound levels being easily measurable, and noise being a ubiquitous problem, quite a lot is known about what noise levels lead to what subjective response, and there is a fair amount of consensus on what are appropriate noise standards (at least insofar as they relate to the average person). Nevertheless, these standards can still be applied and interpreted–used and misused- in very different ways.

This chapter starts with a brief analysis of the component parts of noise standards, and why noise standards look the way that they do. It presents some noise standards and explains the contexts within which they are often used. It then focuses on a case study – the 'Generation 1' expansion of Stansted Airport in southeast England – which exemplifies some of the ways that noise standards are applied in practice in environmental impact assessment.

26.2 Components of Noise Standards

Noise is unwanted sound, and the role of noise standards is to control the degree of the 'unwanted-ness' of the sound. This can take several dimensions.

How loud the sound is. Sound is measured in decibels (dB). These describe sound pressure on a logarithmic scale, so that a change of 3 dB is equivalent to a doubling or halving of the sound pressure. However, in terms of our perception, a change of 3 dB in sound levels is barely detectable by the human ear, and it requi-

Standards and Thresholds for Impact Assessment. Edited by Michael Schmidt, John Glasson, Lars Emmelin and Hendrike Helbron. © 2008 Springer-Verlag

res a change of approximately 10 dB to be perceived as a doubling or halving of loudness.

What kind of sound it is. Sounds vary in terms of frequency (pitch), amplitude (loudness) and duration; whether they contain distinct clatters and thumps, are intermittent or continuous, are rhythmic or not, and/or contain information such as speech or music. Examples are:

- background traffic noise: low hum, roughly consistent throughout the day, not impulsive, not rhythmic, no information
- someone talking on a mobile phone in a quiet public area: comparatively loud, short duration, not rhythmic, information content
- police helicopter: loud, short duration, impulsive, rhythmic, no information

These will have different meanings and 'feel' very different to different listeners, but traffic noise is probably less likely to interrupt concentration than the others; and the helicopter noise will be rather less annoying to someone whose house is in the process of being burgled than to their neighbours. Abrupt changes in sound levels, even if small, can be significant to the listener (DoT 1993).

Who hears the sound? Many sounds are pleasurable to the listener, or intrinsic aspects of a pleasurable experience, for instance those resulting from concerts, sporting events, or riding a motorcycle. Noise levels at such events/activities are often higher than is strictly healthy (as those of us who have left concerts with ringing ears will testify), but typically no noise standards are set for those activities to protect the participant because they are not 'unwanted', provided, of course, that they are not disturbing non-participants.

For many work-related activities, occupational health standards apply. These ensure the health of the listener, and are broadly based on dose-response studies involving the level of noise and how long the worker can listen to it. For others, the worker's ability to concentrate is the basis for the noise standard. In either case, the assumption is that workers will put up with sound – 'unwanted' does not come into question - as long as their health and productivity is not compromised.

Community noise standards, instead, are much lower because they do try to reflect the concept of 'unwanted'. They focus on ensuring a pleasant and healthy living environment, not just a physically bearable one. The aim of these standards is to prevent annoyance, for instance due to people's inability to have a comfortable conversation or to sleep. They are typically based on standard responses by a standard population. However, individuals' responses to different noises vary by the type of noise and the individual. For instance, the World Health Organisation (1999) notes that

> *"Data from a number of sources show that aircraft noise is more annoying than road traffic noise, which, in turn, is more annoying than railway noise (...) Some populations may also be at greater risk [from] the harmful effects of noise. Young children (especially during language acquisition), the blind, and perhaps foetuses are examples of such populations."*

What time of day it is and the listener's expectations. The same level of sound is typically more 'unwanted' at night than during the day, and often also on week-ends than during weekdays. We expect a quieter environment for sleeping and resting than for other activities. Baseline sound levels are also typically lower at night, so that the same sound (e.g. people walking by a building talking loudly) will seem comparatively noisier at night than during the day.

Similarly, people in rural environments would expect lower sound levels than those in a busy city centre. This interplay of location and noise is not often seen in noise standards, though the OECD (1996) recommends different noise levels for urban, suburban and rural areas, and the CPRE's (1995) 'tranquil area' maps define tranquillity as absence of both visually intrusive and noisy activities. It is because of these different parameters that noise standards look the way they do. Figure 26.1 explains the components of noise standards.

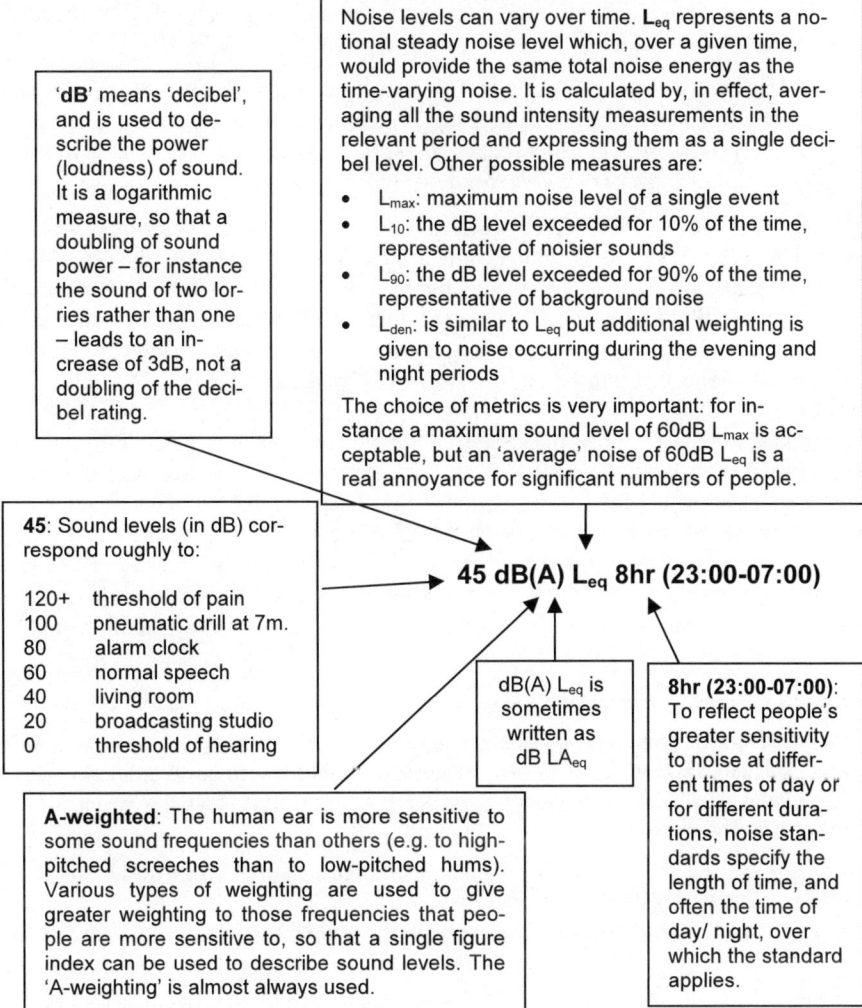

Fig. 26.1. Components of noise standards

26.3 Examples of Noise Standards

Noise standards can be used for a range of regulatory/control purposes. In a work environment, they can specify for how long a worker can use a piece of machinery or what kind of protective equipment they need. For instance, the Australian Government's (2000) national standard for occupational noise is:

"The national standard for exposure to noise in the occupational environment is an eight-hour equivalent continuous A-weighted sound pressure level, LAeq, 8h, of 85dB(A). For peak noise, the national standard is a C-weighted peak sound pressure level, LC, peak, of 140dB(C)".

The European Commission (2003) uses more complex trigger levels, but specifies that action must be taken where levels are exceeded:

"Where noise exposure exceeds the lower exposure action values, the employer shall make individual hearing protectors available to workers; Where noise exposure matches or exceeds the upper exposure action values, individual hearing protectors shall be used."

For new transport infrastructure projects, noise standards can determine whether noise insulation has to be provided for nearby residents, or whether the project can be built at all: Box 26.1 shows an example. For airports, they can constrain the hours of operation and the types of aircraft that are allowed to land. For individual pieces of machinery, including cars and construction equipment, they can limit the amount of sound that can be emitted from that machinery: Table 26.1 shows an example.

Standards such as those discussed above can help to control overall ambient (background) noise, but if many vehicles or aircraft are in use, even if each is relatively quiet, then the total noise can still be unacceptable. Ambient (or community) noise standards deal with the sum total of noises at the receiver's end. They support land use planning decisions about where to locate different activities and can prevent some noisy activities from taking place at certain times. They are considerably lower than the other types of standards, and are typically recommended

---

Box 26.1. Example of traffic noise standard leading to compensation (ODPM 2004)

For residential buildings that are within 300m of a new or altered highway and that were occupied before the highway was first opened to the public; the Highway Authority will be required to undertake sound proofing works if, within 15 years from the opening of the new or altered road

- the traffic noise level at one or more facades will increase by at least 1dB(A) and will be not less than 68dB(A)L10(18hr); and
- noise caused or expected to be caused by traffic using the new or altered road section will contribute at least 1dB(A) to the noise level.

Rooms will be eligible for noise insulation provided they have one or more facades exposed to the noise: living room, dining room, study, bedsitter, bedroom.

levels rather than limits/standards *per se*. Table 26.2 shows the World Health Organisation (WHO) guidelines for community noise; Box 26.2 shows the English Government's Planning Policy Guidance on Noise.

Table 26.1. UNECE (1996) regulations on sound levels for vehicles

Vehicle categories	limit values (dB(A))*
Vehicles used for the carriage of passengers and capable of having not more than nine seats, including the driver's seat	74
Vehicles used for the carriage of passengers having more than nine seats, including the driver's seat, and a maximum authorized mass of more than 3.5 tonnes	
- with an engine power less than 150 kW (ECE)	78
- with an engine power of 150 kW (ECE) or above	80
Vehicles used for the carriage of passengers having more than nine seats, including the driver's seat; vehicles used for the carriage of goods	
- with a maximum authorized mass not exceeding 2 tonnes	
- with a maximum authorized mass greater than 2 tonnes but not exceeding 3.5 tonnes	76 77
Vehicles used for the transport of goods with a maximum authorized mass exceeding 3.5 tonnes	
- with an engine power less than 75 kW (ECE)	
- with an engine power of 75 kW (ECE) or above but less than 150 kW (ECE)	77 78
- with an engine power of 150 kW (ECE) or above	80

*Regulation 51 specifies how the sound levels should be measured. This methodology (though not the standards themselves) is in the process of being revised: the Working Party on Noise recommended amendments in June 2006.

Table 26.2. Example of ambient noise guidelines: World Health Organisation Guidelines for Community Noise

Specific environment	Critical health effect(s)	LA_{eq} (dB)	Time base (hours)	LA_{max}, fast (dB)
Outdoor living area	Serious annoyance, daytime and evening	55	16	-
	Moderate annoyance, daytime and evening	50	16	-
Dwelling, indoors Inside bedrooms	Speech intelligibility and moderate annoyance, daytime and evening	35	16	-
	Sleep disturbance, night-time	30	8	45
Outside bedrooms	Sleep disturbance, window open (outdoor values)	45	8	60
School class rooms and pre-schools, indoors	Speech intelligibility, disturbance of information extraction, message communication	35	during class	-
Pre-school bedrooms, indoors	Sleep disturbance	30	sleeping time	45

Table 26.2. (cont.)

Specific environment	Critical health effect(s)	LA_{eq} (dB)	Time base (hours)	LA_{max}, fast (dB)
School, playground outdoor	Annoyance (external source)	55	during play	-
Hospital, ward rooms, indoors	Sleep disturbance, night-time	30	8	40
	Sleep disturbance, daytime and evenings	30	16	-
Hospitals, treatment rooms, indoors	Interference with rest and recovery	#1		
Industrial, commercial, shopping and traffic areas, indoors and outdoors	Hearing impairment	70	24	110
Ceremonies, festivals and entertainment events	Hearing impairment (patrons:<5 times/year)	100	4	110
Public addresses, indoors and outdoors	Hearing impairment	85	1	110
Music through headphones/ earphones	Hearing impairment (free-field value)	85 (#4)	1	110
Impulse sounds from toys, fireworks and firearms	Hearing impairment (adults)	-	-	140 #2
	Hearing impairment (children)	-	-	120 #2
Outdoors in parkland and conservation areas		#3		

#1: as low as possible.
#2: peak sound pressure (not LA_{max}, fast), measured 100mm from the ear.
#3: existing quiet outdoor areas should be preserved and the ratio of intruding noise to natural background sound should be kept low.
#4: under headphones, adapted to free-field values.

Box 26.2. Example of ambient noise guidelines used for planning purposes: English Planning Policy Guidance Note 24: Planning and Noise

"This guidance introduces the concept of Noise Exposure Categories (NECs), ranging from A-D, to help local planning authorities in their consideration of applications for residential development near transport-related sources. Category A represents the circumstances in which noise is unlikely to be a determining factor, while Category D relates to the situation in which development should normally be refused. Categories B and C deal with situations where noise mitigation measures may make development acceptable. The NEC procedure cannot be used in the reverse context for proposals which would introduce new noise sources into areas of existing residential development."

Box 26.2. (cont.)

Categories For New Dwellings $L_{Aeq,T}$ dB

	Noise Exposure Category			
Noise Source	A	B	C	D
road traffic				
07.00 - 23.00	<55	55 - 63	63 - 72	>72
23.00 - 07.00[1]	<45	45 - 57	57 - 66	>66
rail traffic				
07.00 - 23.00	<55	55 - 66	66 - 74	>74
23.00 - 07.00[1]	<45	45 - 59	59 - 66	>66
air traffic2				
07.00 - 23.00	<57	57 - 66	66 - 72	>72
23.00 - 07.00[1]	<48	48 - 57	57 - 66	>66
mixed sources3				
07.00 - 23.00	<55	55 - 63	63 - 72	>72
23.00 - 07.00[1]	<45	45 - 57	57 - 66	>66

[1] Night-time noise levels (23.00 - 07.00): sites where individual noise events regularly exceed 82 dB LAmax (S time weighting) several times in any hour should be treated as being in NEC C, regardless of the LAeq,8h (except where the LAeq,8h already puts the site in NEC D).

[2] Aircraft noise: daytime values accord with the contour values adopted by the Depart ment for Transport which relate to levels measured 1.2m above open ground. For the same amount of noise energy, contour values can be up to 2 dB(A) higher than those of other sources because of ground reflection effects.

[3] Mixed sources: this refers to any combination of road, rail, air and industrial noise sources. The "mixed source" values are based on the lowest numerical values of the single source limits in the table. The "mixed source" NECs should only be used where no individual noise source is dominant.

26.4 Use of Noise Standards in Practice: Stansted 'Generation 1' Environmental Impact Statement

26.4.1 Context

Although there is considerable consensus over noise standards – evidenced by the fact that many are set at an international level – there is still considerable variation in how they are used (and at times misused) in practice. The example below is not at all atypical, and illustrates two different organisations' views of how noise standards should be applied.

Stansted Airport is the third busiest airport in the UK, and serves London and the east of England. It is owned and operated by BAA. It is located in a rural and quite wealthy part of the country, home to a particularly well-informed and well-resourced pressure group, Stop Stansted Expansion (SSE). Stansted Airport has grown very rapidly, from 5.5 million passengers per annum (mppa) in 1997-8 to 22.3 mppa in 2005-6.

In 2003, Uttlesford District Council granted BAA[1] planning permission to extend the airport's operations up to 25mppa or 241,000 air traffic movements, whichever was reached first. These two operational thresholds were expected to be reached in about 2014/15. In practice, due to the very rapid growth of air travel in the UK, 25mppa is now expected to be hit in 2008. The government's 2003 Air Transport White Paper identified Stansted as being important to the UK's economic growth and leisure demands, and highlighted the importance of making full use of the capacity of Stansted's existing runway.

In April 2006, BAA applied to Uttlesford District Council for a change in the planning permission that would raise the air traffic movements to 264,000 and remove the cap on the number of passengers altogether. BAA believes that this so-called 'Generation 1' proposed development would lead to 'some 35mppa'; Stop Stansted Expansion believes that it could eventually lead to up to 50mppa on the single existing runway. Most of the additional taxiways, aircraft stands, changes to access road, hotels etc. needed to service the additional passengers and cargo already have planning permission, but BAA's planning application noted that some additional development could result if planning permission was granted.

Uttlesford District Council had, in November 2004, put out a scoping report which specified what topics it expected BAA's Generation 1 environmental statement (ES) to cover. For noise impacts, this included: Leq 16 hour daytime noise contours at 44, 47, 50 and 54 dB(A); information about helicopter noise; information on noise impacts on the public realm including parks, markets, places of worship etc.; night noise contours; LA_{max} at specific points; reports on increases in flight movements using the metrics pioneered by the Australian Department of Transport and Regional Services (DOTARS); and ground noise contours for airport activity, surface access movements, and a combination of the two.

- BAA's 2006 proposal was accompanied by a lengthy ES which includedair noise: nine pages in the main ES plus a 43-page 'Volume 2';
- ground noise, i.e. noise from aircraft taxiing, testing of aircraft engines, equipment servicing the aircraft, and traffic to and from the airport: ten pages in the main ES plus a 41-page 'Volume 8'; and
- construction noise: two pages in 'Volume 15' on construction

Since planning permission had already been given for 25mppa, the ES considered this as the baseline, and its impact predictions focused on the difference between what are now considered to be the likely impacts at 25 mppa[2] and those at 35 mppa. At the time of writing this chapter, the Generation 1 proposal was about to be considered at a public inquiry. Stop Stansted Expansion had critically ana-

---

[1] Following a takeover in mid-2006, BAA has become part of the Ferrovial group of companies.

[2] The new predictions for noise impacts at 25 mppa are considerably reduced from those put forward at the time of the 2003 planning application, because of improvements in the noise profile of Stansted's aircraft mix.

lysed BAA's ES, and raised a range of issues regarding the ES's use and interpretation of noise standards. All of the quotes in the tables below are from the 'Generation 1' ES (BAA, 2006) and Stop Stansted Expansion's (2006) response, unless otherwise stated.

26.4.2 What Noise Sources do Standards Apply to?

To listeners/receivers, it is the sum total of noise that matters, and the character of that total noise. For them, separating out different noise sources from the same development is akin to splitting a large development into small components for the purposes of ES: it is a 'salami-slicing' approach that may not fully identify the development's cumulative impacts. On the other hand, for developers, different noise standards and mitigation measures apply to different components of their operations; some noise sources contribute only in very minor ways to overall noise levels; and they don't want to have to mitigate for noise impacts that are not clearly attributable to their proposed development. The table below illustrates these different viewpoints.

BAA	SSE
"The contribution of helicopter traffic is likely to be very small and will not be accounted for in the noise contours. The implications of helicopter movements will, however, be discussed in the assessment" (BAA 2005).	"[The ES] indicates that the average of 3-4 [helicopter] movements per day will continue. Planning Policy Guidance Note 24 [states] that 'Helicopter noise has different characteristics from that from fixed wing aircraft, and is often regarded as more intrusive or more annoying by the general public'."
"There are no infrastructure changes proposed to the existing rail access to the Airport under the 35 mppa case. For this reason, there is no further consideration of airport related rail traffic noise in this assessment."	"[The ES] states that rail access noise is excluded from the assessment. Yet trains are currently running at capacity, so more passengers will surely require more rolling stock and produce more noise"
"The Uttlesford District Council scoping opinion requested that: 'Ground noise contours be prepared for on airport activity, increased surface access movements, and for a combination of both sources... Combined contours would be difficult to interpret and might conceal minor impacts attributable to secondary noise sources. It is therefore not proposed to present combined contours."	"[A] combination of both [airport activity and surface access noise] sources... is important, since noise is noise from wherever it comes."

26.4.3 Who is Covered by the Standard?

Some receivers are more sensitive to noise than others. For instance, the users of schools, hospitals, recording studios and other music venues are generally more likely to be disturbed by a given level of noise than others. This accounts, for instance, for the WHO's lower standards for schools and hospitals. But what counts as a school or a hospital? For instance, are noise sensitive uses that are closely related to a proposed development exempt from the tighter standards?

BAA	SSE
"No further [noise sensitive uses] were identified within any of the remaining contour bands, ie 60 – 72 dB LAeq, except for a doctors surgery (Inflight Hanger 1) which is not considered in this assessment because it is an aviation-related establishment and the Stansted Airport Adult Learning Centre which is supported by [Stansted Airport Limited]."	"The implication that financial support given to an establishment absolves noisemakers from their environmental responsibilities is clearly unacceptable".

26.4.4 What is the Correct Standard?

Despite a generally good understanding of the links between noise levels and annoyance, there is still much room for debate around the edges, particularly about community standards. Different types of noise annoy different people in different ways; different background studies underlie different standards; and different studies and standards can thus be used to support different arguments.

The last example in the table below also shows that even simple factual statements in government documents – in this case about ground reflection effects – can be interpreted differently by different parties.

BAA	SSE
"In the UK, aircraft noise assessment is based on the findings of the 1982 'Aircraft Noise Index Study'(…) From the results of [this] study, the Government currently considers 57 dB LAeq as the onset of significant community annoyance (…) although it is accepted that a small proportion of residents exposed at lower sound levels might still be annoyed (…) The WHO suggests that daytime outdoor noise levels of less than 55 dB LAeq are desirable to avoid significant community annoyance and the upper boundary for PPG24 Noise Exposure Category A is based on this."	"57 dBA Leq is the noise contour identified by the Government as being related to the onset of significant community annoyance. However, other authorities including the World Health Organisation (WHO) put the level lower"
"[S]upplementary analysis has been undertaken to show the distribution in maximum noise levels (LAmax) at specific locations in the vicinity of Stansted Airport. Six representative locations have been chosen to help inform the changes in the LAeq contours".	In separate correspondence, SSE suggest that the Australian DOTARS 'N65' approach should also have been used at these and other locations. N65 contours show all locations subject to more than 10, 20, 50, 100 etc. aircraft movements of more than 65dB per day.

(The ES does not provide LAmax information for night-time air noise. It provides LAmax information for night-time ground noise).

"[T]he WHO's Guidelines for Community Noise state: 'If the noise is not continuous, LAmax or SEL [3] are used to indicate the probability of noise induced awakenings. Effects have been observed at individual LAmax exposures of 45 dB or less. Consequently, it is important to limit the number of noise events with a LAmax exceeding 45 dB. Therefore, the guidelines should be based on a combination of values of 30 dB LAeq, 8h and 45 dB LAmax'".

For ground noise:
"The adopted benchmarks are: 55 LAeq for the day and evening; and 45 LAeq and 65 LAmax for the night-time, all measured outdoors. In the context of ground noise assessment, the precise meanings of the terms minor, moderate and major adverse impacts as defined in this volume take precedence over any alternative definitions.
Major adverse impacts – benchmark exceedances by more than 10 dB except where 35 mppa case sound levels do not exceed 25 mppa case sound levels or do not exceed baseline LA90.
Moderate adverse impacts – benchmark exceedances within the range from 0 up to 10 dB except where 35 mppa case sound levels do not exceed 25 mppa case sound levels or do not exceed baseline LA90.
Minor adverse impacts – benchmarks are not exceeded where 35 mppa case sound levels exceed 25 mppa case sound levels by more than 10 dB and exceed baseline LA90, or where 35 mppa case sound levels exceed 25 mppa case sound levels by more than 3 dB and baseline LA90 by more than 5 dB, or where 35 mppa case sound levels exceed 25
mppa case sound levels and exceed baseline LA90 by more than 10 dB.

"The benchmark system does not tell people what they want to know, which is how much more ground noise will there be compared with now. Firstly a complex set of interlocking exceedance criteria... are required to be assessed; the combinations are apparently designed to minimise the likelihood of the system identifying as adverse any impacts that may arise from a comparison between 2005 and the projected 35mppa levels: i.e. the impact that is of most importance to any resident wishing to assess the effects of expansion. Secondly the decibel increases (in some cases +10 dB) required to trigger the criteria are high considering that it is apparently Leq measurements that are at issue."

"For aircraft noise, PPG24 states that: 'For the same amount of noise energy, contour values can be up to 2 dB(A) higher than those of other sources because of ground reflection effects.' This means that 57 dB LAeq for air noise is considered broadly equivalent to 55 dB LAeq for other noise sources in terms of its use as an indicator of community response to noise."

"[The] rationale appears to be that 2 dBA should be subtracted from any aircraft noise reading obtained by a noise meter as these 2 dBA would relate to noise being reflected off the ground. This ignores the fact that the human observer would also hear this reflected noise and so the full, unadjusted reading is the one that actually reflects the human noise experience. In the light of this, there appears to be no logic in putting the annoyance level up to 57 dBA Leq when 55 dBA Leq is the selected level for rail and road traffic noise. Indeed there are strong arguments for reducing

---

[3] SEL - Sound Exposure Level - is a noise metric designed to take account of both duration and loudness. It does this by expressing the total noise energy experienced during the whole of the noise event as if it had occurred evenly spread over a period of one second. It is therefore equivalent to a one second Leq, relating to a single noise event.

the threshold of aircraft noise annoyance as compared with road and rail. Research by Finegold et al. indicates that, for the same noise measurements, aircraft noise is more annoying than other forms of transportation noise. The EU's Noise Working Group 2 builds on this research and finds that at a noise level of 50 dBA LDen, 19 % of people are annoyed by aircraft noise, compared with 11 % for road traffic noise, and 5 % for rail traffic."

26.4.5 How Significant is Exceedance of a Standard?

Even where there is agreement on which standard to use, this standard can be described and interpreted in different ways. Is exceedance of the standard a 'moderate', 'significant', 'substantial' or 'serious' impact? How is a technical noise standard described in non-technical terms in the ES?

BAA	SSE
"The WHO suggests that daytime outdoor noise levels less than 55 dB LAeq are desirable to avoid significant community annoyance and the upper boundary for PPG24 Noise Exposure Category A is based on this."	"[The ES] misquotes the World Health Organisation Guidelines for Community Noise as stating that daytime outdoor noise levels of less than 55 dBA Leq are desirable to avoid significant community annoyance. In fact (...) the WHO Guidelines use the words 'serious annoyance'(...) the Guidelines go on to state that levels above 50 dBA Leq will cause moderate annoyance."
"Interference with sleep patterns is frequently reported by those living near airports. When assessing the potential for sleep disturbance, a number of studies have been undertaken ant the most recent reported that between 1 in 5 and 1 in 10 respondents found difficulty getting to sleep or being woken early (...) One cited example suggests that the average sleeper was woken by aircraft events on only 1 in 75 occasions when noise events were greater than 80 dB(A)."	"An early standard noise study (Wilson, Noise, HMSO 1963) ironically uses an alarm clock as an example of an 80 dBA noise. Clearly the manufacturers of alarm clocks need to raise their game."

Combined aircraft taxiing and APU [Auxiliary Power Unit] sound levels (LAeq) for 35mppa case compared against baseline background sound levels

Receiver location	day	evening	night
Easterly operation (Runway 05)			
Tye Green	6.8	2.4	4.3
Fullers End	-	-	-
Gaunts End	9.0	7.2	10.5
Molehill Green	9.2	11.7	21.7
Coopers Villas, Takeley	6.5	7.8	8.9

"Very significant increases in ground noise impacts arise even in the 35mppa case, for example 21.7 dBA Leq at Molehill Green above baseline background levels at night (...) BAA defines this as 'moderate adverse' – i.e. the use of soft language to disguise the anguish that would be caused to very many local residents who live near the airport."

Summary of impacts – 35 mppa case

Receiver location	day	evening	night
Easterly operation (Runway 05)			

Farm Cottage, Tye Green	no*	no	no
Appletree House Fullers End	no	no	no
Motts Hall, Gaunts End	no	no	moderate
The Forge, Molehill Green:	minor	minor	moderate
Coopers Villas, Takeley	no	no	no

*no = no impact

"3350 people would experience an increase in air noise of 1 - 2 dB LAeq between the 25 and 35mppa scenarios. PPG24 notes that 'a change of 3 dB(A) is the minimum perceptible under normal conditions'. No population experiences an increase in the LAeq of more than 2dB between the 25mppa and 35mppa cases."

"[T]he PPG quotation is made in reference to decibels in terms of single impact events (for instance an LAmax measurement). However the additional 1-2dB exposure referred to by BAA equally clearly related to an Leq measurement, i.e. a measurement of intermittently occurring sounds over a period of time. The Leq metric factors in the periods of quiet between the intermittent noise events in a process akin to averaging, and a 3dB change in Leq, far from being 'the minimum perceptible', is highly noticeable. Indeed it represents a doubling of the exposure and this can be most simply understood by reference to the fact that if an airport were to double the number of [air transport movements], but retain the same fleet mix proportions, then the Leq measurement at any one location would increase by 3dB."

26.5 Conclusions and Recommendations

Noise standards aim to describe people's subjective responses to sound and control the "unwantedness" of sound. They can account for different types of sound, times of day, and receivers. In part because of this, they are quite complex and technical, and difficult for the layperson to understand and critically analyse.

There is a fair degree of consensus about noise standards. But despite this, the standards are still used in very different ways in practice. The same scenario – for instances changes in operations at an airport – can lead to different uses of almost every aspect of noise standards: what sources and receivers do they apply to? what standards should be used in what situations? what level of exceedance of a standard is significant? how should cumulative noise impacts be measured?

The answer is probably not to change the standards themselves. Nor is it possible to raise general public understanding of such a complex technical subject to the level needed to be able to critically analyse the standards and their use: not many local residents' groups will have the nous and technical sophistication of Stop Stansted Expansion. The answer may be to ensure that local authority planning and environmental health officers have the training and resources needed to critically analyse how noise standards are used, and to provide more detailed advice on what to expect (including pitfalls) in the noise section of an ES.

References

Australian Government, National Occupational Health and Safety Commission (2000) National Standard for Occupational Noise [NOHSC: 1007(2000)], 2nd edn, Internet: http://www.nohsc.gov.au/pdf/standards/Noise_standard_NOHSC1007_2000.pdf, last accessed on 12.02.2007

BAA (2005) Stansted Airport 25+: BAA response to UDC Scoping Opinion. Internet: http://www.uttlesford.gov.uk/stanstedairport/planning+issues/stanstedairport25.pdf

BAA Stansted (2006) Generation 1 Environmental Statement, Stansted Airport, Internet: http://future.stanstedairport.com/main/user/page.phtml?page_id=276, last accessed on 12.02.2007

Council for the Protection of Rural England (1995) Tranquil Areas, London

Department of Transport (1993) Design Guide for Roads and Bridges, vol 11, Section 3 Part 7, Internet: http://www.standardsforhighways.co.uk/dmrb/vol11/section3/11s3p07.pdf, last accessed on 12.02.2007

EC – European Commission (2003) Directive 2003/10/EC 'on the minimum health and safety requirements regarding the exposure of workers to the risks arising from physical agents (noise), Internet: http://www.europa.eu.int/eur-lex/pri/en/oj/dat /2003/l_042/l_04220030215en00380044.pdf, last accessed on 12.02.2007

Finegold LS, Harris CS, Gierke von HE (1994) Community annoyance and sleep disturbance: Updated criteria for assessing the impacts of general transportation noise on people. Noise Control Eng. J 42(1):25-30

Organisation for Economic Co-Operation and Development (OECD) (1996) Pollution Prevention and Control: Environmental Criteria for Sustainable Transport, Paris

ODPM – former Office of the Deputy Prime Minister, now Communities and Local Government (2004) Compulsory Purchase and Compensation: Reducing the Adverse Effects of Public Development: Mitigation Works, Internet: http://www.communities.gov.uk/pub/821/Booklet5ReducingtheAdverseEffectsofPublicDevelopmentMitigationWorks_id1144821.pdf, last accessed on 18.05.2007

Stop Stansted Expansion (2006) Uttlesford District Council Planning application UTT/0717/06/FUL, Applicant: BAA plc and Stansted Airport Ltd, Response on behalf of Stop Stansted Expansion. Takeley, UK, Internet: http://www.stopstanstedexpansion.com/documents/SSE_R1_RESPONSE_UDC_VOLUME_01.pdf,   last   accessed   on 18.05.2007

Therivel R, Breslin M (2001) Noise. In: Morris P, Therivel R (eds) Methods of Environmental Impact Assessment. Spon Press, London, pp 65-82

United Nations Economic Commission for Europe (1996) Agreement concerning the Adoption of Uniform Technical Prescriptions for Wheeled Vehicles, Equipment and Parts which can be fitted and/or be used on wheeled vehicles and the conditions for reciprocal recognition of approvals granted on the basis of these prescriptions, E/ECE/324, E/ECE/TRANS/505, Internet: http://www.unece.org/rans/main/wp29/wp29regs/r051r1e.pdf, last accessed on 12.02.2007

UNECE (2006) Proposal for Draft Amendments to Regulation no 51. Internet: http://www.unece.org/trans/doc/2006/wp29grb/ECE-TRANS-WP29-GRB-200606e.pdf,   last   accessed on 12.02.2007

World Health Organisation (1999) Guidelines for Community Noise, Internet: http://www.who.int/docstore/peh/noise/guidelines2.html, last accessed on 12.02.2007

27 Assessing Environmental Impacts on Human Health – Drinking-Water as an Example

Ingrid Chorus

Federal Environment Agency, Germany

27.1 Introduction

27.1.1 Hazards to Humans – Hazards to the Environment

What hazards do humans face when exposed to water, either through drinking it or through recreational activities? How well do we understand their relative impact on human health?

→ Before reading any further, try to write down 5-7 hazards that come to your mind when you think about drinking water from the kitchen tap in your home town, and when you think about going swimming at your favourite lake or beach. Next, try to rank these in order of priority, putting those you think pose the greatest risk on the top of your list. As third step of this exercise, try to assess which of the hazards you noted also endanger aquatic ecosystems, and which are specific to human health.

Most likely, in trying this you found that you feel uncertain about criteria for ranking hazards according to risks, and in the third step you probably noticed that not all hazards to human health also endanger aquatic ecosystems and *vice versa*. This exercise highlights the demand for objective, scientific criteria for assessing risks caused by different hazards to different target organisms or whole ecosystems. This chapter outlines the current state of the art towards meeting this demand, highlighting the issue with the example of setting guidelines and standards for hazardous agents in drinking-water and discussing how human health targets relate to Environmental Impact Assessment (EIA) for aquatic systems.

27.2 Societal Perception and Scientific Understanding

Our societal perception of risks from hazards in the environment strongly impacts on the questions scientists address.

Standards and Thresholds for Impact Assessment. Edited by Michael Schmidt, John Glasson, Lars Emmelin and Hendrike Helbron. © 2008 Springer-Verlag

Vice versa, our perception has evolved together with the scientific understanding about hazards, and with our increasing control over some of them. For drinking-water, we may distinguish three phases, as follows:

1. Focus on microbial pathogens: In Europe in the late 19th century, the concept of drinking-water hygiene started to develop in consequence of the newly developed understanding of pathogenic micro-organisms from faecal pollution in drinking water as the cause of cholera and typhoid epidemics. In many parts of the world, diarrhoea contracted from faecal pathogens in drinking-water is still a major cause of childhood mortality. Societies in such settings respond by setting a priority on the control and avoidance of faecal pollution. Indeed, where this control has been successfully achieved, public health benefits have been dramatic.

2. Focus on (anthropogenic) chemicals: In the second half of the 20th century, environmental pollution in the wake of industrialisation, urbanisation and motorisation triggered the development of an understanding that chemicals occurring in the environment may harm human health. In 1958 the World Health Organisation (WHO) published its first edition of what at the time was called "International Standards" for drinking-water quality. Besides a discussion of microbial indicators such as *E. coli* for faecal pollution, these had a strong focus on chemicals, for the first time providing upper limits for tolerable concentrations in drinking-water, particularly pesticides. By that time, in many industrialised and urban settings, rather good control of faecal pollution had been achieved for drinking-water. Also, with the discovery of antibiotics infectious disease had lost a fair bit of its scare. Increasing awareness of environmental pollution in the 1970's thus shifted the attention of both the general public and of scientists to chemicals, particularly to those of anthropogenic origin and/or with carcinogenic properties.

Although environmental exposure of humans to chemicals almost never occurs in acutely health-threatening concentrations, long-term prolonged exposure to subacute levels can result in severe health impairment, or in premature death through cancer. This recognition resulted in a widespread perception of chemicals, particularly anthropogenic ones, being equally "bad" as pathogens. In parallel, with an increasing interest in alternative medical and health approaches, a (scientifically not underpinned) distinction emerged between negatively perceived "chemicals" and positively perceived "natural products".

To date, with the exception of drinking-water professionals few people – including scientists – are aware of the fact that human exposure to hazardous chemicals occurs chiefly through food and air and (with the exception of very few substances) almost never primarily through water.

3. Increasing awareness of hazard from natural chemicals: The early 1990's saw a newly developing understanding of the widespread occurrence of toxic cyanobacteria in water, causing a human health risk both for drinking-water and for recreation. Although in most settings massive cyanobacterial blooms are a consequence of anthropogenic eutrophication, cyanobacterial toxins are natural substances. A spontaneous reaction in discussions on the human health risk caused by cyanobacterial toxins often is that these couldn't be nearly as dangerous as anthropogenic chemicals, simply because they are "natural". However, if we the same toxicological criteria that we use for other substances to assess the risk they pose to human health, we find no scientific rationale for this argument. In fact, together with fluoride, arsenic and lead (the latter only in systems where lead is used for plumbing), specific cyanotoxins (i.e. microcystins) are among the few chemicals in water resources that occur widely and frequently in concentrations well above guideline values or standards.

This discussion of risk perception raises the question to which extend the standards and guideline values we currently use provide objective scientific criteria to quantify the human heath risks and to compare the relevance of these different hazards.

27.3 Deriving Guideline Values and Setting Standards for Hazardous Agents in Drinking-Water

The World Health Organisation's (WHO) Guidelines for Drinking-water Quality are the most widely used scientific basis for setting national standards for maximum tolerable concentration of hazardous agents in drinking-water. This is because of the broad scientific basis used in their derivation by international groups of highly acknowledged experts and an extensive process of scientific review. The 3rd edition of these WHO Guidelines (WHO 2004) emphasizes that the overall target is not compliance to a long list of standards, but to protect public health by minimizing illness transmitted through drinking-water. National standards should therefore be set for those agents which are relevant in the respective settings, for which surveillance and enforcement is possible, and at the levels that are locally most adequate to optimize achievement of this overall target.

For chemicals in drinking-water, WHO gives Guideline values (GVs) for 95 different substances. These are set at levels that should not cause harm to human health even if exposure continues for a whole lifetime. Two approaches are used to derive these GVs, depending on the toxic mechanism of the respective substance – one for "threshold chemicals", i.e. those that are assumed to have no effect below a threshold dose, and a very different one for "non-theshold" chemicals.

Threshold chemicals: Whether or not these cause harm depends on the dose. (Table salt, i.e. NaCl is a classical example: we need some, but too much can kill a person). The threshold for harm is usually determined with animal experiments conducted over a time span of usually at least 13 weeks and preferably up to 104 weeks with several groups, each receiving a different dose of the substance to be tested. During and after terminating the experiment, a wide range of parameters ("toxic endpoints") is assessed – e.g. blood serum enzyme levels, changes observed in the structure of organs and tissues. The "No Observed Adverse Effect Level" ("NOAEL") is the highest dose at which none of these show any statistically significant change in that experiment as compared to the control group that –

in the same experiment – was dosed only with the carrier, e.g. drinking water. For reasons of standardization and interlaboratory comparability, the animal species routinely used for such experiments are well defined strains of rats and mice, but also of rabbits, dogs or (rarely) monkeys. The NOAEL as measured in the most sensitive species is the Point of Departure (PoD) to derive a tolerable dose for humans, i.e. a "Tolerable Daily Intake" (TDI) usually intended for safe lifetime exposure and expressed as mg substance per kg bodymass and day. This is derived by a two-step extrapolation, the first of which accounts for the unknown but possible difference in sensitivity (i) between humans and the test animal species and (ii) within the human population between average and sensitive individuals or groups. Both uncertainties or possible differences in sensitivities are accounted for by dividing the PoD by an *extrapolation* factor, usually amounting to 10 for each, i.e. in total for both uncertainties the PoD is divided by 100. In many cases, division by an additional *uncertainty* factor – of up to 10 – is applied to account for qualitative insufficiencies of the available toxicological data, such as lack of data which can be used for assessing effects from lifetime exposure or on offspring, or toxic effects on further important endpoints like reproductivity, neurologic performance or immunologic defense.

Thus, the overall factor applied to the PoD for deriving the TID may reach 1000 or more. In other words, division of the PoD by this overall factor results in a TDI which is 100-1000 times lower than the NOAEL in the experiment chosen as basis for the derivation. The last step in converting the TDI into a permissible concentration in drinking-water is to assess which share of the total exposure to the respective substance is expected to occur through drinking-water in relation to exposure through the sum of other sources (largely air and food). The TDI is then multiplied by this share and the product divided by the maximum volume of water assumed to be consumed (often 2 L, but e.g. in tropical settings possibly more) by an adult person with 60 or 70 kg body mass. This gives the Guideline value or standard in terms of µg/L or mg/L.

It is important to realize that the PoD is an animal dose at which no impact of the substance was observed after many weeks of exposure, while at the next higher dose some (often rather slight) effect occurred. To exemplify, let's use a cyanobacterial toxin – Microcystin-LR. This damages liver tissue by blocking important cellular enzymes (see Kuiper-Goodman et al. 1999 for a summary). This toxin was administered to 4 dose groups of mice (15 animals per group) in a 13-week experiment at 0, 40, 200 or 1000 µg per kg bodymass and day (Fawell et al. 1994). At 200 µf per kg bodymass and day, some of the animals showed slight liver tissue changes, and these injuries were more pronounced at 1000 µg/kg per day. In contrast, at 40 µg/kg per day no effect was observed, and this dose was hence chosen as the PoD. From the NOAEL of 40 µg/kg per day, a provisional TDI was derived at 0.04 µg/kg per day, i.e. using an overall factor of 1000. It is provisional because the overall factor of 1000 includes an uncertainty factor of 10 for lack of information on chronic exposure beyond the 13 weeks of the experiment's duration, including lack of data on reproductive effects and on carcinogenic potency (WHO 1998). Assuming an adult to weigh 70 kg, to drink 2 litres per day and scarcely be exposed to microcystin through other sources (i.e. 80 % of exposure coming from drinking-water), a provisional Guideline-value for Micro-cystin-LR in drinking-water was set at 1 μg/L (WHO 1998).

Using this example, let's discuss what exposure to concentrations somewhat above this Guideline value implies for our risk of illness from Microcystin-LR. What does exposure to a dose between the PoD and the TDI mean? How worried would you be if you had received a dose on the range of – say – 0.4 μg per kg of your bodyweight for some days, or occasionally over a time span of several weeks or months? This would be 10 or 100 times above the TDI for lifelong exposure, but still well below the PoD. Intuitively, most of us would feel that harm is unlikely if a much higher dose in the 13-week animal experiment did not show any damage. The point is, however, that we simply don't know whether humans are more sensitive, whether we happen to be a very sensitive individual within the human population, or whether other perhaps more sensitive endpoints were missed in this 13-week animal study because nobody may have looked for them (e.g. al-lergenicity). Each of the uncertainty and extrapolation factors of 10 that were ap-plied to derive the TDI are scientifically well justified: those for differences in sensitivity between species and between individuals within a species have actually been observed repeatedly with a range of chemicals when testing a number of dif-ferent species with the same substance, or when comparing responses of different individuals within a species. The factor of 10 for the limitations in the data-base reflects repeated observations of NOAELs being reduced by about this range after data became available from longer-term experiments. A dose between the PoD and TDI thus does not *per se* imply a high risk. What it does mean, however, is high uncertainty about whether or not it there is a risk. We know for sure only that below the TDI risk is improbable. From a regulatory point of view, the TDI of a substance is its virtual effect threshold in the most sensitive individuals within a human population assumed to be significantly (10 times) more sensitive than the animal species in which the reference-PoD was measured.

It is important to understand that this widely used method of deriving TDIs and GVs does not *quantify the risk* to human health from exposure to certain levels of a given substance, but rather is a yes-no approach with drinking-water being con-sidered very probably safe if concentrations are below the GV and possibly unsafe if they are higher. The only emerging differentiation is an acknowledgment of TDIs being calculated for lifetime exposure, while in the shorter term, higher ex-posures may be tolerated in certain cases, with maximum concentrations allowable for a short time depending on the mechanism and acute or long-term endpoint of toxicity. In the example of Microcystin-LR discussed above, if the exposure to 0.4 μg/kg and day happened only on one day, indeed we would have little cause for concern, because the TDI of 0.04 μg/kg is safe for lifetime exposure and the corre-sponding toxic endpoint, and this is significantly different from the one for (much higher) dangerous short-term exposure.. Such differentiation between short-term and lifetime exposure may be important in situations where investments are better placed for removing the cause of a problem than for short-term remediation.

Non-threshold chemicals are largely genotoxic carcinogens. For most of these, it is assumed that there is no threshold below which "nothing happens", but that the induction of a mutation has a theoretical chance of causing cancer, with whether or not this happens being a matter of probability. As for the threshold chemicals discussed above, guideline values are determined from animal experiments conducted with long-term exposure. However, here the development of cancer is the most relevant endpoint studied, and rather than using the data to determine a TDI from a NOAEL, the rate of cancer incidence at the effective dose is extrapolated with mathematical models in order to calculate the dose at which not more than 1 additional case of cancer would be expected in a population of 10^5 or 10^6. *Vice versa*, this means that your chances of contracting cancer from exposure to a a dose at the guideline level would be one in 100 000 or a million at up to life long exposure to that dose. This probability is a level agreed by society as acceptable, and where it is set for substances in drinking-water will depend on priorities in relation to other hazards. For example, in settings with poor sanitation resulting in widespread illness and childhood mortality from exposure to untreated sewage, setting very low standards for carcinogens in drinking-water is not likely to improve public health significantly whereas controlling the spreading of infectuous disease by improving sanitation would have a subsantial positive impact.

The values of one in 10^5 or 10^6 are rather rough estimates of risk at low environmental doses. The animal experiments to derive these rates of cancer incidence are not conducted with thousands of animals at a very low dose, but rather with smaller groups, using much higher doses in order to detect statistically significant effects. The ensuing mathematical extrapolation to a lower dose does not take into account that a range of biological processes may be different at a lower dose, such as DNA repair mechanisms, pharmacokinetics (i.e. clearance of substances from – or their accumulation in the body) and immune responses, all of which affect the probability of developing cancer from low-level exposure. Thus, this approach also involves substantial uncertainty. Nonetheless, it is one that does give an estimate of risk rather than a yes-no answer.

Health implications of exposure to levels above the guideline value or standard for a given chemical: For public communication it is important to understand the implications of these approaches for the likelihood of harm from exposure to chemicals at levels somewhat above guideline concentrations: it does not *per se* imply a high likelihood of harm. For non-threshold chemicals, it merely means a correspondingly higher calculated cancer risk – e.g. if the dose was tenfold higher, a risk of one in 10^5 instead of one in 10^6. For threshold chemicals, it means greater uncertainty of whether or not any harm is likely. For both types of chemicals, short-term exposure to a higher dose is less likely to cause harm because guideline values or standards are developed for safe lifetime consumption. If it seems reasonable to assume that the effect of a substance – strictly speaking, its toxic endpoint toxikinetic behaviour - is different for short term exposure as compared to that for chronic or lifetime exposure, guideline values which protect specifically from effects of short term exposure must be derived and applied separately.

Pathogenic micro-organisms: Until the late 1990s no attempt was made to quantify the health impact of pathogen concentrations in water. Rather, the indicator principle has been and continues to be widely (and indeed very successfully) applied. This means that 100 ml of water sample are tested for the presence microorganisms which are rather easy to detect, although they themselves are not pathogenic. If these are found, this indicates recent faecal contamination which very likely implicates the presence of human pathogens, whichever happen to be prevalent in a given population at a given time. The indicator micro-organisms used are *E. coli* and, as supplementary indicator, also thermotolerant coliforms. This principle works a bit like a traffic light: if the light switches to red, this indicates cars are likely to cross the intersection, though they will not necessarily do so. However, the red light does not totally exclude that cars will come anyway and kill you even if the light is green and likewise, sometimes pathogens may present even though neither of the two indicators was detected (e.g. because some pathogens are more resistant to disinfection than the indicators are). Also, a pure presence-absence assessment gives no information about the actual risk of infection and illness.

In response to an increasingly perceived demand for a more differentiated and risk-based approach to setting targets for pathogen occurrence, quantitative microbial risk assessment (QMRA) is being developed (WHO 2004). This assesses the probability of an adverse health effect from (i) the risk of infection in response to exposure to a specific dose of a specific pathogen and (ii) the risk of falling ill after being infected. Dose-response curves differ between pathogens: for some, ingesting only one pathogen involves a specific probability of infection, while for others a higher dose is necessary. Likewise, infection does not always lead to symptoms of illness and in fact, asymptomatic infection is common for many pathogens. The risk of illness is therefore calculated from the probability of infection multiplied by the probability of illness resulting from that infection. A third factor in QMRA is the severity of illness, which can range from slight impairment to death. Together, these three criteria – i.e. (i) risk of infection, (ii) likelihood of illness from infection and (iii) severity of illness – define the impact of a given dose of a specific pathogen on a population.

27.4 Assessing Human Health Risks: Can We Quantify or at Least Prioritise them?

A public health target can be set by defining the number of days of illness from infection considered acceptable per person – or   translated to the impact on a human population – the "Disablilty Adjusted Life Years" (DALY) lost in that population. Figure 27.1 visualises this concept and WHO (2004) explains it in detail. WHO (2004) suggests 10^{-6} DALYs per person per year as tolerable disease burden caused by a given pathogen. For a population of one million, this can mean any of the following examples:

- 1 person dying one year earlier from infectious disease, or
- 2 persons dying half a year earlier, or
- 1 person living for 2 years with 50 % disability, or

- 1 person living for 4 years with 25 % disability, or
- 365 people loosing one day of their healthy active lives by spending it in bed with 40°C fever, i.e. at close to 100 % disability, or others...

Infectivity, likelihood and severity of illness vary between pathogens, and therefore this target translates to rather different acceptable concentrations for different pathogens. For example for Cryptosporidium, it translates to 1 organism in 1 600 liters, for Campylobacter to 1 in 8000 litres and for Rotavirus to 1 in 32 000 liters (WHO 2004). These are very low pathogen concentrations, and obviously hard to analyze. We may therefore dismiss the idea of ensuring safety by regularly monitoring pathogen occurrence. Instead, we need to make sure that the measures and processes controlling their prevention or elimination are safely in place and reliably operating, i.e. to achieve safety through system assessment and process control (see Sect. 27.5). Furthermore and equally obviously, it is impossible for surveillance to cover the wide range of pathogens potentially occurring in water resources used to produce drinking-water. This problem is addressed by focusing system assessment on a few reference pathogens whose behavior in the environment is representative for a group, e.g. the three pathogens mentioned above may be used as representative for Protozoa, Bacteria and Viruses. The DALY approach provides a "common metric" with which we can compare the impact of non-threshold chemicals to that of infectuous pathogens as exemplified with the fictive cases illustrated in Fig. 27.1. DALYs provide a conceptual framework for deve-

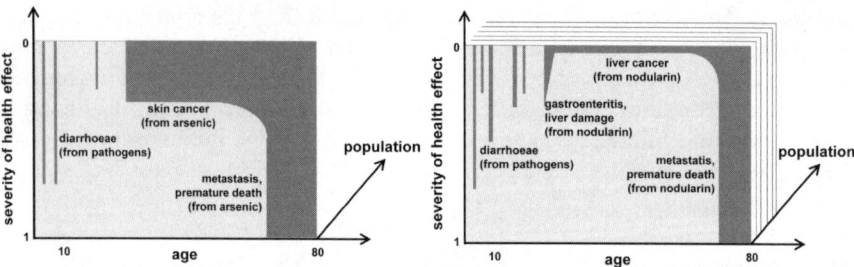

Left-hand panel: individual suffering 2 cases of diarroeae during childhood, disabling her by 70-75 % each, and 1 case as young adult with 30% disability. Skin cancer contracted at age 25 leads to 33 % disability and to premature death at age 40. In total, this individual spends 15 years with 33 % disability and looses 30 years, i.e. a total of about 35 healthy life years.

Right-hand panel: another individual experiencing similar childhood diarrhoea, gastroenteritis from exposure to nodularin as young adult resulting in latent liver cancer and a premature death at age 50, i.e. he looses about 30 healthy life years. The panels behind the one for this individual indicate this assessment for the entire population to estimate population disease burden in terms of DALYs .

In settings with a large infectious disease burden, most of the life years lost would be due to childhood mortality caused by diarrhoea.

Fig. 27.1. Using DALYs (Disability Adjusted Life Years) to compare health risks (adapted from Schmoll, pers. com, reprinted from Chorus 2006)

loping objective scientific criteria for quantifying human heath risks in order to be able to compare the relevance of these different hazards. However, currently the data-base for many of the hazardous agents that may occur in drinking-water needs further development before it can be used for calculating DALYs. For example, for carcinogenic substances a risk of contracting cancer from exposure to a given dose can be estimated, but data for the average life expectancy with a given type of cancer and severity factors for the disability that this cancer causes have not been integrated in order to express this risk in terms of DALYs. Particularly for threshold chemicals, approaches are lacking that look at the consequences of injury from a given chemical, i.e. the data from animal experiments at doses above the NOAEL, and then translate these into an impact on an average individual's and a human population's fitness. Indeed this is likely to be a difficult as it requires estimating the implications that endpoints such as histopathological changes observed in liver tissue or elevated blood serum enzyme levels would have for fitness and life expectancy. Applying the DALY approach to chemicals thus would require a paradigm shift in human toxicology and guideline derivation.

Such developments could, however, benefit from approaches in ecotoxicology, - a discipline which has been investigating effects on the population level for some time. The new WHO Guidelines for Drinking-water Quality (WHO 2004) strongly promote the DALY approach, and in other fields (such as air quality) it is already more widely used. For example, in 2005 German newspaper headlines gave numbers of people estimated to die prematurely due to fine dust particles from fuel emissions. As DALYs are the way forward to reach the aim of a rational, data-based approach to quantifying risks, this need is likely to stimulate research to fill the data gaps.

Meanwhile, expert judgement is used to compare risks from different agents, i.e. for assessing them relative to each other, even if their quantification in absolute terms is not possible. Simple risk matrices comparing the likelihood of a hazard to occur against the severity of its public health impact are a useful tool for such rankings. They typically use between 3 and 5 categories for estimating the likelihood of a hazardous event to occur and the severity of its consequences, as shown in (fig. 27.2).

An important aspect of this approach is that the likelihood of a hazard to occur depends upon local conditions. Thus, risk assessments are always specific to a given setting. A common misconception from imprecise use of the term "risk assessment", is to assign a risk for human health to a specific hazard, ignoring the circumstances of the setting which determine the frequency and concentrations with which a hazardous agent will occur. As risk assessment includes the probability of a hazard to occur, it is not generic, but always setting-specific. Fig. 27.3 gives two simplified examples of risk assessments for specific settings. Obviously, such relative risk assessments have a subjective element, and different groups of experts may come up with slightly divergent assessments. However, it is unlikely that their assessments will be radically different. In spite of their lack of scientific stringency, such assessments have two important benefits: (i) experts' implicit, of ten not even totally conscious assessments are made explicit and amenable to discussion, and (ii) this process improves transparency. For both, it is important that

Probability of occurrence	Severity of consequences				
	Insignificant	Low	Moderate	High	catastro-phic
almost certain	M	M	H	E	E
Likely	M	M	H	E	E
Possible	L	M	H	H	E
Likely	L	L	M	H	E
Rare	L	L	M	H	H

E:*extremely high risk;*   **H**: *high risk;*   **M**: *moderate risk;*   **L**: *low risk*

Fig. 27.2. Example of a matrix for comparing risks from different hazards occurring in a given setting

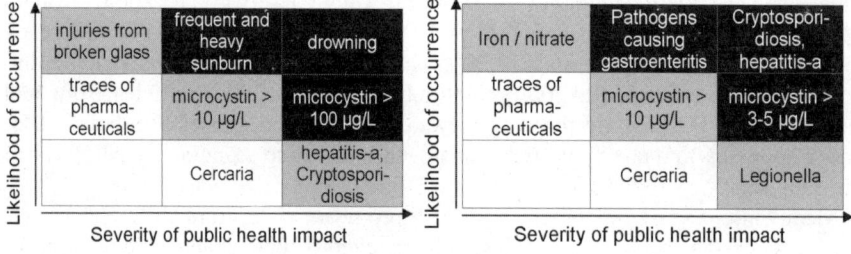

Fig. 27.3. Recreational site X and Drinking-water Supply Y: Two examples of setting specific risk assessments (from Chorus and Niesel 2007)

Recreational site X: the public health risks from drowning ranked clearly above risks from exposure even to high microcystin concentrations (higher than the German guideline value for recreational exposure, which is 100 μg/L; UBA 2003), as drowning obviously is lethal while it is uncertain whether or not short-term exposure even to rather high microcystin concentrations will cause harm. Among the hazards causing less severe impact, the risks from heavy and frequent sunburn were ranked above those from moderate microcystin exposure, because of the potential long-term consequence of skin cancer.

Drinking-water supply system Y: the risks from *Cryptosporidia* and Hepatitis-a Virus ranked higher than those from microcystin exposure to concentrations up to 5 times the provisional WHO Guideline value, because in this setting the former are more likely to occur and they implicate a certain death rate, while the health impact of microcystin exposure to these levels is uncertain. The health risk from *Legionella* received an equally high ranking in this setting (although *Legionella* are assessed as less likely to occur) because their occurrence also implicates a certain death rate. In this setting, iron often causes the water to appear brownish, and this is not ranked in the "green" area as low risk because its unpleasant taste and colour leads users to turn to other – often less safe – water sources. The system's nitrate levels in concentrations up to 100 mg/L in theory could be a more serious hazard, but in this setting they were not ranked in the "red" area (i.e. as extremely high risk) because here, infants are well protected through widespread breast-feeding and an excellent information system for parents.

the outcome of the assessment is documented together with the rationale for the positioning of each hazard in the matrix and for uncertainties of this positioning.

27.5 How do we Effectively Protect Human Health from Environmental Hazards – Process Control

The very low levels for acceptable concentrations of pathogens given above (i.e. in the range of one organism in several thousands of liters of water) highlight the limitation of an approach to safety that relies on monitoring their occurrence in drinking-water. A further caveat of such an approach is the time needed for microbiolgocal testing: in many cases, by the time the laboratory results are available, the water will already have been consumed. This also applies to many chemical analyses. Therefore, safety of drinking-water supply systems is best ensured by stringent process control for key protective measures in the whole supply chain, i.e. from the catchment to the consumers. The WHO (2004) Guidelines for Drinking-water Quality take this into account by recommending that "Water Safety Plans" (WSP) are developed for individual supply systems. These include assessing the risks caused by the hazards that occur in the given supply system and the system's performance in controlling them. Measures are identified that can control the risks, i.e. keep them at sufficiently low to ensure that the health-based targets for drinking-water quality are being met. Monitoring focuses on ensuring that these measures are always in place and functioning.

Such control measures may be very diverse, including, for example, land use planning in the catchment, keeping stock out of a drinking-water protection zone, restricting the use of manure and fertilizer in the catchment or critical parts of it, managing abstraction to avoid intake of cyanobacterial blooms, treatment to remove pathogens and substances, and disinfection. For each control measure, management plans are developed that define an operational monitoring system which will show whether or not the measure operating as it should, i.e. that critical limits of the operational parameter are being met. Operational monitoring has nothing to do with monitoring the concentration of hazardous agents in water. Rather, it addresses parameters such as stock density, farm records of fertiliser and manure use or residual soil nutrient content for control measures in farm management, turbidity at the outlet of filters for control measures in drinking-water treatment, or residual chlorine concentration at specified points in the distribution network if disinfection is a control measure. Supplementary to such operational monitoring, the monitoring of concentration of hazardous agents in water remains important as overall indication that the system is working as it should, but not for day-to-day operational control. In this system, the role of independent surveillance extends beyond occasional "end-product monitoring" of concentrations of hazardous agents in finished drinking-water, by also checking whether the Water Safety Plan of a given drinking-water supply system is adequate and is being adhered to.

In Sect. 27.4 we had seen that the very low target concentrations for some pathogens that typically may occur in water are hard to analyze, and this cannot be done on a regular basis. However, control measures can be validated as to whether they are likely to meet the targets set in order to protect human health. Research in engineering investigates to which extend specific drinking-water treatment technologies remove pathogenic micro-organisms and hazardous chemicals, and environmental research establishes the extent to which they are transported and attenuated in water-body catchments and the aquatic environment. For the purpose of such research, the use of highly sensitive analytical methods to detect even very low concentrations is more feasible than for routine monitoring. Once removal rates of typical hazards have been established for control measures, Water Safety Plans can be designed for the specific setting by combining measures so that they will achieve the removal rate needed in relation to the concentration of the hazard expected, including in the case of specific pollution events.

This approach gives standards and guideline values a new role, shifting the focus from compliance to standards towards control of the processes chosen to render water safe for human use. A similar approach was taken by the new EU Bathing-water Directive: It has relaxed on the number of parameters for which it defines standards (particularly for chemicals), but instead requires a "bathing water profile", which means assessment of the potential for pollution hazardous to human health and its causes (EC 2006).

27.6 Water Quality Targets for Human Health and for Aquatic Ecosystems

In the perception of the general public as well as of many scientists, impacts of water pollution on human health tend to be equated to impacts on aquatic environments, implicitly postulating water from aquatic ecosystems in good ecological condition to be *per se* healthy and wholesome for humans as well. This is roughly true, but there are many exceptions: For example, micro-organisms that are pathogenic to humans are harmless to aquatic organisms and tend to die off or to be predated upon rather quickly. Phosphate that is harmless to humans causes massive eutrophication of aquatic ecosystems (and only under some conditions leads to cyanobacterial blooms that are toxic to humans). Physical measures, such as impoundment of rivers, will drastically alter an aquatic ecosystem, and ecologically this impact is usually rated negatively. However, whether it will cause deterioration or even improvement of the quality of drinking-water abstracted from the system depends on specific conditions of the individual setting. Even direct impacts of hazardous chemicals may be rated very differently, e.g. a low and transient exposure to a neurotoxic pesticide may have no lasting impact on a human swallowing a mouthful of contaminated water when swimming, but for an aquatic crustacean "breathing" water with the same concentration for 24 hours a day, it may mean a reduction of alertness with the consequence of being eaten by predators or being translocated downstream to environmental conditions unsuitable for this species. *Vice versa*, assessing the human health risks from chlorine in relation to those from *Vibrio cholerae* would likely result in chlorination of drinking-water if the risk of contracting cholera cannot be sufficiently controlled with other measures, while for the aquatic environment, chlorine and chlorination by-products would likely be more detrimental than the occurrence of *Vibrio cholerae*.

The DALY discussion in section 27.4 highlighted the current conceptual challenges towards finding a common metric for assessing human health risks from different hazardous agents in water. Criteria for weighing human health risks against risks to aquatic ecosystems are not in view, and indeed developing them would be challenge primarily for social sciences. At the current state of the art it is important to understand that water quality targets for human health and for aquatic ecosystems share a subset, but there are also important differences, as highlighted in Figure 27.4.

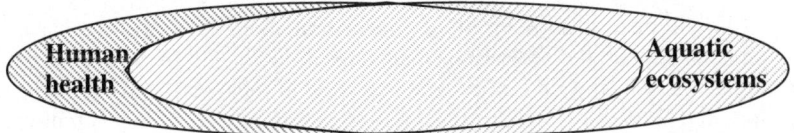

Fig. 27.4. Overlap and differences between water quality targets for human health and aquatic ecosystems

27.7 Conclusions and Recommendations

In conclusion of the considerations outlined above, the following hypotheses are proposed for further discussion:

1. Health risks currently tend to be communicated to the public in a „yes - no" fashion, and in consequence, public risk perception does not reflect real risk priorities.
2. Transparency in communication of the uncertainties of our risk assessments is of critical importance, both for credibility and for recognizing the information gaps that need to be filled most urgently.
3. Setting priorities is important for effective protection of public health, i.e. to avoid misguided investment and in some cases – such as chlorination of drinking-water, or allowing recreational use of water-bodies in spite of mild cyano-bacterial blooms – also for weighing benefits against risks.
4. Though tools need to be further refined, setting priorities for public health on a more rational basis, e.g. with the DALY approach or at least by relative risk ranking, is the way forward.
5. We should strive to do the right things for the right reasons – i.e. the protection of aquatic ecosystems is not always justified by the target of protecting human health. Rather, it is a target in its own right and does not need to „hide" behind the human health argument.

Exercise

Reconsider the drinking-water and lake or beach in your home town that you thought about when starting to read this chapter, and the 5-7 hazards you listed from exposure to that drinking-water and that recreational site. Could you now assess the risks from these hazards more systematically and prioritize, i.e.:

1. Could you estimate the likelihood of these hazards to occur? Or the severity of their impact on health (not only yours, but of your community)? Could you place them in a matrix such as the one in Fig. 27.2?
2. Could you describe (i) the reasons for your assessment and (ii) the uncertainties involved?
3. Would the outcome help find priorities for measures to improve the situation?

Possibly, you will need to look up more information about each of these hazards before you can estimate the severity of their impact and/or the frequency of their occurrence. You can easily find information on most of them on the WHO website that supports the Guidelines for Drinking-water Quality as well as for bathing water (http://www.who.int/water_sanitation_health/dwq/guidelines/en/index.html). This site contains links to background documents on many hazards in water. It might be interesting to see to which extend this exercise changes your perception of relative risks from different hazards, and/or of the uncertainties involved in assessing each.

Acknowledgement: special thanks are due to Hermann Dieter for checking and improving the toxicological discussion and to Oliver Schmoll for many fruitful discussions of the WHO concepts presented here – particularly for his idea of depicting DALYs as shown in Fig. 27.1!

References

Chorus I, Niesel V (2007) Toxic and Bioactive Peptides in Cyanobacteria. PEPCY final report. EU Contract Number Contract Number: QLK4-CT-2002-02634

EC – European Commission (2006) Directive 2006/7/EC of the European Parliament and of the Council of 15 Febuary 2006 concerning the management of bathing water quality and repealing Directive 76/160/EEC

UBA (2003) Empfehlungen zum Schutz von Badenden vor Cyanotoxinen. Bundesgesundheitsblatt – Gesundheitsforschung – Gesundheitsschutz 46:530-538

Chorus I (2005) Toxic Cyanobacteria: Controlling Risks. Lakeline 26 (2):16-23

Kuiper-Goodman T, Falconer IR, Fitzgerald DJ (1999) Human health aspects. In: Chorus I, Bartram J (eds) Toxic cyanobacteria in water: a guide to their public health consequences, monitoring and management. E & FN Spon, London, pp. 114-153

WHO – World Health Organisation (1998): Guidelines for Drinking-water quality, 2nd edn, Addendum to vol 2, Health Criteria and other supporting information. WHO Geneva

WHO – World Health Organisation (2004) Guidelines for Drinking-water Quality, 3rd edn, vol 1, Recommendations. WHO Geneva, 515 p.

28 Management of Dams in Trans-National River Basins – a Preliminary Sustainability Impact Assessment for the Upper Elbe River Basin

Martin Socher[1], Stefan Dornack[1] and Hans Ulrich Sieber[2]

1 Saxon State Ministry for Environment and Agriculture, Dresden
2 Dam Authority of the Free State of Saxony, Pirna

28.1 Introduction

Dams or systems of dams can significantly change the properties of river basins, such as flow, retention or the management of the water resources in general. Furthermore, dams and their management are known to show relevant impacts on the environment, economic activities and social structures in their very neighbourhood, upstream and downstream. Taking into consideration that nowadays numerous dams in Middle Europe are being operated as multifunctional/multipurpose installations, their impact assessment should start with a broadly scoped analysis with regard to the hydrological, environmental, social and economic properties of the particular river basin. This ambitious undertaking is in fact not that easy to exercise while looking at new dams projects. Due to forced environmental impacts up and down stream, social concerns about i.e. resettlement and complex interest and conflict lines between beneficiaries and non- beneficiaries quite often new dams projects lack mainly social acceptance despite of proven positive economic benefits.

However, due to a number of external requirements for changing management/ use patterns, already existing dams also come up with a number of hitherto unknown difficulties with regard to new strategic management schemes. Naturally, an assessment of dams situated in a trans-national river basin is even more challenging taking into account different socio economic structures and environmental conditions in the countries sharing that particular part of the river basin. This situation is without doubt very common in middle Europe due to the fact that a relevant number of major European streams are trans-national rivers such as the Rhine, Danube, Elbe, Tagus and Odra. Of course, from a general point of view, member states of the European Union (EU) have to obey the aquise communau-

Standards and Thresholds for Impact Assessment. Edited by Michael Schmidt, John Glasson, Lars Emmelin and Hendrike Helbron. © 2008 Springer-Verlag

taire which guarantees that the legal frame follows the same stipulations in any of those riparian countries. For individual projects or installations in trans-national river basins the situation is quite complex, with respect to the real impact in the downstream parts of the catchment. New dam projects are eligible to strategic impact assessment and trans-national consultations according to the Espoo Convention which entitles any concerned party to come up with their opinions and problems as part of public hearings or consultations. Looking at already existing dams, incremental changes of management schemes over a longer period of time do not legally require transnational consultations or public hearings, it rather depends on the extend and quality of co-operation among the concerned Member States to settle relevant issues in a sense of common understanding. This situation may alter after the currently discussed and negotiated European Flood Risk Management Directive will become law. However, from a general point of view it still remains necessary to analyse impacts of dams on a sound and generally accepted methodological basis. So far, different institutions such as the "International Commission on Large Dams", the "World Commission on Dams" and "United Nations Environment Programme Dams and Development Project" strive with zeal in order to bring together different systematic approaches about reach and impacts of dams on the environment, the economy and society. Taking into consideration that tremendous results have already been achieved it still makes sense to follow along this line strictly the idea of "Sustainable Development" for the assessment of dams in trans-national river basins on the basis of an accepted methodology which has been developed for a different purpose but has demonstrated its applicability for a variety of cases. Thus, a "Sustainability Impact Assessment" based on the work of Kirkpatrick et al. (1999) may have the flexibility to be adopted for the preliminary assessment of the impact of dams in trans-national river basins.

28.2 The Method of SIA

Environmental Impact Assessment (EIA) has become a legally defined, broadly accepted and widely used instrument for the assessment of projects, programmes and strategies. It is surrounded by a variety of methodological support and secondary modelling so that almost any subject can be thoroughly addressed. In the late 90's of the last century worldwide the need for an applicable methodology for handling the new paradigm of "Sustainable Development" became evident. This has been recognised by the European Commission while preparing the European Institutions for the next round of World Trade Organisation (WTO) negations mainly with countries from other parts of the world, with different state of development. It turned out to be quite reasonable to take advantage of experiences already gathered with the application of EIA methodology and instruments. The contractor (Kirkpatrick et al. 1999) proposed a scheme for a "Sustainability Impact Assessment" (SIA) with almost an equivalent structure as known from the well established EIA. However, for the very beginning it has not been intended to apply this method in other external fields. This has changed and meanwhile SIA became a field of advanced scientific endeavour (further under Kirkpatrick et al. 1999). In an attempt to operationalise SIA for projects, programmes and strategies outside the originally designated fields of interest of the EU, it has been shown in the early years of this century that SIA has an intrinsic capacity to be used for any assessment of the sustainability of projects, programmes or strategies (Socher 2000a, 2000b, 2001, 2002).

The classical Phase One SIA has the following three steps:

- Screening
- Scoping
- Preliminary assessment (with the analysis of mitigation and enhancement measures)

Screening comprises as a first step the identification of the projects, programmes or strategies to be assessed, whereas the scoping step describes the terms of reference such as:

- Specific properties of the case
- Specific possible scenarios
- Criteria by which the significance of the sustainability impacts are to be assessed
- Time horizons over which the impact should be assessed
- If possible casual chain analysis (Kirkpatrick et al. 1999).

Scoping is, furthermore, the stage where SIA core indicators are introduced and defined. This means that economic, social and environmental indicators may get in the further assessment process a connotation as: likely significant impact (v), unlikely significant impact (x) or significance of impact very uncertain (?).

As it is well established in EIA, also in SIA scoping is target- and process- oriented and involves hearing and consultation of concerned parties / stakeholders.

A Phase One SIA comes to a preliminary impact assessment and the proposal for mitigation and enhancement measures whereas a full (or Phase Three) SIA includes the formulation of monitoring and post- evaluation (MPE) proposals.

The aim of this paper is not to run a full SIA with all necessary technical and public requirements met, it is directed towards the demonstration that the methodological frame of a preliminary SIA offers a sound basis for the analysis and assessment of rather complex management issues in shared or trans-national river basins. Hence, the complexity of the Upper Elbe River Basin will be sketched and first results concerning the sustainability of differentiated management schemes will be derived.

28.3 Dam Systems of the Upper Elbe River Basin, the Moldau Cascade and Saxon Dams in Tributaries from the Ore Mountains and the Saxon Switzerland

With a length of 1 094 km and a catchment area of 148 268 km² the Elbe River is on the fourth position of rivers in middle Europe. The catchment area includes parts of the four States Germany, Czech Republic, Austria and Poland. According to geomorphological criteria the Elbe is differentiated between the Upper, the Middle and the Lower Elbe. The Upper Elbe reaches from its spring in the Giant Mountains (Czech Republic) to its entrance in the Northern German Lowlands near Meißen in Saxony. This river section takes a length of 467 km, with a 371 km part in the Czech Republic and 96 km in Saxony. An overview on the major tributaries in the Upper River Elbe Basin is shown in Table 28.1 (ICPE 2005).

Compared with the small tributaries in Saxony by far the major tributaries of the Upper Elbe River are – for their most part – situated in Czech territory. The natural flow conditions in these tributaries and in the Elbe River itself have been increasingly affected in the last centuries by the construction of weirs and especially of dams. Today 139 dams with a total storage capacity of about 2 566 million m³ and a flood storage of about 271 million m³ exist in the Upper Elbe River Basin (see Table 28.2). Only two of these dams are directly built in the Elbe River-Labska dam and Les Kralovstvi dam, about 12 and 54 km beneath the Elbe spring.

The construction of such a large number of dams leads to a significant change of flow conditions. Especially large dams that can level off wide differences of the natural flow even over a period of several years are used to increase the flow in times of low water for different operations like shipping or water supply. Moreover, flood waves can be retained ore reduced in the dam reservoirs. Beside the dams a total number of 28 weirs have significantly modified the character of the Upper Elbe River in the Czech Republic at a length of about 250 km. Due to the small storage capacity upstream of weirs compared with the dams, there is only a

Table 28.1. Major Tributaries of the Upper Elbe River in the Czech Republic and Saxony

Major tributaries of the Upper Elbe River	Length in km	Catchment area in km²
Czech Republic		
Moldau	430	28 090
Eger	305	5 614
Jizera	167	2 193
Orlice	136	2 037
Ploucnice	101	1 194
Saxony		
Vereinigte Weißeritz	67	384
Wesenitz	71	270
Lachsbach	34	270
Gottleuba	34	252
Müglitz	49	214

Table 28.2. Number and storage capacity of dams in the Upper Elbe River Basin (ICPE 2005)

Catchment area	Dams with storage capacity > 0,3 mil. m³	Storage capacity [mil. m³]	Flood retention storage capacity [mil. m³]
Elbe from spring to mouth of Moldau	19	167	45
Moldau	72	1.893	125
Eger	13	400	70
Upper Elbe beneath mouth of Moldau (without Eger)	35	106	31
Sum	139	2.566	271

relatively small impact of those weirs on high or low flow conditions in the Elbe River.

As an essential part of infrastructure dams serve multifold uses and create a special area for flora and fauna at the same time. This leads to conflicts of interests in view of different uses in water management and of the requirements of the European Water Framework Directive (WFD). Regardless of the dam's purpose the reservoir has to achieve the "good ecological potential" according to the goals of the WFD in future. Relating to this requirement dams are to be considered as an integrated part of the belonging catchment area. As their operation is an essential element of the influenced part of the river basin management and flow regulation the different uses of dams have to be sustainable embedded in this process.

Because of the multifunctional use it is an ambitious task to meet different needs for a dam on its own. Especially on dams used for water supply and flood protection controversial requirements have to be accomplished. For drinking water dams the conflict of interests primarily refers to the fill level of the reservoir – since water supply needs a preferably filled reservoir, flood protection in contrast demands a large flood storage volume and therefore a low water level within the reservoir. In search of an optimum several aspects have to be considered: sufficient retaining capacity to ensure water supply with an adequate security of the water quantity, sufficient reserve capacity for adequate security of the water quality, sufficient flood retention capacity to retain the volume of a flood wave or to reduce its maximum for the essential flood protection (Fig. 28.1).

Besides the evident conflict between water supply and flood prevention there are also uses of dams with essentially the same objectives of reservoir management, but which are nevertheless incompatible. The best example for this competition is the use for drinking water supply and for water sports or bathing. Though both uses require best water quality in the reservoir they are almost incompatible because of the immanent risk potential for water quality aligned with the touristic utilisation. In Table 28.3 the main occupancies of dams are contrasted to the primary objectives of water quantity and quality management. It shows compliances and opposites of interests. Several management objectives vary or do not necessarily have to correspond between all uses. Obviously the maximum difference is between drinking water supply and flood protection. These remarks on the dams im-

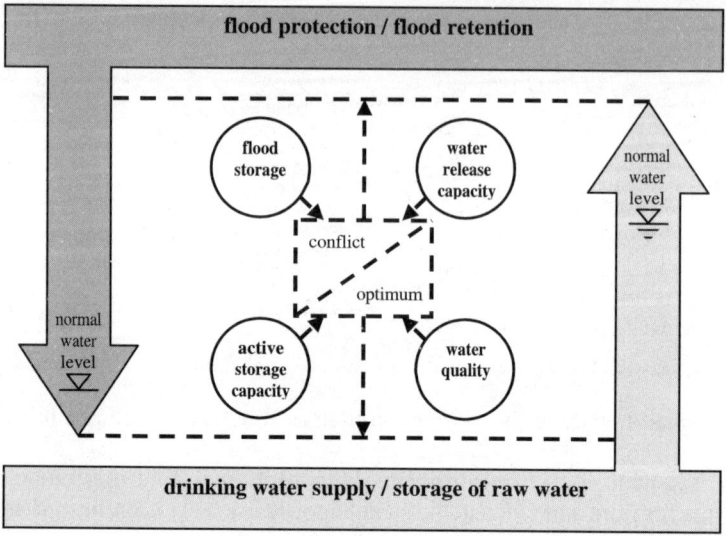

Fig. 28.1. Conflict of interests between water supply and flood prevention with dams

pact on flow conditions in catchment areas under consideration of their multifunctional use of course apply also and in particular for large dams or dam systems in the Upper Elbe River Basin. Here the Moldau Cascade especially with Lipno I and Orlik dam (No. 5 and 8 in Fig. 28.2) is by far the largest dam system with a correspondingly significant effect on the downstream Moldau including the Elbe beneath Melnik. For the Eger River the Nechranice dam has a major importance for this tributary and also an effect to flow conditions in the Elbe beneath (No. 18 in Fig. 28.2).

Table 28.3. Main purposes of dams and their management aims

Main purposes	Management aims				
	High water level	Large water volume in reservoir	High minimum discharge	High requirements on water quality in reservoir	High requirements on water protection in catchment area
Drinking water supply	+	+	-	+	+
Industrial water supply	+	+	-	*	x
Flood protection	-	-	x	x	x
Low-water enhancement	+	+	+	*	x
Power generation	+	x	-	x	x
Tourism / recreation	+	+	-	+	x
nature and landscape protection	*	*	+	*	+

+ corresponding aim   - opposed aim
x aim without concern   * differentiated aims

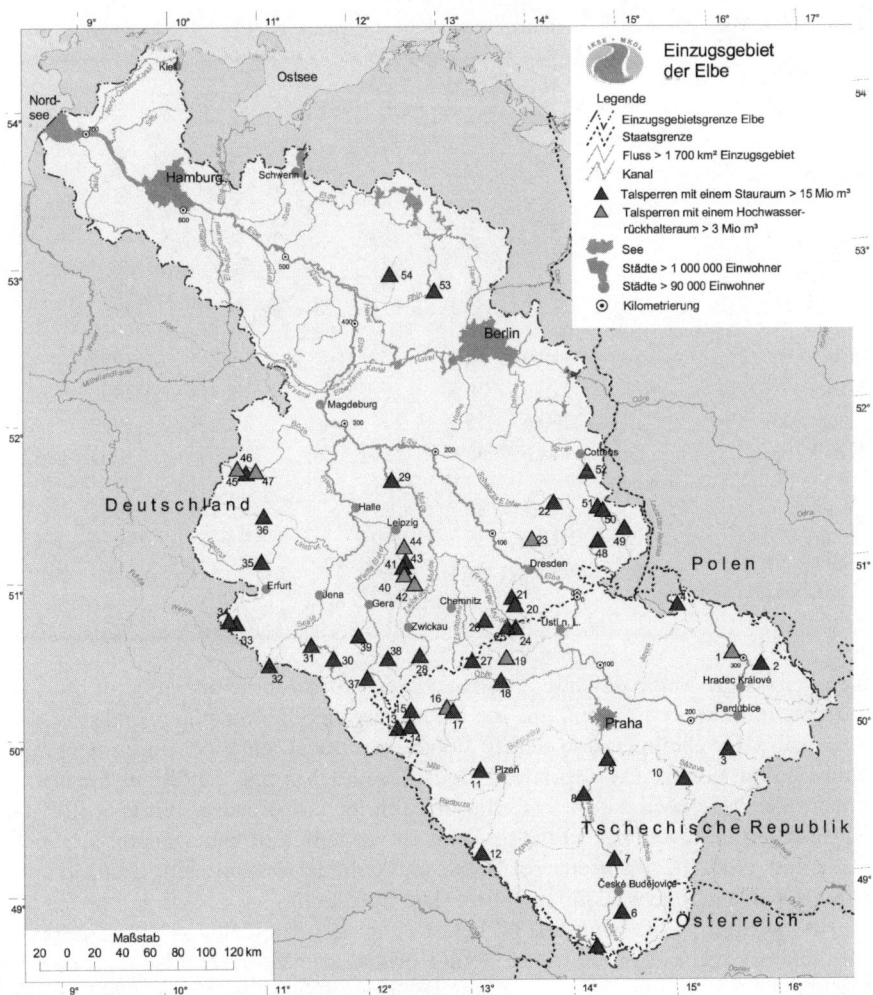

Fig. 28.2. Dams with a total storage capacity > 3 mil. m³ in the Elbe River Basin (state of data December 2003, ICPE 2005)

The Moldau Cascade nowadays consists of 9 dams with a total storage capacity of 1 353 mil. m³ and a flood retention capacity of 95 mil. m³. The main purposes are to impound water to secure flows for power generation and flow augmentation mainly for water abstraction by Prague (Q_{min} = 40 m³/s) and also for navigation. Additional purposes of the cascade include partial protection of Prague against floods (harmless discharge 1500 m³/s). Two dams, Hnevkovice and Korensko, were built in conjunction with water needs of Temelin Nuclear Power Plant. Main parameters of the dams are given in Table 28.4 (ICPE 2005). The water in the Moldau Cascade reservoirs is managed by Vltava River Basin, State Enterprise (Povody) of Moldau River Basin, State Enterprise in Prague, which uses operatio-

Table 28.4. Main parameters of dams in the Moldau Cascade (ICPE 2005)

Name of reservoir/ hydroelectric power station	River km	Mean flow (m3 s-1)	Total reservoir capacity (mil. m3)	Flood retention capacity (mil. m3)	Installed output (MW)	Initial year of operation
Lipno I	329.5	13.16	309.5	33.2	2 x 60	1959
Lipno II	319.1	13.4	1.7	-	1 x 1.5	1957
Hnevkovice	210.4	30.8	21.1	-	2 x 4.7	1992
Korensko	200.4	55.2	2.8	-	2 x 1.9	1992
Orlik (most important dam of the Moldau cascade)	144.6	83.4	716.5	62.1	4 x 91	1961 - 1962
Kamyk	134.7	83.7	12.9	-	4 x 10	1961
Slapy	91.6	84.7	269.3	-	3 x 48	1954 – 1955
Stechovice I	84.4	85.1	10.4	-	2 x 11.25	1943 – 1944
(Stechov. II)			(0,5)	-	(1 x 45)	(1947)
Vrane	71.3	111.0	11.1	-	2 x 6.94	1936

nal rules for the whole cascade. The Povody co-ordinates the operation of individual reservoirs to secure optimum use of the cascade's power plants and it is also responsible for the operation during floods, in critical situations in co-operation with Commission for Protection of Floods. Due to changing conditions the operational rules have been adopted several times in the last decades. While at first the construction of new dams set the need for an adaptation of management schemes, in the last years the occurrence of floods and political development changed constraints and lead to new requirements of dam operation.

The occurrence of August 2002 catastrophic flood initiated research projects for the assessment of influence on flood flows in selected river sites (e.g. Elbe at Decin, Usti and Melnik, Moldau at Prague). The results show effects and limits of flood protection with the existing dams in the Upper Elbe River Basin. The Moldau Cascade for example is most effective in reducing floods whose return period is from 10 to 20 years, while the reduction decreases for return periods outside this range. In contrast, the flood reduction at the Nechranice Reservoir increases in the full range of return between 1 and 100 years because of its sufficiently high retention capacity compared to the flood flow volume (Kasparek et al. 2006).

Possible flood reduction in the Moldau River Basin is an important economic and political problem, which is in discussion since finalisation of the cascade. Now the results of recent research projects on possible flood protection with dams will have to be evaluated under consideration of other purposes and requirements of dams (Kasparek et al. 2006).

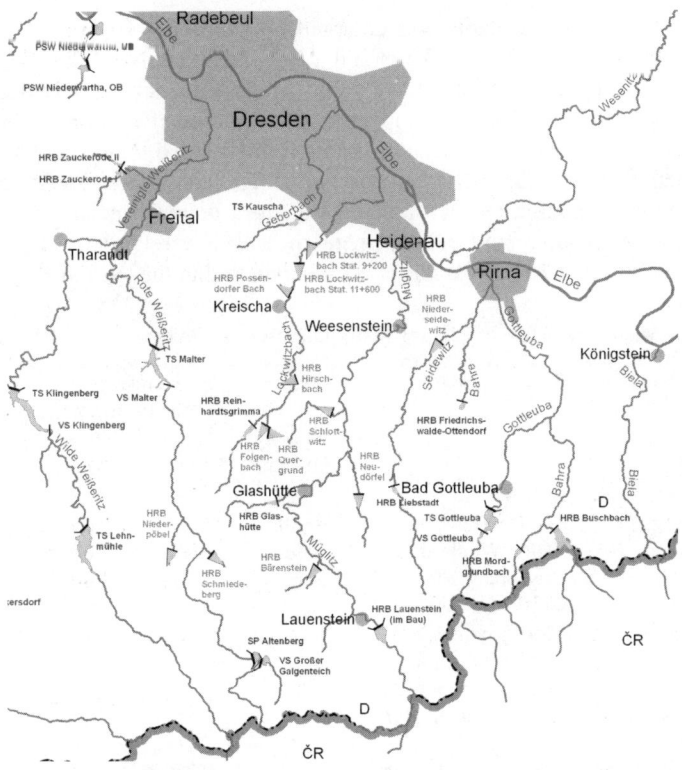

Fig. 28.3. Existing and planned dams and reservoirs in the eastern Ore Mountains in Saxony (state of data July 2005)

The dams at the small tributaries of the Upper Elbe River Basin in Saxony cannot affect the flow conditions in the Elbe River significantly. Beside other functions (mainly drinking water supply) they are nevertheless very important for flood prevention especially in the valleys of the eastern part of the Ore Mountains. Here the construction of dams started at the beginning of the last century in particular to secure regional drinking water supply. Because of the high flood risk in this area the dams were complemented by a system of flood protection reservoirs (without or with very little permanent water storage). An overview on the reservoir system in the eastern ore mountains is given in Fig. 28.3 (red means existing and green planned reservoirs) and in Table 28.5. As a result of the Saxon flood protection concepts for the tributaries in this area it is foreseen to extend the flood protection system by construction of more flood protection reservoirs (see Fig. 28.3).

The major part of flood retention capacity is provided by dams that are mainly used for water supply. For these dams a validation of the main purposes water supply and flood prevention was carried out under consideration of changing conditions like the decreasing consumption of water and the increasing demand on improved flood protection especially after August 2002. As a result the flood retention capacity in these dams was enlarged by additional 10 million m³ taking into account their use for drinking and raw water supply (Table 28.5). This enlargement of flood retention storage volume required a substantial upgrading of three concerned waterworks to compensate the negative effects on water quality. In these cases a change of constraints and its validation led to an adaptation of the management schemes for individual reservoirs. The above mentioned conflict of interests required additional measurements to meet opposite needs of multifunctional dams. Mitigation and enhancement measures were proposed and implemented step by step and individually for each dam within the system.

Table 28.5. Main parameters of existing dams and reservoirs in the eastern Ore Mountains

Name of dam / reservoir	River	Total reservoir capacity (mil. m3)	Flood retention capacity (mil. m3) before 2002	after 2002	Initial year of operation
Malter	Rote Weißeritz	8.78	2.28	5,97	1913
Klingenberg	Wilde Weißeritz	16.38	1.63	2.00	1914
Lehnmühle	Wilde Weißeritz	21.92	1.95	7.00	1931
Buschbach	Buschbach / Bahra / Gottleuba	2.40	2.40	2.40	1964
Mordgrundbach	Mordgrundbach / Gottleuba	1.27	1.15	1.15	1966
Liebstadt	Seidewitz / Gottleuba	1.08	1.01	1.01	1967
Friedrichswalde – Ottendorf	Bahre / Seidewitz / Gottleuba	1.52	1.45	1.45	1970
Reinhardtsgrimma	Lockwitzbach / Elbe	0.38	0.38	0.38	1969
Gottleuba	Gottleuba	12.97	2.00	3.00	1974
Lauenstein	Müglitz	5.22		5.05	2006
Sum		71.9	14.25	29.42	

28.4 Screening and Scoping

Screening and scoping are major prerequisites for preliminary assessment within the frame of a SIA. In order to follow a focused approach relevant scenarios will be identified aiming towards a clear understanding of the role that multifunctional/multipurpose dams play in that particular shared Elbe river basin. It appears to be quite reasonable to limit this assessment to key scenarios, without neglecting that a number of mixed and different scenarios could be possible. This self constraint requires, however, an overview of and considerations about the ongoing discussions among the riparian states, mainly within the frame of the "International Commission for the Protection of the Elbe River" (ICPE) which consists of delegations from Czech Republic and the Federal Republic of Germany and guests from Poland and Austria. From a very general point of view three scenarios are conceivable:

1. *Baseline Scenario:* This scenario can be described as business as usual, with independent operation of dams in the Czech and German part of the river basin, with, however, strong reliance on mutual exchange of relevant information and the establishment of an information platform on both sides.
2. *Breakthrough scenario:* This scenario sets out strong bilateral agreements among the two countries sharing the catchment area and integration into an organisational setting on trans-national basis.
3. *Advanced scenario:* This scenario takes into consideration the implementation of the currently discussed European Directive on the Assessment and Management of Flood Risks where mutual acceptance of up- and downstream measures will become a legal requirement in trans-national river basins.

The definition of core indicators is crucial for a SIA, for the purpose of this paper those will be target indicators to indicate the final impact of measures within the scenarios. With a view to focus on the function of already existing dams in shared/ trans-national river basins the core indicators will be defined as follows:

- *Economic indicators:* Contribution to Gross Domestic Product (GDP), fixed capital formation, employment
- *Social indicators:* Income generation, health, social stability
- *Environmental indicators:* Environmental quality (compartment water), natural resource stocks, renewable resources

The preliminary assessment rests on a wide range of technical analysis and modelling in Czech Republic and Saxony as described in an overview in recent documents of the International Commission for the Protection of the Elbe River (ICPE 2006).

28.5 Preliminary Assessment

The preliminary assessment is the first relevant result of a SIA. For the aim of this work it is necessary to take into account the whole life cycle of a dam and its technical infrastructure. Since the assessed dams already exist and are still far beyond an overall lifetime of approximately 100 years (and more), only two phases – operation and maintenance – will be addressed, whereas planning, construction and decommissioning will not be considered of being relevant for this work. Furthermore, it is given that both countries co-operate bilaterally und within the frame of existing institutions such as the ICPE and it can be extrapolated that both countries will transpose and implement the "European Directive on the Assessment and Management of Flood Risks" (Council of the European Union 2006).

Baseline Scenario

After the tremendous flood in 2002, flood protection policy in Germany as a whole, Saxony in particular and in the Czech Republic has changed considerably.

As shown in section 28.3 storage capacities in dams for retaining flood waves have been increased and management schemes were revised with a view to bring different use patterns together without neglecting traditional uses such as power generation, tourism, drinking water supply and maintaining downstream minimum flows. In both countries information platforms in the internet have been established in order to meet the information needs of administrative bodies, companies and citizens (Saxonian State Department for Environment and Geology 2005; Ministry of Agriculture of the Czech Republic 2005). Changes are communicated through the ICPE and actual properties of the Moldau cascade can be easily extracted from the internet. In fact, this baseline scenario comes rather close to actual co-operation patterns between the Czech Republic and Saxony. Incremental changes in multifunctional use patterns of dams improve also incrementally positive impacts on the core indicators, an expected decrease in power generation will be compensated by an increased avoidance of flood damages downstream the Moldau cascade, the Czech part of the Elbe and Saxon tributaries to the Elbe. Social and environmental indicators will be positively affected as well.

Breakthrough Scenario

This scenario requires far reaching ties among German/ Saxon and Czech authorities and a series of strong bilateral agreements among Czech Republic and Germany in order to bolster necessary changes of dam management schemes in due time. This scenario might be an option for states embedded in a different political and infrastructural surrounding; it requires standing bodies and detailed case by case trans-national benefit cost analysis for any change in dam management. It is surely not an appropriate scenario for Member States of the EU, where already a broad range of legal and technical instruments exist to smoothly solve bilateral problems. Hence, this scenario does not show more significant impacts on the core sustainability indicators than the baseline scenario in our case.

Advanced Scenario

This scenario is based on the assumption that in the coming year the proposed "Flood Risk Management Directive" (Council of the European Union 2006) will come into force and Member States such as Czech Republic and Germany will transpose and implement this directive. One focus of this directive is on flood risk management for trans-national river basins where the main elements.

- preliminary risk assessment
- risk/ hazard mapping
- flood risk management plans and
- upstream – downstream solidarity principle

have to be obeyed. For the Elbe River Basin this Directive will have a reliable basis on already existing organisational structures such as ICPE and its technical working groups, information platforms in the internet and further bilateral co-operation. Due to strong relation with the Water Framework Directive core sustainability indicators will all have positive impacts.

River Basin as a Whole

The Elbe River is formed by the confluence of the Elbe River from its headwaters in the Giant Mountains and the Moldau River from the Bohemian Forest. For the Upper Elbe River no further main tributaries are of importance, more than 200 km downstream from the Czech border the first relevant German main tributary reaches the Elbe. With respect to the reach and impact of dam management in the Upper Elbe River Catchment it can be concluded that significant positive impacts of dams are restricted to the Czech and Saxon part of the river basin.

The Preliminary Assessment gives the following results for the key indicators projected on the three scenarios (Table 28.6).

Table 28.6. Scenario based sustainability impact assessment for multifunctional dams in the Upper Elbe River Basin

Impact on	Significance of Impacts								
	Baseline Scenario (I)			Breakthrough Scenario (II)			Advanced Scenario (III)		
	A	B	C	A	B	C	A	B	C
Czech Republic	+1	+1	+1	+1	+1	+1	+2	+2	+2
Free State of Saxony	+1	+1	+1	+1	+1	+1	+2	+2	+1
River Basin as a whole	0	0	0	0	0	0	+1	+1	+1

A - economic indicators
B - social indicators
C - ecological indicators
0 - no significant impact compared with starting point
1 - slight impact

2 - significant impact
+ - positive impact
- - negative impact
-/+ - effects vary over time
± - positive and negative impacts/overall effect unclear or contextual ambiguous

28.6 Conclusions and Recommendations

Multifunctional dams can play an important role in terms of sustainable development on a river basin level. This paper aimed at an assessment whether such dams may add to improve the situation of flood protection while keeping multi-functionality as a main principle of dam management. SIA has been used as an instrument for relevant scenario analysis. As a result of the preliminary assessment a baseline scenario which will later merge into an advanced European scenario turns out to be of high relevance for the sustainability of multifunctional/multipurpose dams in a trans-national/ shared river basin. Environmental Assessment is a key process and methodology for sustainable dam management. Together with a rather qualitative SIA a wider picture can be drawn, thus positively rendering environmental impacts and social acceptance towards the function of dams in a complex river basin situation. Here, the Water Framework Directive together with the coming Flood Risk Management Directive will form an European basis for further enhancing international co-operation and standard setting mainly with a view on trans-national river basins.

References

Council of the European Union (2006) Proposal for a Directive of the European Parliament and the Council of the Assessment and Management of Flood Risks (Doc 11027/07), Brussels 2006

ICPE – International Commission for the Protection of the Elbe River (2006) Erster Bericht über die Erfüllung des „Aktionsplans Hochwasserschutz Elbe im Zeitraum 2003 bis 2005", Magdeburg

ICPE (2005) Die Elbe und ihr Einzugsgebiet. Ein geographisch-hydrologischer und wasserwirtschaftlicher Überblick, Magdeburg, August 2005

Kašpárek L, Novický O, Jeníček M, Buchtela S (2006) Influence of large reservoirs in the Elbe River Basin on reduction of floods, Ministry of the Environment of the Czech Republic, Masaryk Water Research Institute, ICPE Documents Prague March 2006

Kirkpatrik C, Lee N, Morrissey O (1999) (Institute for Development Policy and Management, Univ. Manchester) WTO New Round - Sustainability Impact Assessment Study, Phase One Report, Manchester, Nottingham 1999, see also http://trade.ec.europa.eu/doclib/docs/2005/february/tradoc_112353.pdf, last accessed on 10.05.2007

Ministry of Agriculture of the Czech Republik (2005) The Water Management Information Portal, 2005, http://www.voda.mze.cz/en/, last accessed on 23.04.2007

Saxonian State Department for Environment and Geology (2005) State Flood Centre 2005. http://www.umwelt.sachsen.de/lfug/wasser_hwz.html, last accessed on 23.04.2007

Sieber H-U, Glasebach H-J (2000) The importance of dams for infrastructure and landscape conservation in Saxony", International Commission on Large Dams, XX. Congress, Question 77 / Report 16, Beijing, http://www.talsperrenkomitee.de/german_research/index.cgi/page/article/article_id/5, last accessed on 23.04.2007

Sieber H-U (2003) Talsperren als multifunktionale Anlagen – Konkurrenz zwischen Rohwasserbereitstellung und Hochwasserschutz am Beispiel der sächsischen Trinkwassertalsperren. Vortrag zur ATV-DVWK Tagung am 16.10.2003

Socher M (2000a) Das Konzept des Sustainability Impact Assessment der Europäischen Kommission - mehr als ein neuer methodischer Ansatz?" TA Datenbank Nachrichten 1 (9), FZK, ITAS Karlsruhe avaliable under: http://www.itas.fzk.de/deu/tadn/tadn001/soch00a.htm last accessed on 23.04.2007

Socher M (2000b) Sustainability Impact Assessment – Erste Erfahrungen und ein Ausblick zur Nachhaltigkeitsprüfung" TA Datenbank Nachrichten 2(9), FZK, ITAS Karlsruhe http://www.itas.fzk.de/deu/tadn/tadn002/soch00a.htm, last accessed on 23.04.2007

Socher M (2001) Brennstoffzellen auf dem Prüfstand der Nachhaltigkeit. Energiewirtschaftliche Tagesfragen, 51(1/2)

Socher M (2002) Nachhaltigkeitsprüfung von Wasserkraftanlagen – Erste Ergebnisse für Sächsische Gewässer. Energiewirtschaftliche Tagesfragen, 52(4)

Part IV – Emerging Issues

Impact assessment must be an adaptive and evolving process, in response to both the development of policy and techniques, and to trends in and knowledge of emerging types of development and of environmental receptors. In this context, Section IV seeks to provide at least a partial insight to some emerging issues.

Chapter 29 (Amoah and Ertel) introduces, respectively, the increasingly significant world of nanotechnology. It has the potential to generate important impacts on the physical and socio-economic environments. Whilst there is some evidence of the introduction of standards and thresholds, the current coverage by EIA processes is shown to be only marginal at best.

Chapter 30 (Kratz) provides a less conventional perspective on the impact of the pharmaceuticals industry, by considering the serious ecotoxicological risks to human health from human pharmaceuticals, via their transmission through the waste/drinking water systems. Chapter 31 (Stepniewski) also takes a different perspective on the pollution of the media of water, air and soil, in terms of environmental oxygenology, and the impact of dioxygen and trioxygen concentrations.

Finally, Chapter 32 (Helbron and Schmidt) provides an insight to evolving methodology and impact assessment processes, highlighting the development of quantitative threshold values for Strategic Environmental Assessment, with a particular focus on land consumption and soil sealing in a case study in Saxony in Germany. The chapter clearly reveals the valuable potential of a structured and quantitative approach to SEA, as a means of better balancing development and environment at the strategic level

29 Environmental Impact of Nano Technology on Human Health

Amoah Benedicta and Jürgen Ertel

Department of Industrial Sustainability, Brandenburg University of Technology (BTU), Cottbus, Germany

29.1 Introduction

Technological deployment has aided humankind immeasurably in raising living standards, controlling disease, and expanding our resource base. Successful use of new and efficient technologies is an important component of most future sustainability strategies. As such, research aims to develop innovative compounds and products that can significantly enhance product performances and inevitably the quality of life for consumers. This is what research into Nanotechnology hopes to achieve.

Currently there is no precise definition for nanotechnology; however, one appropriate definition is that of the National Research Council which defines nanotechnology as research and technology development at the atomic, molecular, or macromolecular levels, in the length of approximately 1-100 nm (nanometre) range, to provide a fundamental understanding of phenomena and materials at the nano-scale, and to create and use structures, devices, and systems that have novel properties and functions because of their small size. A nanometre is the current standard unit of length in the metric system equal to one billionth of a metre with symbol being nm.

The novel and differentiating properties and functions are developed at a critical length scale of matter, typically under 100 nm (National Research Council 2002). In theory nanoparticles can be produced from nearly any chemical; however, most nanoparticles that are currently in use today have been made from transition metals, silicon, carbon (single-walled carbon nanotubes; fullerenes), and metal oxides (zinc dioxide and titanium dioxide). In many cases engineered nanoparticles exist as nano crystals composed of a number of compounds such as silicon and metals (quantum dots) (Murray et al. 2000).

Although the world is striving to achieve sustainable development and nanotechnology is a promising mode of reaching this target, the world cannot wait

Standards and Thresholds for Impact Assessment. Edited by Michael Schmidt, John Glasson, Lars Emmelin and Hendrike Helbron. © 2008 Springer-Verlag

until it is hit by a crisis before appropriate action is taken. The health and environmental impacts that arise from the deployment of new technological innovations such as nanotechnology have to be taken into consideration since there are no specific regulations, standards or thresholds governing this technology.

29.2 Research Approach

A systematic analysis and evaluation of nanotechnology was carried out based on literature sources (journals, articles and books) to establish the state of the existing logistical systems with regard to materials and processes, applications and environmental benefits and associated health problems. In understanding the existing relationship between nanotechnology and the environment, a systems analysis approach was employed. Associated health problems are reviewed and a proposed environmental impact assessment option is given.

29.3 Application of Nanotechnology

The transition of nanotechnology research into manufactured products is limited but some products have moved relatively quickly into the market place and already are having significant impact. According to Theodore et al. (2005) examples of nanotechnology that are in actual commercial use, under serious investigation or on the verge of commercialisation include:

- Semiconductor chips and other microelectronics applications
- High surface-to-volume catalysts, which promote chemical reactions more efficiently and selectively
- Ceramics, lighter-weight alloys, metal oxides, and other metallic compounds
- Coatings, paints, plastics, fillers, and food-packaging applications
- Polymer-composite materials, including tyres, with improved mechanical properties
- Transparent composite materials, such as sunscreens containing nano size titanium dioxide and zinc oxide particles
- Use in fuel cells, battery electrodes, communications applications, photographic film developing, and gas sensors
- Nano barcodes
- Tips for scanning probe microscopes
- Purification of pharmaceuticals and enzymes
- Diagnostic and drug delivery systems, including specific targeting of cancer cells (not yet commercialised)
- Sensing of pollutants, pH, and chemical warfare agents (not yet commercialised)

- Ultraviolet light (UV)-activated catalysts for treatment of environmental contaminants (not yet commercialised)
- Removal of environmental contaminants from various media, including in situ remediation of pesticides, polychlorinated biphenyls (PCBs), and chlorinated organic solvents, such as trichloroethylene (TCE) (not yet commercialised)
- Post treatment of contaminated soils, sediments, and solid wastes (not yet commercialised)
- Absorption of contaminants for air and water pollution control, in a manner said to be vastly superior to activated carbon (not yet commercialised)
- Chelating agents for polymer-supported ultra filtration (not yet commercialised)
- Oil-water separation (not yet commercialised)
- Destruction of bacteria (including anthrax) (not yet commercialised)
- Purification of drinking water, without the need for chlorination (not yet commercialised)

29.4 Impact of Nanotechnology on Human Health

Human contact with nano-materials can take various forms such as inhalation, ingestion and absorption through the skin. These have the potential to cause damage to the human body in different ways. In many cases engineered nano-particles are believed to have negative health impact. According to Maynard et al. (2003) the manufactured nano-particle called single-wall carbon nano-tubes (SWCNTs) have negative health impact on humans and he observed that the bio-persistence of SWCNTs may be a significant occupational safety issue since chronic exposures to low levels of SWCNTs could be associated with severe adverse health effects to the respiratory tract. Oxidative stress on keratinocytes and defensive cells in humans was observed by Shvedova et al. (2003). Researchers from NASA/Johnson Space Center reported that studies on effects of nanotubes on the lungs of rats produced a toxic response (Hogan 2003). Auto-immune diseases were found to result from wear debris generated by orthopaedic implants and it was observed that patients with such implants have a statistically significant rise in the incidence of auto-immune diseases (Akisue 2002).

Cadmium selenide nano-particles (quantum dots) have the ability to break down in the human body, potentially causing cadmium poisoning (Mullins 2004). Industrial workers who breathe particulate matter (i.e. silica dust) develop fibrosis in their lungs, and other respiratory problems leading to tissue damage (Maloney 1998). Another study also indicated that gold nano-particles can move across the placenta from mother to foetus (Wootliff 2004). Titanium dioxide/zinc oxide nano-particles from sunscreen are found to cause free radicals in skin cells, damaging DNA (Salinaro et al. 1997).

A study conducted on the largemouth bass (fish) exposed to water containing buckyballs at concentrations of 500 parts per billion (the concentration level is comparable to pollutant levels commonly found in port waters) identified "severe" damage to brain tissue in the form of "lipid per oxidation," leading to the destruction of cell membranes, after only 48 hours of exposure (Oberdörster et al. 2005).

Nano particles are very different chemically from the original parent material and they have very different properties that might make them more harmful, therefore in-depth research should be conducted on them. Since sufficient evidence is not yet available with reference to the behaviour of nano-particles in the environment, caution should be taken to ensure that nano-particles are not released into the environment. Current experimental evidence is insufficient to reach conclusions about the effect of nano-particles on human health, however it is recommended that modification should be made on the nano-particles that are being used such as coating them to ensure that they are safely biocompatible.

The international community needs to formulate a legally binding mechanism to govern the products of new technologies in order to minimize early detrimental environmental effects. Past industrial failures associated with bio-uptake and accumulation had grave costly environmental consequences as was the case with the introduction of earlier chemical substances (such as DDT), therefore it is essential that much research should be conducted. Technology should focus on particle mediation transport, potential for bio-assimilation, adsorption, biological uptake, material transformations and the fate of nano-structures in all mediums.

A central (worldwide) research organization should be set up and all engaging in research and manufacture of nano-particles should automatically become members. This organization should have a central database accessible to all, in order to serve as a watchdog to monitor nanotechnology developments and assess potential risks associated with the utilization of this technology. This central research organization should work in collaboration with research grant councils to establish a grant program to promote research in this area. A standard nomenclature should be developed and specific legislation particularly concerning this technology must be introduced.

Since the manufacture of nano-materials is not widespread and has no standard nomenclature it is essential that researchers indicate in their proposals which nano-materials would be used and where they will be obtained. Industry, researchers, environmental organizations, politicians and government, should work together on the risk evaluation of this technology. Environmental impact assessment should be applied on products manufactured by nano-technology because most of the chemicals used during production are not inert. As such a safety assessment is vital for the whole life cycle of the product and there is the need to comply with general legal requirements for the individual chemicals used during production.

In the meantime health and environmental risk can be controlled with laws such as those relating to general occupational health and safety, as well as to consumer goods. Other countries can follow the example of Germany where the following regulations are equally important: the EU novel food regulation, food and preservative legislation, medical product legislation, pharmacy legislation, labour legislation, chemistry legislation and air pollution regulations. Other regulations that concern the individual chemical substances should be adhered to (e.g. Environmental Protection Agency laws, occupational health and safety agency laws and the toxic substance control act). Table 29.1 summarises potential impacts and sensitivity groups.

Table 29.1. Potential impact with corresponding sensitivity groups

Sensitivity group	Potential impacts
Neonate = 0 to 1 month	Birth defects, Auto-immune diseases
Pediatric = 1 month to 12 years	Auto-immune diseases, Detrimental effects to respiratory tract, Metal poisoning and tissue damage
Adolescent = 13 to 18 years	Detrimental effects to respiratory tract, toxic response, Auto-immune diseases Metal poisoning and tissue damage,
Adult = 19 to 64 years	Toxic response, Auto-immune diseases, Detrimental effects to respiratory tract, Metal poisoning and tissue damage
Geriatric = 65+ years	Toxic response, Auto-immune diseases, Detrimental effects to respiratory tract, Metal poisoning and tissue damage

29.5 Technical Suitability of Nanotechnology

The suitability of nanotechnology is shown in figure 29.1 by means of a casual loop diagram. The suitability of this technology depends on the intrinsic properties of its particle size and the application of the technology. The nano-particle size and high efficiency of material use and recovery promote the use of this technology since this conserves resources and leads to sustainability. The efficiency with which the technology is handled relies on the availability of trained personnel and the behaviour of emitted nano-particles. The behaviour of emitted nano-particles promotes the possibility of substitution of environmentally harmful substances.

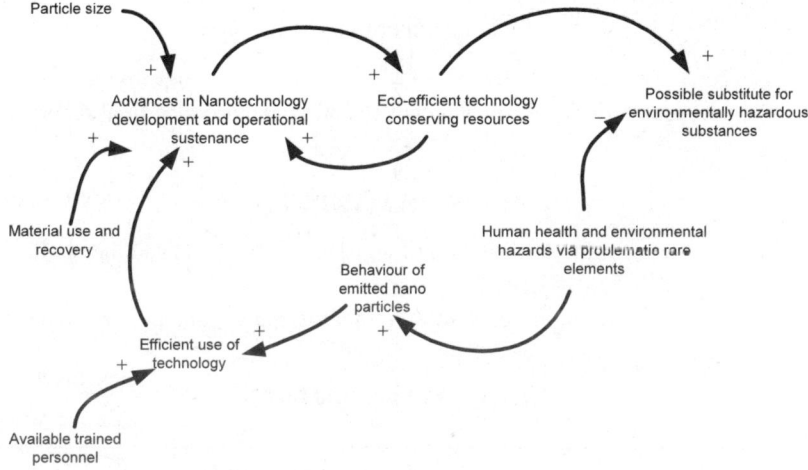

Fig. 29.1. Casual loop diagram for the suitability of nanotechnology

29.6 The Need for Environmental Impact Assessment

Environmental impact assessment (EIA) is a process, having the ultimate objective of providing decision makers with an indication of the likely environmental consequences of a proposed activity. This refers to the evaluation of the effects likely to arise from a major project significantly affecting the natural and man-made environment and supplies decision makers with an indication of the likely consequence of their action. It also aids the identification of the likely effects at an early stage thus improving the quality of both the project planning and decision making (Wood 1996). Fig. 29.2 illustrates how regulation can come into force in line with the EIA process. The above serves as a driving force for the need for environmental impact assessment in nanotechnologies- relating to economy, quality, costs, time, environmental safety, health impacts, technology innovation, re sources, patents, politics, and social acceptance. The potential environmental impacts (for nanotechnology and chemical processes) can be calculated by means of the potential environmental impact equation (PEI) (Diniz et al. 2003). Here the most attractive alternatives with regard to environmental protection are arrived at.

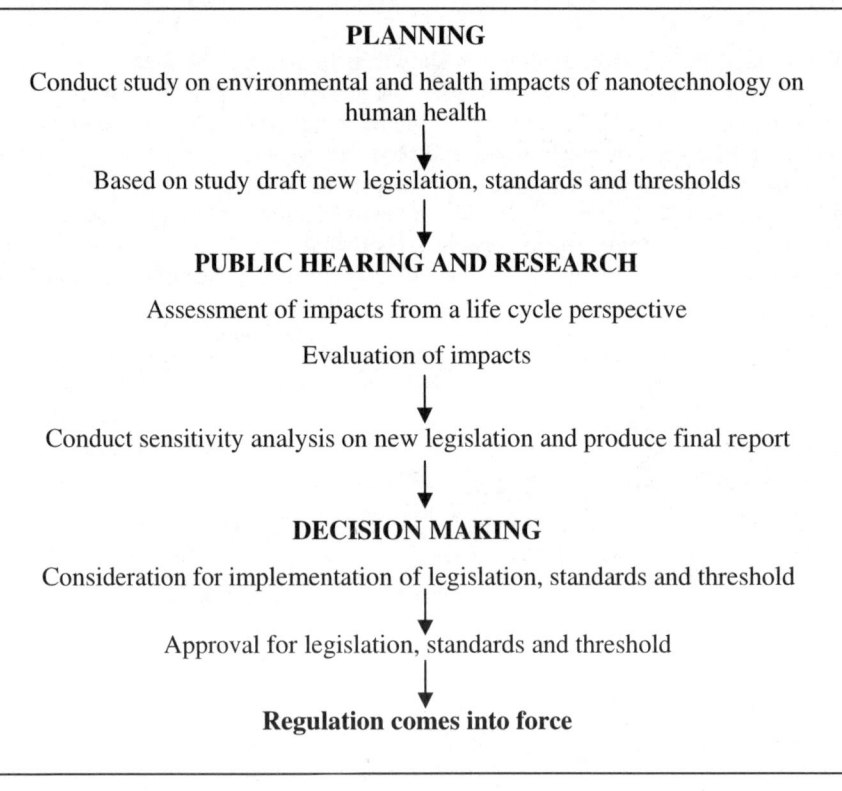

Fig. 29.2. Pathway towards nanotechnology legislation through EIA

(PEI) = Potential Environmental Impact

(PEI) i = Σ (Mass Indices of Component j) * (Impact categories) *or*

(PEI) i = Σ (Weighting Factors of Component j) * (Impact category)

Further work in EIA of chemical processes with regards to nanotechnology should seek to integrate economic and safety aspects to find the best compromise solution.

29.7 Conclusions and Recommendations

Environmental impact assessment should be applied on nanotechnology products since nano-particles are very different chemically from the original parent material and they have very diverse properties that might make them more harmful. Since sufficient evidence is not yet available with reference to the behaviour of nano-particles in the environment, caution should be taken to ensure that nano-particles are not released into the environment. Current experimental evidence is insufficient to reach conclusions about the effect of nano-particles on human health, however it is recommended that modification should be made on the nano-particles that are being used such as coating them to ensure that they are safely biocompatible.

The international community must formulate a legally-binding mechanism to govern the products of new technologies in order to minimize early detrimental environmental effects. Past industrial failures associated with bio-uptake and accumulation had grave and costly environmental consequences as was the case with the introduction of earlier chemical substances (such as DDT) therefore it is essential that much research should be conducted on this technology and should focus on particle mediation transport, potential for bio-assimilation, adsorption, biological uptake, material transformations and fate of nano-structures in all mediums.

A central research organization should be set up and all engaging in research and manufacture of nano-particles should automatically become members. This organization should have central database accessible to all, in order to serve as a watchdog to monitor nanotechnology developments and assess potential risks associated with the utilization of this technology. It should work in collaboration with grant councils to establish a grant program to promote research in this area.

A standard nomenclature should be developed and specific legislation introduced. Since the manufacture of nano-materials is not widespread and has no standard nomenclature it is essential that researchers indicate in their proposals which nano-materials would be used and where they will be obtained.

Industry, researchers, environmental organizations, politicians and government, should work together on the risk evaluation of this technology. Environmental impact assessment should be applied on products manufactured by nanotechnology because most of the chemicals used during production are not inert, therefore a safety assessment is vital along for the life cycle of the product and there is the need to comply with general legal requirements for the individual chemicals used during production. In the meantime health and environmental risk can be controlled with existing laws such as those relating to health and safety at work, consumer goods and medication. Other regulations that concern the individual chemical substances should be adhered to e.g. Environmental protection Agency laws, occupational health and safety agency laws and the toxic substance control act.

References

Akisue T (2002) Journal of biomedical materials research (2002) 59(3):507

Diniz da Costa, JC, RJ Pagan (2006) Sustainability metrics for coal power generation in Australia. Journal of process safety and environmental protection 84(B2):143–149

Hogan J (2003) How safe is nanotech? Special Report on Nano Pollution, New Scientist, vol 177, no 2388:14

Maloney WJ (1998) Journal of biomedical materials research 41(3):371

Mullins J (2004) Safety concerns over injectable quantum dots, New Scientist, vol 181, no 2436:10

Murray CB, Kagan, CR, Bawendi MG (2000) Synthesis and characterization of monodisperse nanocrystals and close-packed nanocrystal assemblies. Annu Rev Mater Sci 30:545-610

National Research Council (2002) Small Wonders, Endless Frontiers: A Review of the National Nanotechnology Initiative. Washington, DC, National Academy Press, p. 12

Oberdorster G, Sharp Z, Elder AP (2005) Translocation of inhaled ultrafine particles to the brain. Inhal Toxicol 2004, 16:437–45

Salinaro A, Dunford R, Cai L, Serpone N, Horikoshi S, Hidaka H, Knowland J (1997) Chemical oxidation and DNA damage catalysed by inorganic sunscreen ingredients. FEBS Letters, vol 418, no 1-2, 24:87-90

Sampson MT (2004) Type of buckyball shown to cause brain damage in fish. Eurekalert, Internet address: www.eurekalert.org, last accessed on 10.05.2007

Shvedova AA, Murray AR, Kisin ER, Schwegler-Berry D, Kagan VE, Gandelsman VZ, Castranova V (2003) Exposure to Carbon Nanotube Material: Evidence of Exposure-Induced Oxidant Stress in Human Keratinocyte and Bronchial Epithelial Cells. Free Radical Research 37:97

Theodore L, Kunz RG (2005) Nanotechnology: Environmental Implications and Solutions. John Wiley & Sons

Wood (1996) Environmental Impact Assessment: A Comparative Review. Addison-Wesley Publishing, Harlow

Wootliff B (2004) Nanoparticles Might Move from Mom to Foetus, Small Times, Internet address: www.smalltimes.com, last accessed on 10.05.2007

30 Ecotoxicological Risk of Human Pharmaceuticals in Brandenburg Surface Waters?

Werner Kratz

Department of Ecology, Nature Conservation and Water Economics, unit Ö3, Environmental Survey and Ecotoxicology, Brandenburg State Office for Environment (LUA), Potsdam, Germany

30.1 Introduction

According to the definition of the German medicine law (AMG 2001), pharmaceuticals are, amongst other things, intended to heal, to mitigate or prevent in their application in the human or animal body, diseases, suffering, body damage or pathological burdens. The undisputed positive purpose of pharmaceuticals and the question of cost budgeting has, in the recent past, limited the room to critically consider, from an environmental perspective, the possible unwanted side effects arising from medicament consumption in Germany. In the German Federal state in 1999 alone the five most important and available pain and rheumatism drugs had approximately 39 tonnes of active substances which, in pharmaceuticals for humans, contained 2.1 tonnes of active antibiotic substances (Abbas and Kratz 2000; BLAC 2001). These biologically high-activity materials are partly converted after intake in the human body and separated afterwards into original and converted forms (BLAC 2001). Within the sewage waters, these residues arrive at the sewage treatment plant, at which point they are often undesirably, incompletely mitigated or held back in surface waters (BLAC 2001).

Many unwanted effects of medicament arrears in surface waters are transferable: eg. the emergence and transmission of resistance of genes to antibiotics, the impairment of the microbiology though disinfectants, a variation in the life rhythms of organisms as well as the food web relations by anaesthetics, the reduction of the fertility and a change of sex conditions by hormones as well as hereditary good-changing and reproduction-toxic effects by cytostatic drugs (BLAC 2001)). The questions of whether and to what extent medicine residues can cause such effects in Brandenburg's surfaces waters, are addressed in the following sections, where the individual stages of the risk assessment are described and applied

Standards and Thresholds for Impact Assessment. Edited by Michael Schmidt, John Glasson, Lars Emmelin and Hendrike Helbron. © 2008 Springer-Verlag

with regards to the active substances. At the same time it must be stated here that the data on the ecotoxiclogical effect of many substances is insufficient and the sparse information that does exist on the environmental behaviour of many of the medical active agents refers only to the first affect (BLAC 2001). Although the EMEA (2003) has developed guidance on environmental risk assessment for pharmaceuticals, much more ecotoxicological effects data is needed, especially in the context of the implementation of the EU Water Framework Directive over the next few years.

30.2 For Which Pharmaceutical Active Substances Can a Risk Assessment be Accomplished?

Altogether 58 active agents and two substances are identified, on the basis of the following criteria:

- Consumption quantity, i.e. greater than 250 kg of active agent consumption per year in the State of Brandenburg (evaluation for 1999 through the LUA Brandenburg)
- General active agents of specific effect such as antibiotics, hormones, cytostatic drugs
- Active agents , which are not contained in the first two groups, but were proven in the meantime in the Brandenburg running waters monitoring program by the Federation/Land Committee for Chemical Security (BLAC, WG Pharmaceuticals in the Environment)

30.3 Which Concentrations are to be Expected from these Active Substances in Brandenburg's Surface Waters and Which are Already Proven?

The expected concentration of the pharmaceutical active substances in the surface waters of Brandenburg (Predicted Environmental Concentration (PEC BRB)), was determined with reference to the following influential factors:

- the annual consumption quantity of the pharmaceutical in Brandenburg
- the active agent share eliminated unchanged by the person
- a sewage share of 150 litres per head and day
- the active agent share held back in the sewage treatment plant
- a supposed 10-fold thinning by the sewage treatment plants flow to the surface waters

The assumption was made of a local and temporal uniform distribution of the active substances in waters, and the analytically determined active substances (the maximum measured concentration determined in each case) was put as environmental concentration (PEC BRB). A composition of the 30 active substances with a PEC BRB of 0.1 µg/l is presented in figure 30.1. For the eight substances marked in grey the maximum concentrations were analytically determined.

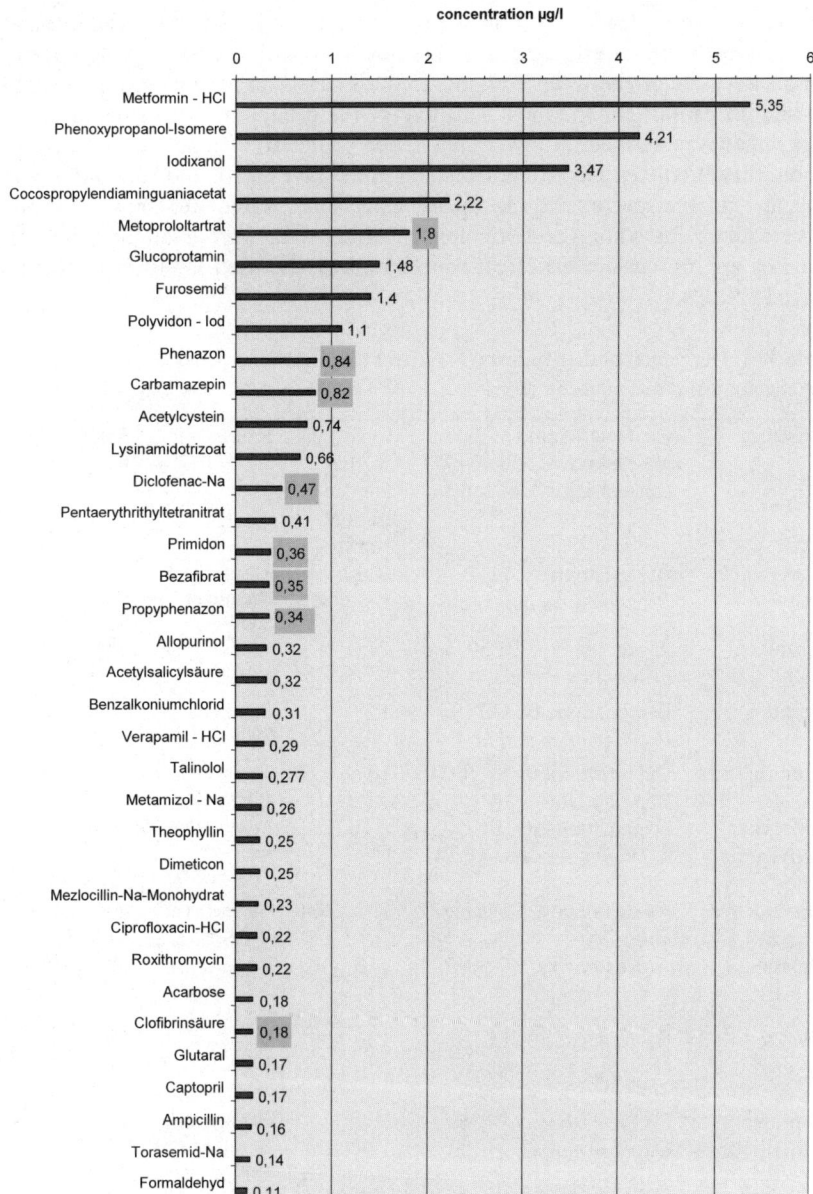

Fig. 30.1. Pharmaceutical Substances with a Predicted Environmental Concentration in Brandenburg (PEC BRB)> 0.1 µg/l

30.4 Which Active Agent Concentrations Adversely Affect the Aquatic Ecosystem?

For each active substance the effects on aquatic organisms (algae, crustacea, fish and bacteria) were investigated. For the most compatible active agent concentration a safety factor (between 100 and 25 000) based on the many imponderables and simplifications for the most sensitive of the groups of organisms in the surfaces water was derived. A lower step of the so determined active agent concentration, that Predicted No Effect Concentration (PNEC), should rule out any risk of danger for the species and the aquatic ecosystem. Table 30.1 shows for the 40 active agents, for whom ecotoxicological effect data were available, the corresponding appropriate lowest effect concentrations, the used safety factors and the derived PNECs.

Table 30.1. Derivation of the Predicted No Effect Concentrations (PNEC) from the ecotoxicological effect concentrations

Substance	Characterization of the lowest known effective concentration	lowest effective concentration [µg/l]	Source	Safety Factor	PNEC [µg/l]
Acetylsalicylacid	Bacteriatoxicity, EC 0, *Pseudomonas putida*	8000	Merck KgaA Darmstadt 2000	200	40
Benzalkoniumchloride	Algae toxicity, EC 50, *Scenedesmus subspicatus*	24	Gartiser et al. 1999	200	0,12
Ciprofloxacin-HCl	Bacteriatoxicity, EC 10, *Pseudomonas putida*	1,8	Bayer AG 2001	100	0,018
Clofibrinacid	Daphnia toxicity, NOEC Reproduction,	10	Ternes et al. 1999	100	0,1
Dodecylbis-propylentriamine	Daphnia toxicity, EC 50, 24 h, *Daphnia magna*	< 1000	Lonza AG Basel 1997	200	< 5
Formaldehyde	Algae toxicity, *Scenedesmus*	300 - 500	Gartiser et al. 1999	200	1,5
Glutaral	Algae toxicity, EC 50, 72 h	610	Dr. Theodor Schuchardt & Co 1999	200	3,05
Glyoxal	Bacteriatoxicity, EC 50 Growth, *Pseudomonas putida*	133 700	Gartiser et al. 1999	200	668
Ibuprofen	Daphnia toxicity, NOEC, *Daphnia magna*	3000	Halling-Sørensen et al. 1997	100	30
Iodixanol	Daphnia toxicity, EC 50	> 2500 000	Nycomed A Buchler GmbH & Co KG 2001	200	> 12500

Table 30.1. (cont.)

Paracetamol	Daphnia toxicity, EC 50, 48 h, *Daphnia magna*	9200	Halling Sørensen et al. 1997)	200	46
Acarbose	Daphnia toxicity, EC 0, *Daphnia magna*	> 1000 000	Bayer AG 1999	500	> 2000
Cyclophos-phamide	Fish toxicity, NOEC 96 h, *Salmo gairdneri*	> 984 000	Stuer-Lauridsen et al. 2000	500	>1970
Diclofenac-Na	Daphnia toxicity, EC 0, 24 h, *Daphnia magna*	18 000	BIOCHEMIE Ges. mbH. Kundl 1999	500	36
Ethinylestra-diol	Fish toxicity, LOEC Plasmavitellogenin 10 d, 16,5 °C, Rain bow trout	0,0001	Kümmerer 2001	500	0,000 0002
Glucoprotami-ne	Daphnia toxicity, EC 50	500	Gartiser et al. 1999	1 000	0,5
Ifosfamid	Daphnia toxicity, NOEC 48 h, *Daphnia magna*	100 000	Schecker et al. 1998	500	200
Lidocain-HCl	Fish toxicity, LC 50, 96 h, *Zebra danio*	106 000	Astra Zeneca GmbH 2000a	1 000	106
Mezlocillin-Na-Monohydrate	Daphnia toxicity, *Daphnia magna*	20 000	Bayer AG 2001	1 000	20
Pentaerythri-thyltetranitrate	Bacteriatoxicity, EC 50, 30 min, *Vibrio fischeri*	14 500	Rippen 1990	1 000	14,5
Prilocain-HCl	Daphnia toxicity, EC 50, 48 h, *Daphnia magna*	61 000	Astra Zeneca GmbH 2000b	1 000	61
Verapamil-HCl	Fish toxicity, LC 0, 48 h,	4 600	Knoll GmbH Ludwigshafen 1997	1 000	4,6
5-Fluorouracil	Algae toxicity, EC 50, Growth, *Desmodesmus subspicatus*	21 300	Hanisch et al. 2002	5 000	4,3
Carbamazepin	Algae toxicity, EC 50, *Desmodesmus subspicatus*	85 000	Hanisch et al. 2002	5 000	17
Furosemid	Fish toxicity, LC 50, 96 h,	> 500 000	Hoechst Marion Roussel 1999a	5 000	> 100
Laurylpro-pylendiamine	Fish toxicity, LC 50, 96 h, *Zebra danio*	100	Hoechst Marion Roussel 1999a	5 000	0,02
Metamizol-Na	Fish toxicity, LC 50, 96 h,	> 500 000	Clariant GmbH Frankfurt M.2000	5 000	> 100
Metformin-HCl	Daphnia toxicity, EC 50, 48 h, *Daphnia magna*.	60 000	Hoechst Marion Roussel 1999b	5 000	12
Naproxen	Daphnia toxicity, EC 50, 24 h, *Daphnia magna*.	140 000	Kümmerer 2001	5 000	28

Table 30.1. (cont.)

Norethis-teron	Daphnia toxicity, EC 50, 48 h, *Daphnia magna*	> 4600	Schering AG Berlin 2000	5 000	> 0,92
Pentoxifyllin	Fish toxicity, LC 50, 96 h, *Zebra danio*	100 000	Aventis Pharma 2000	5 000	20
Polyvidon-Iod	Fish toxicity, NOEC, *Leuciscus idus*	4 600	BASF AG 2001	2 500	1, 84
Propyphena-zon	Fish toxicity, LC 50, 96 h, *Zebra danio*	220 000	Hoechst Marion Roussel 1999c	5 000	44
Clarithromy-cin	Bacteriatoxicity, EC 50, Growth, *Enterecoccus faecalis*	151	Alexy et al. 2001	25 000	0,006
Cocospropy-lendiamin-guaniacetat	Fish toxicity, LC 50, 96 h, *Zebra danio*	< 1 000	Dr. Schuhmacher GmbH 1995	25 000	< 0,04
Naftidrofu-rylhydroge-noxalat	Daphnia toxicity, EC 50, 48 h	20 000	Merck Lipha s.a. France 1997	25 000	0,8
Phenazon	Fish toxicity, LC 50, 96 h	> 500 000	Kraemer & Martin Pharma Handels GmbH 1998	25 000	20
Propranolol-HCl	Daphnia toxicity, LC 50, *Daphnia magna*	2 700	Kümmerer 2001	25 000	0,108
Roxithromy-cin	Fish toxicity, LC 50, 96 h	> 100 000	Hoechst Marion Roussel 1999d	25 000	4
Theophyllin	Daphnia toxicity, EC 50, 24 h	155 000	Kümmerer 2001	25 000	6,2

[a] The grey gradation of the column ranges within the table refers to the different ecotoxicological database concerning the trophical level in the aquatic ecosystem. The transition from white to the darkest grey tone corresponds to the availability of ecotoxicological effects data between four and one trophical levels.

One observation found is that the active agent groups with the highest toxicity to water organisms (PNEC <1 µg/l), are the sexual hormone Ethinylestradiol, the antibiotic Ciprofloxacin-HCl, the blood fat reducermetabolit Clofibrinacid and the disinfectant active agent Benzalkoniumchloride and Glucoprotamin. The Daphnia represents, with the Benzalkoniumchloride, the algae and with the Ciprofloxacin HCl, the bacteria, the most sensitive organisms in the aquatic system in the cases of the Clofibrinacid and Glucoprotamin.

There must however be an assumption of the possible strong underestimatimation of the data of the actual ecotoxicity resulting from the usual tests to acute fish, Daphnia- and algae toxicity, because chronic effects are almost not considered. In the second group with a PNEC between 1 µg/l and 100 µg/l the analgesic Acetylsalicylacid, Paracetamol, and Ibuprofen and Diclofenac-Na, the disinfectant active agents formaldehyde, Dodecylbispro-pylendiamin, Glutaral and the localanästeti-cum Prilocain-HCl are found. In addition Mezlocillin-Na Monohydrate, through hemorrhage means Pentaerythrityl tetranitrate the antihypertonikum Verapamil HCl belong to this antibiotic group.

The majority of the most sensitive organisms in the aquatic ecological system seem to be in the Daphnia. Only acetylsalicylic acid and Pentaerythrithyltetranitrat (lowest effective concentration for bacteria) as well as formaldehyde and Glutaral (algae) and Verapamil HCl (fish) deviate from this trend. For the cytostatic drugs belonging to the third group (PNEC >100 µg/l) Ifosfamid and cyclophosphamide, the X-raycontrast media Iodixanol and Antdiabeticum the Acarbose must be pro-ceeded according to the past data situation from a rather small (acute) aquatic tox-icity. The same applies to localanaestetikum   Lidocain-HCl and the disinfectant Glyoxal. The cytostatic drugs must however be possibly counted on a smaller PNEC, if with humans and with the mammal proven chronic-toxic effects (cancze-rogenity, mutagenicity, reproduction toxicity), they should play a role also in the aquatic ecological system.

30.5 For Which Pharmaceutical Active Substance Does an Ecotoxicological Risk Exist?

Due to ecotoxicological effects, the environmental risk is to be accepted for the pharmaceuticals, with which the measured Environmental Concentrations in Brandenburg (PEC BRB) are in the range or in excess of the No Effect Concen-trations (PNEC) (with consideration of safety factors determined). The active sub-stances, to which this criterion applies, are perceptible in figures 30.2 and 30.3 by those bars, to the right of PEC BRB: PNEC-ratio of 1 representing lines stand. Furthermore probably also the active agents with a PEC-BRB must be considered because of the existing imponderables: PNEC-ratio between 1.0 and 0.1.

The pharmaceutical active substances, for which an environmental risk for the aquatic ecological system can be accepted based on proven effects, include the sexual hormone Ethinylestradiol (the active pill substance), the antibiotics Cipro-floxacin and Clarithromycin, the Antidiabeticum Metformin as well as the conver-sion product of some blood fat countering agents, Clofibrinacid and six disinfec-tants. In addition other examined active agents were tested to establish criteria for possible environment relevance. These criteria were:

- a high biotic accumulation ability
- a heavy biological degradability and therewith connected high persistence
- a small purification effect though sewage plants, as well as
- in human or animals, connected carcinogenic behaviour, hereditary, reproduc-tion toxicity or the hormone system influencing substances

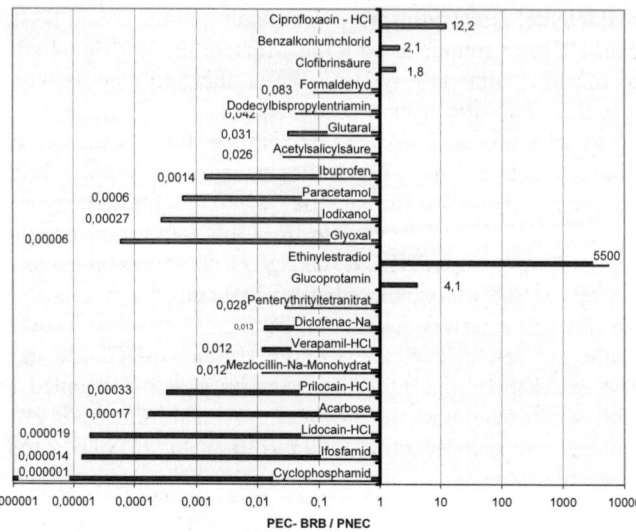

Fig. 30.2. PEC/PNEC relation (impact data for 2 and 3 trophical levels)

Table 30.2. Substances that fall directly outside criteria of environmental relevance

Substance	Substance Group	Number of trophical levels with known eco-toxicoloical effect data
Bezafibrat	Blood fat reducer	0
Fenofibrinacid	Blood fat reducer (metabolite)	0
Indometacin	antirheumaticum	0
Levonorgestrel	Sexual hormone	0
Lysinamidotrizoat	X-ray contrast media	0
Medroxyprogesteron	Sexual hormone	0
Primidon	Antiepileptic	0
Naftidrofurylhydrogenoxalat	Blood penetration compound	1
Carbamazepin	Antiepileptic	2
5-Fluoruoracil	Cytostaticum	2
Metamizol-Na	Analgetic	2
Naproxen	Analgetic	2
Norethisteron	Sexual hormone	2
Propyphenazon	Analgetic	2
Cyclophosphamide	Cytostaticum	3
Ifosfamid	Cytostaticum	3
Lidocain-HCl	Localanesteticum	3
Pentaerythrityltetranitrate	Coronar compound	3
Prilocain-HCl	Localanesteticum	3
Verapamil-HCl	Antihypertonicum	3
Iodixanol	X-ray contrast media	4

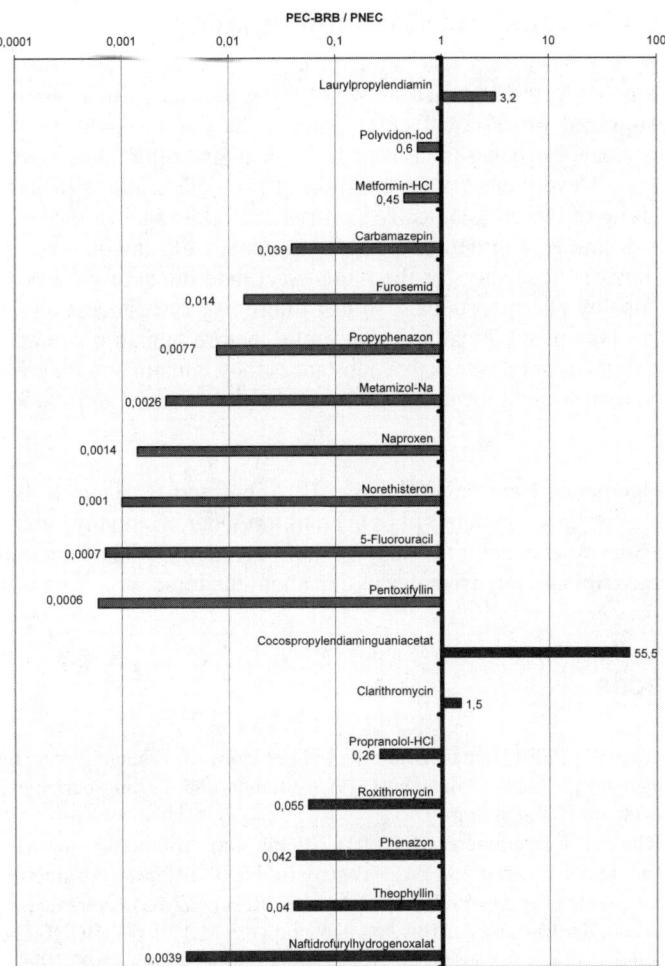

PEC-BRB / PNEC

Fig. 30.3. PEC/PNEC relation (impact data for 1 and 2 trophical levels)

Active agents, that fulfil at least two of the characteristics mentioned, should be likewise regarded despite the absence of an effect presently for water organisms and for the aquatic ecosystem. These active agents are listed, equipped with the associated trophy level, with ecological risk data, in Table 30.2. On the other hand, Table 30.2 is not for the active substances with a PEC BRB, contained: PNEC ≪ 1 as well as for the cough pharmaceuticals, Acetylcystein, the charge means Allopurinol, the hemorrhage, means Isosorbiddinitrate and the beta blocker Metoprololtartrat as the environmental relevance is low. For a further ten active agents (Ampicillin, Captopril, Dimeticon, Phenoxypropa-nol-Isomere, Piperacil-lin-Na, Prednisolon, Sulbactam-Na, Sultamicillin, Talinolol, Torasemid-Na) no statement is currently possible to the environmental behaviour due to a gap in our current knowledge.

30.6 Conclusions and Recommendations

It is not to be expected and also must not to be demanded that pharmaceuticals, with a recognized or assumed environmental damage, should be immediately withdrawn, since the benefits derived from their use often clearly outweigh the disadvantages. Nevertheless a sensitisation of physicians and patients for a more aware handling of these biologically high-risk materials against the environmental dangers is desirable. Furthermore the development of suitable ecotoxicological testing systems is necessary for the more exact determination of a possible environmental risk by pharmaceuticals, so that future risk assessments on a more solid database can take place. Finally, in the future, before human pharmaceuticals are allowed on to the market, the active substances they contain and their behaviour in the environment should be examined against the current and future states of knowledge.

Acknowledgements: I am grateful to Dr. B. Abbas and B. Hanisch, Brandenburg State Office for Environment (LUA), Frankfurt/Oder, Germany, unit Ö3, Environmental Survey and Ecotoxicology, for the inventory of the literature, support by the manuscript and intensive discussion about the topic.

References

Abbas B, Kratz W (2000) Humanarzneimittel in der Umwelt. Erhebung von Humanarznei-mittelmengen im Land Brandenburg 1999. Studien und Tagungsberichte 25, ed Landesumweltamt Brandenburg

Alexy R, Kümpel T, Kümmerer K (2001) Effekte von Antibiotika auf das Bakterien-wachstum in der Umwelt. (SETAC -Tagung Berlin 10.09.2001), Abstract-Band

AMG (2001) Gesetz über den Verkehr mit Arzneimitteln (AZMG) vom 24.08.1976 (BGBL I S. 2445) in der Fassung der Bekanntmachung vom 11.12.1998 (BGBL I S. 3586), zuletzt geändert durch 9. Gesetz zur Änderung des AZMG vom 26.07.1999 (BGBL I S. 1666) last changed on 26.07.2000

Astra Zeneca GmbH (2000a) Sicherheitsdatenblatt Lidocaine Hydrochloride

Astra Zeneca GmbH (2000b) Sicherheitsdatenblatt Prilocaine Hydrocloride

Aventis Pharma (2000) Sicherheitsdatenblatt Pentoxyfyllin

BASF AG (2001) Sicherheitsdatenblatt PVP-Iod

Bayer AG (1999) Sicherheitsdatenblatt Acarbose 050192/09

Bayer AG (2001) Sicherheitsdatenblatt Baypen p.i./Mezlocillin

Bayer AG (2001) Sicherheitsdatenblatt Ciprofloxacin-Hydrochlorid

BIOCHEMIE Ges. m.b.H., Kundl (1999) Sicherheitsdatenblatt Diclofenac-Natrium

BLAC – Bund-Länderausschuss für Chemikaliensicherheit (ed) (2001) Untersuchungspro-gramm Arzneimittel in der Umwelt. Senatsverwaltung Hamburg Hygiene Institut. 94 pp

Clariant GmbH Frankfurt / M. (2000) Sicherheitsdatenblatt Genamin LAP 100 D

Dr. Schuhmacher GmbH (1995) Sicherheitsdatenblatt Cocospropylen-1,5-bis-guanidini-umacetat

Dr. Theodor Schuchardt & Co (1999) Sicherheitsdatenblatt Glutardialdehyd (25 %-ige Lösung zur Synthese)

EMEA (2003) The European Agency for the Evaluation of Medicinal Products. London. CPMP/SWP/4447/00 11 June 2003, 19 pp

Gartiser S, Stiene G, Hartmann A, Zipperle J (1999) Umweltverträgliche Desinfektionsmittel in Krankenhausabwässern. Hydrotox GmbH Freiburg im Auftrag des Umweltbundesamtes. FKZ 29727526, 106 pp

Halling-Sørensen B, Nielsen Nors S, Lansky PF, Ingerselv F, Holten Lützhøft HC, Jørgensen SE (1997) Occurrence, Fate and Effects of Pharmaceutical Substances in the Environment – A Review. ed The Royal Danish School of Pharmacy, 357–391

Hanisch B, Abbas B, Kratz W (2002) Ökotoxikologische Bewertung von Humanarzneimitteln in aquatischen Ökosystemen. Studien und Tagungsberichte 39, ed LUA Brandenburg, Potsdam

Hoechst Marion Roussel Deutschland GmbH (1999a) Sicherheitsdatenblatt Furosemid

Hoechst Marion Roussel (1999b) Sicherheitsdatenblatt Metamizol-Na Monohydrat

Hoechst Marion Roussel (1999c) Roxithromycin. Safety Data Sheet

Hoechst Marion Roussel (1999d) Sicherheitsdatenblatt Propyphenazon. Version 1.1

Knoll GmbH Ludwigshafen (1997) Sicherheitsdatenblatt Verapamil-HCl

Kraemer & Martin Pharma Handels GmbH (1998): Sicherheitsdatenblatt Phenazon

Kümmerer K (Hrsg.) (2001) Pharmaceuticals in the Environment - Fate, Effects and Risks. Springer Verlag Berlin, 265 p

Lonza AG Basel (1997) Sicherheitsdatenblatt Lonzabac 12.100

Merck KgaA Darmstadt (2000) Sicherheitsdatenblatt Acetylsalicylacid

Merck Lipha (1997) Material Safety Data Sheet Metforminhydrochlorid

Merck Lipha s.a. France (1997) Safety Data Sheet, Naftidrofuryl Oxalate, ed No 2

Nycomed Amersham Buchler GmbH & Co. KG (2001) Umweltdatenblatt Visipaque ® Iodixanol

Rippen G (ed) (1990) Handbuch Umweltchemikalien. Stoffdaten-Prüfverfahren-Vorschriften. ecomed Verlag Landsberg - Loseblatt-Ausgabe, 3rd ed (4-8)

Schecker J, Al-Ahmad A, Bauer M, Zellmann H, Kümmerer K (1998) Elimination des Zytostatikums Ifosfamid während der simulierten Zersetzung von Hausmüll im Labormaßstab. UWSF-Z. Umweltchemie und Ökotoxikologie 10 (6):339–344

Schering AG Berlin (2000) Sicherheitsdatenblatt Norethisteron

Stuer-Lauridsen F, Birkved M, Hansen LP, Holten Lützhøft HC, Halling-Sørensen B (2000) Environmental risk assessment of human pharmaceuticals in Denmark after normal therapeutic use. Chemosphere 40, 783-793

Ternes Th, Hirsch R, Stumpf M, Eggert T, Schuppert B, Haberer K (1999) Nachweis und Screening von Arzneimittelrückständen, Diagnostika und Antiseptika in der aquatischen Umwelt. Abschlussbericht des ESWE-Institutes für Wasserforschung und Wassertechnologie GmbH zum Forschungsvorhaben 02WU9567/3 des BMBF 234 pp

WFD – Water Framework Directive (2000) Directive 2000/60/EC of the European Parliament and of the Council of 23 October 2000 established a framework for Community action in the field of water policy, OJ L 327 of 22.12.2000

31 Environmental Oxygenology and Related Thresholds and Standards

Witold Stepniewski and Agnieszka Rozej

Department of Land Surface Protection Engineering, Lublin University of Technology

31.1 Introduction

Oxygenology has been defined as the scientific discipline related to the presence and the role of oxygen in nature on earth (Stepniewski and Stepniewska 1998; Stepniewski et al. 2005). It is a branch of environmental sciences comprising issues of oxygen generation, absorption, turnover, storage, transport, functions and measurement in the environment.

Identification of oxygenology as a separate branch of science is justified not only by the unique role of oxygen as the most abundant component of the lithosphere and as an exceptional component, on the cosmic scale, of the atmosphere of our planet, but also by the need for a holistic approach to the oxygen related problems faced in aquatic, wetland and dry-land ecosystems because of their common nature and structure.

Oxygen plays an essential role in the life of all macro- and micro-organisms, as well as in the biochemical and chemical processes occurring in the environment. Having limited our interest only to earth's oxygenology (e.g. without oxygenology related to other planets), and more precisely only to the contemporary oxygenology (without entering into the paleo-oxygenology related to previous geological periods) such sub-branches as atmospheric oxygenology, aquatic oxygenology, lithospheric oxygenology and bio-oxygenology can be distinguished.

Atmospheric oxygenology is related to the presence of dioxygen and ozone in the troposphere and stratosphere. So within the atmospheric oxygenology we can distinguish tropospheric, stratospheric, and mesospheric oxygenology as well as ozonology. Within hydro-oxygenology or aquatic oxygenology such research areas as ocean oxygenology, marine oxygenology, limno-oxygenology (oxygenology of lakes), and potamic oxygenology (i.e. oxygenology of rivers) have been distinguished. Oxygenology of lithosphere relates mainly to the pedosphere being its most active part. Here we can define some special research fields such as:

Standards and Thresholds for Impact Assessment. Edited by Michael Schmidt, John Glasson, Lars Emmelin and Hendrike Helbron. © 2008 Springer-Verlag

- oxygenology of wetlands (natural, agricultural and constructed wetlands)
- oxygenology of drylands (natural and agricultural)
- oxygenology of anthropogenic systems (landfills, recultivated areas, waste water treatment plants, storages of agricultural materials, etc.)

Bio-oxygenology has been defined as the oxygenology of living organisms i.e. oxygenology of biota. Within bio-oxygenology, such areas as microbial oxygenology, phyto-oxygenology (oxygenology of plants), and zoological oxygenology (zoo-oxygenology) and human oxygenology have been distinguished.

The influence of a development on human health is one of the elements of environmental impact assessment. The article 47 of Act of Change of Polish environmental protection law and of other acts from 18 May 2005 states that the EIA procedures must estimate and evaluate direct and indirect influences of a development on the environment, health and conditions of human life (Dz.U. 1135, 954, 2005). In this context only ozone threshold values are listed as permissible for human health and activity. Dioxygen (O_2) is considered as natural component of the atmosphere and is not taken into consideration directly in the constructions on the ground.

There are only few standards and thresholds related to oxygenological issues in the environment. They concern mainly indoor and outdoor air, and to a much lower extent the hydrosphere and the soil. In this chapter the oxygenological thresholds and standards are treated in a somewhat different way compared to the commonly used approach to environmental impact assessment. From one side dioxygen is necessary for all aerobic organisms. So the attention has to be paid to keep the desirable dioxygen level for humans. But in the case of trioxygen (e.g. ozone) the approach is similar to other toxic substances. An open problem and a curious example of environmental impact assessment is the production of air cleaners producing ozone. Such devices do not generate ozone at the place of production but the environmental effect is manifest far away at a place of their use.

31.2 Thresholds and Standards Related to Dioxygen (O_2)

Outdoor air

The oxygen content in standard outdoor air is treated as a reference for indoor air quality. The standard composition of the dry atmosphere at the earth's surface is presented in Table 31.1. The composition of standard atmosphere is presented in more detail as the US Standard Atmosphere (1976).

Indoor Air

The indoor air is the air inside the dwellings and public utility buildings. More specific cases of such closed spaces are mines, spaceships, aircrafts and submarines.

Table 31.1. Typical composition of dry natural atmospheric air (ISO 2533)

Component	% of mass	% of volume
Oxygen (O_2)	23.14	20.947 6
Nitrogen (N_2)	75.52	78.084
Argon (Ar)	1.288	0.934
Carbon dioxide (CO_2)	0.048	0.031 4
Hydrogen (H)	0.000 003	0.000 05
Neon (Ne)	0.001 27	0.001 818
Helium (He)	0.000 073	0.000 524
Krypton (Kr)	0.000 33	0.000 114
Ksenon (Xe)	0.000 039	0.000 008 7

Dwellings and public utility buildings

Standards related to the internal atmosphere are constructed from the point of view of human respiration in closed spaces. But even in such situations the regulations usually do not give concrete threshold oxygen concentration values but only an indirect indicator, such as for example number of air exchanges in a room during a definite period of time or a number of cubic meters of air exchanged per person and hour. The exchange rate is calculated in this way such that the composition of the indoor air should be similar to that of the outdoor air. For example the Polish Norm: Ventilation in dwelling and public utility buildings. Specifications. (PN-83/B-03430) recommends: >20 m^3 h^{-1} for each person of outdoor air for public utility buildings, 30 m^3 h^{-1} when smoking is allowed, 15 m^3 h^{-1} for each child in nursery schools. The recommended flows of ventilation air in particular closed spaces are as follows:

- for kitchen: 70 m^3 h^{-1}
- For flat for three persons: 30 m^3 h^{-1}
- For flat for more than 3 persons: 50 m^3 h^{-1}
- For bathroom: 50m^3 h^{-1}.

Sometimes carbon dioxide concentration is used as an indicator of the effectiveness of the ventilation system. According to American Society of Heating, Refrigerating, and Air Conditioning Engineers (ASHRAE 2004) the carbon dioxide level should not be higher by more than 700 ppm above the outdoor air concentration.

Specific Closed Spaces: Mines, Submarines, Aircrafts

Mines

Mines are specific constructions where the underground mining adits are characterised by higher air pressure (1074, 1140, and 1207 hPa at depths of 500, 1000 and 1500 m, respectively) compared to that in the atmosphere at the sea level being 1013 hPa (US Standard Atmosphere 1976). Due to this the partial pressure of oxygen could sometimes increase to the level higher than that in the atmospheric air. On the other hand due to the evolution of methane containing mine gas from

Table 31.2. Threshold values of oxygen and carbon dioxide content in specific closed spaces

Closed space	Minimum concentration of O_2 [%]	Maxmum concentration of CO_2 [%]	Reference
Mines	19.5	0.5	Walle and Jennings 2001
Wells, sewers, tanks	18.0	-	Dz.U. 03.169.1650
Submarines	18.0	1.0	Grzesikowski 1998

the rocks as well as oxygen consumption by humans the partial pressure of oxygen may be substantially lowered. According to 'The handbook of safety and health in the small scale surface mines' of The International Labour Organization, if the length of mining adits exceeds 6m the input of fresh air should be assured by natural or pressurised ventilation (Walle and Jennings 2001). The same document gives 19.5 % O_2 as a minimum oxygen concentration and 0.5 % CO_2 as a maximum carbon dioxide concentration for human safety in the mine atmosphere (Table 31.2). Even in the international standard for soil sampling (ISO 10381-3:2001) there is a warning that "where the investigation requires entry into deep excavations or confined spaces, particularly those below ground level, the build-up of explosive and/or toxic gases and the formation of an atmosphere which is deficient in oxygen is a possibility. An atmosphere deficient in oxygen even by a small amount (1 %) can be fatal".

Sewers, wells, tanks, and vessels

According to the Ordinance of the Polish Ministry of Labour and Social Policy (1997) for working without any protection of the respiratory system inside containers, wells, sewers and other closed spaces, the entrance to which is through a narrow manhole, the minimum recommended oxygen concentration is 18 %.

Submarines

The construction of the first submarine is connected indispensably with the discovery of oxygen at the very beginning of the XVII century (Lane 2002) by Polish alchemist Michael Sendivogius Polonus who explained the method of producing it to his Dutch fellow, physician Cornelius Drobbel. According to Lane (2002) Drobbel demonstrated the role of oxygen in 1621 by construction of the world's first wooden submarine that stayed under water for three hours with twelve oarsmen. Thus submarines are the first examples of practical application of oxygenology to completely closed spaces.

Since that time much has changed. The present day submarines are equipped with an efficient system of ventilation and the system of regeneration of air (absorption of carbon dioxide and replacing it by oxygen) enabling long stay (sometimes several months in case of atomic submarines) under water. The ventilation system ensures the exchange of air in the entire submarine within several minutes when on the surface of water. The system of ventilation and regeneration usually maintains oxygen concentration above 18% and that of carbon dioxide below 1 % (see Table 31.2).

Aircraft

Traveling by aircrafts is connected with people staying, sometimes for several hours, in a medium characterised by reduced air pressure and, due to this, poorer in the oxygen which is necessary for respiration. Already before World War II closed cabins were used in warplanes and the pilot was provided with air enriched in oxygen. At present aircraft are prepared for dehermetization if problems happens at a height above 14 000 ft (e.g. of 4 200 m). Under such conditions oxygen masks will be delivered automatically to passengers and the aircraft lowers its height to the level below 3 000 m, where the underpressure does not present a direct threat for human life.

The current Ordinance of the Minister of Infrastructure of Poland (5 November 2004) on safety of aircraft exploitation (Dz. U. 262.2609) says that in the passenger aircrafts the carrier shall not fly the machine at a height above 10 000 ft as long as it is not equipped with additional devices to store and deliver the required amounts of oxygen. The crew members of helicopters fulfilling duties important from the point of view of safety should use oxygen delivered continuously if the pressure altitude is above 10 000 ft for more then 30 min, and always when the pressure altitude exceeds 13 000 ft.

The same Ordinance states that the carrier shall not use aircraft with pressure cabin, on which there is an obligation for the aircrew for flights above 25000 ft unless it is equipped with a source of additional oxygen for passengers who, because of physiological reasons, will require delivering of oxygen in the case of cabin decompression. The amount of oxygen shall be calculated assuming an average coefficient of flow of at least three litres per minute per person. The carrier has to assure after decompression of the cabin at pressure altitude above 8000 ft (2 500 m) but not higher than 15 000 ft a supply of oxygen for at least 2 % of passengers, and not less than for one person (Dz. U. 262.2609 JAR-OPS 1.760).

Besides the additional oxygen the aircraft should be equipped with first-aid oxygen for those passengers, who after exhaustion of the additional oxygen still feel the hypoxic problem (Dz. U. 262.2609 IEM OPS 1.760). For the calculation of the first-aid oxygen the carrier should take into account that the amount of oxygen should be sufficient to satisfy the hypoxic problem of all passengers, at the cabin pressure height exceeding 15 000 ft and for some of the passengers at cabin pressure height within the range 10 000-15 000 ft.

31.3 Thresholds and Standards Related to Thrioxygen (O₃)

Outdoor air

Normal ozone content in the troposphere (being the result of its transport from stratosphere) is about 20 ppb (parts per billion) (Wayne 2002). Troposphere ozone is formed only under the conditions of contamination of air by NO_2 happening mainly in the urban air. Polluted atmosphere for example in Los Angeles contains about 2 ppm O_3. Ozone in the outdoor air is a subject of interest because of its effect on humans and on plants due to generation of free radicals in tissues (Manahan 1993). Human exposure to ozone irritates and inflames the lining of the respiratory system, causes such symptoms as coughing, chest tightness and shortness of breath. Air containing 1ppm of ozone has a distinct odour and causes headache. Persons especially vulnerable include children and those who suffer from respiratory diseases such as asthma (Stepniewski et al. 2005).

Polish thresholds for ozone are established by the Ordinance of the Minister of Environment (6 June 2002) on permissible levels of substances in the air (Dz.U. 87.796). The threshold value of ozone concentration for outdoor air from the point of view of human health protection is 120□g O_3 m⁻³ (Table 31.3). The same threshold has been accepted in the EC Directives 1999/30/EC and 2000/69/EC. The acceptable maximum frequency of exceeding the above threshold value is 25 days a year. The threshold O_3 level when the society should be informed about the risk of the occurrence of alarm levels is 180 □g m⁻³. The alarm level is 240 □g m⁻³ (for 1 hour averages).

The threshold value for the protection of plants is averaged for the vegetation period. It is defined as a sum of 1-hour mean ozone concentrations above the threshold of 20 μg m⁻³ over a defined time period. For agricultural crops the time period is defined as daylight hours between 8.00 and 20.00 CET for the months from May to July and the critical level is set to 24000 μg m⁻³ till the end of 2009 and 18000 μg m⁻³ later. For the forest trees in Switzerland the time period stretches over the months from April to September (Zierl 2002). The specificity of oxygen che-

Table 31.3. Thresholds values of ozone concentration in outdoor and indoor air according to Polish regulations

Air	Averaging period	Permissible level of O_3 concentration [μg m⁻³]	Reference
Indoor air	8 h	150	Dz.U. 02.217.1833
Outdoor air for human health protection	8 h	120 (180 warning value and 240 alarm value)	1999/30/EC 2000/69/EC Dz.U. 87.796
Outdoor air for plant protection	Vegetation period (1V-31VII)	24 000[1] 18 000[2]	Dz.U. 87.796

[1] until 31.12.2009
[2] after 01.01.2010

Table 31.4. Reactive oxygen species (After Bartosz 1995)

Name	Structure
Singlet oxygen	$^1O_2$
Ozone	O_3
Hydroperoxyl radical	HO_2
Superoxide radical anion	$O_2^{\cdot-}$
Hydrogen peroxide	H_2O_2
Hydroxyl radical	$^\cdot OH$
Nitric oxide	$NO^\cdot$
Nitric dioxide	$NO_2^\cdot$
Peroxynitrous acid	O=N-OOH
Peroxynitrite	O=N-OO$^-$
Hypochlorous acid	ClOH
Hypochlorite	ClO$^-$
Hypobromous acid	BrOH
Hypobromite	BrO$^-$
Hypoiodous acid	IOH
Hypoiodite	IO$^-$
Hypothiocyanous acid	S=C=N-OH
Hypothiocyanite	S=C=N-O$^-$
Alkoxyl radical	RO$^\cdot$
Peroxyl radical	ROO$^\cdot$
Peroxide (=hydroperoxide)	ROOH
Acyloxyl radical	RCOO$^\cdot$
Acylperoxyl radical	RCOOO$^\cdot$
Aryloxyl radical	ArO$^\cdot$
Arylperoxyl radical	ArOO$^\cdot$
Semiquinone radical	H-Ch$^\cdot$
Semiquinone radical anion	Ch$^{\cdot-}$
Epoxide	-COC-

mistry is characterised by the presence of highly reactive forms of oxygen, which are called "reactive oxygen species" (ROS) presented in Table 31.4 (Halliwell and Gutteridge 1999; Mittler 2002). These molecules can influence the cell structure and thus functioning of living organisms. All organisms living under toxic conditions possess mechanisms protecting the cells against ROS. In the situation of uncontrolled ROS production and non sufficient scavenging, the problem of oxygen toxicity appears.

Indoor air

The presence of ozone in the indoor air is connected with different devices generating ozone at a working place. The maximum permissible concentration of ozone for 8 hours exposure each working day in Poland, expressed as weighted average, is $150 \mu g\ m^{-3}$ (Dz.U. 02.217.1833).

Another problem involves the production of different types of ozone-generating indoor "air purifiers" for which the maximum permissible ozone concentration is 70 ppb after 8 hours and 90 ppb after one hour of ozone generator use. It has been found that ozone generators can often produce indoor ozone levels several times the state outdoor health standard (Mason et al. 2000).

A special case is the application of ozone, preferred to other oxidizers as more friendly for the environment, to different technological processes e.g. conditioning of cooling waters, decolourization and deodorization of industrial waste waters and disinfection of water in swimming pools. It is recommended usually within the best available techniques (BAT) approach (Council Directive 96/61/EC).

For example the installation for disinfection of swimming pool water consists of several devices such as an ozone generator, a device for ozone mixing with water, a reaction vessel, and an ozone destructor. The disinfection reaction time is 3 minutes. According to DIN 19627 the ozone concentration in the air introduced to water should be above 20 g per normalised cubic meter, while the output air should be obligatorily directed to the ozone destructor. The maximum ozone concentration in the air introduced to the destructor (due to explosion risk) is 4 gm^{-3}.

31.4 Water

Water quality is related, among others, to oxygen saturation degree. Usually in the acts related to water quality two values are given: the desirable and permissible

Table 31.5. Recommended and permissible values of oxygen concentration and biochemical oxygen demand (BOD) for different surface water classes

Destination of water	Indicator	Recommended value	Permissible value	Reference
Surface water for drinking water acquisition	O_2	A1 >70 A2 >50 A3 >30		75/440/EEC Dz. U. 204.1728
	BOD_5	A1 <3 A2 <5 A3 <7		75/440/EEC Dz. U. 204.1728
Bathing waters	O_2	0-120		76/160/EEC
	O_2	80-120	>80	Dz. U. 183.1530
	BOD_5	<6	<6	Dz. U. 183.1530
	d O_2	>5		Dz. U. 183.1530
Water for crustaces living	O_2	≥ 80	≥ 70	79/923/EEC
Freshwater for fish (salmon)	d O_2	$50\% \geq 9$ $100\% \geq 7$	$50\% \geq 9$	78/659/EEC Dz. U. 176.1455
	BOD_5	≤ 3		78/659/EEC Dz. U. 176.1455
Freshwater for fish (carp)	d O_2	$50\% \geq 8$ $100\% \geq 5$	$50\% \geq 7$	78/659/EEC Dz. U. 176.1455
	BOD_5	6		78/659/EEC Dz. U. 176.1455

O_2 - Oxygen saturation rate [% O_2]; BOD_5 - Biochemical Oxygen Demand [mg O_2/l]; d O_2 - Dissolved Oxygen [mg O_2/l]

thresholds, as well as their determination method. For instance the classification of surface waters for drinking water production into A1, A2 and A3 categories is based on 44 different indicators of which one is oxygen concentration. The values are presented in Table 31.5. The values, based on European Directives, can differ among the Member States. For example Polish regulations are more stringent compared to the EU thresholds. This is manifested by increased number of parameters of characterisation, elevation of some thresholds and, first of all, by the changes of the limiting values for particular categories of water quality. In the EU Directive the permissible values are always higher than the recommended ones, which means that the conditions of evaluation are less restrictive. Also in the case of bathing water (characterized by 15 indicators) and water for fish farming (14 indicators) oxygen is included and the approach is similar.

31.5 Soil

Wetland soils are submerged and are generally anoxic. The oxygen penetrates from the surface and creates a thin (several mm deep) oxic soil layer. The second way of O_2 migration into the deeper soil layer is through root systems with vesicular tissue (aerenchyma) (Conrad 1996).

The depth of the oxic surface layer is a result of equilibrium between the influx of O_2 from the atmosphere and its consumption by soil microflora and roots. The oxygen is usually depleted rapidly, especially in the presence of significant amounts of organic matter, if O_2 transport from the atmosphere ceases. With the increase of the distance from soil surface or from the root surface the redox potential (Eh) stratification occurs. The different redox zones are characterized by the presence of different groups of micro-organisms such as aerobic, facultative aerobic, micro-aerophilic, and finally strictly anaerobic. They use successively various electron acceptors for their respiration: O_2, NO_3^- (at Eh values lower than 350 mV), Mn^{4+} (at Eh<200 mV), Fe^{3+} (at Eh<100 mV), SO_4^{2-} (at Eh around<-150 mV) and CO_2 (at Eh<-350 mV) (Le Mer and Roger 2001). Although redox potential value of 300 mV is generally accepted as the limit between oxidized and reduced conditions (Glinski and Stepniewski 1985) there is no official standard for

Table 31.6. Indicators of soil oxygenation (Glinski and Stepniewski 1985) and their threshold values suggested by authors as sufficient for adequate oxygenation of soil

Indicator	Value sufficient for good aeration
Air-filled porosity	>0.25 m^3 m^{-3}
Relative gas diffusion coefficient (D/Do)	>0.02
Soil air composition	$>10\%$ O_2; $<10\%$ CO_2
Oxygen diffusion rate (ODR)	$>70\mu$g m^{-2}s^{-1}
Redox potential	> 300 mV

D- gas diffusion coefficient in soil, Do – the diffusion coefficient of the same gas in the free atmosphere at the same temperature and pressure conditions.

that. There is only a standard for the method of determination of redox potential (ISO 11271:2003).

The complexity of interactions within the atmosphere - soil- plant continuum is such that although there are many indicators describing oxygen status in soil, none of them separately reflects the situation satisfactorily. The indicators currently in use are: air-filled porosity, air permeability, gas diffusion coefficient, respiration rate, soil air composition, oxygen diffusion rate (ODR), redox potential and some other indicators (Glinski and Stepniewski 1985). The threshold values suggested by us for some of them as sufficient for adequate oxygenation of soil are presented in Table 31.6.

Aerobic Transformations of Xenobiotics

Chemical pollutants released into the biosphere are extraordinarily diversified. The biological impact of many pollutants operating together can be stronger than the effect of individual chemical factors. Chemical or biochemical transformations of pollutants in the environment are possible to lead to formation of more stable, more toxic and more carcinogenic chemicals than the initial one. Metabolism of xenobiotics in living organisms is called detoxication. In view of the possibility to increase toxic effect this kind of process should be called biotransformation. Biotransformation reactions are divided into four types: oxidation, reduction, degradation and conjugation. Reactions of oxidation, in particular, often lead to the formation of more dangerous substances than the initial ones (Ostroumov 1992).

Some examples of such processes are presented below:

- oxidation of amines e.g. during reaction between diethylamine and nitrite in the acidic medium in the stomach, carcinogenic diethylnitrosoamine is produced
- oxidation of aromatic amines through N-hydroxylation can lead to formation of carcinogenic products, especially during hydroxylation in –orto position
- epoxidation of aromatic compounds causes generation of strongly toxic, carcinogenic and mutagenic epoxides (e.g. during microsomal and microbial oxidation of such pesticides as heptachlor and aldrine toxic epoxides of these compounds are produced)
- oxidation or oxidative substitution of sulphur: e.g. insecticide paration during microsomal and microbial oxidation in soil is transformed into paraoxon, being also an insecticide, but two times more toxic for mammals than paration.

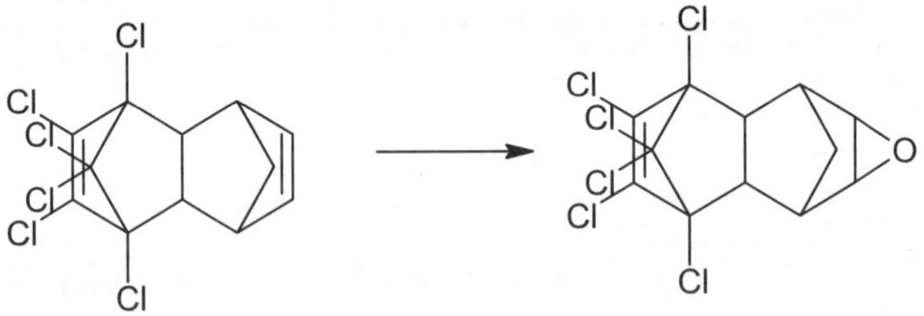

Fig. 31.1. Oxidation of aldrine and formation toxic epoxide (dieldrine)

Methods of laboratory testing for biodegradation of organic chemicals in soil under aerobic conditions are the subject of Polish and European Standards (ISO 11266:1994; PN-ISO 11266:1997).

31.6 Conclusions and Recommendations

Two forms of oxygen, dioxygen (O_2) and trioxygen (i.e. ozone (O_3)) are at the centre of interest while considering oxygenological thresholds and standards for environmental impact assessment. It should be noted that these standards are not sufficiently elaborated and require further studies. Oxygen in the form of dioxygen O_2 at almost atmospheric concentrations is indispensable for normal functioning of humans and these levels for dwellings, public utility buildings, mines as well as for closed spaces such as submarines, aircrafts are assured in the pertinent standards. Requirements for diving mammals as well as those for plants and microorganisms may be lower, but they are less recognized and not included yet in the up to date normalization acts. In both cases further specification is needed. Also the toxicity of over-atmospheric concentrations of oxygen requires further attention. Consideration of oxygen as a criterion of environment use is not practised yet. That means that industrial plants have integrated permits including, for example, limits for water use or limits for carbon dioxide or dust emission but no limits for oxygen use as this source is considered renewable or unlimited. The problem of oxygen limitation occurs only in the closed spaces such as submarines, aircrafts, spaceships, or eventually future colonies on other planets. Ozone or trioxygen is treated as a chemical agent for water sterilization or treatment as well as for air purification. Because of harmful ozone effect on humans and plants there are some thresholds connected with ozone concentration limits in indoor and outdoor air. It should be stressed that despite its toxicity ozone is preferred over other chemicals as more friendly for the environment. Reactive oxygen species influence the chemical reactions in the environment and cell structures of living organisms. They can cause toxic effect on people, animals and plant tissues. Establishing their thresholds or standards requires further investigations.

References

Act of Change of environmental law protection and other acts of 18 may 2005 (Dz.U. 1135, 954, 2005)

ASHRAE Standard 62.1-2004. Ventilation for Acceptable Indoor Air Quality.

Bartosz G (1995) Druga Twarz Tlenu (The Other Face of Oxygen - in Polish). Wyd. Naukowe PWN, Warszawa

Conrad R (1996) Soil microorganisms as controllers of atmospheric trace gases (H_2, CO, CH_4, OCS, N_2O, and NO). Microbiol Rev 60(4):609-640

Council Directive 1999/30/EC of 22 April 1999 relating to limit values for sulfur dioxide, nitrogen dioxide and oxides of nitrogen, particulate matter and lead in ambient air

Council Directive 2000/69/EC of the European Parliament and the Council of 18 November 2000 relating to limit values for benzene and carbon monoxide in ambient air

Council Directive 75/440/EEC of 16 June 1975 related to surface water intended for the abstraction of drinking water

Council Directive 76/160/EEC of 8 December 1975 related to bathing water quality

Council Directive 78/659/EEC of 18 July 1978 related to protection or improvement of waters supporting fish life

Council Directive 79/923/EEC of 30 October 1979 related to shellfish water quality

Council Directive 96/61/EC of 24 September 1996 related to Integrated Pollution Prevention and Control

DIN 19627   Ozonerzeugungsanlagen zur Wasseraufbereitung

Glinski J, Stepniewski W (1985) Soil Aeration and its Role for Plants. CRC Press Inc Boca Raton, Florida

Grzesikowski T (1998) Middle size submarines, series 613, ORP Orzel II (in Polish). The Seas, Merchant Vessels and Naval Ships, 2:3-19, (Published by Magnum-X, Warsaw)

Halliwell B, Gutteridge JMC (1999) Free Radicals in Biology and Medicine, third ed. Oxford University Press

ISO 10381-3:2001. Soil quality - sampling – Part 3: Guidance on safety

ISO 11266:1994.Soil quality - Guidance on laboratory testing for biodegradation of organic chemicals in soil under aerobic conditions

ISO 11271:2003. Soil quality – Determination of redox potential – Field method

ISO 2533:1975. Standard Atmosphere

Lane N (2002) Oxygen. The Module that Made the World. Oxford University Press

Le Mer J, Roger P (2001) Production, oxidation, emission and consumption of methane by soils: A Review. Eur J Soil Biol 37:25-50

Manahan SE (1993) Fundamental of environmental chemistry. Lewis Publ. Boca Raton

Mason MA (2000) Characterisation of ozone emissions from air cleaners equipped with ozone generators and sensor and feedback control circuitry. Engineering Solutions to Indoor Air Quality Programs Symposium, Research Triangle Park, NC VIP-98, AWMA, July, pp 254-269

Mittler R (2002) Oxidative stress, antioxidants and stress tolerance. Trends Plant Sci. 7:405-410

Ordinance of Minister of Environment of Poland of 4 October 2002 on requirements to be met by surface waters likely to support fish life under natural conditions (in Polish). Dz.U. 176.1455

Ordinance of Minister of Health of Poland of 16 October 2002 on requirements for bathing water quality (in Polish). Dz. U. 183.1530

Ordinance of Minister of Health of Poland of 19 November 2002 on requirements concerning quality of drinking water for humans (in Polish). Dz. U. 204.1728

Ordinance of the Minister of Environment of 6 June 2002 on permissible levels of substances in air (In Polish). Dz.U. 87.796

Ordinance of the Minister of Infrastructure of Poland of 5 November 2004 on safety of aircraft exploitation (In Polish). Dz. U. 262.2609 – JAR-OPS 1.760, IEM OPS 1.760

Ordinance of the Minister of Labour and Social Policy of 26 September 1997 on the general regulations on work safety and hygiene (In Polish). Dz.U. 03.169.1650

Ordinance of the Minister of Labour and Social Policy of 29 November 2002 on the maximal permissible concentrations and intensities of harmful factors for health in the work environments (In Polish). Dz.U. 02.217.1833

Ostroumov SA (1992) Introduction into the biochemical ecology (In Polish). PWN, Warsaw, pp 159-162

PN-83/B-03430 Ventilation in dwelling and public utility buildings. Specifications

PN-ISO 11266:1997 Soil quality - Guidance on laboratory testing for biodegradation of organic chemicals in soil under aerobic conditions

Stepniewski W, Stepniewska Z (1998) Oxygenology as a new discipline in the environmental sciences – a proposal for discussion. Internat. Agrophysics 12:53-55

Stepniewski W, Stepniewska Z, Bennicelli RP, Glinski J (2005) Oxygenology in Outline. Institute of Agrophysics PAS, Lublin

US Standard Atmosphere: National and Atmospheric Administration, National Aeronautics and Space Administration and the US Air force. (1976) US Government Printing Office, Washington D.C. 20402

Walle M, Jennings N (2001) Safety and Health in Small – Scale Surface Mines. A Handbook. International Labor Organization

Wayne R (2002) Chemistry of Atmospheres: An Introduction to the Chemistry of the Atmospheres of Earth, the Planets, and their Satellites. Third Edition, Oxford

Zierl B (2002) Relation between crown condition and ozone and its dependence on environmental factors. Environmental Pollution 119: 55-68

32 Quantitative Threshold Values for Strategic Environmental Assessment

Hendrike Helbron and Michael Schmidt

Department of Environmental Planning, Brandenburg University of Technology (BTU) Cottbus, Germany

32.1 Introduction

This chapter gives an insight into the methodology and derivation of assessment thresholds during an INTERREG IIIA pilot project on a transborder SEA of the regional plan of Upper Lusatia-Lower Silesia (OL-NS) in Saxony (called Trans-SEA) (see Figure 32.1; Helbron and Schmidt 2007; regional planning authority of OL-NS: http://www.rpvolns.homepage.t-online.de/frame3.htm).

The derivation of quantitative threshold values for strategic environmental assessment in regional land use planning with a scale of 1: 200 000 (here SEA-REP) is still an evolving process in most EU Member States after their national implementation of the SEA Directive (2001/42/EC) in 2004 and later (see for instance Schmidt et al. 2005). For a programmatic SEA assessment, methodologies have to be adapted for instance from sustainability appraisals, ecological risk analysis and Environmental Impact Assessment (Project EIA) and have to be modified in order to meet the requirements of the higher more abstract spatial planning level.

Environmental quality objectives (EQO) and environmental quality standards (EQS) specify or refer to certain qualities of resources, potentials or functions defined in material, spatial or temporal terms, which are to be preserved or attained in specific situations. They may be scientifically, legally or politically defined and expand on general environmental policy and environmental planning objectives. Derived from these objectives, environmental quality standards give specific values for day-to-day environmental policy and address certain parameters and indicators, measuring methods and conditions (UBA 1995). They are necessary in order to determine assessment thresholds to classify the conflict intensity on the adversely affected individual area. As the regional plan is a comprehensive spatial plan, including all land uses of the region, SEA in this sector has a high potential to contribute to an optimisation of the plan and thus to a sustainable development in the region. Limit values influence the efficiency of SEA-REP to a great extent.

Standards and Thresholds for Impact Assessment. Edited by Michael Schmidt, John Glasson, Lars Emmelin and Hendrike Helbron. © 2008 Springer-Verlag

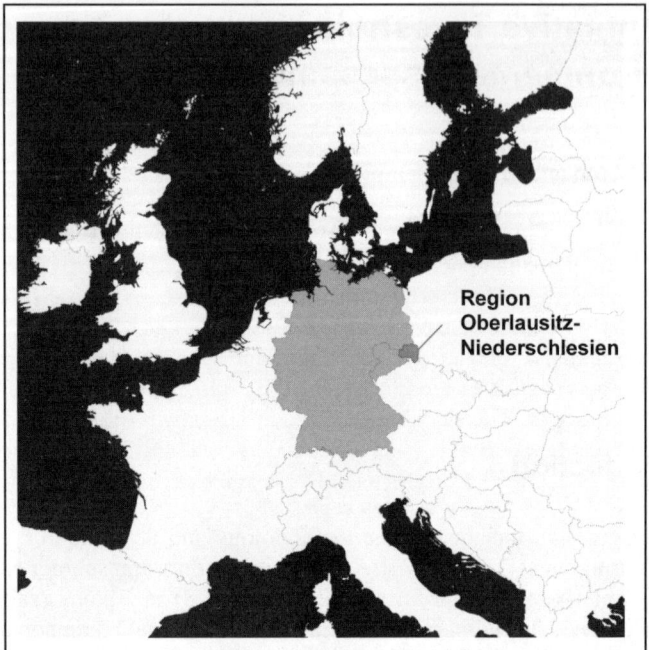

Fig. 32.1. Study Area Upper Lusatia-Lower Silesia (changed from RPV OL-NS 2007)

32.2 Assessment Methodology

The methods and techniques used in SEA can be drawn from two main sources: those that are already used in project level environmental impact assessment (EIA) and can be adopted for the use at more strategic levels of assessment (e.g. for scoping and impact identification, describing baseline decisions, predicting impacts from multiple sources, etc.) and those which can be adopted from policy analysis and planning studies (e.g. scenario and simulation analysis, site selection and suitability analysis and evaluation techniques) (Lee and Wood 1995). The methods used in this pilot study were borrowed from the traditional methods of German landscape planning – a more or less independent area-wide sector planning, *spatial potential analysis* and EIA, and *ecological risk assessment* leading to a trans-nationally approved, baseline-led and indicator-based impact assessment approach.

In Germany, in contrast to most other EU Member States, landscape framework plans (LRP) can provide a major part of the essential area-wide environmental information at regional level (Scholles and von Haaren 2005). Landscape planning as a more or less independent sector planning delivers a variety of objectives and assessed data on nature and landscape issues. These are to be considered in the weighting process of the regional plan and may lead to a very comprehensive approach to SEA. A compromise has to be found between what is adequate for the level of detail of regional planning and what is applicable in practice.

32.2.1 Environmental State Indicators

Twenty nine environmental state indicators were selected (Table 32.1) which adequately cover the environmental media for EIA (Federal EIA Act) and which were enlarged in the SEA Directive (Annex I, f). These include human health, fauna/flora and biodiversity, soil, water (groundwater and water bodies), climate and air, landscape and recreation, cultural and other material assets), spatially inclusive and comprehensive at regional planning level, and partly include initial pollution and impacts. The choice of the environmental components was firstly undertaken in an interdependent approach from necessary data due to environmental objectives, i.e. desired end-states, followed by a continuous revision of the environmental indicator set.

Table 32.1. Environmental components and state indicators applied in TransSEA (modified from Helbron and Schmidt 2007)

Code	Environmental component	State indicator
Environmental media human health		
HH 1	Noise pollution in settlement areas	Land uses of settlement areas to be protected against noise with priority
HH 2	Pollution in settlement areas	Land uses of settlement areas to be protected against pollution with priority
Environmental Media Fauna, Flora, Biodiversity		
FFB 1	Biotope types and habitats	5 Assessment classes of the biotope and land use types after Bastian (1994), protection states of biotopes after Saxon Nature Conservation Act (SächsNatSchG)
FFB 2	Protected species	Importance of the areas for protected species after Annex II or IV Habitats Directive, Annex I Birds Directive, red list of Germany or red list of Saxony
FFB 3	Biotope connection network	Biotope connection areas and elements with transborder, Saxon or regional importance
FFB 4	Protected areas	Importance of national parks, nature reserves and biosphere reserves in zones
FFB 5	FFH areas and SPA	Importance of areas protected after Habitats Directive and SPA
Environmental Media Soil		
So 1	Natural productivity	Five productivity classes (F classes) of the soil concept map
So 2	Storage and regulation function	Five buffer classes (P classes) of the soil concept map
So 3	Biotic habitat function	Site class V of the soil concept map (special sites, which are moist, dry or poor in nutrients)

Table 32.1. (cont.)

Code	Environmental component	State indicator
So 4	Erosion risk	Erosion risk through water and wind in five classes, erosion protection forest
So 5	Contaminated soils	Urgency of need for action on contaminated sites in five classes
So 6	Unsealed area	Largely unsealed areas (< 25 %) or rest areas, which are not within the sealing classes of (26-50 %, 51-75 %, 76-100 %); brownfield sites according to data of lower planning levels, aerial data and field trips
Environmental media Groundwater		
Gw 1	Groundwater development rate	Groundwater development rate in three classes in mm/a in groundwater catchment area
Gw 2	Protection of Groundwater against pollution	Protection potential of groundwater cover in three classes
Gw 3	Groundwater level below ground	Average groundwater level below ground < 2 m, in connection with biotopes depending on groundwater and average groundwater level >= 2 m in connection with woodland and forest biotopes depending on groundwater after CIR and selective biotope map
Gw 4	Drinking water catchment areas	Importance of drinking water catchment area in protection zones
Environmental media water bodies		
Wa 1	Biological structure of rivers and streams	Biological structure in seven classes
Wa 2	Flood plains and retention areas	Flood plains and retention areas; nature-close river sections with biological structure 1-2 or protection after Art. 26 SächsNatSchG
Wa 3	Water quality of rivers and streams	Water quality in seven classes
Wa 4	Barrierless flow of rivers and streams	Importance of rivers and streams concerning their barrierless flow in categories I and II
Environmental media air and climate		
AC 1	Bioclimatic Condition	Need for preservation of open areas from climatic perspective in three classes; forests > 4 ha for fresh air production; climate protection forest
AC 2	Highly polluted areas	Potential highly polluted areas in valleys and settlement areas in dependence on traffic volume and industry/commerce; immission protection forest
Environmental media landscape		
La 1	Landscape character	Landscape character quality in three classes
La 2	Areas for recreation in proximity of central places	Suitability of areas > 4 ha for recreation in proximity of settlement areas of central places
La 3	Unfragmented areas	Importance of unfragmented areas related to their size in three classes

Table 32.1. (cont.)

Code	Environmental component	State indicator
La 4	Protected areas for recreation and open areas in need for protection against noise	Importance of biosphere reserves, landscape protection areas, nature parks, recreation forests

Environmental media cultural and other material assets

Code	Environmental component	State indicator
Cu 1	Constructed cultural and material assets, monuments	Protected cultural heritage and regionally important cultural monuments and other constructions
Cu 2	Landscapes, elements, soils with archive function	Natural monuments, protected landscapes and elements, archaeological soils and geotopes

32.2.2 Impact Factors and Indicators

Within the scoping step, it was determined in the project team that those designations and contents of the regional plan require an SEA, which are site-specific and set the framework for project EIA (e.g. new housing areas) according to the EIA Directive 85/337/EEC (listed in the Annexes I and II) on the lower planning level or which require an assessment pursuant to Article 6 or 7 of the Habitats and Birds Directive (SEA Directive Art. 3 para 2 a, b).

The most relevant regional plan designations (partly integrated into the regional plan from other related sector plans) likely to cause significant impacts are:

- new urban settlements
- industrial and commercial sites
- routes for transport infrastructure (integrated from sector plan)
- areas for the exploitation of near-surface non renewable resources (selection from a map, which presents several deposits of resources)
- areas for wind energy power plants (designation criteria of regional planning)
- sites for recreation on formerly used mining areas (integrated from restoration structure plans and regional development concepts)
- sites for the extension of forests

The coarse database and less detailed scale (scale 1: 100 000 up to 1: 200 000) on the strategic level of regional planning do not allow predictions on intensity of land uses in terms of settlement density or exact decibels for noise pollution. Thus spatially relevant *impact factors,* which have to be considered for each individual site-specific designation, were defined. These can be divided into adverse effects within the regional borders and effects with transborder relevance.

At the regional planning level it is generally not known which parts of areas or which percentage of a specific designated area is affected by soil sealing, soil excavation and land use change. Therefore several regional designations can include

Box 32.2. Impact factors with relevance for SEA-REP (modified from Helbron and Schmidt 2007)

Of regional relevance on the individual site:

Land consumption*	LC	Soil abstraction or soil sealing
Land use change#	LU	Change of function of the area without soil removal or sealing; e.g. afforestation, recreation
Fragmentation/Barriers	FB	Separation of functions, effects on accessibility of areas

Additionally of regional and transborder relevance on impact zone:

Change of Groundwater level	GW	Groundwater level increase or decrease
Flooding	F	Anthropogenic caused flooding of areas
Noise	N	Potential exceedance of noise standards or general increase of noise level
Pollution	P	Potential exceedance of pollution standards or increase of pollution risk
Visual Impacts	VI	Impacts on landscape character (change of peculiarity, variety and natural charctersitics)

* The impact factor land consumption includes all impacts, which lead to an irreversible destruction or removal of the soil such as soil sealing and soil degradation/excavation.

\# The impact factor land use change defines an alteration of the utilization and/or vegetation of the area without accompanying irreversible soil loss such as afforestation on formerly used arable land.

land consumption and at the same time land use change, but these two impact factors cannot overlay on an identical area. *Impact indicators* combine the impact factors with the environmental component (e.g. land consumption on an agricultural land with high productivity of the soil) (see also Chap. 17).

32.3 Environmental Quality Objectives (EQO) and Environmental Quality Standards (EQS)

Environmental quality objectives and standards are needed for the selection of indicators and threshold values in SEA. The core module in SEA-REP is the indicator set, which is needed to predict and assess the impacts of the regional plan which are likely to occur. This component of SEA methodology has to be examined in sufficient detail in order to overcome technical difficulties.

In a first step a detailed *spatial potential analysis* of data on land cover or land-use type, in the form of existing surveys gathered in maps and databases, was carried out in cooperation with the Saxon Board for Environment and Geology (LfUG) – with reference to Annex II No. 2 of the SEA Directive, the legal basis for a comprehensive analysis of the environmental status quo. It is crucial to evaluate the available environmental baseline data, as SEA is very much dependant on its applicability. The study area of OL-NS and the spatial regional objectives in a scale down to 1: 200 000 m dictate the baseline data requirements.

Figure 32.2 shows that set environmental standards in the EU and Germany mainly protect the environmental media of soil, air and water against pollutants, but there is a lack of regional environmental objectives for the protection of the land and resources against progressing consumption. This is not a new problem, but was recognised before the early 1990s (UBA 1993). Although 'taboo zones' in the form of legally-binding protected areas and buffer zones (minimum distances of harmful land uses to sensitive areas) approved among regional planners do exist, the countryside lacks an area-wide "ecologisation", as regional environmental objectives are missing. For example, how much noise should be allowed for quiet recreation? How noisy is our countryside? What share of soil sealed areas in our region is sustainable?

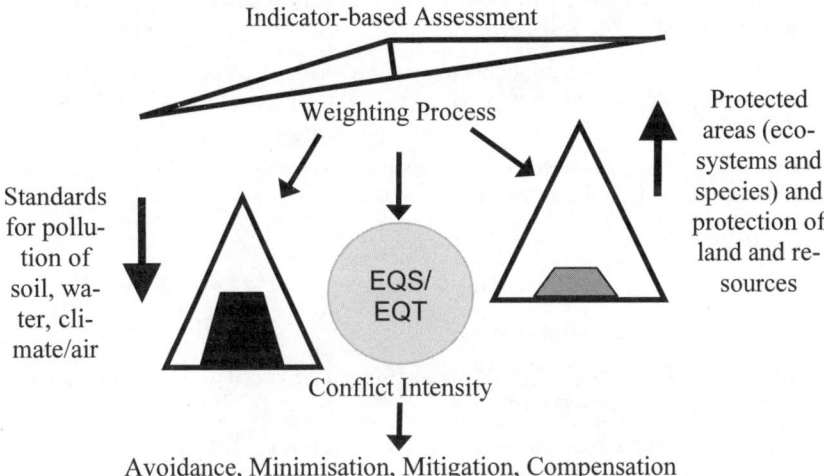

Fig. 32.2. "Unbalanced" situation of set standards for environmental components

The German Enquête Commission proposed in 1998 the action objective, to achieve by 2010 a reduction of the conversion rate of undeveloped areas into settlement and transport areas by 10 % of the rate of the years 1993 to 1995. A further 15 % of Germany's territory shall be reserved for nature conservation. The urban and transport areas in Saxony increased after 1990 by approximately 8 hectare per day. The entire area sealed through housing development made up more than 11.6 % of Saxony in 2004. The consumption of the natural soil resource increased rapidly (LfUG 2006). The real sealed (>50 %) share of this area is approximately more than 45 %. Only existing standards and objectives were applied in SEA-REP. The following Box 32.1 and Table 32.2 give examples of applied EQT, EQS and threshold values in TransSEA. The threshold values are thus regionalised for the region of OL-NS and cannot be transferred without modification to another region. The values must be regionalised, as the current environmental state is the basis for the formulation of state indicators.

Box 32.1. Environmental state indicators, EQS/strong legally-binding protected areas and precaution-oriented guidance values (I) and EQT/objectives on environmental activity (II) (extract, not complete) of TransSEA (modified from Helbron and Schmidt 2007)

SOIL:

So 6: Unsealed soils (excluded are urban, industrial and transport areas)

I) No legally-binding standard exists

II) Sustainable and considerate use of soil with soil sealing as low as possible; new soil sealing shall be minimised, desealing (Art. 1 a Abs. 1 BauGB; Art. 5 BBodSchG)

Reduction of the change of undeveloped areas in urban and transport areas (by 2010 to 10 % of the rate of 1993-1995) (Enquete Commission 1998)

Recycling of brownfield land has priority in sublocal and communal spatially-relevant plans before land consumption of greenfields especially in unzoned open land of the countryside (LRP 1.3-1, SMI 2003).

The growth of urban and transport area in Saxony shall be reduced to a quarter of the current growth by 2020 under consideration of a shrinking population until 2020 (LRP 1.3-1, SMI 2003)''

Slowing down the land consumption by urban and transport development in Germany to 30 ha/d by 2020 (German Government 2002).

WATER BODIES:

Wa 3: River water quality in 7 classes

I) No legally-binding standard exists

II) Achievment of the biological quality class II for all water bodies (Environmental report of the German Government 1994)

Streams in a state worse than quality class II and still water bodies, whose water condition does not satisfy the designated requirements for their use, shall be remediated step-by-step (LEP SMI 2003).

AIR /CLIMATE:

AC 1: Bioclimate: forest areas > 4 ha; air corridors and air production areas with relevance for settlement areas

I) No legally-binding standard exists

II) The function of areas with relevance for the climate in settlement areas shall be secured (Objective 4.5.1 LEP; LRP 1.4-1, SMI 2003).

Preservation of large-scale fresh air and cold air production areas in unzoned, open spaces on the basis of a qualitative assessment with the help of criteria such as cold air flow higher than 100 m^3/s, a minimum distance of 500 m to emittants, roads with less than 10 000 vehicles/day, an initial pollution of less than 25,0 μg $NO2/m^3$ and good surface-near aeration conditions (average wind speed of over 3,0 m/s, frequency of inversion less than 220 days per year) (LRP 1.4-1, SMI 2003 SMI 2003).

32.4 Quantitative Threshold Values and Conflict Intensity Classes

The standards and objectives influence the determination of quantitative threshold values and conflict classes for SEA-REP. In the evaluation it has to be concluded whether or not a predicted effect will be environmentally significant: In order to do so, it is necessary to analyse how far the current situation of the environment is from any established threshold or environmental objective. Examples from Germany for threshold values are noise exposure limits for residential areas according to the Federal Immissions Control Act; an objective is for example the designation and protection of 10 % of the territory of the Länder with the purpose of biotope connection according to the Federal Nature Conservation Act Art. 3 para. 1). Environmental objectives specific for the region mainly derive from the Saxon development plan LEP of 2003 (SMI 2003). With the help of the environmental indicator set the potential environmental impacts of individual site developments are

Matrix 32.1. Allocation of impact factors to the obligatory regional plan objectives requiring SEA (extract of only obligatory contents) (modified from Helbron and Schmidt 2007)

Regional Plan Objectives Requiring SEA / Impact Factor	Site-Specific			Site-specific/Impact zone (transborder relevance)					
	Land consumption	Land use change	Fragmentation/Barriers	Change of groundwater level	Flooding	Pollution with noxious substances	Noise pollution	Visual impacts	Positive environmental impacts
Priority and reservation area (VRG, VBG) for measures of the technical flood prevention	X	X	X	X	X			X	X
Priority and reservation area for a controlled use of local resources	X		X	X		X	X	X	
Designation of lignite coal deposits in the open-cast mining areas of Nochten and Reichwalde as priority areas for lignite coal mining	X		X	X		X	X	X	
Priority and reservation area for afforestation		X	X					X	X
Securing the spatial conditions for a future use of lignite	X		X	X		X	X	X	
Securing the spatial conditions for a use of wind energy as terminal planning	X	X	X				X	X	
Priority and reservation area water resources				X					X
(…)									

estimated and assessed in relation to the environmental value of the respective affected area following the ecological risk assessment method. The selected state indicators are cross-linked with the impact indicators acting on the individual sites (see Matrix 32.1 and 32.2).

Matrix 32.2. Environmental components potentially affected by impact factors (extract of human, flora/fauna/biodiversity and soil) (modified from Helbron and Schmidt 2007)

Impact Factors / Environmental Components			Site-Specific		Site-specific/Impact zone (transborder relevance)						
			Land consumption	Land use change	Fragmentation/Barriers	Change of groundwater level	Flooding	Pollution with noxious substances	Noise pollution	Visual impacts.	Positive environmental impacts
Human Health	Hu 1	Noise pollution in settlement areas							X		
	Hu 2	Pollution with noxious substances in settlement areas						X			
Flora, Fauna, Biodiversity	FFB 1	Biotope types	X	X	X	X	X	X	X		X
	FFB 2	Red List Species and rare species	X	X	X	X	X	X	X		
	FFB 3	Biotope connection areas	X	X	X						X
	FFB 4	Protected areas (national park, nature reserve, bio-	X	X	X	X	X	X	X		
	FFB 5	FFH-areas and special protection areas (SPA)	X	X	X	X	X	X	X		
Soil	So 1	Natural Productivity	X	X		X	X	X			
	So 2	Storage and buffer function	X	X		X		X			
	So 3	Biotic habitat function	X	X		X		X			
	So 4	Erosion risk		X							
	So 5	Contaminated areas	X	X				X			
	So 6	Unsealed areas	X								
(...)											

Two types of spatial impacts can be distinguished:

• *Impact on an environmental component at an individual site* represents an intersection of a designated area and the environmental value. It characterises a direct nuisance (in the sense of loss) of the environmental function's quality.

- An *impact on an environmental component at an impact zone* represents an intersection of an *impact zone* (potential scope of significant environmental impacts) and the environmental importance (evaluated). This is an indirect effect on environmental functions through pollution with noxious substances, noise or visually adverse effects (Helbron and Schmidt 2005).

The conflicts on the site are generally quantified in square kilometres or hectares. The result is a range of conflict intensities determined by the environmental value of the site and the *impact indicators*. The assessment of every single site is necessary to be able to compare site alternatives and find the least possible environmental conflict. The conflicts were ranked into classes.

The first class (1) *'High Conflict'* categorises areas with a high potential for conflicts. This covers firstly (1.a)) *restrictions* in the form of legal exclusion zones, where adverse effects are prohibited by law such as national nature reserves or according to environmental legally binding standards. With additional impacts on a specific site there is a risk that threshold values and environmental objectives will be exceeded (respectively 'broken'). Secondly (1.b)) areas with a high ecological or social value of nature or landscape where significant effects are likely to happen are categorised as high conflicts.

The second class (2) *'Medium Conflict'* characterises a medium potential for conflicts between impacts and the environmental value of the area. This medium rank is necessary in order to steer *'High conflict'* developments towards environmentally more resistant locations and/or away from sensitive areas according to the precautionary principle (comparison of alternative sites).

The third class (3) *'Low conflict'* categorises conflicts without significance due to the inferior importance of the state of the environmental component on the specific site.

Each impact may lead to a decrease of the environmental potential of a site and thus to an 'upgrade' in the conflict classification. Alternative site selections have to be considered especially in the case of *'High conflicts'*. The final result should be sites or strategic options with the least possible negative effect on the environment and the best compromise between conflicting objectives. Additionally mitigation and compensation measures have to be documented in the environmental report. The indicator's function is to describe and assess the necessary environmental baseline. Thus a strong linkage with environmental data, environmental thresholds and objectives was created. The indicator set was used in initial scoping exercises to determine issues that could be affected by the regional plan of Upper Lusatia-Lower Silesia. This stage, which was undertaken in cooperation and under consultation with Polish and Czech authorities, is especially important since it ensures that the baseline study is not limited to already existing information, but may also act as a driver for broader data assembly.

In the SEA report all derived conflict maps, prepared with the help of a geographic information system (GIS), are documented. The following examples of threshold values create the limits for the determined conflict intensity classes (Table 32.2). Table 32.3 presents an example of a checklist for SEA-REP, which includes state and impact indicators, threshold values and classes of the conflict intensity. For each environmental component a checklist was created, which contains all relevant information for SEA-REP.

Table 32.2. Threshold values for SEA-REP (examples) (modified from Helbron and Schmidt 2007)

1.a High Conflict with Restriction	1.b High Conflict without Restriction	2. Medium Conflict	3. Low Conflict
FFB 4 Entire area of NP, NSG, BR	–	–	No negative impacts on protected areas
So 2 No legally-binding standard exists	Special importance: buffer classes 5 and 4	Medium importance: buffer class 3	Inferior importance: buffer classes 2 and 1
AC 1 No legally-binding standard exists	Forest areas > 4 ha; cold and fresh air corridors and production areas of relevance for settlement areas	Cold and fresh air areas/corridors in the catchment area of highly polluted areas, relevant for settlement areas	Cold and fresh air areas and corridors not relevant for settlement areas

Table 32.3. Example of state indicator, impact indicators and conflict intensity classes in a checklist (modified from Helbron and Schmidt 2007)

So 6	Checklist Unsealed Area (Extract)
Indicators	
State Indicator: Unsealed areas with soil sealing share of < 25 %	
Impact indicator: Land Consumption of areas with soil sealing share of < 25 %	

Assessment of the Environmental State	
Class of Environmental state	
SI= Special Importance	Areas with soil sealing share of < 25 %
GI = General Importance	-
II = Inferior Importance	Areas with soil sealing share of > 50 %

Assessment of the Environmental Impacts on the Site

Conflict Intensity	Lay-out	Effects of Impact factors (impact indicators) on individual sites with importance according to the assessment of the environmental state (SI, GI, II)
		Land Consumption
1 a. High Conflict with Restriction		Currently no legally-binding standard exists
1 b. High Conflict without Restriction		SI
2. Medium Conflict		-
3. Low Conflict		II
Positive Environmental Impacts		Created frame for desealing at lower land use planning level (tiering)

32.5 Conclusions and Recommendations

Land consumption is a progressing environmental impact in Germany and Saxony, even in regions with a shrinking population (see Storch, chapter 19). SEA-REP is a decision-making instrument, which has a potential to optimise the regional plan in favour of the environment. The feasibility of a baseline-led SEA method depends to a great extent on the availability and adequacy of environmental objectives, standards and data. If the aim of an efficient consideration of alternatives at regional planning level is to be achieved, further environmental objectives have to be derived for the prevention of the environment and human health against environmental harm.

Regional planning has to take over the responsibility and strengthen its competencies, especially in the field of implementation of environmental objectives to reduce and finally stop land consumption and recycle brownfield land instead. Currently there is still a lack of limit values for soil sealing at regional level. In future all possibilities especially to reduce land consumption and other negative effects on the environmental media have to be used. The step of the derivation of assessment thresholds and conflict classes is crucial in SEA, as the results directly influence decision-making. Not only should legally-binding restrictions be complied with but, in an ideal situation, general designations assessed as high conflicts should also be avoided.

SEA is an essential tool to take on this challenge of a movement towards more environmentally-friendly regions in the EU. In order to achieve efficient SEA-REP in EU Member States in the future, the following issues should be addressed in policy making and selection of methodologies:

- in order to overcome the unbalanced situation concerning legally-binding standards (i.e. stricter set limits of urban sprawl and consumption of resources), some key indicators have to be defined, which shall function as an "emergency alarm" in case of proceeding developments
- additional environmental quality objectives and standards have to be set at EU and national levels for the long-term protection of soil, land and resources especially against consumption; these have to be regionalised
- a consideration of the prevention of the environment and human health against environmental harm can be promoted with the determination of a range of medium conflict intensity
- a transparent identification of environmental conflicts strengthens the position of environmental concerns in the weighting/balancing processes of regional planning, so that an 'overweighting' in favour of economic concerns becomes more difficult
- more research and expertise is needed in the selection of adequate indicators and the derivation of efficient assessment thresholds and classes.

As an incentive for using SEA's potential at regional planning level certificates could be handed out to especially environmentally-friendly European regions.

References

Bastian O, Schreiber K.-F. (1994) A tiered biotope assessment in local landscape planning. (Eine gestufte Biotopbewertung in der örtlichen Landschaftsplanung), Dresden

Enquête Commission (1998) Final report 'Protection of human being and the environment'. Abschlußbericht der Enquete-Kommission "Schutz des Menschen und der Umwelt - Ziele und Rahmenbedingungen einer nachhaltig zukunftsverträglichen Entwicklung" des 13. Deutschen Bundestages: Konzept Nachhaltigkeit. Vom Leitbild zur Umsetzung. Hrsg.: Deutscher Bundestag, Referat Öffentlichkeitsarbeit

German Government (2002) National Sustainability Strategy of the German Government 'Perspectives for Germany'

Helbron H, Schmidt M (2005) Part 1: Development of an indicator catalogue for the environmental assessment in regional planning of Saxony (indicator set/criteria). In: Bölitz D, Helbron H, Schmidt M, Reinke M (2005) First interim report on Interreg IIIA project 'Strategic environmental assessment for regional planning – development of a trans-national assessment and practice concept for Saxony, Poland and the Czech Republic, 27.05.2005, pp. 2-45 (in German) Internet address: http://www.rpvolns.home page.t-online.de/frame1.htm, last accessed 06.12.2006

Helbron H, Schmidt M (2007) Indicator system as instrument for the prediction and assessment of the environmental state and environmental impacts of regional plans. In: final report of pilot project on „Strategic environmental assessment for regional planning – development of a transnational assessment and practice concept for Saxony, Poland and Czech Republic". July 2004 until October 2006, Saxony (forthcoming)

Lee N, Wood C (1995) Strategic Environmental Assessment. EIA Leaflet series No 13, EIA centre, University of Manchester, Internet address: http://www.art.man.ac.uk/EIA/ publications/leafletseries/leaflet13/, last accessed 29.08.2005

LfUG (2006) Soil protection and land consumption (Bodenschutz und Flächenverbrauch) Internet address: http://www.umwelt.sachsen.de/lfug/boden_13402.html, last accessed 06.12.2006

RPV OL-NS – Regional planning authority Upper Lusatia-Lower Silesia (2007) Regionaler Planungsverband Oberlausitz-Niederschlesien. Internet address: http://www.rpvolns.homepage.t-online.de/frame3.htm, last accessed 20.02-2007

Schmidt M, João E, Albrecht E (eds) (2005) Implementing strategic environmental assessment. Springer, Berlin

Scholles F, von Haaren C (2005) Co-ordination of SEA and landscape planning. In: Schmidt M et al. (eds.) Implementing Strategic Environmental Assessment, Springer, Berlin, pp.557-566

SMI – State Ministry of Interior of the Free state of Saxony (2003) Saxon land use development plan (Landesentwicklungsplan Sachsen, LEP 2003), from Government resolved on 16.12.2003 and published on 31.12.2003 (GVBl. Nr.19/2003)

UBA (1995) Glossary of space-oriented environmental planning. Internet address: http://www.umweltdaten.de/rup/glossar/deutsch.pdf, last accessed 06.12.2006

UBA (ed) (1993) Internationaler Vergleich von Verfahren zur Festlegung von Umweltstandards (International comparison of procedure for the setting of environmental standards, Erich Schmidt publisher, 259 p., UBA-reports 3/93, Berlin

Part V – Implementation

Implementation of standards and thresholds with environmental assessment as a tool faces a number of practical and theoretical problems. Standards and thresholds may vary from legally binding, quantified environmental quality norms to "soft standards" such as non-quantifiable goals for future environmental quality (Chap. 36). Implementation and thus EA follow up will mean very different things depending on the type of standards and thresholds. We are thus dealing with a range of tools and instruments for environmental governance. The implementation of standards covers a wide range of topics and sciences dealing with all aspects of standards from the underlying ecology and health sciences to policy research. This section deals with a more limited aspect of implementation: issues related to EA follow up and the power of standards and thresholds as tools in environmental governance.

There are many important aspects of standards in relation to EIA and SEA from "thresholds of concern" and standards for requiring an assessment to be made to standards relating to environmental quality which define both acceptability of projects or plans or define needs for mitigation. Regardless of type there seems to be an increasing consensus concerning an "implementation deficit". Cherp (Chap. 34) attributes this to two principal factors: uncertainty and lack of commitment. The degree of enforcement of standards may be considered to be at one end of the commitment scale. It presupposes follow up and monitoring and the quality of this obviously influences implementation. While monitoring is not a sufficient precondition for enforcement it is a necessary one. It should thus be of major concern that the literature indicates that EA follow up in general and monitoring in particular may be the weakest link in the EA chain. The weaknesses are present both in follow up of environmental assessment and in monitoring of environmental impacts (Wood 2003).

If follow up is lacking then projects may undergo changes of such a total magnitude that initial EIA predictions become invalidated. If the ambitions to actually carry out mitigation measures suggested in the EIS are low then there may be an "implementation gap" acerbated by the fact that impacts of a changing project or plan are not predicted and described in the original assessment.

In well defined, infrastructure projects rationalist models of decision making and follow up are theoretically feasible. As Bartsch and Westerveld (Chap. 33) point out SMART indicators – specific, measurable, acceptable, relevant and time specified – can be defined and used.

Proposals for mitigation is in practice often a more important outcome of EIA than production of alternative solutions. Bartsch and Westerveld offer a number of practical remedies to the problems and failures of mitigation monitoring. They could be summarised as aspects of project integration. For short term projects or projects with a well defined implementation it is necessary to let EIA mitigation findings be directly integrated into post decision project plans. For longer term implementation environmental management plans (EMP) and environmental management systems (EMS) offer possible instruments for SEA follow up (Chap. 34).

As Glasson points out (Chap. 1) one important function of standards may be to give reliable signals of future requirements. Thresholds and standards should be "realistic" (Chap. 10) in order to be implemented. Defining realism is part of the value base or political element of setting standards. As Cherp notes (Chap. 34) follow up and thus ultimately implementation depends among other things on commitment. The degree of commitment will influence perceptions of realism of standards especially if they have considerable economic implications or are in conflict with the primary goals of an organisation, firm or sector agency. There is an inherent tension between setting standards and thresholds scientifically related to "Nature's limits" or health impacts on the one hand and compromising to achieve "realism". Environmental quality norms are in fact often presented as scientific alternatives to the compromises inherent in other tools such as "best available technology". One way of handling the problem of short term realism versus long term sustainability is to set successions of increasingly stringent standards over time. This gives actors time to adapt which may positively influence both perception of realism and commitment. The question of whether different types of standards and thresholds in fact give reliable signals about future environmental quality is an important aspect of reliability and legitimacy of standards (Chap. 36).

Uncertainty is inherent in all aspects of EA. Tenøy (Chap. 35) discusses different forms of uncertainty which influence the possibility both of informed decision making based on EA and of meaningful follow up. Her cases demonstrate empirically an experience that many practitioners and researchers have: the successive loss of information concerning uncertainty as documents move from scientific background to material directly aimed at decision making. If uncertainty is not revealed to the decision maker rational decisions may become difficult as alternatives or mitigation measures may be presented as more distinctly different or certain than the predictions actually show them to be. Furthermore Rolf (2006) notes that conceptual uncertainty may be the most intractable form in many environmental issues. In contrast to this studies of EA effectiveness point to the profession seeing epistemic uncertainty as the main problem (Emmelin 1998). As the case of Sweden demonstrates (Chap. 36) conceptual uncertainty can come from an unclear mixture of binding environmental quality norms, scientifically set thresholds and more diffuse environmental goals. Lack of recognition of uncertainty presents a special problem in relation to legally binding environmental quality norms. Violation or even the possibility of violation of such norms may give rise to drastic requirements for mitigation or entirely prohibit a project of plan. Unrecognised uncertainty concerning whether norms are in fact in danger of being violated may undermine the entire legitimacy of decisions and systems of norms (Chap. 36).

A lack of clarity concerning the practical, theoretical and methodological differences between EIA and SEA is problematic in follow up and implementation of standards. In higher level SEA outright prediction of levels of pollutants is often not meaningful. Instead the role of SEA may be to guide lower level EIA and follow up (Handbook on SEA for cohesion policy). Commitment to this may be low if environmental concern is seen to conflict with sector goals or economic growth (Chap. 36). Tiering – i.e. a degree of consistency between higher level goals and lower level implementation – can benefit from coordination between SEA and environmental management in both private and public sectors (Chap. 34).

Apart from the need to foster commitment the single most important issue in implementation may be to increase the quality of EA follow up so as to create learning systems. At present the lack of or deficiencies in follow up in many systems (Wood 2003) is a major obstacle to development of environmental assessment systems and thus to it's effectiveness as an instrument of implementation of environmental standards and thresholds.

References

Emmelin L (1998) Evaluating Environmental Impact Assessment Systems– Part 1: Theoretical and Methodological Considerations. Scandinavian Housing and Planning Research 15(129):148

Handbook on SEA for Cohesion Policy 2007-2013. February 2006. Greening Regional Development Programmes Network. GRDP Project Team. Environment Agency. Exeter EX2 7LQ UK (www.grdp.org)

Rolf B (2006) Decision support tools and two types of uncertainty reduction. pp 134–11 in: Emmelin L (ed) (2006) Effective tools for environmental assessment – critical reflections on concepts and practice. BTH research Report 2006:3

Wood C (2003) Environmental Impact Assessment: A Comparative Review. 2nd edn, Prentice Hall

33 A Method to Monitor the Implementation of Mitigation Measures in Infrastructure Projects – Exemplified with a Project in the Republic of Yemen

Reinhart Bartsch and Dirk Hein Westerveld

GFA Group, Maghreb and Middle East Department

33.1 Introduction

Monitoring deals with the collection, analysis and use of information to support informed decision-making (EC 2004). Monitoring is usually an internal management responsibility and as a rule ongoing. It is carried out to check progress, take remedial action and to update plans. In the following, monitoring is applied to the implementation of measures to mitigate negative environmental impacts of a project. A project is understood as a series of activities aimed at bringing about a specified objective within a defined time-period and budget (EC 2004). Infrastructure projects are such to improve transportation, energy and water supply (household, irrigation), as e.g. roads bridges, dams, canals etc.

Mitigation measures are identified in the process of the Environmental Impact Assessment (EIA) of a project and eventually recorded in an Environmental Impact Statement (EIS). They target on adverse impacts of a project that are relevant, considerable and cannot be avoided or prevented. Of those impacts, which cannot be circumvented or evaded, there are such that require a compensation, e.g. if a dam floods a house, and those that can be mitigated, i.e. lessened, eased, moderated or taken the edge of.

Mitigation measures in the context of infrastructure projects are either structural, i.e. investments into additional construction works and/or non-structural, i.e. managerial measures. Monitoring mitigation measures means here the ongoing observation of defined project implementation activities to prevent or lessen the environmentally undesired impacts of an infrastructure project. In the following, a method is presented to monitor mitigation measures in infrastructure projects. This method bases on the Project Cycle Management approach as well as the Logical

Standards and Thresholds for Impact Assessment. Edited by Michael Schmidt, John Glasson, Lars Emmelin and Hendrike Helbron. © 2008 Springer-Verlag

Framework and ZOPP planning methods (GTZ 1997). The method has proved to be useful in the practice of project management, mainly for project progress monitoring, especially in development cooperation. Its data requirements are reasonable, i.e. *neither too simple nor too complicated* to collect to deliver useful information.

To demonstrate the application of the method for infrastructure – i.e. large-scale – projects, a real project is used. The project is simplified to allow a general discussion of the method proposed. In the following, it will be briefly described. Then, the monitoring method is presented. Some critical remarks on the method conclude this paper.

33.2 The Infrastructure Project

To illustrate the monitoring method, the Al Maneen Irrigation Scheme will be used. This irrigation project is located in the Republic of Yemen at the Governorate of Marib, close to the town of the same name, in the eastern part of Yemen, bordering on the Empty Quarter desert. At Marib, the floodwaters of the central mountains of Yemen reach the plain through a narrow gorge, used since the times of the Queen of Saba (thought to have lived around 1800 BC) to store the water through a weir. This weir, one of the most famous and important works of the pre-historic world, broke finally in the sixth century A.D. (Daum 1988 p.26) to be reconstructed by the grant of a wealthy Dubai Sultan whose family originates from this region, only in 1986.

A dam was constructed, but there was initially no irrigation scheme to make use of the water. This was then constructed on the left side of the dam, financed by the Arab Bank and other regional banking institutions. On the right side of the dam, efforts to rehabilitate the old irrigation areas were made by the European Commission Food Security Programme (Westerveld 2004), first by building a weir to raise the water level and then in a second phase by constructing the primary and secondary canal system[1]. A community development component was added to this infrastructure-building project, focusing especially on the empowerment of women.

There is a historic weir structure at the site of the new weir, one of the oldest monuments in the region, dating from 1800 BC and representing an important archaeological and cultural heritage (Deutsches Archäologisches Institut 2003). This was to be protected from construction works.

---

[1] This was in former times the right hand garden of the Queen of Sheba or Saba. The Queen had a weir barring the water coming from the mountains into the plain to irrigate two gardens. As the Prophet Mohammed said: "For the natives of Sheba this was indeed a sign in their dwelling place: a garden on their left and a garden on their right. We said to them: 'Eat of what your Lord has given you and render thanks to Him. Pleasant is your land and forgiving is your Lord.' But they gave no heed" (Koran 1978, Sura 23, 15ff).

The beneficiaries of the irrigation scheme are farmers, farm workers and the women of two conflicting tribes, inhabiting the right and the left bank of the Wadi respectively. The objective[2] of the Al Maneen irrigation project is expected to be an increase in agricultural production, raising the farm family incomes and reducing food insecurity.

The negative environmental impacts of the project where identified as the following:

- Fertile soils could be destroyed by the canal and weir construction works.
- The water quality of the Wadi and its aquifer could deteriorate through the expanded agricultural activities and the increase in household water supply.
- Valuable cultural heritage might be destroyed and made unattractive for tourist visits.

33.3 The Monitoring Method

The method presented in this paper is based on the LogFrame[3] approach. The first analytical step of this planning method is the problem analysis. It establishes cause-effect relationships between problems – here these are understood as negative environmental effects of a project. The problem analysis is then transformed into an objective analysis, i.e. a hierarchy of objectives. The LogFrame planning method's main output then is a LogFrame Matrix or Project Planning Matrix (PPM). This matrix consists of the project's descriptive summary,[4] i.e. the overall objective and project purpose, results, activities, indicators, sources of verification for the indicators and assumptions related to the descriptive summary (EC 2004 p62) (see Table 33.1).

The activities are planned to achieve the respective results and the respective results are planned to achieve the project purpose. The project purpose is then to contribute to the achievement of the overall objective or the impact.[5] The project management is responsible for the achievement of the activities and thereby for meeting the results as these are planned to be sufficient to achieve the results. The indicators derived for the objectives, results and activities, which are part of this hierarchy, are used as thresholds to be monitored. The monitoring activity it-

---

[2] The terminology of the PCM approach and EIA differ. To avoid misunderstandings, the EIA terminology is used in this article. The terminologies could be harmonized, if EIA had used a cause-effect analysis and presentation of impacts. This is, however, not done but instead impacts are listed regardless of their position in the cause-effect hierarchy.

[3] The logical framework approach or 'Zielorientierte Projektplanung' (ZOPP) is a part of Project Cycle Management (EC 2004). The monitoring method based on indicators of the LogFrame Matrix was developed by Salzer 2002 and refined by GTZ 2004.

[4] In the following the terminology of EC 2004 is used.

[5] In monitoring there is an attribution gap, which is located between the outcome and the impact and is due to the fact that the outcome only contributes to the impact, but does not necessarily evoke it (see Salzer 2002).

self is understood as the subsequent deployment of workforce in quality and quantity defined by the planning.

As a first step of the method the negative effects anticipated (based e.g. on the Environmental Impact Assessment, EIA, of a project and eventually recorded in an Environmental Impact Statement, EIS) are used to establish a problem analysis. This defines the logical hierarchical position of a problem/negative environmental effect. A chain of cause and effects could look like the example in Figure 33.1. The relationship between project objectives and the mitigation needs is given by the correct application of the LogFrame logic, i.e. in the Project LogFrame Matrix and the matrix established for the mitigation measures, the hierarchical position should be the same. The procedure is to do this together with all stakeholders; in the case of infrastructure projects these are the beneficiaries, the construction company, the supervising and implementing agencies as well as the funding entity. The definition of indicators for the subsequently established mitigation measure LogFrame Matrix[6], to be used as thresholds to be monitored, is the second step of the monitoring method presented here. Indicators describe the objective, result or activity in operationally measurable terms of quantity, quality and time (QQT) (2004 p80).

Table 33.1. Information contained in a LogFrame Matrix (EC 2004)

Project Description	Indicators	Source of Verification	Assumptions
Overall objective: The broad impact to which the project contributes – at a national or sector level	Measures the extent to which a contribution to the overall objective has been made.	Sources of information and methods used to collect and report it.	
Project purpose: The outcome at the end of the project expressed as the expected benefit to the target groups	Helps answer the question "How will we know if the purpose has been achieved?	Sources of information and methods used to collect and report it	Factors outside project management's control that may impact on the purpose-objective linkage
Results: The direct/tangible results the project delivers.	Helps answer the question "How will we know if the results have been delivered?	Sources of information and methods used to collect and report it	Factors outside project management's control that may impact on the result-purpose linkage
Activities: The tasks that need to be carried out to deliver the planned results	Sometimes a summary of resources/means is provided in this box		Factors outside project management's control that may impact on the activity-result linkage

---

[6] The LogFrame method is too comprehensive to be presented here completely.

Indicators should be SMART after EC (2004 p81):

Specific to the objective it is supposed to measure
Measurable (either quantitatively or qualitatively)
Available at an acceptable cost
Relevant to the information needs of managers
Time-bound - so we know when we can expect the achievement to happen

If indicators for the objective, purpose and results of our irrigation project were formulated, they could look like shown in Table 33.2, second column. The indicators/thresholds defined determine the costs of implementing the monitoring activity.

The descriptive summary of a LogFrame Matrix developed specifically for the mitigation measures of a project, could in the case given look like the one presented in Table 2. The Table does not show all the activities a mitigation plan for the Al Maneen project would include; examples were chosen to describe the method only.

To complete the Project Planning Matrix, assumptions are formulated and the sources of verification (SOV) for the indicators defined. The SOV should specify:

- How the information is collected?
- Who should collect the information?
- When/in which intervals the information should be collected?

Logically, this determines the quantity and quality of manpower and equipment etc. required to monitor the indicators.

When all the information generated is filled into the LogFrame Matrix and the LogFrame presented in Table 33.2 is obtained.

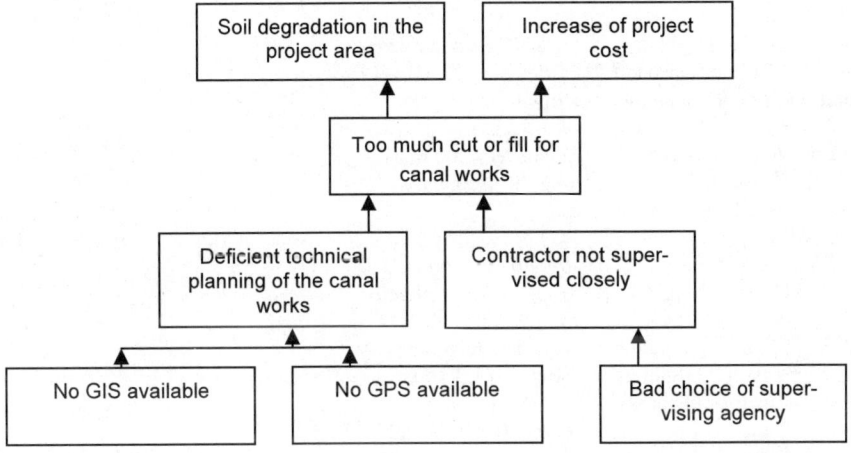

Figure 33.1. An example of cause-effect relationships in a problem analysis

Table 33.2. Mitigation Measure LogFrame Matrix Al Maneen Scheme

Descriptive Summary	Objectively Verifiable Indicator (OVI)s	Source of Verification (SOV)	Assumptions
Overall objective: Environment of the Maneen area is conserved	IOO: Quality of water and soil is maintained for 10 years on the level of 2004.	Water and soil tests	
Project purpose: Adverse impacts of constructing the irrigation project are reduced	IPP: Population of scheme area benefits of scheme construction without negative effects in a period of 5 years after scheme construction beginning.	Available water is used by the farmers	APP1: Food security aid continues as politically desired APP2: Population growth in the project area does not exceed national growth rate
Results: 1) Destruction of fertile soils by the canal and weir construction works is minimised. 2) Agricultural activities and the increase in household water supply do not affect the Wadi water quality adversely 3) Adverse impacts of the project on cultural heritage are impeded.	IR2: Water quality of the aquifer measured at two village wells has the same quality in 2009 as in 2004 regarding salt content and harmful residues. IR3: Historic weir is preserved on scientific standard until 6.2005.	Water tests	AR1: EC Technical Assistance continues AR2: Project proposals submitted by implementing agencies are properly documented and presented on time AR4: Project proposals submitted by implementing agencies are technically and economically correct
Activities: A 1.1: Purchase GPS A 1.2: Minimise canal cut and fill through planning A 1.3: Supervise contractor of canal works A 2.1: Establish drainage A 2.2: Establish extension service A 3.1: Secure funding A 3.2: Design site protection A 3.3: Secure cooperation of Ministry of Antiquities and German Archaeological Institute A 3.4: Secure historic site	IA 1.3: The contractor carries out cut and fill with an accuracy of +/- 0.1 cubic meter (5%) every 200 m along the secondary canal IA 3.4: Historic weir structure is protected by a Rip-Rap covered wall (6.2005), the site is cleaned of flood damages (8.2005), presentation signs are set up (10.2005) and the site is included in tourist guide (12.2005)		

33.4 Result: The Mitigation Measure Monitoring Plan

The establishment of a procedure to monitor mitigation measures or the Mitigation Plan[7], is the third and final step of the method. They are established on the basis of the indicators defined, using the hierarchy of the LogFrame Matrix. Indicators are formulated for objectives, purpose and result levels. For monitoring purposes they can also be formulated for the activity (see Table 33.2) and assumption levels.

All the indicators considered relevant for monitoring purposes are then inserted into a monitoring plan (see Salzer 2002). The consideration of relevancy can be shown for Activity 1.3 (see also Figure 33.1). It is not needed to monitor the purchase of a GPS. To monitoring of the proper project planning should also not be part of the mitigation measures plan. Relevant is the monitoring of the activities of the building contractor, who may aim to maximise cut or fill.

The mitigation measures monitoring plan should be set up according to the requirements and the nature of the respective project. It should include exact time indications, not just the year and should assign responsibilities for the verification and the corrective measures. In our case it could look like the monitoring plan given in Table 33.3 and 33.4 showing examples of two activities monitored by the help of an indicator.

Table 33.3. Al Maneen Irrigation Scheme Project Monitoring Plan

Source of Verification	Time of verification	Value to be achieved/ Threshold +/- 5%	Respon sible for verification	Value achieved	Difference	Corrective measures to be taken	Responsible for corrective measures
Supervision report Payment certificate	1.2. 2005	At 50 m CM cut = 0.59 m2 CM fill = 3.63 m2	Supervision	-	-	Correct fill and cut	Construction company
-	1.5. 2005	At 450 m CM cut = 0.55 m2 CM fill = 2.99 m2	Supervision	-	-	-	-
-	1.10. 2005	At 900 m CM cut = 0.39 m2 CM fill = 4.47 m2	Supervision	-	-	-	-

Indicator A. 1.3: The contractor carries out cut and fill with an accuracy of +/- 5% every 200 m along the secondary canal

---

[7] This plan identifies the responsibilities, schedules and budgets of the environmental and social management measures to be implemented for a certain project.

Table 33.4. Indicator A 3.4: Historic weir structure is protected by a Rip-Rap covered wall (6.2005), the site is cleaned of flood damages (8.2005), presentation signs are setup (10.2005) and the site is included in tourist guide (12.2005)

Sources of Verification	Time of verification	Values to be achieved	Responsible for verification	Value achieved	Difference	Corrective measures to be taken	Responsible for corrective measures
Supervision report	6. 2005	300 meters of protective earth wall 750 meters Rip-Rap cover 50 meters of stone masonry	Mott-McDonalds	-	-	-	ERADA
Project visit of project engineer	8.2005	All rubble has been removed Stones refitted where possible	Dr. Iris Gerlach	Rubble removed, stones not refitted	Stones not refitted	Staff of the Sirvah excavation refits stones	German Archaeological Institute, ERADA
Report of German Archaeological Institute Sana'a	10.2005 12.2005	Sign set up Archaeological Guide published	Rural engineer (Westerveld)	Sign set up Guide not yet published	Guide not published	No action required	-

33.5 Conclusions and Recommendations

If the method is followed through it results in a mitigation plan and a monitoring plan for the implementation of these measures. The establishment of SMART indicators is, of course, essential because indicators not fulfilling these requirements are use-less in practice. The established monitoring system should be practical, affordable budget wise and efficient enough to achieve the targeted goals.

So, which are the drawbacks of the method? Or why is it not used more commonly? The drawbacks experienced are the following and explain why the system is not as popular as it should be theoretically:

- Perfection makes the system crash. Using the problem analysis, too many indica-tors are formulated and monitored, in our example e.g. the costs of unnecessary cut or fill or the agricultural area lost. At the beginning of the monitoring fever, projects were thought to have a person exclusively dedicated to the monitoring, the updating of the planning and the follow up of corrective measures. This proved to be overkill and resulted in giving up monitoring altogether in many cases. In the case of the Al Maneen Irrigation Project the Assistant Engineer was charged with the cost monitoring, with included the control of the Interim Payment Certificates where the amounts of cut and fill where registered. The Engineer monitored the overall progress of the project and the works on the cultural heritage site. More was not monitored.

- Planners and especially the implementers did not make use of the monitor-ing because the results came to late or were inaccurate. In the case of the project presented, the payment certificates gave the cut and fill figures of two to three months before. The monitoring thus resulted in admonishing the contractor "not to fail again". Participatory monitoring, a remedy sug-gested, is time consuming. Even more time consuming or/and expensive is generally the survey to establish the baseline values required for the monitoring of the indicators. In case of the project presented, water and soil analysis were not conducted. Most probably – given the framework conditions of Yemen – these values will never be monitored.
- Also, the method proposed requires knowledge of the LogFrame approach if it were to be applied logically. The high costs to provide this knowledge can be studied in the case of the GTZ, a case of failure to provide ade-quate training in time can be seen at the European Commission's struggle with the introduction of this planning method.

The solution to these problems is tentatively:

- To restrict the number of the indicators monitored by a monitoring plan to the absolutely essential. The means of verification should be practically accessible, not only theoretically. Surveys e.g. to verify a situation are al-most never carried out in practice. Data, which other entities would have to supply, should be excluded as sources.
- Special monitors are not useful. It is the engineers or economists that are responsible of the project implementation that should do the monitoring. It can be delegated to the supervision consultancy, but the monitoring of the monitoring or evaluation of the monitoring etc., should be strictly avoided.
- Formalisation should be avoided. The establishment of indicators or the moni-toring plan should suffice as a rule in projects up to € 5 million. Re-ports on the monitoring lead to "cheating" and deduct working time from implementation tasks.
- The monitoring time frame should not be outside the project implementation period as ex post evaluations are extremely rare in practice.

References

Bartsch R (2004) The Al Maneen Irrigation Scheme, Feasibility study, FSMU, Sana'a

Daum W (ed) 1988 Die Königin von Saba, Stuttgart

Deutsches Archäologisches Institut (2003) Orient Abteilung, Außenstelle Sana'a: Hefte zur Kulturgeschichte des Jemen, 25 Years Excavations and Research in Yemen, 1978-2003, Sana'a

EC – European Commission (2004) aid delivery methods. vol 1, Project Cycle Management Guidelines

GTZ – Deutsche Gesellschaft für Technische Zusammenarbeit (1997) GmbH: ZOPP, Objectives-oriented Project Planning, Eschborn

GTZ (1999) project cycle management, Eschborn

GTZ (2004) results-based monitoring, Guidelines for Technical Cooperation projects and Programmes, Eschborn

Lagemann J (2004) monitoring and evaluation of natural resource management programmes

Salzer W (2002) impact monitoring, training materials, Mombasa 2002

Westerveld DH (2004) Marib Dam project, Al Maneen Diversion Weir, Primary and Secondary Irrigation canals, FSMU, Sana'a

34 The Role of Environmental Management Systems in Enforcing Standards and Thresholds in the Context of EIA Follow-Up

Aleh Cherp

Department of Environmental Sciences and policy, Central European University, Budapest, Hungary and International Institute for Industrial Environmental Economics at Lund University, Sweden

34.1 Introduction

The effectiveness of Environmental Assessment (EA) depends on its ability to effect change in the way human activities impact the environment. Unfortunately, environmental professionals are all too familiar with a gap between protecting the environment *'on paper'* (e.g. in the EIS and related documents) and destroying it *'on the ground'* where the activities undergoing EA are implemented. To bridge this gap, EIA follow up contains the management component defined as *"making decisions and taking appropriate action in response to issues arising from monitoring and evaluation activities"* (Morrison-Saunder et al. 2003). Environmental Management Plans (EMPs) and Environmental Management Systems (EMS) are the main management tools which can improve EIA effectiveness during follow-up. This section considers the rationale for and the experience of using EMPs and EMS in the context of EIA and SEA with particularly attention to the role of standards and thresholds.

The causes of the discrepancy between EIA findings and recommendations on the one hand and the actual practices and their impacts on the other hand are pervasive and diverse, but can be roughly divided into two groups: *uncertainty* and *commitment*. Uncertainties are inherent in the Environmental Assessment process. First of all, there are always uncertainties in predicting environmental impacts. These result from our lack of understanding of how complex socio-environmental systems work, lack of data, time and other resources to construct accurate prediction models, as well as the inherently 'chaotic; and 'emergent' nature of many phenomena we are dealing with. Secondly, there is uncertainty regarding the way the planned activity will be implemented. There may be change in such contextual factors as technology, budgets, politics, or the output linked to the market de-

Standards and Thresholds for Impact Assessment. Edited by Michael Schmidt, John Glasson, Lars Emmelin and Hendrike Helbron. © 2008 Springer-Verlag

mands for final products. Few projects, to say nothing of plans and programmes, operate as originally planned. The significance of these two categories of uncertainties has been widely documented in the EIA literature starting from the early 1980s, which essentially concluded that it is typically a very minor proportion of environmental impacts which can be verified as 'accurately predicted' within the EIA process (see a summary of this evidence in Noble and Storey (2005)). Uncertainty is also affected by the constant change in monitoring techniques. Finally, there is uncertainty regarding the standards themselves. These naturally change with time as attitudes, knowledge and capacity of those who set standards and identify thresholds change. Some of the standards, for example, those involving the notion of the Best Available Technique (BAT)[1] are by definition meant to change with time. Various locally-specific thresholds are also bound to change as the social and eco-systems framing them evolve.

This puts EA – a forward-looking, predictive and deliberative tool – in a precarious situation where it seeks to base its recommendations on validating compliance of uncertain impacts from an inprecisely defined activity with ever-changing standards and evolving thresholds. On top of these difficulties there is another group of causes potentially making EA conclusions less effective: that of commitment. In most cases, proponents of proposed activities have all the incentives to present EA findings in such a way as to secure necessary permits and approvals. This usually implies expressing and justifying both the intention and feasibility to comply with relevant standards and thresholds. When it comes to implementing the proposed activities the commitment of the developer to the environment may, for many reasons, diminish. It may turn out that respecting standards and thresholds is not technologically or financially feasible. The institutional framework and ownership may transform so that both the pressure for and attitudes to compliance decrease. In case of projects funded by international financial institutions such as the World Bank this problem has been well documented, for example, by Goodland and Mercier (1999) who noted that while EIAs are developed and endorsed by the World Bank and central governments of lending countries, the actual implementation is often relegated to provincial authorities who lack capacity to implement EIA recommendations. In this respect, Goodland and Mercier (1999) note that "the quality of Environmental Assessment report is necessary but not sufficient for successful EIA" (p.7) and conclude that "EIA without political commitment is inadequate, risky and wasteful" (p.14). This commitment problem is not specific to projects funded by the World Bank. For example, in a suite of 1118 mitigation proposals from 41 EISs Marshall (2001)) found that only in 700 (62 %) was the proponent viewed as being committed to their implementation, whilst the remaining 418 (38 %) were expressed in such a way that the proponent could not be held to be committed to their implementation. The author of that re-

---

[1] BAT-based standards are, for example, used in the permitting systems guided by the European Union's Integrated Pollution Prevention and Control (IPPC) Directive (COUNCIL DIRECTIVE 96/61/EC of 24 September 1996 concerning integrated pollution prevention and control (OJ L 257, 10.10.1996, p. 26)).

search concludes that *"mitigation is of little value if it remains as a series of [non-implemented] proposals in an EIS."*

The lack of commitment combined with uncertainty creates especially prob-lematic conditions:

> *Uncertainty is a key criterion in determining the need for follow-up, but the EA review process is designed to discourage discussion of uncertainties, which in turn may distort the resources allocated to follow-up. The priority of the pro-ponent is to get their project approved. Accordingly, they will be reluctant to concede to decision-makers any uncertainties associated with project design, schedule or implementation methods or outcomes during the preapproval stage* (Noble and Storey 2005)

Already in the first decade following the passage of the US NEPA, the EA com-munity noted that uncertainty can render EA findings less relevant and the lack of commitment – less actionable. A variety of suggestions ranging from pragmatic adjustment to rethinking of the EA approach have emerged from these observa-tions. Perhaps the most radical re-conceptualization of EIA to address the issue of uncertainty was proposed by Holling in 1978 under the name Adaptive Environ-mental Assessment and Management (AEAM) in the book with the same title Holling (1978). AEAM proceeds from the fact that environmental impacts cannot be accurately predicted and locally-specific thresholds cannot be meaningfully de-fined in advance. Instead of focusing on impact prediction, Holling proposes to set up interactive and adaptive systems which will start with however imperfect mod-els of the environment-development interactions and then – through continuous monitoring, trial-and-error interventions and learning – gradually improve these models, thus accumulating knowledge and capacity for minimizing unwanted ef-fects in a longer-term perspective.

If Holling primarily deals with the issue of *uncertainty*, Goodland and Mercier (1999) explain how the World Bank addresses the issue of *commitment* in their ar-ticle "The Evolution of Environmental Assessment in the World Bank: from 'Ap-proval' to Results". These authors point out that prior to the 1990s EIA findings had often stayed on paper due to three main reasons: the lack of political will to protect the environment; weak institutional capacity for environmental manage-ment and the lack of financial resources specifically designated for implementing mitigation measures. Goodland and Mercier argue for shifting the emphasis of the EIA process from preparing an EIS (with focus on predicting environmental im-pacts) to preparing an EMP (with focus on presenting planned environmental pro-tection activities). Instead of making an EMP an insignificant annex to an EIS, they propose the opposite: to make the EMP the main document and use the EIS primarily to justify environmental management activities proposed in the EMP with reference to expected environmental impacts. This approach has been used by the World Bank in the last decade for environmentally significant activities where EMPs were considered part of lending contracts.

AEAM and the World Bank system are similar in that they consider 'environ-mental management' as central in addressing both uncertainty and commitment - the main causes of the gap between EIA statements and reality. Noble and Storey

(2005) also remark that "recent practice experience suggests that advances in EA have been greater in terms of the development of impact management and mitigation approaches than in the development of predictive techniques and methods" concluding that:

> given the difficulties associated with predictive accuracy, the potential benefits of follow-up for the environment generally, and to proponents specifically, a broader argument about EA tasks and EA 'good practice' can be made to reallocate resources from providing extensive baseline analyses and sophisticated predictions towards better management and follow-up of effects.

'Environmental Management' is a very broad concept with many aspects and definitions. For the purposes of this chapter we will focus on two environmental management instruments: Environmental Management Plans and Environmental Management Systems. The difference between the two is not always clear-cut, although many view an EMP as a formal document and an EMS as a set of both documents and processes for implementing and (re)formulating EMPs. The relationship between an EIA, an EMP and an EMS in the most simplified form can be depicted as in figure 34.1 (see similar diagram in Marshall 2001). It shows that EIA findings and recommendations are reflected in an EMP, which, in turn, contributes to the planning component of an EMS.

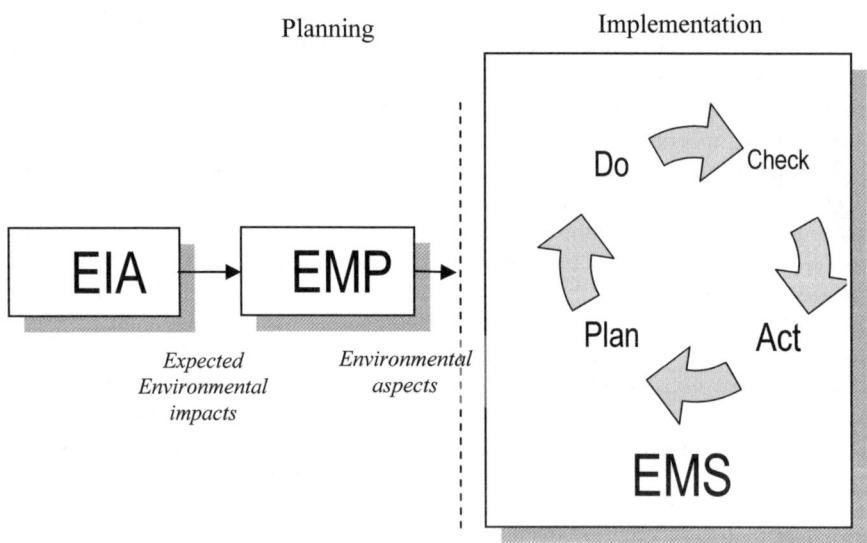

Figure 34.1. A simplified relationship between EIA, EMP and EMS

34.2 EMPs and EMSs

Environmental Management Plans are prepared at the planning stage in parallel with preparation of Environmental Impact Statements. An EMP may either be incorporated in the EIS or submitted as a separate document. An EMP describes the proposed management systems and the monitoring and auditing arrangements required to ensure both the proper implementation of agreed mitigation measures and the verification of predicted environmental impacts. In particular, an EMP may include specification of technical and institutional measures and arrangements such as budgets, timelines and responsibilities designed to ensure compliance with relevant environmental standards. An EMP should form part of the documentation used for consultation purposes and decision making and should be used by the competent authorities when specifying conditions to be met by the developer when implementing the project. Although EA follow-up is required in many systems it rarely incorporates the management component such as preparation of EMPs. In this regard, few exceptions deserve particular attention.

One is Canada, where Environmental Protection Plans (EPPs) are prepared on regular basis at both provincial (e.g. Saskatchewan) and federal-level EAs (Noble and Storey 2005). Another example is Western Australia where common EIA practice includes the setting of conditions in which an environmental objective is specified but the proponent is not directed as to how they should meet this objective. This often leads to consent decisions requiring the production of EMPs in the post-decision stage of projects, but prior to project construction and operation. The EMPs provide details of mitigation and monitoring measures to satisfy EIA objectives. (Morrison-Saunder et al. 2003). Monitoring and reporting on the extent to which environmental objectives have been met is mandatory for proponents. However, methods for project and environmental management are not prescribed, leading to an adaptive management approach (Morrison-Saunders and Arts 2004).

Finally, in the already mentioned World Bank system EMPs are essential elements of EA reports for Category A (i.e. the most environmental significant) projects whereas for many Category B (less significant) projects, the EA process may result in a management plan only. To prepare a management plan, the borrower and its EA design team (a) identify the set of responses to potentially adverse impacts; (b) determine requirements for ensuring that those responses are made effectively and in a timely manner; and (c) describe the means for meeting those requirements. EMPs in the World Bank focus on mitigation, monitoring, and capacity strengthening. Each proposed mitigation measure is linked to budgets and schedules, thus, enhancing commitment. Loan covenants commonly specify that EMP shall be implemented on an agreed schedule by an agreed budget (Goodland and Mercier 1999; www.worldbank.org).

It is important to ensure that the EMP is not just 'another paper' disconnected from the actual activities. According to Goodland and Mercier (1999), this is best ensured by incorporating EMP's budget in the total project cost before loan signing. They consider a separate budget for EMP a warning signal that environmental protection is considered merely as an 'add-on' to the 'main' project operation.

Another effective means of internalizing commitments made in the EIA and the EMP is Environmental Management System (EMS) commonly defined as:

> *A management approach which enables an organization to identify, monitor and control its environmental aspects. EMS is a part of the overall management system that includes organizational structure, planning activities, responsibilities, practices, procedures, processes and resources for developing, implementing, achieving, reviewing and maintaining the environmental policy.*

The definitions mentions *environmental aspects* which include not only direct and indirect impacts, as may be identified in the EA, but also environmental objectives and targets as may be defined in the EMP. Thus, environmental aspects link the three management tools: EIA, EMP and EMS.

The definition also refers to EMS as part of the overall management system, thus stressing its aim to mainstream environmental issues into all organizational activities, not just operations of the environmental (or EIA) unit.

In principle, any organization which deals with environmental effects of its operations may be said to have some sort of an EMS even if only informal and rudimentary. However, an important trend under the last decade has been increasing standardization and formalization of organizational EMSs so that it is now universally accepted that a functional EMS should follow certain universal principles. Such principles are for example, reflected in the international EMS standard ISO 14 001 first issued in 1996 and updated in 2004[2]. Hundreds of thousands of organizations worldwide have certified their EMS to comply with ISO 14001 and many more are using the standard without formal certification. The majority of these are private corporations, but a growing number are public utilities and authorities. For example, about one-half of Swedish municipalities have formal EMSs (although only a small portion of these are certified) (Emilsson and Hjelm 2002).

An ISO 14001-compliant EMS is based on a four-element Plan-Do-Check-Act (PDCA) cycle as shown in figure 34.2. Adhering to this cycle offers a possibility of continuous (albeit incremental) improvement, systematic approach and adaptation (cf. with 'adaptive management' in AEAM). The 'Plan' stage is naturally linked to the EIA and the EMP. At the next – 'Do' – stage the planned measures and arrangements are implemented. Such arrangements may also include continuous monitoring and operation of 'early warning' systems that detect deviations from standard operating conditions or other circumstances that may lead to violation of accepted standards and thresholds.

At the 'Check' stage the operation of the EMS is subject to a regular and systematic assessment. Such assessment may take form of an EMS audit (actually re-

---

[2] Another widely known EMS standard is the Eco-Management and Auditing Scheme (EMAS) issued by the European Commission in 1993 and updated in 1996. EMAS is primarily used in Europe and not at all as widely as ISO 14001. In the second version of EMAS its EMS requirements are identical with those of ISO 14001 (EMAS also contains other requirements notably for legislative compliance and for public reporting). Thus, we focus the subsequent discussion on ISO 14001.

quired by ISO 14001 and guided by another ISO standard (ISO 19001:2002)) and use the so-called EASO (Environmental Assessment of Sites and Organizations) (ISO 14015:2001). The EMS 'Check' stage is strongly linked to the evaluation component of EA follow-up. The results of such audits should normally be documented alongside recommendations on what, if anything, should be improved for the next PDCA cycle. Making the actual improvements, including modifications to organizational routines, monitoring arrangements, budgets, etc, makes up the last – 'Act' – element of the cycle.

34.3 SEA Follow-Up: Similarities and Differences with the Project-Level

The importance of SEA follow up and especially its management component is dictated by the same factors – uncertainty and commitment – that provide rationale for the project level EIA follow up. At the level of strategic initiatives, however, these factors impose even greater challenges. Implementation arenas for strategic initiatives are much less controlled by their proponents than in case of individual projects. That is why the link between formulation and implementation is much weaker at the strategic level. SEA follow-up is needed to extend the influence of SEA from merely promoting 'green rhetoric' in policies, plans and programmes (PPPs) to safeguarding environmentally sound patterns of activities arising from these PPPs.

Though the need for SEA follow-up was noted already in early SEA publications (e.g. Lee and Walsh 1992; Therivel and Partidario 1996), there has been only a handful of recent research papers on this topic[3], most notably Partidario and Arts (2005). In both research and practical guidance, SEA follow-up has also been less emphasised than other SEA elements and its monitoring component received more attention than evaluation, management and communication. This lack of attention to the management component of SEA follow-up has two reasons. First, the very concept of SEA is largely based on the idea to influence "strategic decisions" i.e. those made "higher" in the "decision-making hierarchy" and affecting the whole range of "lower-level" decisions and "implementation" activities. SEA traditionally attempts to make its contribution 'as early as possible' rather than dealing with 'later' stages of planning and implementation. Secondly, the very nature of many strategic initiatives hinders SEA follow-up. The link between formulation and implementation of such initiatives is often complex and sometimes very weak. In cases when institutional frameworks for formulation of strategic initiatives cease to exist after the initiative has been endorsed, SEA follow-up looses its 'organizational anchoring' or 'ownership'.

Effective SEA follow-up should untangle this complex relationship between formulation and implementation of strategies learning from policy and organizational studies, particularly from corporate strategy formation theories. Most of such theo-

---

[3] See Partidario and Fischer 2004; Persson and Nilsson 2006 for a review.

ries offer significant evidence that actual implementation often dramatically differs from formally conceived plans even in successful strategies. Moreover, in so-called 'emergent' strategies action does not follow formal decisions, but instead decisions articulate learning gained from action. In this interpretation, implementation is more important than formulation (Mintzberg et al. 1998). The SEA thinking has also evolved to become more receptive to this concept of 'emergent' strategies (see an overview of this evolution in Cherp et al. (forthc.).

Despite many similarities in approaches to SEA and EIA follow-up there are two important differences. The first concerns more profound, in case of SEA, shift of focus from *impacts* of strategic initiatives to their assumptions, objectives and implementation mechanisms. For example, the 'multi-track approach' to the evaluation element of SEA follow-up proposed by Partidario and Arts (2005) aims to validate underlying assumptions, evaluate goal-achievement and performance of the strategic initiative and activities linked to it. This echoes the general shift of focus in EA follow-up (also at the project level) from verifying 'what was expected' to 'what was wanted', from impacts to objectives, as suggested by Noble and Storey (2005). Such shift of focus is significant for the way standards and thresholds are dealt with as explained in the next section.

The second difference between EIA and SEA follow-up – more important for the management component – concerns implementation mechanisms to which they are linked. EIA follow-up and project implementation are often undertaken by the same project proponent who has been responsible for the project proposal and EIA. Implementation, in most cases, means adhering to the plans drawn at the pre-decision stage. In case of SEA, implementation may assume more complex and diverse forms depending upon the nature of the strategic initiative in question. It may be connected to decisions and actions of many different actors, not just the proponent of the initiative. In order to meaningfully facilitate sustainable development, SEA follow-up should affect most significant of these decisions and actions, which may be classified in the following four types (Fig. 34.2).

I.   Decisions on revising and amending the strategic initiative itself. For example, land use plans and many development plans undergo periodic reviews and renewals.
II. Actions directly prescribed in the strategic initiative and often implemented, funded, or supervised by the proponent. For example, a road improvement programme may presume construction of bicycle lanes. This type most closely resembles implementation of individual projects.
III. Decisions and actions implemented by other actors but controlled by the strategic initiative through formal frameworks. For example, a land-use plan may guide developments in particular zones.
IV. All other decisions and actions, which are affected by strategic initiative. For example, a national energy policy may influence consumer and investor behaviour without directly controlling it. Such effects are mainly achieved through price signals, 'soft' incentives and other information-based or similar indirect mechanisms. Many strategies are adopted because they set agendas, articulate commitments, or promote certain principles. Their implementation

occurs through information, communication and learning rather than through any specific development activities.

The prominence of different types of actions and decisions depends upon the nature of the strategic initiative. Types II-III management and control responses may be more relevant for plans and programmes, Type IV actions and decisions might be especially relevant for policies which sometimes are more exercises in rhetoric rather than direct action-forcing mechanisms. Type I is relevant for all types of strategic initiatives undergoing revision cycles. It should also be noted that all four types overlap to a certain degree.

Various management approaches are needed for ensuring that SEA findings and recommendations are reflected in all these decisions and actions. In case of Type I decisions, the proponent of the strategic initiative will be most likely also responsible for its revision and amendment. The original SEA findings as well as the results of SEA follow-up (which should ideally be timed so as to correspond to the revisions cycle) may in principle be directly linked to the renewal decisions, in the same way SEA is used at other 'evaluative moments' of planning processes. In case of Type II decisions, management tools used within SEA follow-up may be similar to those used in project-level EIA as discussed in the previous section. The role of SEA follow-up in Type II and III actions may be related to the concept of *tiering*. Tiering (Lee and Wood 1987) focuses on decisions that are 'nested' within

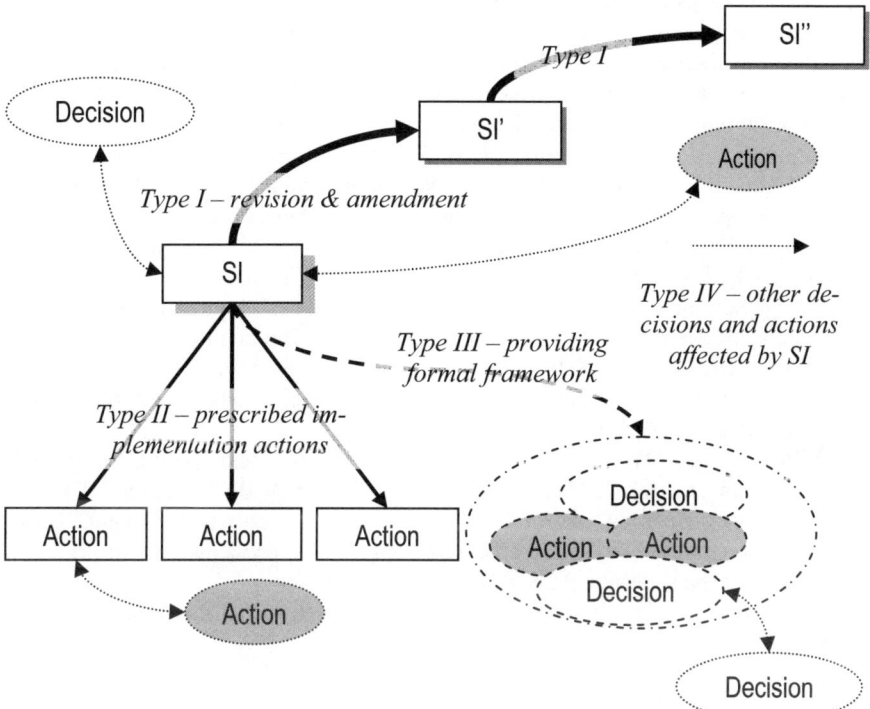

Figure 34.2. Types of 'implementation activities' for a strategic initiative (SI)

or hierarchically linked to the strategic initiative. The mechanism of tiering suggests that SEA follow-up should shape such decisions by linking their Environmental Assessments (SEAs or EIAs) to the original ('higher-level') SEA findings and recommendations. The concept of tiering features in most SEA texts and principles, but its empirical evidence on its practical implementation remains, so far, scarce. Tiering is naturally more effective in formal hierarchically organized planning systems, such as transport or waste management planning (Fischer 2003).

Many actions and decisions relevant to implementation of strategic initiatives belong to Type IV, i.e. lack a priori known, formal and easily traceable links with the original initiative. This is very well described by the 'splash' metaphor by Partidario and Arts (2005), which means that a strategic initiative may affect decisions and actions at the same, lower or higher levels across sectors and administrative jurisdictions. The question is: how can these actions and decisions be possibly tracked and shaped by SEA follow-up?

The problem of influencing Type IV actions and decisions is closely related to the challenge of institutional ownership of SEA follow-up. Actors behind Type IV decisions are rarely 'owners' of the original SEA. Thus, their participation in the SEA follow-up should be assured by specific organizational, communication or other arrangements. In certain cases such arrangements may be provided by an EMS much as in case of EIA follow-up (Marshall and Arts (2005). If the project-level EIA links with EMS for a company-developer, the SEA for public sector strategic initiatives may be linked to EMS in public authorities.

So far, this approach seems to be not much used in the public sector and it is merely at the stage of discussion and testing (for example, in the framework of the MiSt research programme in Sweden (www.mist-sea.se, project SEAMLESS). Despite certain potential, it may face serious obstacles. The first obstacle concerns the nature of EMS in (local) authorities. Worldwide research has shown that authorities most often employ EMS to deal with their own affairs (use of energy, paper, production of waste etcetera), but not with their plans, policies or programmes (which are subject to SEA). (This observation was made in Sweden (Naturvårdsverket 2004), New Zealand (Cockrean 2000), Japan (Srinivas and Yashiro 1999), and the Netherlands) In other words, EMS in authorities is not *strategic* to serve the purpose of SEA follow-up (Cherp 2004). The second challenge is that appropriate management responses within SEA follow-up may be needed outside of the organizational boundaries from those in which the original strategic initiative was adopted (and the SEA conducted).

EMPs may also play a role in facilitating the management component of SEA follow-up. For example, EMPs are likely to be encouraged (or mandated) for all large European cities as part of the EU strategy on urban environment. These EMPs which contain both management and assessment elements may in principle link SEA of urban master plans and daily implementation of environmental significant urban activities. It is too early to say to which extent this link will be articulated.

34.4 The Role of Standards and Thresholds

Within the management component of EA follow-up, environmental thresholds and standards assume an important and complex role which is somewhat distinct from their role at other stages of the EA process. Firstly, we are dealing with actual rather than planned or predicted phenomena, which naturally gives us more certainty in determining whether a particular standard or threshold is met. Secondly, these standards and thresholds assume a very high degree of *specificity*. They relate to certain components of given ecosystems and/or to specific facilities or activities. Thirdly, the standards and thresholds in this context are often management- rather than science-oriented, in particular relating to targets and objectives rather than on impacts which are normally difficult to trace and verify.

> *It is managing the real, rather than the predicted, impacts that matters. Predictive accuracy thus becomes of secondary importance to the effectiveness of management measures. (...) In a constantly changing environment, the larger question of outcome objectives is of greater importance than predicted impacts when designing and implementing follow-up programs... Rather than allocating time and resources to address the probability and the likelihood of an impact materializing as predicted, in our view, the primary objective of follow-up should be ensuring whether what were intended as project outcomes were in fact realized.*

(...) any approach to follow-up that is science-, rather than business-focused, is unlikely to be well received (...) In our view, the primary objective of follow-up should be the management function (Noble and Storey 2005).

Finally, the standards and thresholds used at this stage should be treated as dynamic rather than static because they are more likely to change over relatively longer time periods when the activities in question are implemented[4].

This contrasts standards and thresholds used at the EA follow-up stage (particularly in management) from those that guide earlier EA stages when the standards and thresholds used are more generic (universal), objective, science-based, impact-oriented, and static. In essence the whole EA process (including its follow-up stage) is not only about applying (pre-existing) standards and thresholds, but also about defining and clarifying them in relation to specific circumstances. Such si-

---

[4] For example, in a composting facility in the Netherlands, a key issue were odour emissions. The original permit awarded to the composting company drew on an odour policy that was based on the perceptibility of the odour in residential areas. Since the permit was awarded, the odour policy had been altered to also incorporate the nuisance level of odour. Following public complaints about odour, a number of investigations and modifications of the air treatment system have occurred. To accommodate both the changes in policy and the technical alterations of the plant has required a flexible permitting system on behalf of the EIA regulators as well as a succession of follow-up investigations and evaluations (van Vliet 2000).

Table 34.1. Key characteristics of standards and thresholds in the context of management plans and systems

Standards used at early stages of the EA process	Management element of the EA follow-up
Generic, universal, e.g. 'minimizing emissions of greenhouse gases (GHGs)'	Site- and activity-specific, e.g. "limit GHG emissions from the incinerator to xxx ton/day of CO2 equivalent"
Science-oriented, e.g. "existence of minimum viable population of bats", "concentration of heavy metals in fish below dangerous levels"	Management-oriented, e.g. "restricted activities during night hours", Emission Limit Values for discharges of pollutants
Impact-oriented, e.g. "no adverse impact on health of local population"	Objective-oriented e.g. "provide high standard health services to local population"
Static: e.g. WHO standards for air quality	Dynamic: e.g. BAT

multaneous application and clarification of standards echoes the process of refining 'sustainability criteria' in Sustainability Impact Assessments as described by Gibson (2006). At the implementation stage the standards are reflecting not only the knowledge about ecosystems, but also social, institutional and other factors. The summary of these differences is presented in Table 34.1.

These characteristics define an opportunity, but also a challenge and a threat. The opportunity is that enforcing such management- and objective- oriented dynamic and place-specific standards may make EA more effective in achieving environmental objectives even in situations when no impacts can be precisely predicted. For example, Noble and Storey (2005) describe the case of Hibernia construction project in Canada where several of the key impact predictions proved to be incorrect, but the ultimate outcomes of the project met the intended objectives. Similarly Morrison-Saunder and Bailey (1999) note that in Western Australia predictions expressed in vague and qualitative terms were equally likely to have management actions associated with them as the more scientific, quantitative impact predictions.

34.5 Conclusions and Recommendations

The challenge is in meaningfully linking such standards with global and universal environmental policy objectives, and focus them on strategic and significant issues. A common criticism aimed at EMS (and most "management tools" in general) is that by their very nature they can best address issues which can be dealt with by incremental change, 'continuous improvement'. There is a threat that standards defined in this way would be most convenient for developers, possibly improving somewhat their environmental performance but unable to address 'strategic' environmental issues or induce 'genuine' change in organizational activities towards radically better environmental performance. They may also fail to address genuinely unexpected and unpredicted impacts and circumstances. Although

EMSs and EMPs are good in programming existing commitment into effective actions they are not likely to foster commitment in those situations where it does not exist from the beginning. One may argue that formal and independent 'external' EIA process with its rigid universal standards is needed precisely to foster (or sometimes force) such a commitment which can hopefully be internalized through more flexible and management-oriented follow-up systems. This echoes a more general observation familiar to scholars of EMS that such 'voluntary' and 'flexible' management tools operate best in a well-defined and transparent legal and market context where the threat of legal sanctions or market risks ensure that EMS lead to better performance.

References

Cherp A., Watt A, Vinichenko V (forthcoming) SEA and Strategy Formation Theories." Environmental Impact Assessment Review x(xx): xxx-xxx

Emilsson S, Hjelm O (2002) Mapping Environmental Management Systems Initiatives in Swedish Local Authorities - a national survey." Corporate Social Responsibility and Environmental Management 9:107-115

Fischer TB (2003) Strategic environmental assessment in post-modern times. Environmental Impact Assessment Review 23:155-170

Gibson RB (2006) Sustainability assessment: basic components of a practical approach." Impact Assessment and Project Appraisal 24(3):170-182

Goodland R, Mercier J-R (1999). The Evolution of Environmental Assessment in the World Bank: from Assessment to Results. Washington, DC, Environment Department, The World Bank

Holling CS (1978) Adaptive Environmental Assessment and Management. New York, Wiley

Lee N, Walsh F (1992) Strategic environment assessment: an overview. Project Appraisal 7:126-136

Marshall R. (2001) Mitigation linkage: EIA follow up through the application of EMPs in transmission construction projects. Impact Assessment in the Urban Context, Annual Conference of the International Association for Impact Assessment, Cartagena, Colombia, IAIA

Mintzberg H, Ahlstrand B, Lampel J (1998) Strategy Safari. New York, Prentice Hall

Morrison-Saunder A., Arts J, Baker J (2003) Lessons from practice: towards successful follow-up. Impact Assessment and Project Appraisal 21(1):43-56

Morrison-Saunder A, Bailey J (1999) Exploring the EIA/Environmental Management Relationship. Journal of Environmental Management 24(3):281-295

Morrison-Saunder A., Baker J, Arts J (2003). Lessons from practice: towards successful follow-up. Impact Assessment and Project Appraisal 21(1):43-56

Morrison-Saunders A, Arts J (2004) Exploring the Dimensions of EIA Follow-up. Impact Assessment for Industrial Development: Whose Business Is It? 24th annual meeting of the International Assosiation for Impact Assessment, Vancouver, Canada, IAIA

Noble B, Storey K (2005) Towards increasing utility of follow-up in Canadian EIA. Environmental Impact Assessment Review 25:163-180

Partidario MR, Arts J (2005) Exploring the concept of strategic environmental assessment follow-up. Impact Assessment and Project Appraisal 23(3):246-257

Therivel R, Partidario MR (1996) The Practice of Strategic Environmental Assessment. London, Earthscan

35 Consequences of EIA Prediction Uncertainty on Mitigation, Follow-Up and Post-Auditing

Aud Tennøy

Institute of Transport Economics (TØI) – Norwegian Centre for Transport Research, Oslo, Norway

35.1 Introduction

This paper presents a discussion about how conditions for efficient and effective mitigation, follow-up and post-auditing are influenced by uncertainty in EIA predictions, lack of communication about such uncertainty and lack of transparency in prediction processes. As a conclusion, it is discussed how better communication about uncertainty and more transparent prediction processes can improve the conditions for efficient and effective follow-up and post-auditing, and thereby also for protection of the environment.

The Environmental Impact Assessment (EIA) process consists of a number of steps; screening, scoping, prediction, evaluation and mitigation design, in the pre-decision stage, and mitigation, follow-up and post-auditing in the post-decision stage (during project-implementation and when projects are in operation). The pre-decision steps of EIA should help decision-makers decide if the detected and predicted magnitude and significance of environmental and other impacts are tolerable and within given standards and thresholds, e.g. if the project should be allowed or not, and which mitigation measures that must be implemented in order to reduce such impacts. In the post-decision steps of EIA, mitigation is implemented, while follow-up should help decide if the project, included mitigation measures, is performing as predicted with respect to effects on environmental or other variables decided important in the pre-decision stages, and if additional mitigation measures should be implemented in order to reduce appearing negative impacts. Post-auditing is undertaken in order to enable the effectiveness of particular forecasting techniques to be tested and thus to improve further practice by learning from experience.

In EIA literature, there is a broad understanding that too little weight is put on the post-decision stages of EIA (see among others Marshall 2005; Glasson et al. 2005; Arts et al. 2001; Wood 1995).

Standards and Thresholds for Impact Assessment. Edited by Michael Schmidt, John Glasson, Lars Emmelin and Hendrike Helbron. © 2008 Springer-Verlag

A common complaint is about EIA being just pro forma paper exercises, rather than the tool to achieve sound environmental management it could be. If the post-decision steps of EIA are not carried out, there is no guarantee that the project is implemented as approved or that agreed mitigation measures are implemented, there is no monitoring to detect unpredicted impacts and call for additional mitigation measures if environmental thresholds and standards are violated, and there are no feedback systems to ensure learning from experience.

There are several reasons why prediction uncertainty in EIA is of importance in implementing standards and thresholds. Uncertainty in scoping and prediction of impacts may mean that violation of environmental quality norms and standards may be missed. With binding quality norms this may at worst mean that project permits are inappropriate. Conversely an uncertain prediction may seem to indicate a violation of a standard or norm leading to demands for costly mitigation or at worst rejection of an alternative or of a project. Furthermore the uncertainty in EIA predictions affects follow up and monitoring related to goals for future environmental quality. Reasonable decisions concerning mitigation and environmental permits need to take uncertainty into account in a systematic and rational way. The potential clash between uncertain predictions and the precision signaled by standards and thresholds underscores the importance of communication about prediction uncertainty in EIA to decision makers and the public.

In a study by Tennøy et al. (2006), it was concluded that uncertainty is almost unavoidable in EIA predictions, and that better communication about uncertainty, and more transparent prediction processes may improve EIA as a decision-making tool. This paper will take the discussion a step further, asking how uncertainty in EIA predictions, lack of communication about this uncertainty and lack of transparency in prediction processes are affecting the conditions for efficient and effective EIA mitigation, follow-up and post-auditing, and thereby also the environmental risks. As a conclusion, we will discuss how better communication about uncertainty and more transparent prediction processes can improve the conditions for efficient and effective follow-up and post-auditing, and thereby also for protection of the environment.

35.2 Uncertainty and Handling of Uncertainty in EIA Predictions

The object of EIA predictions is to identify the magnitude and other dimensions of change in the environment with a project or action, in comparison with the situation without that project or action (Glasson et al. 2005). The question to be answered is "what happens if...?" Once impacts have been predicted, the next step is evaluation, where the relative significance of impacts is assessed, in order to equip the decision-makers with information enabling them to judge if they consider the impacts to be acceptable or not. The question here is "what are the consequences of...?" This also includes judging and indicating whether the impacts are likely to violate standards or thresholds given in laws, guidelines etc. Part of the prediction and evaluation processes is to discuss whether negative impacts and consequences can be mitigated, and to set up a mitigation design

EIAs are, in this way, supposed to serve as tools for decision-making by informing decision-makers about the possible or likely impacts of the activity about which a decision is to be reached. This means, that in order to be a good tool, an EIA should forecast with a high, or at least sufficient, degree of accuracy what will happen in the future if an activity is implemented. Most authors recognize the unavoidable uncertainty inherent in EIA predictions.

Tennøy et al. (2006) asked, in a study including 22 Norwegian cases with predictions about transport effects, geohydrological effects and air pollution effects, if decision-makers are adequately informed about EIA prediction uncertainties, as well as input data and assumptions underlying such predictions. The magnitude of prediction uncertainties were examined, as well as the reasons why predictions are uncertain, in order to say something about the importance of decision-makers being informed about prediction uncertainties and prediction input. Based on the findings, the paper presents a discussion on how EIA can be improved as a decision-aiding tool, by improving the communication of uncertainty in EIA predictions and by making the prediction processes more transparent.

EIA Predictions Appear more Certain than they are

It was found that 42 % of the predictions in the study were deemed accurate, 29 % nearly accurate and 29 % inaccurate[1] (Tennøy et al. 2006). These results were in accordance with numerous other studies (Flyvbjerg et al. 2003, Teigland 2000, Wood et al. 2000, Dipper et al. 1998, Buckley 1992, Bisset and Tomlinson 1988). As for the transport cases 24 % of the predictions were deemed accurate, 41 % nearly accurate and 35 % inaccurate. Since there is clear evidence that EIA predictions are uncertain, one would expect this to be expressed throughout EIA processes. The study showed this not to be the case. In 43 % of the documents reviewed[2] (prediction-documents, EISs[3] and decision-documents), uncertainty

---

[1] The definitions established by Wood et al. (2000) are used to discuss accuracy. 'Accurate' is defined as within 10 % of the predicted level, and 'nearly accurate' as within 11–25 %.

[2] The data sources referred here are documents produced during the planning processes, mainly prediction-documents (background documents or technical reports of different kinds, presenting the impact predictions made prior to the EIS), EISs (where they exist) and decision-documents (mainly the planning office's report to the committee for cases at municipal level and Royal Propositions for cases at national level). Monitoring data during the construction phase was gathered from the road- and railway-authorities, while post implementation measurements were collected from project specific before- and after studies. All documents reviewed are dated between 1990 and 2004.

[3] The findings of the EIA process are presented in a document known as an Environmental Impact Statement (EIS) which is considered by the competent authority in the process of determining the development consent application (Dipper et al. 1998). We use the term EIS for the report, and the term EIA for the process.

associated with predictions was not mentioned. In 23 % of the documents, uncertainty was suggested but not referred to as uncertainty, in 13 % of the documents uncertainty was indicated but not explained or discussed, and in 21 % of the documents it was explained or discussed to some degree. Even if the professionals making predictions express uncertainties in their reports, this does not necessarily reach decision-makers. While uncertainty was discussed or indicated in 50 % of the prediction-documents, it was discussed or indicated in 18 % of the EISs and in 33 % of the decision-documents. This lack of information about uncertainty in predictions is described by other authors as well (Geneletti et al. 2003; Dipper et al. 1998; Andrews 1988). It was concluded that EIA predictions are uncertain, but that decision-makers are not made aware of the prediction uncertainty. EIA predictions thus appear more certain than they are.

This can also be seen as a parallel to environmental standards and threshold being given as very precise number, signalling high degree of safety in decision making and follow up, which contrasts the uncertainty we know exists in EIA predictions.

Prediction Processes are not Transparent

If EIA is to be an open and transparent process leading to informed decision-making, the EIA predictions must be transparent. The study by Tennøy et al (2006) found this not always to be the case. Data and assumptions used in the predictions were well accounted for in 39 % of the documents in the study. In 24 % of the documents, they were partly accounted for, and in 37 % they were not accounted for at all. Input data and assumptions were well accounted for in 70 % of the prediction-documents, but only in 25 % of the EISs and in 17 % of the decision-documents. There was a general lack of discussion of the input data used and the assumptions built into the models, i.e. why these data and assumptions (and not others) were chosen, and the impact of using this set of data and assumptions (compared with other sets) on the results. This lack of discussion make it difficult for others to realise that quite different results could have been reached if other input data and assumptions had been chosen. Other studies have shown similar results (Barker and Wood 1999, De Jongh 1988). It was concluded that decision-makers get only limited access to information about the input data and assumptions used in the predictions. The prediction process is thus not particularly open or transparent.

Prediction Performance can be Improved, but can EIA Predictions ever be Objective and Accurate?

Reasons for uncertainty in EIAs are multiple and complex and appear throughout the process (Glasson et al. 2005; Teigland 2000; De Jongh 1988). In the screening and scoping processes subjective considerations must be made, often on the basis of insufficient knowledge. The dynamics of projects often results in changes in the project between the time when predictions are made and the time a project is implemented[4]. Uncertainties in the predictions themselves occur because of model errors, errors in baseline data, input data and assumptions. Predictions often tell what will be the situation several years ahead in time, which adds to the uncertainty in many different ways. In addition there are the unknown cumulative effects, the discussion about whether and how to include long-term effects (e.g. consideration for future generations), and all the unpredictable events that can impact a project. With all these possible sources of uncertainty, it is hard to envisage that EIA predictions can ever be accurate.

The many subjective considerations necessary in EIA predictions make bias possible. Decisions must be made about which alternatives to assess, which processes to include in the model, how to describe the model reality (cause-effects), which data and assumptions to use in constructing and using a model, how to interpret the results etc. This means that the knowledge and ideas of people involved in the process, including their personal and professional values, world view, background and experience, as well as the culture, views and understandings at their workplace or in the system they are working within, inevitably influence the results (Flyvbjerg et al. 2002, Emmelin and Kleven 1999, Emmelin 1998, Wachs 1990).

The study by Tennøy et al. (2006) found evidence for all four explanations for deviation between predictions and measurements after project implementations which were checked (change in project; model errors; error in data or assumptions; and bias). The study was too small to serve as anything but an example of the multitude and complexity of the reasons for uncertainty in EIA predictions. Nonetheless, based on these findings as well as on the literature on the topic, it was found that reasons for uncertainty in EIA predictions are complex and manifold, and include subjective decisions that open for bias. In spite of several approaches to improve EIA prediction performance, it was concluded that EIA predictions probably never will become accurate and objective.

If focusing on EIA as a decision-aiding tool, the combination of uncertainty in EIA predictions, the failure to communicate that uncertainty and the lack of transparency in the predictions entail several disadvantages: Decisions resulting in unwanted environmental consequences may be made on the basis of erroneous information; democratic influences on decisions may be impeded by a lack of information; information valuable to discussions may fail to be brought to light; and EIA may lose credibility among decision-makers and the public. Since accurate and objective EIA predictions are an almost unobtainable goal, it was concluded, better communication about prediction uncertainties and more transparent prediction processes may be more realistic objectives.

Below, this discussion is taken a step further, asking how uncertainty in EIA predictions, lack of communication about this uncertainty and lack of transparency in prediction processes are affecting the conditions for efficient and effective EIA mitigation, follow-up and post-auditing, and thereby also the environmental risks.

---

[4] In Norwegian legislation, a new EIA is required only if the changes in the project are of such a magnitude that the changes themselves would have required an EIA.

But first we present the understandings of mitigation, follow-up and post-auditing which are used in the further discussion.

35.3 Mitigation, Follow-Up and Post-Auditing

As described, the pre-decision steps of EIA should help decision-makers decide if the predicted magnitude and significance of the impacts are tolerable and within given thresholds and environmental standards, and thus if mitigation measures could reduce negative impacts to a tolerable level or if the project should be rejected. Setting up a mitigation design to reduce negative impacts is (or should be) part of this process. The post-decision steps of EIA, when the project and its mitigation measures are under implementation and later in operation, are meant to make sure that the project and mitigation measures are implemented in accordance with the approval, that the real impacts are within tolerable or agreed limits, and that action (mitigation measures) is taken if the impacts exceed these limits.

Mitigation Measures – a Principal Aim of the EIA Process

Mitigation is defined in EC Directive 97/11 as "measures envisaged in order to avoid, reduce, and if possible, remedy significant adverse effects" (CEC 1997). Mitigation is inherent in all aspects of the EIA system (Glasson et al. 2005). Wood (1995) states, that the mitigation of environmental impacts is the principal aim of the EIA process. He argues that the main purpose of EIA is to allow the proposed development to proceed, while reducing its impact to an acceptable level. The secondary purpose of EIA is to prevent unsuitable developments by demonstrating that certain impacts cannot be mitigated to the point of acceptability or affordability. With reference to the Department of Environment (1994), Wood classifies mitigation measures into avoidance (using an alternative approach to eliminate an impact), reduction (lessening the severity of an impact) and remedy (which may involve some enhancement or compensation). It is underlined that mitigation measures could also have negative impacts that need to be identified, predicted and evaluated. Sánchez and Gallardo (2005) argue that the effectiveness of EIA depends largely on fully implementing cost-effective mitigation and other management measures to prevent significant environmental degradation. This is recognised in various EIA systems, Wood (1995) found that mitigation of action impacts must be considered at the various stages of the EIA process in all the seven EIA systems he examined (United States, California, United Kingdom, The Netherlands, Canada, Australia and New Zealand).

Follow-up – Getting a Grip on Uncertainties

Morrison-Saunders and Arts (2004) define EIA follow-up as "The monitoring and evaluation of the impacts of a project or a plan (that has been subject to EIA) for management of, and communication about, the environmental performance of that project or plan". EIA follow-up is useful and necessary because it provides information about the real consequences of an activity as they occur, as well as the opportunity to mitigate or prevent negative effects on the environment and to make sure that impacts are within given or agreed standards and thresholds (Marshal et al. 2005; Wood 1995; Bisset and Tomlinson 1988). Arts et al. (2001) add the learning opportunity to improve further EIA practice.

Monitoring is often defined as the repetitive measuring and recording of physical, social and economic variables associated with development impacts (e.g. traffic flows, air quality, noise, employment levels or fish stocks), and comparison of the collected data with standards, thresholds, predictions or expectations (Arts et al. 2001, Wood 1995, Glasson et al. 2005, Bisset and Tomlinson 1988). Implementation monitoring, checking that the project is implemented in accordance with the approval and that the mitigation measures correspond with those required, is a form of monitoring. *Evaluation* is the periodic or single appraisal of conformity with pre-defined criteria, in EIA often forecasts and commitments made in the pre-decision stages of EIA, quality standards or thresholds of concern (Arts et al. 2001; Wood 1995). *Management* is in Arts et al.'s (2001) definition about making decision and taking appropriate action in response to issues arising from monitoring and evaluation activities, while *communication* is about informing stakeholders about the results of EIA follow-up.

In the further discussion we will use the term follow-up for activities focused on monitoring and managing impacts in the implementation and operational phases of projects.

Post-Auditing – Learning from Experience

Post-auditing is often described as comparing impacts predicted in EIA with those that actually occur after implementation, in order to assess whether the impact prediction performs satisfactorily (Wood et al. 2000; Dipper et al. 1998; Wood 1995; Glasson et al. 2005; Bisset and Tomlinson 1988; Sadler 1988). The aim is to enable the effectiveness of particular forecasting techniques to be tested and thus to improve further practice.

35.4 Consequences of EIA Uncertainty on Mitigation, Follow-Up and Post-Auditing

EIA Uncertainty is a Main Rationale for Follow-Up and Post-Auditing

There are two different lines of argumentation for this. The line of argumentation for *follow-up* says that since we know that the pre-decision EIA process results in inaccurate and uncertain answers about what will be the impacts of a project, for a number of reasons, follow-up must be undertaken in order to make sure that it is the real, not only the predicted impacts, that are dealt with (Marshall et al. 2005; Arts et al. 2001; Glasson et al. 2005; Wood 1995; Bisset and Tomlinson 1988).

The line of argumentation for *post-auditing* says that if we are to improve EIA prediction performance, we must undertake post-auditing in order to be able to learn from experience by i.a.; to highlight the types and categories of impacts that tend to be predicted less accurately than others; to stimulate gradual improvement in prediction techniques used to make predictions; to provide baseline information for future EIAs (Marshall et al 2005; Flyvbjerg et al. 2003; Wood et al. 2000; Teigland 2000; Dipper et al. 1998; Glasson et al. 2005; Glasson 1994; Bisset and Tomlinson 1988; Sadler 1988). An argument for EIA follow-up and post-auditing, mentioned by some authors, is to *prove* the quality and accuracy of EIAs.

Uncertainty in EIA predictions poses somewhat different challenges on EIA follow-up and post-auditing, and we will therefore discuss them separately. Problems caused by pre-decision predictions for post-auditing seem to be more frequently and thoroughly discussed in the literature than for follow-up. For that reason, we will treat post-auditing more lightly than follow-up.

Challenges for EIA Follow-Up

EIA follow-up concerned with handling EIA prediction uncertainties in the post-decision stages of the project is mainly focused on monitoring the impact of a project or activity, in order to reduce or hinder unwanted or unexpected environmental impacts, or impacts that exceed environmental quality standards or thresholds of concern, through mitigation. Glasson et al. (2005) say, with reference to Lee and Wood (1980), that a primary requirement is to focus monitoring activity only on those environmental parameters expected to experience a significant impact, together with those parameters for which the assessment methodology or basic data were not so well established as desired. When discussing these requirements in relation to the earlier described knowledge about EIA prediction uncertainty, lack of communication about such uncertainty and lack of transparency in the prediction process, several challenges emerge: important impacts may be missed or severely underestimated; effective (that it should provide relevant information about implementation and impacts) and efficient (that needless monitoring and mitigation is not undertaken) follow-up is hard to set up, and; unnecessary or contra productive mitigation measures may be applied because of mistrust to the predictions.

The prediction uncertainty itself may lead to a situation where an impact is underestimated, giving the impression that the impact is far below thresholds, environmental standards or other maximum levels, and thus that neither monitoring nor mitigation measures are necessary. The same goes for other reasons for uncertainty in EIA, like missing out on important impacts in scoping, or when changes in the project from prediction to implementation[i] occur. Arts et al. (2001) say that the degree of uncertainty is a criterion for determining whether follow-up is necessary. But, if the uncertainties are not communicated, as was the case in a majority of the EISs and decision documents in the study carried out by Tennøy et al. (2006), the ones planning follow-up cannot know if uncertainty exists, what its magnitude is etc. And, if the prediction process is not transparent, i.e. data, assumptions, models and underlying theories are not well accounted for, they will not be able to do their own judgements about possible underestimated environmental impacts hidden in the prediction uncertainties, or about important impacts being missed out. Without uncertainty descriptions and transparent predictions, there are small chances that faults made in early stages of the EIA will be discovered later in the process, for instance by the ones setting up follow-up. This could lead to a situation where possibly serious environmental impacts are not monitored or mitigated and thereby to unexpected or unwanted negative impacts on the environment, exceeding important environmental quality standards and thresholds, and to implementation of costly and unplanned mitigation measures. This could be illustrated by the case of the 14 kilometre railway tunnel Romeriksporten, discussed in Tennøy et al. (2006). The geohydrological effects and secondary impacts of these effects on surface water, groundwater and peat lands were missed out on in the pre-decision EIA. This was later explained by lack of knowledge about multidisciplinary cause-effect relationships (model error) and by misjudgement in the scoping process. Neither monitoring nor action programmes to mitigate such effects were set up. Consequently, the drainage effects caused great problems to nature, buildings and to the construction of the tunnel, as well as the need for costly mitigation measures.

When screening and scoping for setting up efficient and effective EIA follow-up, uncertainty may lead to lack of follow-up of important impacts because of underestimation, as described above, or to unnecessary follow-up because of overestimation. Lack of communication about uncertainties makes it hard to know where extra resources should be put to work to find out which impacts or variables are uncertain and should be monitored for that reason. Predictions given as accurate numbers may give the impression that the predictions are more certain than they are, the same goes for standards and thresholds in the cases where they actually are given as objectives. Non transparent prediction processes remove the possibility to go back to previous work to check out how the predictions were done. This problem could be extra troublesome if there are changes in the project from prediction to implementation, and thus that new considerations are needed in order to determine which monitoring and mitigation is needed. If working in complex or sensitive areas, where strict or absolute environmental standards and thresholds are applied, and/ or with new technology or with projects that of other reasons involve risks, severe known prediction uncertainties could result in uncertain recommendations about expensive mitigation measures or to reject the project. This could lead to reactions from project owners asking which level of certainty is adequate for putting such burdens on them, or from environmentalists, neighbours etc. asking the same question with respect to burdens on the environment or the neighbourhood. We see that uncertainties in EIA predictions, lack of communication about uncertainties and lack of transparency in prediction processes can result in severe difficulties setting up efficient and effective follow-up.

Another problem may be that unnecessary or even contra productive mitigation measures are applied because of mistrust in the predictions. In the case of relocating *The National Hospital* in Norway from a central to a less central location, as described in Tennøy et al. (2006), a new tram line was supposed to secure 50 % public transport share of all transport to and from the hospital. It appeared from the documents in the study that a 50 % public transport share was more an objective than a prediction, and that several of the stakeholders did not believe in its probability. Since the planners did not dare believe the prediction (or objective), the parking space was designed with more parking capacity than would have been needed with a 50 % public transport share. The ample availability of parking spaces probably contributed to the failure of the prediction or objective of 50 % public transport share, since good access to parking space tends to make people choose cars as their preferred mode of transportation. Perceived prediction uncertainty among important stakeholders thus resulted in counter-productive mitigation measures (extra parking spaces) that contributed to failure of the predictions (or the objective).

Challenges for Post-Auditing

Reasons for uncertainty in EIA predictions are manifold and complex. While the reason why uncertainties appear is not a major issue in EIA follow-up, this is more so in post-auditing. In post-auditing, a principal aim is to test the forecasting techniques. Other reasons for uncertainty and inaccuracy than faults in these techniques make it hard to do rigorous post-audits in such a way that the techniques could be improved. This means that uncertainties and inaccuracies occurring because of e.g. errors done in screening and scoping, change in project from prediction to implementation (mitigation included), and unknown and unpredicted effects, are merely disturbing elements. The optimum would be if only uncertainties in the predictions themselves, because of error in baseline data, input data, assumptions, models, underlying theories etc. occurred.

The problems for post-auditing is not only that uncertainties and inaccuracies occur for several and often non-relevant reasons (for post-auditing). Lack of communication about uncertainty and not at least lack of transparency in prediction processes may ruin the possibilities of doing good and meaningful post-auditing. If it is not known how the predictions are made, i.e. if data, assumptions, models and underlying theories are not well accounted for, it is hard to envisage that good post-auditing may take place. This could be illustrated by the three distinct different types of structural errors that can occur in models of the environment, described by De Jongh (1988): *process errors* – the model simplifies the reality by assuming that only certain processes are important, and by including only those; *functional errors* – the model is unable to describe the actual processes of cause and effect, and; *use of the model outside its range of circumstances* – the model is not valid for the problem at hand. If to improve theories, models and techniques used in EIA predictions, it would be necessary to know what kind of error that has occurred, and this requires a thorough description of how the predictions are made. We also know that the many subjective considerations necessary in EIA mean that the knowledge and ideas of people involved in the process inevitably influence the predictions and the results (Tennøy et al. 2006, Flyvbjerg et al. 2003, Emmelin 1998). This enhances the need for transparent predictions in order to enable post-auditing.

Vagueness, Precision and Intervals

An interesting discussion exists about vagueness or imprecision in EIA predictions (Dipper et al. 1998, Wood 1995, De Jongh 1988). If to be auditable, some say, EIA predictions should be framed as precise, falsifiable hypothesis. They are "better quantitative and wrong than qualitative and untestable" (Duinker 1985). Others have opined that, while precision is desirable when appropriate, the value of EIA predictions does not depend of their strict auditability (Munro 1987). He also argues that it is more important to be able to predict whether an environmental factor is likely to move into a range which holds serious implications for e.g. the ecosystem, or we will add, to violate given environmental standards and thresholds. Even an imprecise prediction can lead to appropriate mitigation. Tennøy et al. (2006) criticised EIA predictions to appear more certain than they are, among others because predictions are given as precise numbers and without discussions about uncertainty. In their study, it was found that all predictions given as intervals (geohydrological effects, air pollution effects) were deemed accurate, while predictions given as precise numbers (these were all predictions about transport effects) were deemed accurate in 24 % of the cases. In addition to the advantage of correct predictions, this direct indication of the uncertainty inherent in predictions was seen as a way of communicating uncertainty. Another strong argument for doing and presenting predictions as intervals rather than as precise values is that predicting and presenting minimum and maximum numbers for effects and impacts will tell more clearly than an average value, if there is a danger that impacts will move into ranges that are in conflict with commitments, quality standards or thresholds of concern. We also believe that predictions done and presented as intervals are just as auditable as precise numbers or average values. Still, the usefulness of post-auditing for calibration of forecasting techniques may be reduced if the intervals are given to wide.

Consequences of Pre-Decision EIA Uncertainty

As we have seen, EIA prediction uncertainty, lack of communication about such uncertainty and lack of transparency in the prediction process, pose several challenges to EIA follow-up and post-auditing. For follow-up, the main challenges are that efficient and effective follow-up is hard to set up, that unnecessary or contra productive mitigation measures may be applied and, most important, that unpredicted, unwanted and unmitigated environmental damage may occur. For post-auditing, uncertainty not related to the forecast techniques etc. that are to be assessed, are disturbing the post-audit, while lack of transparency in prediction processes may ruin the possibilities of doing good and meaningful post-auditing. Even if Sánchez and Gallardo (2005) argue that the discussion about effectiveness of EIA has shifted from accuracy of impact predictions to our ability to effectively prevent significant environmental degradation and to fully implement cost-effective mitigation and other management measures, we will point at the importance of post-auditing in order to achieve accurate predictions.

The main purpose of EIA is often defined as to generate information on various changes that may occur in the environment in response to the implementation of a particular proposed activity, and to be an aid to decision-making by informing decision-makers about the possible or likely impacts of the proposed activity (Norwegian Ministry of the Environment 2003; Wathern 1990). If predictions are not accurate enough, and if they are missing out on important impacts, EIA is not serving its purpose. Decision makers and other stakeholders would not necessarily be informed about possible consequences for the environment and the risks for violating environmental standards and thresholds, and they may e.g. approve projects that they would not have approved if they knew about those consequences, both for environmental reasons and for reasons concerning mitigation costs. In addition, not all effects, impacts or consequences can be mitigated. Once the project is built and in operation, and the consequences appear, there is in reality not much to do about it. The effects of urban road construction on traffic growth and the impact of this on global environment, local environment, health and well-being etc. is an example of this. If the traffic growth is higher than predicted (and wanted, or set as a threshold) because of a new road, one could of course mitigate this by closing down a lane or using heavy road tolls in order to reduce the traffic levels. In reality this will not happen. In such cases, it is important that decision makers and other stakeholder are aware of the possible effects and impacts of the project in advance of the decision, since mitigation in reality is not an option.

It becomes clear that uncertainty in EIA predictions, lack of communication about such uncertainty and lack of transparency in prediction processes poses different challenges on follow-up and post-auditing, but that the consequence in both cases is an increased risk of unpredicted, unconsidered and unwanted environmental impacts, that can or will not in reality be mitigated.

35.5 Conclusions and Recommendations

Uncertainty poses challenges to follow-up and post-auditing, and thereby risks for environmental damage. Of course, accurate and objective predictions would reduce these problems. Tennøy et al. (2006) argued that EIA predictions hardly ever can be fully accurate and objective, because of the many and complex reasons for uncertainty and the many subjective decisions to be made. They argue for giving more emphasis to communication of uncertainty and to transparency in EIA predictions, especially in EISs and decision-documents. This could lead to a situation where although EIA predictions continue to be uncertain, decision-makers and other stakeholders are aware of this fact. Transparent predictions may put decision-makers and other stakeholders in a better position to judge the predictions and, possibly, to oppose them. This could furnish a better basis for informed decision-making, and improve EIA as a decision-aiding tool. We will argue that such communication and transparency may help improving follow-up and post-auditing as well.

Good communication about uncertainties throughout the planning and decision processes means in Tennøy et al. (2006) that uncertainties are expressed and clarified in prediction-documents, as well as in EISs and decision-documents. Which predictions are uncertain, the ways in which they are uncertain, the possible magnitude of the uncertainty and, most importantly in pre-decision EIA, the likely implications of the uncertainties, must be discussed? Stating predictions as intervals rather than precise numbers (as previously discussed) may be one way to do this.

Transparent predictions imply that input data, assumptions, methods, models and theories involved are described. The reasons why these particular data, assumptions, models, methods and theories are used should be explained, as should the implications of the choices made. It is important to point out that other conclusions might have been reached if another set of input data, assumptions, methods, and models were chosen. Often, this will be a question of presenting opposing theories, and then discussing the reasons for one's choices.

Good communication about uncertainties and transparent prediction processes may help improving the conditions for follow-up and post-auditing as well. *Firstly,* good communication about uncertainty would be of great help in the struggle to design efficient and effective follow-up. By knowing which predictions are most uncertain, the possible magnitude and implications of uncertainties, the ones setting up follow-up programmes will know if and which environmental standards and thresholds are likely to be violated, and can do this in a far more efficient and effective way than if they don't have this information. A procedural approach to communication of uncertainty, as discussed by Glasson et al. (2005), is to require a particular uncertainty report as a step in the EIA process. Geneletti et al. (2003) argue that relatively simple uncertainty analyses would increase the awareness of decision-makers by calling their attention to the problematic aspects of the procedure. Both these approaches would be useful for follow-up as well.

Secondly, through communication about uncertainties and transparent predictions, the problem whether to require expensive mitigation measures or to reject projects on the basis of severe uncertain predictions could be laid out for the politicians, neighbours, environmentalists and other stakeholders to discuss and decide, rather than bureaucrats and scientists trying to decide on scientific basis questions that in reality are value-based and political.

Thirdly, transparent predictions would obviously make EIA predictions far more auditable. It would prepare the ground for learning from experience, since it would be possible to see what others really have done.

Fourthly, transparency would be useful for follow-up as well, especial where e.g. changes in the project make new uncertainty considerations necessary, in order to set up relevant monitoring and mitigation.

Fifthly, communication about uncertainty and transparency in prediction processes could contribute to more relevant discussions involving more stakeholders than at present, active public participation included, both in pre- and post-decision stages of EIA. This could bring together more knowledge, and also raise awareness of possible impacts and mitigation possibilities not previously considered. This is also relevant for discussions on if the preciseness of environmental standards and thresholds are giving the impression of EIA and the following decision-making being less uncertain than we know it can be.

And *lastly,* an effect of the previous five points may be less uncertainty in EIA predictions, which in itself would reduce the challenges posed on follow-up and post-auditing, and thereby possibly lead to more effective and efficient follow-up and post-auditing.

This demonstrates that better communication about uncertainty in EIA predictions and more transparent prediction processes will improve the conditions for follow-up and post auditing, and thereby contribute to reducing the risk of violating environmental standards and thresholds and damaging the environment, without having done conscious decisions about this or without being aware of it.

References

Andrews R N L (1988) Environmental impact assessment and risk assessment: learning from each other. In: Wathern P (ed) Environmental Impact Assessment. Theory and Practice. Routledge, London

Arts J, Caldwell P, Morrison-Saunders A (2001) Environmental impact assessment follow-up: good practice and further discussions – findings from a workshop at the IAIA 2000 conference. In Impact Assessment and Project Appraisal 19(3)

Barker A, Wood C (1999) An evaluation of EIA system performance in eight EU countries. Environmental Impact Assessment Review 19(4):387-404

Bisset R, Tomlinson P (1988) Monitoring and auditing of impacts, in Peter Wathern (ed) Environmental Impact Assessment. Theory and Practice. Routledge, London

Buckley R (1992) How Accurate Are Environmental Impact Predictions? Ambio, 20(3-4) May 1991

CEC – Commission of the European Communities (1997) Council Directive 97/11/EC amending Directive 85/337/EEC on the assessment of certain public and private projects on the environment. Official Journal L73/5, 3 March

De Jongh P (1988) Uncertainty in EIA. In Peter Wathern (ed.) Environmental Impact Assessment. Theory and Practice. Routledge, London

Department of Environment (1994) Guide on Preparing Environmental Statements for Planning Projects. Consultation Draft, DOE, London

Dipper B, Jones C, Wood C (1998) Monitoring and Post-auditing in Environmental Impact Assessment: A Review. Journal of Environmental Planning and Management, volume 41(6):731-748

Duinker PN (1985) Forecasting environmental impacts: better quantitative and wrong than qualitative and untestable. In: Sadler B (ed) Audit and Evaluation in Environmental Assessment and Management: Canadian and International Experience, Proceedings of the Conference on Follow-up/ Audit of EIA Results

Emmelin L (1998) Evaluating Environmental Impact Assessment – Part 2: Professional Culture as an Aid in Understanding Implementation. Scandinavian Housing and Planning Research 15:187-209

Emmelin L, Kleven T (1999) A paradigm of Environmental Bureaucracy? Attitudes, thought styles, and world views in the Norwegian environmental administration. NIBR's Pluss Series 5-99

Flyvbjerg B, Bruzelius N, Rothengatter W (2003) Megaprojects and risk. An anatomy of ambition. Cambridge University Press

Flyvbjerg B, Holm MS, Buhl S (2002) Underestimating Costs in Public Works Projects. Error or Lie? Journal of the American Planning Association 68(3):pages 279-295

Geneletti D, Beinat E, Chung CF, Fabbri AG, Scholten H-J (2003) Accounting for uncertainty factors in biodiversity impact assessment: lessons from a case study. Environmental Impact Assessment Review 23 (2003):471–487

Glasson J, Therivel R, Chadwick A (2005) Introduction to Environmental Impact Assessment, 3rd edition. Routledge

Glasson J (1994) Life after the decision: The Importance of Monitoring in EIA. Built Environment 20(4)

Lee N, Wood C (1980) Methods of environmental impact assessment for use in project appraisal and physical planning, Occasional Paper no. 7, University of Manchester

Marshal R, Arts J, Morrison-Saunders A (2005) International principles for best practice EIA follow-up. Impact Assessment and Project Appraisal 23(3)

Morrison-Saunders A, Arts J (eds) (2004) Assessing Impact: Handbook of EIA and SEA Follow-up. Earthscan James and James, London

Munro DA. (1987) Learning from experience: auditing environmental impact assessment. In B. Sadler (ed) Audit and Evaluation in Environmental Assessment and Management: Canadian and International Experience, Proceedings of the Conference on Follow-up/ Audit of EIA Results

Norwegian Ministry of the Environment (2003) Environmental impact assessment. Oslo, Norway

Sadler B (1988) The evaluation of assessment: post-EIS research and process development. In Peter Wathern (ed) Environmental Impact Assessment. Theory and Practice. Routledge, London

Sánchez LE, Casteli Figueireo AL (2005): On the successful implementation of mitigation measures. In Impact Assessment and Project Appraisal, volume 23, number 3

Teigland J (2000) Impact Assessments as Policy and Learning Instrument. Why Effect Predictions Fail, and How Relevance and Reliability can be Improved. Ph.D. thesis 2000 Roskilde University

Tennøy A (2003) Prediksjoner og usikkerhet i trafikkfaglige rapporter i KU. (Predictions and uncertainty in reports on traffic-related issues in EIA. With an English summary) NIBR report 2003:13

Tennøy A, Kværner J, Gjerstad KI (2006) Uncertainty in environmental impact assessment predictions – the need for better communication and more transparency. Impact Assessment and Project Appraisal 24(1):45–56

Wachs M (1990) Ethics and Advocacy in Forecasting for Public Policy. Business & professional ethics journal, vol.9, nos. 1 & 2.

Wathern P (1990) An introductory guide to EIA. In Peter Wathern (ed) Environmental Impact Assessment. Theory and Practice. (Routledge, London)

Wood C (1995) Environmental Impact Assessment. A Comparative Review. Longman

Wood C, Dipper B, Jones C (2000) Auditing the Assessments of the Environmental Impacts of Planning Projects. Journal of Environmental Planning and Management, 43(1):23–47

36 Environmental Quality Standards as a Tool in Environmental Governance – the Case of Sweden

Lars Emmelin[1] and Peggy Lerman[2]

1 Spatial planning, Blekinge Institute of Technology, Karlskrona, Sweden
2 Lagtolken PL AB (publ), Nättraby, Sweden

36.1 Introduction

In this article we examine the role of environmental quality standards (EQS) and thresholds as a tool in environmental governance and the strive for sustainable development in Sweden. Such standards, as opposed to emission standards, are a relatively new addition to the environmental policy tools in Sweden. The role of EQS is ultimately to operationalise an approach that is described in Sweden's National Strategy for Sustainable Development: "The society of the future must be formed within the limits set by nature, environment and human health (…)" (Gov Bill 2003/04:129) The point of departure is thus environmental quality in determining acceptability of development proposals, programmes and plans and industrial projects rather than emissions. Sweden has introduced EQS of different kinds as tools for governance and the effectiveness of different types can thus be examined. Standards, especially the National Environmental Objectives structure with its system of quantified targets is accorded a prominent role in policy.

Lundqvist (2004) argues that Sweden is a good case study for "ecologically rational governance" and of "ecological modernization". This entails solving environmental problems and working towards sustainability within a framework of market economies and an assumed win-win situation for ecology and economy (Hajer 1995) with increased eco-efficiency as a main objective (Dobers 1997).

Integrating environmental concern with the aid of stringent legislation is part of "ecological modernization". An important tool in "ecological modernization" is environmental policy integration and thus the use of management by objectives,

Standards and Thresholds for Impact Assessment. Edited by Michael Schmidt, John Glasson, Lars Emmelin and Hendrike Helbron. © 2008 Springer-Verlag

MBO (Emmelin and Kleven 1999)[1].The introduction of EQS in Sweden is part of the process of control of diffuse and multiple sources of pollution after the first phase of clean up of point sources (Lundqvist 2004). As opposed to permit processes and EIA/SEA their purpose is also to handle on-going land use and to give incentives for strategies or specific action programmes that should lead to better environmental quality. Cumulative impacts, that are notoriously difficult to handle in EIA and permit processes should thus be tackled by EQS. EQS is a tool to drive technological development and force environmentally better alternatives in programming, planning and project design. EQS is thus a tool for "ecological modernization". Standards have so far been developed for chemicals and particulate matter in air, chemicals in water and to some extent for noise. Standards for biological parameters have also been discussed. The National Environmental Objectives are a form of EQSof particular interest.

Lundqvist's extensive study of the case of Sweden deals almost exclusively with the policy level. Good policy may be a prerequisite for environmentally rational governance but it is necessary to evaluate the implementation structures as well as the policy level (Emmelin 1998a). We have previously argued that the piecemeal and unstructured system of EIA and the minimalist introduction of SEA in Sweden casts doubt on the effectiveness and efficiency of the implementation level (Emmelin and Lerman 2004; Emmelin 1998b).

In this article we examine EQS as a tool in environmentally rational governance with special reference to their role in environmental assessment. Four aspects of standards and thresholds are crucial to implementation:

- Are they constructed so as to fulfil the object of operationalising limits set by "nature, environment and human health"?
- Do they give clear signals concerning demand for future environmental quality?
- Can they function in a system of management by objectives?
- How do they function within the environmental assessment system?

For this examination we need first to introduce salient features of the Swedish EA system and the types of standards to be applied within that system. We then discuss some of the theoretical aspects of governance with standards, especially at the interface of spatial planning and environmental management. Our discussion is evidence based in a number of studies that we have carried out on tools for environmental governance, especially the Swedish EIA system and the National Environmental Objectives for the Centre for Regional Development Planning, Blekinge Institute of Technology (Emmelin and Lerman 2004a, b; Emmelin 2005), on environmental governance for the National Committee on Public Sector Responsibility (Emmelin and Lerman 2006) and on environmental quality norms for the Swedish Environment Protection Agency (Lerman 2004)

---

[1] Although both "ecologically rational governance" and "ecological modernization" have become accepted terms covering a much wider area of policy than the strict use of the term "ecological" implies we prefer to use "environmentally rational governance" for the policy and policy tools discussed in this article.

36.2 Types of Environmental Quality Standards

There are basically three types of EQSin the Swedish system:

First, the *binding standard* which in principle set a fixed and measurable limit or threshold. It is equivalent to legislation and can, in legal terms, be called a *norm*. *Environmental quality norms, EQN* can be likened to a stop light, and is supposed to stop any project or plan that will pass a threshold defined by the standard. Sweden has at present relatively few of this type. The prime examples, with important consequences for spatial planning, are the standards for nitrous oxides and for particulate matter, PM10, in ambient air.

The second type, a standard giving *a guiding threshold value,* can be regarded as an imperative for action, but is not necessarily an immediate hindrance for a project or a plan. A programme, planning decision or project permit may violate the standard if this is balanced by being in other respects in the public interests. One example of such a standard i the one on ground level ozone discussed below.

EQSof these kinds, are usually quantifications of the worst level of an environmental parameter that can be accepted, with regard to impacts on health and environmental quality. Such a standard should thus not be mistaken for goals for good environmental quality. It is in fact not even an accepted level of pollution. The risk of violating a standard, both a binding norm and a guiding threshold, is sufficient cause to demand mitigation measures and for a demand for better alternatives to be elucidated in an EIA or SEA.

The third type is the softest focusing on good – or at least acceptable - environment rather than the bad and unacceptable. It can be described as an *objective* operationalised – quantified and time set – as a *goal or target value*. A *soft standard* can, just as the types above, be used as an indication that a project or a plan is in line with regulations in the sense of being in line with general societal ambitions for environmental quality. The goal value by itself can not stop a project or a plan on their own, but may be a part of the over all evaluation for example of public versus private interests. They can add weight in the application of other, legally binding demands. The prime example of international interest in the Swedish system is the *National Environmental Objectives, NEO* discussed below.

A simple distinction between the roles of softer standards and formal environmental quality norms in EIA is that both can be used in impact assessments but play different roles: the former being a general inspiration to search for good alternatives while standards weed out the undesirable alternatives as well.

36.3 The Swedish Implementation System

To discuss the Swedish case we need to set out some basic elements of the Swedish implementation system for environmental quality standards. First, some of the peculiarities of the Swedish EA system need to be briefly mentioned. Here we will also discuss the issue of thresholds for the demand for EIA or SEA. Second, we will elaborate on the three types of EQSdiscussed above and their role in the

Swedish system. We will place special emphasis on the National Environmental Objectives.

The Swedish EA System

We have analysed the Swedish environmental assessment system, i.e. the system of EIA and SEA regulation and implementation, elsewhere (Emmelin and Lerman 2004 a, b). The major problems that make us conclude that the Swedish EA system is of low effectiveness are:

- it is complex, overlapping and contradictory in regulation and terminology
- around 50 different pieces of legislation are involved
- implementation of EU directives has been explicitly minimalist which is a partial cause of several of the other problems listed here
- a very large number of assessments or "semi-assessments" are made annually precluding effective quality control or follow up
- certification of consultants or other quality control of the actors or any formalised or systematic control of EA assessment quality at all, is lacking
- follow up is inadequate or non existent; to some degree this is a function of the large number of assessments but also of divided responsibility between national agencies and lack of priority
- screening and scoping tends to be a mixed up process with ad hoc determination of outcomes – rather than formalised case-by-case screening and systematic scoping
- official guidance deals with formal aspects rather than substantive and methodological problems
- official guidance has consistently lagged behind introduction of legislation. At present no handbook on SEA exists more than two years after the directive was introduced into Swedish legislation
- no clear distinction between project level EIA and different forms of SEA is made perpetuating the "EIA illusion of SEA" making the lack of substantive guidance even more problematic

Follow up and monitoring is crucial to implementation of EQS using the EA system as a tool. Some comments on monitoring will be made below. Regarding EIA follow up the situation in Sweden is not favourable to standards implementation. Using Wood's criteria for EIA follow up gives a rather bleak result for Sweden. For monitoring and auditing of impacts (Wood 2003:245), which is an acknowledged weakness in almost all systems, Sweden comes out at the low end compared to any of the systems reviewed by Wood. This is mainly because of lack of formal provisions and guidance on monitoring. The exception is in the environmental permit system where monitoring is regularly included; however this normally refers to monitoring of compliance with emission levels rather than any impact related monitoring. For EIA system monitoring (Wood 2003 p.302) the result is definitely at the very low end in that the basic criteria for monitoring system function are not met. There is no available overview of EIAs made both because of the large number and because of the fragmented system with many agencies and some 50 different pieces of legislation involved. No attempt at systematic collection of information or experiences is made and no provisions for regular review are made.

The Swedish EA-system is essentially externally driven. Changes and revisions have been made as a result of EU directives. The declared minimalist introduction of directives into Swedish law in combination with the lack of follow up makes for a system with minimal learning and development capacity. The lack of EIA auditing is serious from the point of legitimacy. As Wood (2003 p.243) points out auditing provides public reassurance in the sense of accountability about impact management. It also gives industry and government an opportunity to demonstrate competence to the public (Buckley 1991).

One type of standard in EIA which is in principle different from EQS is the *threshold for requiring an EIA* to be made. Thresholds of this type do not determine the outcome of planning or permit processes but merely activates environmental assessment of some sort as a tool in decision making. This is in contrast to EQS which relate to degrees of acceptability of alternatives in planning and EIA. Thresholds may be size of investment, installed capacity, power, voltage etc of projects. The EIA-directive (85/337/EEC) has a number of such thresholds in Annex I. Annex II gives member states the obligation to set thresholds for a number of activities. Implementation of Annex II in national legislation shows the importance of national contexts for setting thresholds.

The problem of implementation of Annex II is the national setting of thresholds. Essentially the setting of thresholds is a balance between catching the significant environmentally problematic projects or plans and keeping numbers of assessment manageable. Differences may be the function of variations in ambition, a desire to minimise the number of assessments, the environmental or industrial context or of perceptions of environmental quality issues. The high threshold for mandatory requirement of an EIA of pig farms in Denmark compared to for example Greece (Fig. 1.4 in Chap. 1 by Glasson) is arguably a function of a combination of several factors. With the prevalence of pig farming a lower threshold would result in a very large number of EIAs. This may not be needed in Denmark in view of available, well developed technology for handling of wastes including energy recovery, and the large areas of arable land to spread manure on. With low thresholds a large number of EIAs will be produced. This dilemma is sometimes handled by producing "mini-EIAs" (Bond and Wathern 1999). The Swedish experience seems to be a strong argument against this.

In Sweden no thresholds are set for Annex II projects. Nor are there any systematic exclusion criteria[2] such as those initially set up in Denmark (Bond and Wathern 1999). Instead the Annex III criteria are to be used in the ad hoc determination of whether a full EIA needs to be made. One outcome of the lack of thresholds for Annex II projects is the very large number of EIAs made in Sweden, many in the form of "mini-EIAs". This might be thought to be conducive of good handling of cumulative effects. A recent study however shows that incitements to

---

[2] For examples of exclusion criteria see figure 1.4 in article by Glasson, this volume.

even a minimal assessment of cumulative effects are lacking; in fact it seems to be a more or less unknown concept to Swedish EIA practitioners (Oscarsson 2006). The reason for not setting screening thresholds for projects according to Annex II in the Swedish system is a rather curious reasoning concerning equal treatment. Since the impact of a project can not be determined with certainty before an assessment has been made all projects should in principle undergo EA. This reasoning undermines the whole idea of case based screening seemingly without reflection (Government Bill 1990/91:90 s 172, Government Bill 1997/98:45 part 2 s 278, Government Bill 2004/05:129 s 53) The solution to the problem of thus getting a very large number of EIAs was to introduce the concept of "small EIA" as distinct from a full scale EIA, when the probability of a significant impact has been determined in advance. The logic of this seems unclear. Thus the large number of assessments and the occurrence of "mini EIA" in Sweden is less a result of a conscious decision on thresholds or a high ambition than the result of muddled thinking.

Environmental Quality Norms

As noted in the introduction environmental quality norms (EQN) is a relatively recent introduction in the Swedish arsenal of instruments for environmental governance. They were introduced with the Environmental Code in 1998. There are EQN for nitrous oxides, sulphur dioxide, benzene, ozone and particles in air but also less precise for noise. A norm of a completely different kind is the quality norm for sea water for aquaculture of mussels which contains values for a very large number of parameters.

Since EQN refer to a state in the environment they are not directly aimed at any single actor or activity affecting the environment. They have to be operationalised with regulatory instruments addressing such actors and activities. Tools are needed in the form of strategies and action programmes for their implementation (Gipperth and Michanek 2001). One important task of such tools is to distribute responsibility for and the burden of implementation in a reasonable manner for example with regards to effectiveness, cost efficiency or equity. There are three important tools related to EQN: first the action programmes, second the various types of strategies, programmes and not least spatial plans and third, environmental assessment including project EIA especially in permit processes and SEA.

If an environmental quality norm is in danger of being violated an "action programme" has to be drawn up by a responsible authority. This programme may be binding to many different actors and actions. Generally it moves many decisions in spatial planning from the sphere of political decision making to the expert/administrative sphere. The comprehensive municipal plans should indicate how municipalities propose to observe the norms. The risk of violating an environmental quality norm gives the municipality an obligation to monitor that particular environmental parameter.

In sum the EQN serve in a limited number of cases of pollution both as stop lights to unsuitable development and as signals for addressing these problems. Their effectiveness will however depend on implementation structures, not least the way in which they are used in EA and in the action programmes they trigger. Methods of prediction in EA need to be applied in order to relate EA to EQN and the handling of uncertainty in predictions in EA is vital to the legitimacy of application of EQN. A crucial role will be played by follow up and monitoring. An evaluation of experiences of EQN in the field of air pollution (Lerman 2004) has pointed to a number of problems that we will briefly summarise. Several of the points are of relevance to our discussion below of management by objectives and the possibility of using norms for this form of governance.

The norms have taken a long time to become known to relevant agencies and experts which naturally so far has affected their effective use. Part of the explanation for this may be that - as is an unfortunate but common practice with new environmental legislation and instruments in Sweden – official guidance comes late and is based on early trial and error rather than on collated international scientific or administrative experiences.[3] Furthermore the regulations are incomplete and complex. While decision making in road planning may take into account mitigation at other sources than the road, this is legally not possible in other types of physical planning. Another question is where levels are measured and taken into account. Should air quality be measured where people have their noses which is relevant to health issues or on roof tops which is usually the case in order to protect equipment from being vandalised. Should measurements and exposure modelling refer to larger areas where people actually move around or should standards have to be met at points of maximum concentration such as a road tunnel entrances where no one spends time?

The function of EQN as stop light is unclear to many central actors and the function of moving decisions from democratically elected, political assemblies to the national expert agencies probably not fully appreciated or the consequences understood. How EQNs affect municipal spatial planning seems poorly understood except by a few large municipalities, where norms on air quality have come into severe clashes with building plans. One problem these have faced concerns the strive for denser urban development which is a good solution with regard to efficient transportation and thus to global warming, but gives a poorer ventilation climate which may lead to EQNs being exceeded, at least until new technology reduces pollution from cars.

So far the main impact of EQNs seems to be in the environmental permit system under the Environmental Code, where permits have been denied due to pollution above the given level. In spatial planning there are no cases of detailed plans refuted because of the risk of violation of a norm. The reason is usually with reference to expected future technical development and the uncertainty of the actual effects of the plan. This problem of causality is discussed below. The use of the EQN in follow up and monitoring is minimal.

---

[3] The most recent example of this is the fact that official guidance on SEA is not yet published more than two and a half years after the introduction of the SEA directive into Swedish law. The practice goes back at least to the introduction of legislation on nature conservation in the 60's when methods of classifying and inventorying conservation objects were developed ad hoc by the regional authorities (Emmelin 1983b).

The different practise in spatial planning and the environmental permit system may be linked to different traditions in decision making. Planning is generally based on an overall assessment of several conflicting interests, weighing them all in search of the "optimal solution". To get a permit on the other hand, the applicant has to prove that no interests are unacceptably harmed. Acceptability normally depends on the cost in comparison with the level of damage. With this last approach, it is logical to see EQN as a stop or barrier to the extent of weighing allowed. The damage level defined by the EQN is non-negotiable. If violated then the applicant has not proved compliance with all interests concerned. But with planning, there are several different interests, on both sides of the scales at the same time. A stop for one of them makes less sense, if it leads to a worse solution for all the other interests concerned.

With the introduction of new EU norms the situation with regards both to compliance with norms and the monitoring seems problematic. Levels of particulate matter (PM10) in ambient air in a very large number of Swedish urban areas regularly exceed norms in winter. The exact situation is not well known since estimates are based on sampling and model calculations. About half of the municipalities do not carry out regular monitoring of these levels. As with other introductions of EU instruments the situation in Sweden is characterised by lack of strategic planning.

A problem that EQN shares with other types of standards is the setting of suitable geographic systems boundaries. Lerman (2004) notes that the EQN work towards "point fulfilment" or compliance in small geographic areas, typically as fulfilment at a measuring point. In summary our conclusion is that those EQN that are violated and therefore should be actively used are few. In practice EQN have so far had limited effect on spatial planning in Sweden. The importance of EQN for environmental permit processes is so far much greater.

Guiding Threshold Values

The application of binding norms (EQN) in Sweden is thus limited but when it comes to EQS as guiding values it is practically non existent. One reason is that the guiding values are few and new. They derive mainly from the Water Framework Directive – Directive 2000/60/EC of the European Parliament and of the Council of 23 October 2000 establishing a framework for Community action in the field of water policy – which is still being implemented.

The only guiding value for air is for ozone in ambient air. There are in fact two values: one relating to effects on human health and one to effects on plants. The values are exceeded in the south of Sweden, mainly due to long distance transport of pollution. The guiding EQN relating to health is recommended to be fulfilled by 2010 but has so far not been applied either in permitting or planning. The EQN relating to plants is first a five year mean during the growing season and compliance "should be sought" during the period 2010 to 2019 and second a lower level to be applied after 2019. The complexity of these last EQN values makes it highly doubtful what signals are actually sent to planning or to EIA. This is especially so since the official information stresses the dependence on long range transport.

The Swedish National Environmental Objectives

The Swedish Parliament has set a series of National Environmental Objectives, NEO. They represent the main part of the third type of EQS in Sweden discussed above, the goals and target values for good environment.

The NEOs are 16 with a variable number of "sub-objectives" under each objective – see table 1.[4] These "sub-objectives" are targets that in principle should be quantified, have a time frame within which they are to be reached and if relevant a geographic delimitation. The NEOs can roughly be divided and characterised into two categories: "the scientific pollution objectives" and the "utopian landscape quality objectives" (Emmelin 2005). The former can also be characterised as "inverted problems" in that they are positive wordings of identified pollution problems, for example Objective no 3 "Natural acidification only".

Box 36.1. The Swedish National Environmental Objectives, NEO. The "utopian landscape goals" are marked in italics. The rest are the "inverted problems"[5]

1. Reduced Climate Impact	10. *A Balanced Marine Environment, Flourishing Coastal Areas and Archipelagos*
2. Clean Air	
3. Natural Acidification Only	11. *Thriving Wetlands*
4. A Non-Toxic Environment	12. *Sustainable Forests*
5. A Protective Ozone Layer	13. *A Varied Agricultural Landscape*
6. A Safe Radiation Environment	14. *A Magnificent Mountain Landscape*
7. Zero Eutrophication	15. *A Good Built Environment*
8. *Flourishing Lakes and Streams*	16. A Rich Diversity of Plant and Animal Life
9. Good-Quality Groundwater	

The confusion concerning the distinction between strict norms and other types of EQS in Sweden is considerable. While the targets are worded in a manner resembling "hard standards" – quantified, time set and using such phrases as "shall be achieved" – they are formally in fact only "soft standards". The wording of several of the sub-objectives is an illustration of the problem of clashes of political rhetoric with the strict practices of lawmaking. The various types of standards for ground level ozone may serve as an illustration of this confusion, especially the relationship between formal norms and guiding values and goals and targets. The standard for ground level ozone is given under the regulations for environmental quality norms and is formally termed an EQN. However it is formulated as a guiding value "to be sought" complied with at a future date. Furthermore the EQN for

---

[4] The formal name is "environmental quality goals" but they are generally referred to as "environmental goals". Lundqvist (2004) gave them the English name that we use here.

[5] A presentation in English is found on http://miljomal.nu/english/english.php It is interesting to note that the English wording in one case is different from the Swedish. The objective "A Non-toxic Environment" is literally in Swedish "An environment free from poisons". This misleading rhetoric has been criticised (Emmelin 2005).

ozone relating to health effects is duplicated as a target value within the system of NEOs. This double function is confusing. A guiding EQS normally gives the worst level accepted, not a recommended good value to strive for. This mixed use of guiding norms and targets makes an already complex system incomprehensible. The wording of the norm and target acerbates the problem. The text in the norm is as noted above "should be sought" complied with i.e. hardly a wording implying a norm and the date is "after 2010". The target on the other hand is worded "shall not be exceeded (...) in 2010". Thus the target is worded as if it were a norm while the norm is worded as a recommended or guiding value! As we have noted elsewhere the rhetoric of the NEOs tend to give an erroneous impression of formal status as norms (Emmelin and Lerman 2006).

The sub-objectives or targets of the "scientific pollution objectives" are in principle possible to operationalise with a time limit for reaching a quantified environmental quality standard in the hard sense, for example those relating to concentrations of pollutants in ambient air, or with a time frame within which all projects must comply with a given emission standard.

The time element in the targets however precludes in principle their use at any one time as "hard standards", at least as long as the point in time has not been passed. More important is that the general legal framework makes it impossible for the goals to act as "hard standards" that stop a project or a plan – a standard can be a recommendation or binding but not both at the same time. This nature of the NEO seems to be unclear to many actors at local and regional as well as national level who argue that being in conflict with a goal in it self should stop a decision. Application in the courts is however clear. The NEO can be used to give added weight on the scales in a weighting. They can not by themselves stop a project or plan. (Svea Environmental Court of Appeal, case M 9983-04) This is really a necessity since the objectives and their sub-objectives is a mixed lot and there may be conflicts between them in any given case i.e. fulfilling one may put a project in direct conflict with another. There are neither any priorities put on any of them nor any other mechanisms for resolving such conflicts within the NEO system as such. Thus their use in weighting is the only one compatible with any degree of predictability and consistency.

Operationalising the "utopian landscape quality objectives" is a more complex problem because of a combination of dimensions of both space and time. For example the allocation of a national target such as "not more than 10 per cent of the national park area should be affected by aircraft noise" presents an obvious problem in the absence of any mechanism such as cap and trade. In practice they can probably do no more than serve as qualitative guides to problems in planning.

It should be noted that about a quarter of the targets are in fact not related to environmental quality except in an indirect way. These targets are to have strategies or action programmes for certain target factors or even non target factors that may be more or less closely related to the respective goals. Reaching such targets says little about environmental quality except in a very indirect way. This has bearings for monitoring as will be discussed below.

Monitoring the NEOs

Monitoring progress towards objectives is central to MBO. For environmentally rational governance transparency and accountability is central to democratic legitimacy. The NEOs have a system of reporting progress towards the targets and overall objectives. This is managed by a special Environmental Objectives Council. Lundqvist (2004 p.69) claims that "The democratic self-binding is further strengthened by the elaborate system of monitoring and self binding." He concludes that "(…) it enables the general public to make autonomous judgement about progress towards sustainability, and thus to hold political and administrative decision-makers accountable for the success or failure of the MBO strategy". Unfortunately this far reaching and encouraging conclusion is based on the set up of the system and the policy intentions rather than on an analysis of the implementation or on the theoretical basis for the reporting. The Swedish National Audit Office in a review of the reporting reaches far less encouraging results noting for example that the relevance of indicators to the targets and goals is doubtful. This is of course a general problem of indicators. In order to be precise and measurable i.e. have high reliability their relevance or validity to a complex environmental situation may be limited. The review also notes that there are several systems of indicators used internationally and that the NEOs are not co-ordinated with the international systems used for reporting for example to the OECD and the EEA. (Riksrevisionen 2005)

Progress towards the objectives is reported in a number of ways to the public or to Parliament. The objectives and targets are summarised with a system of "smileys" presumed to make the reporting understandable and attractive. Progress is simplified into three classes: "achieved or will be achieved", "progress uncertain" and "not achieved or achievable" signified by a happy, neutral or sad smiley. The monitoring of indicators and targets is done by a large number of actors, largely as part of environmental monitoring, normal gathering of statistics and in the reporting by sectoral agencies. Here it must suffice to note a number of reasons why the monitoring and reporting system in principle may not be able to fulfil the functions that the Environmental Objectives Council claims and Lundqvist attributes to the system:

- The problem of relation of targets to objectives. Objectives are very general, rhetorical or intuitive in nature. Targets can not logically or consistently be derived from the very broad, rhetorical or perhaps even poetical, formulation of objectives. Rather, they are designed to operationalise i.e. fill the objectives with concrete meaning. Targets are a mixture of scientific thresholds and standards on the one hand and on the other hand highly value laden indicators of amenity and minor technicalities.
- About a quarter of the targets are, as noted above, in fact to have action programmes or strategies for some target but also, somewhat confusingly for non targets, produced by a certain date. The strategy or action programme seems to be a proxy target in that environmental quality will be affected only if and when an action programme is actually implemented.

- The illusory nature of some objectives precludes in principle validity of targets. The objective "Poison free environment" has no literal, substantive meaning. Progress towards the targets, which are conventional pollution targets, does not mean progress towards the illusion. Thus the question arises concerning accountability for fostering a counterproductive world view.
- The problem of relevance of indicators to targets. Indicators are specific and detailed which gives a problem of aggregating indicators to a meaningful index mirroring the target. Precision and reliability in indicators is high while validity may be low or doubtful.
- The problem of addition. In indicating the overall progress towards an objective the targets are weighted. This weighting is not a function of priorities in the target structure. The weighting and judgement process is described in some detail and an effort made to make it reliable in the sense of making it independent of the expert making the judgement. Weighting is both logical and necessary with a structure of targets and indicators of variable importance, reliability and validity but it adds to the problems of transparency.
- The conditional nature of classifying achievement. In many cases the judgement concerning whether and objective or a target will be achieved rests on assumptions such as "if decisions are passed", "if action programmes are implemented" etc.
- Achievability as a function of target setting. Should politicians be held accountable for setting unrealistic targets or for not achieving them?
- NEOs do explicitly not address sustainability but merely the ecological or environmental component. To claim that progress towards sustainability is monitored when some environmental targets may in fact preclude social or economic development in some landscapes is misleading.[6]

NEOs have a formally limited function as standards or thresholds. They are more in the nature of environmental policy rhetoric. Their function as "self-binding for the future" (Lundqvist 2004:64) or guidance is severely limited even given the assumption that MBO can function in environmental management and in spatial planning within a format of political decision making.

36.4 Some Theoretical Considerations

To discuss the problems of implementing EQS and their use in environmental assessment we need to briefly delve into some theoretical considerations. We will introduce two paradigms that aid in understanding the conflicts within environmental governance between spatial planning and environmental management.

---

[6] For a discussion see Emmelin and Lerman (2004a).

Environmental Management and Spatial Planning – Two Paradigms of Governance

The need for sound scientific knowledge means that scientific expertise holds a key position in environmental policy. Legitimacy in the environmental paradigm is seen as stemming from scientific quality of the underlying information and the principles. Hajer (1995) has shown how "discourse coalitions" based on different understandings and framing of an environmental problem operated e.g. in the "acid rain issue". Emmelin and Kleven (1999) demonstrate how vague concepts may give rise to a consensus illusion that does not have the form of a discourse coalition i.e. does not extend from the rhetoric of the policy level to practical measures in administration. This reasoning can be extended to the problem of making operational standards from general goals such as the Swedish NEOs.

The Swedish system for environmental governance can roughly be said to contain two principal elements: environmental management and spatial planning with their respective sets of legislation – the Environmental Code and the Planning and Building Act – administrations and the constituent professions and professional cultures. It is useful to distinguish between two paradigms governing the respective elements.

The *"environmentalist paradigm"* springs out of the natural sciences. A decision is legitimate if it rests on sound scientific evidence. Expert knowledge and central overview is critical to "correct" decisions; indeed the notion of "correct decisions" in cases of conflicts of interest is one important figure of thought in the paradigm. Nature as a reference base in the sense of such figures of thought as "natural" or "pristine" ecosystems and "natural conditions". These figures of thought reach into the pollution and environmental health discourses and are not confined to nature conservation. Preservation of natural states is another figure of thought, often complemented with the notion of "restoration" to "original" or "undisturbed" conditions underlying the conservation discourse. The paradigm leads to regulation taking its point of departure in nature and "natural states". The limits to what nature can tolerate is an important concept in the Swedish environmental quality objectives. The notion of "natural" regions as opposed to administratively determined regions as in the "water directive" (2000/60/EG) stems from this paradigm; being "natural" they are somehow superior to other regional divisions.

The basis for the *"plan paradigm"* is that governance of changes in land use and natural resource management should rest on the weighting or balancing of legitimate but not necessarily compatible interests. A central conflict of interest is thus the one between public and private interests in land use. A decision is seen as good and legitimate if it is reached in a process where interests are explicit and weighted. Although methods may vary over a wide scale from strictly rationalist to deliberative the ultimate decisions in spatial planning are political. Their proximate legitimacy is a claim to "fairness" and their ultimate legitimacy is democratic decision making.

These two paradigms can be illustrated as a function of two dimensions. One is the administrative cum geographic of central versus local. The other is the poles of decision rationality defined by Sager (1990; 1994) as between "calculating" and "communicative". The paradigms are basic to respectively the Environmental Code and the Planning and Building Act. Many of the problems and complexities of Scandinavian planning and environmental management can be analysed in terms of the tensions between the two paradigms (Emmelin and Kleven 1999; Emmelin and Lerman 2006; 2004a) The two paradigms are also of use in understanding differences in perceptions of the role of environmental assessment and how this in turn influences implementation of directives and national legislation. (Emmelin and Lerman 2004b; Emmelin 1998a).

36.5 Discussion: Factors Influencing the Effectiveness of Standards and Thresholds in Environmentally Rational Governance

In the introduction we posed four questions about effectiveness of standards and thresholds in Sweden as tools for environmentally rational governance. The questions concern the Swedish systems but are of a general nature related to function of standards and thresholds in governance. The Swedish case should give some general insights into the problems of implementing environmental policy by standards and thresholds. Before attempting to answer these questions we will however discuss some aspects of standards and thresholds in the light of the two paradigms introduced above.

Standards and Thresholds Seen in the Light of the Paradigms

Claiming that democratic accountability and legitimacy entails a communicative rather than instrumental rationality and scientific legitimation (Dryzek 1990: 54) brings the issue of standards and thresholds into the professional and legislative interface of the two paradigms discussed above. The issue of "scientisation" stemming in general in Swedish environmental policy (Lundqvist 2004: 87ff) and in particular from the environmentalist paradigm is interesting in the context of the National Environmental Objectives. The NEOs give a partly spurious impression of having a scientific basis. As noted above especially the group that we call "utopian landscape" objectives in fact are largely time and culture dependent values attached to conservation and amenity. Among these are "purist" or "wildernist" landscape values (Stankey et al. 1999; Heberlein 1973) which are couched in scientific terms as limits to what Nature can sustain rather than what conservation interests can sustain. Such objectives for planning are of course entirely legitimate as part of environmental policy. The problem is that the NEOs are supposed to define "what Nature can sustain" or limits set by "nature, environment and human health". The high legitimacy of scientific argument is thus extended to arguments given a scientific rhetorical dressing.

The two paradigms can be simplified to positions of two opposing, weberian ideal types. The environmentalist paradigm with a base in preservation combined with the precautionary principle and a focus on environmental problems seen as threats, and until recently without goals for environmental governance easily leads to a position of general opposition to all forms of change. This is particularly clear at local level and results in the NIMBY syndrome ("not in my back yard"), EIA becomes a tool for opposing change rather than governing change. Measurable change in environmental parameters becomes an argument against development with little regard for the problem of significance and importance; this is the confusion of "effect" with "impact" (Munn 1979; Emmelin 1996). The paradigm is essentially expert and top down. The rationalism of environmental assessment is founded in this view of planning and management.

The "plan-paradigm" on the other hand in its extreme form leads to a strong belief in the power of planning to shape not only the physical environment but also to fill it with social and economic content. In this view of planning sustainability appraisal is seen as possible and meaningful since planning is attributed with the power to determine development and negotiate the trade-offs between sustainability factors. Planning is depicted as "proactive" whereas impact assessment, permit processes and monitoring is seen as "reactive". This view to a large extent disregards the fact that the state in many cases is not the leading actor of change, having instead to the regulating and reactive role. (Emmelin and Lerman 2006; 2004a) In the Swedish context the plan-paradigm is especially clearly expressed in the spatial planning system which has only one binding level, the local, with limited guidance or steering from central or regional levels.

The view of sustainability promoted in the two paradigms in their ideal-type form is very different. In the environmentalist paradigm ecological sustainability is given priority over social and economic. Sustainability is essentially seen as a state to be achieved. Seeing sustainability as a state stems from the notion that the totality of our environmental problems can be "solved" rather than be temporarily and provisionally "resolved". This state is in several of the NEOs described as a state of balance in Nature which is a pervasive figure of thought often found in ecologist literature (Emmelin 1983a). The notion of sustainability as a steady state is expressed in the NEOs, the ultimate aim of which is to solve the environmental problems within the span of one generation. In the preamble to the NEOs it is claimed that a sustainable society will then be achieved. In relation to EQS it might seem problematic, as noted above, with goals for a future sustainable state that in several cases is identical or close to EQNs which set a minimum standard for certain environmental variables. In the "plan-paradigm" sustainability is seen as a process of continuous negotiation of the three components. The committee revising the Planning and Building Act noted that the Act with its emphasis on weighting of public and private interests is a suitable instrument for this process of negotiating sustainability.

The development of legislation on planning and environment protection in Sweden in the last decades can be seen as an attempt to handle the contradictions and tensions between these two paradigms. To a considerable extent this is done by giving the "environmentalist" paradigm an overarching role. An example of this is the introduction of binding environmental quality norms that essentially function to set the framework and degrees of freedom within which the weighting of interests by planning may take place. The system of National Environmental

Objectives is another example of this. The NEOs have the added problem, noted above, that several of the objectives and many of the targets are not scientifically based standards for environmental or public health, but rather the values of the professions of the "environmentalist" paradigm.

The Effectiveness of Standards and Thresholds in Sweden

In the introduction we posed four questions. It is now time to examine these in the light of our material and discussions. The questions concerning the Swedish system of standards and thresholds are:

- Are they constructed so as to fulfil the object of operationalising limits set by "nature, environment and human health"?
- Can they function in a system of management by objectives?
- Do they give clear signals concerning demand for future environmental quality?
- How do they function within, the environmental assessment system?

Operationalising Limits of "Nature, Environment and Human Health"

The simple answer may be that the EQN largely do operationalise limits since the majority relate to health and ecological effects of pollutants. The implementation problems are related to other aspects such as knowledge, the problems of applying strict norms in conditions of uncertainty etc. The role of EQN and the NEOs in conflicts between the two paradigms has been noted. Spatial planning becomes an arena for this professional and administrative contest with ensuing problems of implementation. We will here discuss aspects of implementation: the problem of EQS in relation to the "precautionary principle", the problems of uncertainty and prediction in applying EQN and the problem of applying EQS spatially.

With soft standards and especially the NEOs the case is less clear. The NEOs are claimed to determine "Nature's limits". However as noted several of the objectives and a large number of the targets can not be described as scientific nor can they in many cases sustain a claim that outcomes are irreversible. Rather obviously all the problems discussed here concerning margins of safety and uncertainty for EQN apply to the NEOs if they are implemented as "pseudo norms".

Standards and thresholds signal certainty and validity, but have several built in uncertainties. The uncertainties are of two kinds. First there is an uncertainty inherent in the setting of a standard. Second there are several uncertainties in implementation concerning for example prediction of impacts that may violate the standard, the relevance of measuring and monitoring when EQS is to be applied to larger areas, the modelling of impact prediction etc. Furthermore there is a problem of trade off within a larger geographic area: can thresholds be exceeded at certain points if the larger area average is acceptable?

In setting standards and thresholds both types and degrees of uncertainty have to be handled. One way of doing this is to introduce margins of safety in the standards. The "precautionary principle" is considered to be a cornerstone of good ecological governance. It defines "a range of outcomes that are impermissible, namely those that can not be altered in the future" (Barry 1999: 225) Strict EQNs attempt to determine and quantify such levels, often based on a dose-response relationship between an environmental parameter and health or ecological impacts. Application of margins of safety in standards focuses a central problem of all strategic decision making: the element of risk taking discussed below in connection with EA. Here we want to point to the problem of the clash between integrated margins of safety and the character of EQN as non-negotiable stop rules. There are at least two aspects to this problem. First the problem of determining *acceptable risk* and second the problem of *causality*.

The concept of "acceptable risk" begs the question of determination of "acceptability" with such complex philosophical and moral implications as intra- and inter-generational equity, both of which are central to sustainable development. (Harrison 2000). We will not go into this problem but simply note the value based nature of standards and thresholds which make them essentially political in nature. The mode of coping with the non-scientific nature of aspects of ecologically rational governance is to introduce the criteria of democratic accountability and legitimacy of decisions (Jansen et al. 1998; Lundqvist 2004). This is sometimes seen as one of the cornerstones of social sustainability (Eckerberg and Lafferty 1997) but needs to be discussed also within the ecological sustainability discourse. The practical way of coping is that those standards and thresholds that have the character of legally binding norms are in Sweden set by Parliament.

Setting binding norms thus curtails further negotiation and weighting. This means that while a general, expert based weighting has been introduced into the standard and legitimized by parliamentary decision local weighting and consideration of acceptability is prohibited or severely curtailed. While this may be an effective way to handle the scientifically based norms relating to health and ecological function it becomes highly problematic in the case of the soft standards. Arguably the utopian landscape objectives and their amenity targets could better be negotiated at regional or local level as part of the weighting of spatial planning than decided by central decree. Recent controversies over aircraft noise in the mountains and over conservation in the coastal environment illustrate that local opinion does not necessarily accept centrally decided amenity and nuisance targets or "pseudo-norms". The discussion by Therivel and Bennett (Chap. 26) on the use and misuse of noise standards may be a case in point. There is however also the opposite problem. Soft standards may leave operationalisation or interpretation to lower levels or to administration in a way that may threaten legitimacy. This can happen if the soft standards are presented as national objectives and targets but at the same time can be seen to be applied in a highly variable manner in different regions or municipalities.

The problem of causality is a major uncertainty in implementing standards in spatial planning. The problem of making predictions from plans is considerable, especially at a higher geographic level such as the comprehensive development plans of municipalities or the regional development programmes. The impacts will largely be a function of factors in implementation of the plans that are difficult to foresee. This is recognised in SEA with the use of scenarios that depict different outcomes of plans and programmes (Emmelin 1996). If an EQN is violated in any one of the scenarios the problem of assessing the likelihood of that scenario being realised arises. SEA of local development plans in Sweden is supposed to predict the impacts of the plan being fully developed and with the maximum activity that the plan allows. However the Planning and Building Act explicitly prohibits local development plans from being overly detailed and precisely regulatory. This built in flexibility is in direct conflict with a need for precision demanded by strict EQN. The fundamental problem is one of imbalance between the inherent uncertainty in predicting plan outcomes and the precision of EQS. With EQN the problem is all the more acute because of their binding nature.

There seems to be a lack of clarity in some standards and threshold as to whether they are in fact either thresholds for a given environmental parameter that is directly related to problems of health or ecological integrity or rather indicators of the state of the environment in more general terms. The problem here is one of causality. Levels of certain air pollutants such as nitrous oxides, particles etc are with a reasonable certainty causally connected to human health via does- response relationships. Actions causing negative development of such environmental factors may thus be directly seen as violations of the norms or standards. Bioindicators on the other hand indicate a state of the environment where the causes may be more or less uncertain, complex and cumulative. Such indicators may be essential in monitoring as warning systems. Biodiversity may in many cases be an example of this problem. They are however difficult to apply ex ante in planning or in environmental assessment. If causality is difficult to establish then it becomes difficult to determine whether any given activity will in fact affect the development of the indicator. If such indicators are used as standards the problem of predicting if a violation will occur will limit their usefulness.

In Sweden a biological quality standard has been discussed: a standard for lowest quality of the marine environment indicated by the minimum width of the coastal band of wrack (*Fucus vesiculosus*). This standard would give an aggregate but complex indication of environmental quality; in fact it probably integrates so many factors that it might rightly be termed an index rather than an indicator. The validity for any specific environmental problem is doubtful. To tie this indicator to any particular emission may be highly problematic. Thus it becomes complicated to make a causal connection between any emission parameter in a project or even more so of a plan. As an indicator signalling the need for caution such a standard may have a function but as a binding norm to be met it seems to have a very large number of problems in implementation.

Management by Objectives, MBO

Environmental policy integration i.e. integrating environmental concern into all sectors and getting sectors to work towards common environmental goals is a central element of "environmentally rational governance". The role of standards and thresholds is to force or coerce actors and planners to work for environmental quality defined by the standards, whether this be a minimum acceptable standard or a desirable goal. This is a much more ambitious and complex undertaking than just complying with emission standards or making use of "best available technology". New alternatives and technologies must actively be sought and violation of certain types of standards may in principle never be permissible.

From the standard literature on MBO a number of simple criteria for effectiveness can be drawn. In the simplest form five sets of criteria are recognised (Emmelin 2005). It is necessary for actors that are to be managed by objectives to:

- understand the objectives
- accept objectives and any derived sub-objectives
- wish to apply the objectives
- understand and accept the consequences of implementation

A somewhat more complex requirement is the contradictory need for both a degree of stability in a goal structure and at the same time a degree of adaptive flexibility.

Legitimacy and thus compliance is a complex interaction of these conditions being fulfilled to a sufficient degree. The two paradigms will lay stress on different criteria. "Scientisation" – which is part of the "environmentalist" paradigm – aims to reduce the importance of "wish to apply" by framing objectives as scientific imperatives.

Legally binding norms can derive their legitimacy from a combination of their scientific basis and democratic legitimacy derived from being decided by parliament. If objectives are open to interpretation the degree to which the above conditions are fulfilled is likely to influence implementation in any multicriteria decision situation. Objectives that are open to highly variable interpretations are obviously problematic from this point.[7] Many of the NEOs are in this category; especially so the "utopian landscape goals".

The criterion "understand and accept the consequences" entails a number of complex issues of which we will deal with only one: the problem of causality and prediction in EIA. Emission standards are in principle reasonably straightforward in this respect.

The conditions for effectiveness of MBO will apply in different degrees to different types of standards. In theory the more binding standards should depend for their effectiveness less on willingness to apply them or on understanding their consequences. However lack of acceptance of understanding the application of binding norms will in practice limit their effectiveness. The problems of prediction and of operationalisation are much the same.

The role of action programmes is complicated in relation to MBO. The central idea of MBO is to allow actors at lower levels to choose both their strategies and programmes of application and the tools and methods of doing so. The problem with EQS is that they are aimed at no specific actors but merely signal demands for environmental quality, be it desirable or minimum acceptable quality. In other words: by being aimed at everybody they are in fact aimed at nobody in particular.

---

[7] A concept such as "sustainable development" is a case in point, being however useful on a high level of policy rhetoric to create a consensus from which discussion of more concrete objectives such as standards and thresholds can follow (see e.g. Emmelin and Kleven 1999).

Complying with emission standards is easily understandable: they target any activity and actor emitting a substance. With plans and programmes the case may be that an actor is not even able to predict activities that may be in conflict with a given EQS as noted above concerning causality. Even if an actor is aware that activities may come into conflict with an EQS the relative importance of a particular activity may not be apparent. The action programmes are supposed to clarify this. When however they also prescribe specific actions that need to be taken they conflict with the central tenet of MBO.

Do Standards and Thresholds Give Clear Signals about the Future

One important potential function of standards and thresholds is to give clear signals concerning investments in environmentally friendly technology (Glasson, this volume). The binding EQN can probably have an increasing importance in this role although that requires present implementation and wide understanding of them as noted above. One important mechanism is the "action programmes" that should be made if a norm is violated or risks being violated. Such action programmes point out actors with special responsibility for taking action. This is a legal/administrative mode of handling the problem of who is addressed by standards.

In a recent case the Environmental Court of Appeal upheld a demand on a waste collection company to invest in a gradual change in its' vehicle fleet towards less polluting trucks in order to comply with the EQN concerning nitrous oxides in ambient air in the future. The EQN would be violated and this condition was imposed on the company as a prerequisite for a permit to expand operations. This is a strong support of the role of EQN action programmes as an instrument of promoting future change and giving signals of the need to plan and invest in cleaner technology.

In the case of soft standards, and especially the National Environmental Objectives the case is much less certain. Their function in management by objectives can be questioned as noted above. However we have also noted that one set of NEOs are in fact inverse, pollution problems, the "scientific NEOs", and to a considerable degree these are operationalised as targets much like standards and thresholds in character. As noted above the NEOs are an expression of political ambition rather than standards in a conventional sense. In principle the "scientific NEOs" could be seen as quantified signals of environmental quality that is to be reached at set times in the future. Unlike the EQN the NEOs do not require binding action programmes if an objective or one of its targets can not be reached. Without action programmes for individual targets they lack address; the signal is not aimed at any specific actors and no indication is given concerning how the target is to be achieved. Thus it is highly doubtful whether even the pollutin related, "scientific" NEOs send a sufficiently credible and clear signal concerning the need for investment. However a recent decision by the Environmental Court of Appeal is interesting in this respect. The Court decided that an individual second home in the Stockholm archipelago could not have a water toilet, but needed to install some form of composting system. The argument was that the cumulative effect of permitting further installation of water toilets, even with individual waste water treatment, would threaten the target 7 "Zero Eutrophication" in the coastal environment. The NEO target is thus used as a strong indication of what can be considered a reasonable level of "public interest". Second homes and household waste water are addressed in the form of local objectives set by the municipality. The Environmental Court of Appeal however reasoned in more general terms concerning releases of nutrients in the Baltic coastal areas. It is also worth noting that the reasoning is clearly related to cumulative effects: if one permit is given now it will be difficult to refuse future permits and it is this potential cumulative effect that will be unacceptable, not the individual releases.

Standards and Thresholds in the EA-System

The forth of our initial questions relates to the function of standards within the environmental assessment system of EIA and SEA. We have previously argued that the piecemeal and unstructured system of EIA and the minimalist introduction of SEA in Sweden casts doubt about the effectiveness and efficiency of the implementation level (Emmelin and Lerman 2006; 2004a, b; Emmelin 1998b). In our discussion above of the Swedish EA system we have pointed both to general deficiencies of the system and some directly related to standards and thresholds. Notable is the lack of screening thresholds which burdens the system with a very large number of assessments and mini-EIAs. The weaknesses in follow up are remarkable even in this internationally rather bleak field. Unfortunately the system of standards in Sweden also has been made imprecise and confusing with the introduction of different types of standards with varying formal strength. It can therefore be argued that the Swedish system of EQShas so many problems in implementation that this casts doubt on their effectiveness in environmentally rational governance (Emmelin and Lerman 2006). Instead of support that EQN could have brought to EA in Sweden, it seems to have increased the confusion.

EQS, and especially binding EQN, makes the problem of prediction of impacts acute. It is important to predict if EQN are being violated, the risk of doing so triggers action programmes with far reaching consequences for many actors in society, permits can be refused based on uncertain predictions etc. Noting that EA fundamentally is concerned with prediction (Glasson et al. 1999) it is interesting to note that prediction does not seem to merit any discussion in Swedish guidance on EIA and SEA so far (Emmelin 2007)

With EQN the problems of handling uncertainty in SEA become acute. First, because the prediction of actual emissions may be difficult or not even meaningful. Second there is the problem of relating the seemingly precise nature of standards with the uncertainty of prediction. The application of the precautionary principle is one way of doing this as noted above. Assigning a different role to SEA than to EIA and recognising the need to make assessments at appropriate levels is an important part of SEA development, breaking out of the "EIA illusion of SEA". The Swedish legislation which lumps together EIA and SEA does little to help here. The role of SEA as a tool for assessing other aspects than the environmental impacts at programme and higher plan levels has so far not been recognised in

Swedish SEA. The use of SEA to assess and ensure mechanisms for getting proper EA done at the appropriate level and for guidance or integration of environmental concern in the implementation stages have so far not been prominent. An attempt was made in the SEAs for the 2007-2013 Objective 2 programmes (Lerman and Emmelin, forthcoming)

To function in EA, there must be environmental objectives for all relevant sectors. Otherwise the formulation of alternatives, evaluations and assessments will be unbalanced and may lead to biased and badly founded decision. The relation between the environmental objectives and other objectives in society must also be addressed, either politically for a nation or region or by the decision maker. The ad hoc approach leads to legally important questions of predictability and equal treatment.

The two paradigms used here as analytical tools have a bearing on EA and EQS also. We have noted that EQS and especially the binding EQN are an attempt by the environmentalist paradigm to set outer limits to the scope of weighting of interests of the plan-paradigm. In a study of professional and administrative cultures in the environmental and planning administrations in the Nordic countries it was noted that those working in central and regional environmental administrations tended to see results of EA as binding to decision makers. The role of EA was generally considered to be to find "the environmentally best alternative" rather than a "balanced" material for decision making (Emmelin and Kleven 1999).

NEOs as TOC?

Threshold of concern, TOC, is a threshold tool for screening that focuses on risks of impact rather than on the given characteristics of a project or plan. The risk of passing a threshold is a warning signal to those assessing a project or plan. With the Swedish system of assessing a very large number of projects and plans TOC checklists might be a tool to make the distinction between "mini EIA" and full scale EIA clearer and more efficient. No such tool has been developed. This is in line with a general tendency in Sweden of concentrating advice and official guidance on formal rather than substantive issues in assessment. A recent, as yet unpublished, proposal for new national EA guidelines from the Swedish Environment Protection Agency unfortunately has the same formal focus.

The NEOs can however be seen as a form of TOC in that they are supposed to signal the need for EIA or for special attention to the respective problem areas. There are three problems with their use as TOC. The first is that the time element in the targets makes the determination of whether any given programme, plan or project in fact comes into conflict with or endangers the objective becomes rather subjective or at least full of uncertainties. The second is the geographical delimitation of targets with a national, aggregate quantification, as in "not more than 10 % of the national park area" being disturbed by aircraft noise. The third is the interpretation of such wording as "shall not increase" as in the target "emissions of water borne nitrogen shall not increase". Is this an absolute injunction to be taken literally or is there room for a weighting of an increase against planned decreases from other sources or for reasoning concerning what is a "significant increase"?

NEOs and EA

Swedish EA guidelines state that the NEOs should normally be included in the specification of relevant environmental issues demanded in both EIA and SEA. The NEOs have been used as indicators of environmental policy in recent SEAs of the 2007-2013 structural fund programmes in Sweden. Some regions have used the regional follow up of NEOs to indicate themes of environmental concern or conflict with regional development which would need attention in programme implementation, especially in the project funding decisions. A study to follow these processes is now being planned. Previous experiences of implementation of the general sustainability rhetoric of programmes in funding decisions is daunting (Emmelin and Nilsson 2006)

A pilot study of the use of NEOs in road EIA which is arguably the best structured and managed part of the Swedish EA system concludes that NEOs are mentioned in the introductory, general descriptive sections but can not be seen to influence either the generation or choice of concrete alternatives (Ekmark 2005)

The NEOs are sometimes used instead of a description of impacts, the conclusion in the EA being that goals are not being violated. The conclusion gives no information of the actual impacts however. This use of goals in EA must be contradicting the very purpose of the assessments; to describe impacts.

References

Bond AJ, Wathern P (1999) Environmental assessment in the European Union. In: Petts J (ed) Handbook of Environmental Impact Assessment vol 2 Environmental Impact Assessment in practice: impact and limitations. Blackwell Science, London, pp. 223-248

Buckley R (1991) Auditing the precision and accuracy of environmental impact predictions in Australia. Environmental Monitoring and Assessment 18:1-23

Dobers P (1997) Organising strategies of environmental control. Stockholm. Nerenius and Santérus

Eckerberg K, Lafferty WM (1997) Comparative perspectives on evaluation and explanation. In: Lafferty WM, Eckerberg K (eds) From earth summit to local forum. Studies of Local Agenda 21 in Europe. Oslo: ProSus, pp. 267-93

Ekmark T (2005) Miljömålens faktiska betydelse i den fysiska planeringen – exempel vägplanering. Masters thesis in spatial planning. BTH Karlskrona

Emmelin L (2007) Att förutsäga miljöpåverkan av planer och program. In: Wallentinus H-G (ed) MKB Perspektiv på miljökonsekvensbeskrivning. studentlit. Lund, pp. 201–230

Emmelin L, Lerman P (2006) Styrning av markanvändning och miljö. SOU Ansvarskommitténs skriftserie. Available from http://www.sou.gov.se/ansvar/skriftserien.htm

Emmelin L (2005) "Att synas utan att verka – miljömålen som symbolpolitik?" i Lars J. Lundgren, Edman J (red.) Konflikter, samarbete, resultat. Perspektiv på svensk miljöpolitik. Festskrift till Valfrid Paulsson. Brottby: Kassandra

Emmelin L, Lerman P (2004b) Problems of a minimalist implementation – the case of Sweden. In: Schmidt M, João E, Albrecht E (eds) Implementing Strategic Environmental Assessment. Springer, Berlin

Emmelin L, Lerman P (2004a) "Miljöregler – hinder för utveckling och god miljö?" BTH Research Reports 2004:09. Available from: www.bth.se/fou

Emmelin L, Kleven T (1999) A paradigm of Environmental Bureaucracy? Attitudes, thought styles, and world views in the Norwegian environmental adminstration. NIBR's Pluss Series 5-99

Emmelin L (1998a) Evaluating Environmental Impact Assessment Systems – Part 1: Theoretical and Methodological Considerations. Scandinavian housing and planning research 15(129):148

Emmelin L (1998b) Evaluating Environmental Impact Assessment – Part 2: Professional Culture as an Aid in Understanding Implementation. Scandinavian housing and planning research 15:187-209

Emmelin L (1996) Landscape Impact Analysis: a systematic approach to landscape impacts of policy. Landscape Research, vol 21, no 1 1996

Emmelin L (1983) Planering med ekologisk grundsyn. Naturresurs och miljökommittén. Bakgrundsmaterial n 13

Gipperth L, Michanek G (2000) Utveckling av miljökvalitetsnormer som rättsligt instrument. Naturvårdsverket Rapport 5138

Government Bill 2004/05:129 p. 53

Government Bill 1997/98:45 part 2 p. 278

Government Bill 1990/91:90 p. 172

Government Paper 2003/04:129 A Swedish Strategy for Sustainable Development

Hajer MA (1995) The politics of Environmental Discourse. Ecological Modernization and the Policy Process. Oxford University Press

Harrison NE (2000) Constructing Sustainable Development State University of New York

Heberlein T (1973) Social psychological assumptions of user attitude surveys: The case of the wildernism scale. Journal of Leisure Research, 5:18-33

Jansen A-I, Osland O, Hanf K (1998) Environmental challenges and institutional changes. An interpretation of the development of environmental policy in Western Europe', In: Hanf K, Jansen A-I (eds) Governance and Environmental Quality. Environmental administration, policy and politics in Western Europe. London: Longman pp. 277–325

Lerman P (2004) Miljökvalitetsnormer som styrmedel Uppföljning och utvärdering av tillämpningen av miljökvalitetsnormer för luftkvalitet under perioden 1999-2003 Rap. 5375

Lundqvist L-J (2004) Sweden and ecological governance. Manchester University Press

Munn R E (1979) Environmental impact analysis. Principles and procedures. 2 ed SCOPE Report no 5. Toronto: Wiley

Oscarsson A (2006) Lack of incitement in the Swedish EIA/SEA process to include cumulative effects. In: Emmelin L (ed) (2006) Effective tools for environmental assessment – critical reflections on concepts and practice. BTH research Report 2006:3, p. 92–115.

Riksrevisionen (2005) Miljömålsrapporteringen – för mycket och för lite. RiR 2005:1

Sager T (1990) Communicate or Calculate: Planning Theory and Social Science Concepts in a Contingency Perspective. Nordplan Dissertation

Sager T (1994) Communicative Planning Theory. Avebury

Stankey GH, McCool SF, Clark RN, Brown PJ (1999) Institutional and Organisational Challenges to Managing Natural Resources for Recreation: A Social Learning Model. In: Jackson EL, Burton TL (1999) Leisure Studies. Prospects for the Twenty-first Century. College State, Pennsylvania, Venture Publishing

Wood C (2003) Environmental Impact Assessment: A Comparative Review. 2nd edn, Prentice Hall

Subject Index

Printing: Krips bv, Meppel, The Netherlands
Binding: Stürtz, Würzburg, Germany